Die Wahrheit ist das Kind der Zeit, nicht der Autoritaet.
Unsere Unwissenheit ist unendlich, tragen wir einen
Kubikmillimeter ab!
(from B. Brecht, in "Leben des Galilei")

There is no end to this wonderful world of experimental discovery
and mental constructions of reality as new facts become known.
That is why physicists have more fun than most people.
(Miklos Gyulassy)

To Luca Lorenzo Strozzi Rigamonti, with hope (A.R.)

To Gegia, Enri, Cate and Dario (P.C.)

Structure of Matter

Attilio Rigamonti   Pietro Carretta

Structure of Matter

An Introductory Course with Problems and Solutions

 Springer

ATTILIO RIGAMONTI
PIETRO CARRETTA
Dipartimento di Fisica "A. Volta", Università degli Studi di Pavia

Library of Congress Control Number: 2007923018

ISBN 978-88-470-0559-4 Springer Milan Berlin Heidelberg New York

Springer is a part of Springer Science+Business Media
springer.com

© Springer-Verlag Italia 2007

Printed in Italy

Cover design: Simona Colombo, Milano
Typeset by the authors using a Springer Macro package
Printing and binding: Grafiche Porpora, Segrate, Milano

Printed on acid-free paper

Preface

Intended Audience, Approach and Presentation

This text is intended for a course of about fifteen weeks for undergraduate students. It arises from the adaptation and the amendments to a text for a full-year course in Structure of Matter, written by one of the authors (A.R.) about thirty years ago. At that time only a few (if any) textbooks having the suited form for introduction to basic quantum properties of atoms, molecules and crystals in a comprehensive and interrelated way, were available. Along the last twenty years many excellent books pursuing the aforementioned aim have been published (some of them are listed at the end of this preface). Still there are reasons, in our opinion, to attempt a further text devoted to the quantum roots of condensed matter properties. A practical aspect in this regard involves the organization of the studies in Physics, after the huge scientific outburst of the various topics of fundamental and technological character in recent decades. In most Universities there is now a first period of three or four years, common to all the students and devoted to elementary aspects, followed by a more advanced program in rather specialized fields of Physics. The difficult task is to provide a common and formative introduction in the first period still suitable as a basis for building up more advanced courses and to bridge the large area between elementary physics and the topics pertaining to research activities. The present attempt towards a readable book, hopefully presenting those desired characteristics, essentially is based on a mixture of simplified institutional theory with solved problems. The hope, in this way, is to provide physical insights, basic culture and motivation, without deteriorating the possibility of more advanced subsequent learning.

Organization

Structure of Matter is such a wide field that a first task to undertake is how to confine an introductory text. The present status of that discipline represents a key construction of the scientific knowledge, possibly equated only by the unitary description of the electromagnetic phenomena. Even by limiting attention only to the conventional parts of the condensed matter, namely atoms, molecules and crystals, still we are left with an ample field. For instance, semiconductors or superconductors, the electric and magnetic properties of the matter and its interaction with the electromagnetic radiation, the microscopic mechanisms underlying solid-state devices as well as masers and lasers, are to be considered as belonging to the field of structure of matter (without mentioning the "artificial" matter involving systems such as nanostructures, photonic crystals or special materials obtained by subtle manipulations of atoms by means of special techniques). In this text the choice has been to limit the attention to key concepts and to the most typical aspects of atoms (Chapters 1-5), molecules (Chapters 7-10) and of crystalline solids (Chapters 11-14), looking at the basic "structural" aspects without dealing with the properties that originate from them. This choice is exemplified by referring to crystals: electronic states and quantum motions of the ions have been described without going into the details regarding the numerous and relevant properties related to these aspects. Only in a few particularly illustrative cases favoring better understanding or comprehensive view, derivation of some related properties has been given (examples are some thermodynamical properties due to nuclear motions in molecules and crystals or some of the electric or magnetic properties). Chapter 6 has the particular aim to lead the reader to an illustrative overview of quantum behaviors of angular momenta and magnetic moments, with an introduction to spin statistics, magnetic resonance and spin motions and a mention to spin thermodynamics, through the description of the adiabatic demagnetization used in order to approach the zero-temperature condition. All along the text emphasis is given to the role of spectroscopic experiments giving access to the quantum properties by means of electromagnetic radiation. In the spirit to limit the attention to key arguments, frequent referring is given to the electric dipole moment and to selection rules, rather than to other aspects of the many experiments of spectroscopic character used to explore the matter at microscopic level. Other unifying concepts present along the text are the ones embedded in statistical physics and thermal excitations, as it is necessary in view of the many-body character of condensed matter in equilibrium with a thermal reservoir.

Prerequisite, appendices and problems

Along the text the use of quantum mechanics, although continuous, only involves the basic background that the reader should have achieved in undergraduate courses. The knowledge in statistical physics is the one based on the Boltzmann, Fermi-Dirac and Bose-Einstein statistical distributions, with the relationships of thermodynamical quantities to the partition function (some of the problems work as proper recall, particularly for the statistical physics of paramagnets or for the black-body radiation). Finally the reader is assumed to have knowledge of classical electromagnetism and classical Hamiltonian mechanics. Appendices are intended to provide ad hoc recalls, in some cases applied to appropriate systems or to phenomena useful for illustration. The Gaussian cgs emu units are used. The problems should be considered entangled to the formal presentation of the arguments, being designed as an intrinsic part of the pathway the student should move by in order to grasp the key concepts. Some of the problems are simple applications of the equations and in these cases the solutions are only sketched. Other problems are basic building blocks and possibly expansions of the formal description. Then the various steps of the solution are presented in some detail. The aim of the mélange intuition-theory-exercises pursued in the text is to favor the acquisition of the basic knowledge in the wide and wonderful field of the condensed matter, emphasizing how phenomenological properties originate from the microscopic, quantum features of the nature.

It should be obvious that a book of this size can present only a minute fraction of the present knowledge in the field. If the reader could achieve even an elementary understanding of the atoms, the molecules and the crystals, how they are affected by electric and magnetic fields, how they interact with electromagnetic radiation and respond to thermal excitation, the book will have fulfilled its purpose.

The fundamental blocks of the physical world are thought to be the subnuclear elementary particles. However the beauty of the natural world rather originates from the architectural construction of the blocks occurring in the matter. Ortega Y Gasset wrote "If you wish to admire the beauty of a cathedral you have to respect for distance. If you go too close, you just see a brick". Furthermore, one could claim that the world of condensed matter more easily allows one to achieve a private discovery of phenomena. In this respect let us report what Edward Purcell wrote in his Nobel lecture: "To see the world for a moment as something rich and strange is the private reward of many a discovery".

Acknowledgments

The authors wish to acknowledge Giacomo Mauro D'Ariano, who has inspired and solved several problems and provided enlightening remarks with his collaboration to the former course "Structure of Matter" given by one of us (A.R.), along two decades. Acknowledgments for suggestions or indirect contributions through discussions or comments are due to A. Balzarotti, A. Barone, G. Benedek, G. Bonera, M. Bornatici, F. Borsa, G. Caglioti, R. Cantelli, L. Colombo, M. Corti, A. Debernardi, A. Lascialfari, D. Magnani, F. Miglietta, G. Onida, G. Pastori Parravicini, E. Reguzzoni, S. Romano, G. Senatore, J. Spalek, V. Tognetti, A.A. Varlamov.

N. Papinutto is acknowledged for his help in preparing several figures. The problems have been revised by Dr. D. Magnani and Dr. G. Ventura, when students. The authors anticipate their gratitude to other students who, through vigilance and desire of learning will found errors and didactic mistakes.

Dr. M. Medici is gratefully thanked for her careful revision of the typed text.

This book has been written while receiving inspiration from a number of text books dealing with particular items or from problems and exercises suggested or solved in them. The texts reported below are not recalled as a real "further-reading list", since it would be too ample and possibly useless. The list is more an acknowledgment of the suggestions received when seeking inspiration, information or advices.

A. Abragam, *L'effet Mossbauer et ses applications a l'etude des champs internes,* Gordon and Breach (1964).

M. Alonso and E.J. Finn, *Fundamental University Physics Vol.III- Quantum and Statistical Physics*, Addison Wesley (1973).

D.J. Amit and Y. Verbin, *Statistical Physics - An Introductory course,* World Scientific (1999).

N.W. Ashcroft and N.D. Mermin, *Solid State Physics*, Holt, Rinehart and Winston (1976).

P.W. Atkins and R.S. Friedman, *Molecular Quantum Mechanics*, Oxford University Press, Oxford (1997).

A. Balzarotti, M. Cini, M. Fanfoni, *Atomi, Molecole e Solidi. Esercizi risolti*, Springer Verlag (2004).

F. Bassani e U.M. Grassano, *Fisica dello Stato Solido*, Bollati Boringhieri (2000).

J.S. Blakemore, *Solid State Physics*, W.B. Saunders Co. (1974).

S. Blundell, *Magnetism in Condensed Matter*, Oxford Master Series in Condensed Matter Physics, Oxford U.P. (2001).

S. Boffi, *Da Laplace a Heisenberg*, La Goliardica Pavese (1992).

B.H. Bransden and C.J. Joachain, *Physics of atoms and molecules*, Prentice Hall (2002).

D. Budker, D.F. Kimball and D.P. De Mille, *Atomic Physics - An Exploration Through Problems and Solutions*, Oxford University Press (2004).

G. Burns, *Solid State Physics*, Academic Press, Inc. (1985).

G. Caglioti, *Introduzione alla Fisica dei Materiali*, Zanichelli (1974).

B. Cagnac and J.C. Pebay - Peyroula, *Physique atomique, tome 2*, Dunod Universit, Paris (1971).

P. Caldirola, *Istituzioni di Fisica Teorica*, Editrice Viscontea, Milano (1960).

M. Cini, *Corso di fisica atomica e molecolare*, Edizioni Nuova Cultura (1992).

L. Colombo, *Elementi di Struttura della Materia*, Hoepli (2002).

E.U. Condon and G.H. Shortley, *The Theory of Atomic Spectra*, Cambridge University Press, London (1959).

C.A. Coulson, *Valence*, Oxford Clarendon Press (1953).

J.A. Cronin, D.F. Greenberg, V.L. Telegdi, *University of Chicago Graduate Problems in Physics*, Addison-Wesley (1967).

G.M. D'Ariano, *Esercizi di Struttura della Materia*, La Goliardica Pavese (1989).

J.P. Dahl, *Introduction to the Quantum World of Atoms and Molecules*, World Scientific (2001).

W. Demtröder, *Molecular Physics*, Wiley-VCH (2005).

W. Demtröder, *Atoms, Molecules and Photons*, Springer Verlag (2006).

R.N. Dixon, *Spectroscopy and Structure*, Methuen and Co LTD London (1965).

R. Eisberg and R. Resnick, *Quantum Physics of Atoms, Molecules, Solids, Nuclei and Particles*, J. Wiley and Sons (1985).

H. Eyring, J. Walter and G.E. Kimball, *Quantum Chemistry*, J. Wiley, New York (1950).

R.P. Feynman, R.B. Leighton and M. Sands, *The Feynman Lectures an Physics Vol. III*, Addison Wesley, Palo Alto (1965).

R. Fieschi e R. De Renzi, *Struttura della Materia*, La Nuova Italia Scientifica, Roma (1995).

A.P. French and E.F. Taylor, *An Introduction to Quantum Physics*, The M.I.T. Introductory Physics Series, Van Nostrand Reinhold (UK)(1986).

R. Gautreau and W. Savin, *Theory and Problems of Modern Physics*, (Schaum's series in Science) Mc Graw-Hill Book Company (1978).

H. Goldstein, *Classical Mechanics*, Addison-Wesley (1965).

H.J. Goldsmid (Editor), *Problems in Solid State Physics*, Pion Limited (London, 1972).

D.L. Goodstein, *States of Matter*, Dover Publications Inc. (1985).

G. Grosso and G. Pastori Parravicini, *Solid State Physics*, Academic Press (2000).

A.P. Guimares, *Magnetism and Magnetic Resonance in Solids*, J. Wiley and Sons (1998).

H. Haken and H.C. Wolf, *Atomic and Quantum Physics*, Springer Verlag Berlin (1987).

H. Haken and H.C. Wolf, *Molecular Physics and Elements of Quantum Chemistry*, Springer Verlag Berlin (2004).

G. Herzberg, *Molecular Spectra and Molecular Structure*, Vol. I, II and III, D. Van Nostrand, New York (1964-1966, reprint 1988-1991).

J.R. Hook and H.E. Hall, *Solid State Physics*, J. Wiley and Sons (1999).

H. Ibach and H. Lüth, *Solid State Physics: an Introduction to Theory and Experiments*, Springer Verlag (1990).

C.S. Johnson and L.G. Pedersen, *Quantum Chemistry and Physics*, Addison - Wesley (1977).

C. Kittel, *Elementary Statistical Physics*, J. Wiley and Sons (1958).

C. Kittel, *Introduction to Solid State Physics*, J. Wiley and Sons (1956,1968).

J.D. Mc Gervey, *Quantum Mechanics - Concepts and Applications*, Academic Press, New York (1995).

L. Mihály and M.C. Martin, *Solid State Physics - Problems and Solutions*, J. Wiley (1996).

M.A. Morrison, T.L. Estle and N.F. Lane, *Quantum States of Atoms, Molecules and Solids*, Prentice - Hall Inc. New Jersey (1976).

E.M. Purcell, *Electricity and Magnetism, Berkley Physics Course Vol.2*, Mc Graw-Hill (1965).

A. Rigamonti, *Introduzione alla Struttura della Materia*, La Goliardica Pavese (1977).

M.N. Rudden and J. Wilson, *Elements of Solid State Physics*, J. Wiley and Sons (1996).

M. Roncadelli, *Aspetti Astrofisici della Materia Oscura*, Bibliopolis, Napoli (2004).

H. Semat, *Introduction to Atomic and Nuclear Physics*, Chapman and Hall LTD (1962).

J.C. Slater, *Quantum Theory of Matter*, Mc Graw-Hill, New York (1968).

C.P. Slichter, *Principles of Magnetic Resonance*, Springer Verlag Berlin (1990).

S. Svanberg, *Atomic and Molecular Spectroscopy*, Springer Verlag Berlin (2003).

D. Tabor, *Gases, liquids and solids*, Cambridge University Press (1993).

P.L. Taylor and O. Heinonen, *A Quantum Approach to Condensed Matter Physics*, Cambridge University Press (2002).

M.A. Wahab, *Solid State Physics (Second Edition)*, Alpha Science International Ltd. (2005).

S. Weinberg, *The first three minutes: a modern view of the origin of the universe*, Amazon (2005).

M. White, *Quantum Theory of Magnetism*, McGraw-Hill (1970).

J.M. Ziman, *Principles of the Theory of Solids*, Cambridge University Press (1964).

Pavia,   *Attilio Rigamonti*
January 2007   *Pietro Carretta*

Contents

1

Atoms: general aspects

Topics

Central field approximation
Effective potential and one-electron eigenfunctions
Special atoms (hydrogenic, muonic, Rydberg)
Magnetic moments and spin-orbit interaction
Electromagnetic radiation, matter and transitions
Two-levels systems and related aspects

The aim of this and of the following three Chapters is the derivation of the main quantum properties of the atoms and the description of their behavior in magnetic and electric fields. We shall begin in the assumption of point-charge nucleus with mass much larger than the electron mass and by taking into account only the Coulomb energy. Other interaction terms, of magnetic origin as well as the relativistic effects, will be initially disregarded.

In the light of the **central field approximation** it is appropriate to recall the results pertaining to one-electron atoms, namely the **hydrogenic atoms** ($\S$1.4). When dealing with the properties of typical multi-electron atoms, such as **alkali atoms** or **helium atom** (Chapter 2) one shall realize that relevant modifications to that simplified framework are actually required. These are, for instance, the inclusion of the **spin-orbit interaction** (recalled at $\S$1.6) and the effects due to the **exchange degeneracy** ($\S$1.3, discussed in detail at $\S$2.2).

The properties of a useful reference model, the two-levels system, and some aspects of the electromagnetic radiation in interaction with matter, are recalled in Appendices and/or in *ad-hoc* problems at the end of the Chapter (Final Problems, F.I).

1.1 Central field approximation

The wave function $\psi(\mathbf{r}_1, \mathbf{r}_2, .., \mathbf{r}_N)$ describing the stationary state of the N electrons in the atom follows from the Schrödinger equation

$$\left[\frac{-\hbar^2}{2m}\sum_i \nabla_i^2 - \sum_i \frac{Ze^2}{r_i} + \sum_{i\neq j}{}' \frac{e^2}{r_{ij}}\right]\psi(\mathbf{r}_1, \mathbf{r}_2, .., \mathbf{r}_N) = E\psi(\mathbf{r}_1, \mathbf{r}_2, .., \mathbf{r}_N)$$

$$\left[T_e + V_{ne} + V_{ee}\right]\psi(\mathbf{r}_1, \mathbf{r}_2, .., \mathbf{r}_N) = E\psi(\mathbf{r}_1, \mathbf{r}_2, .., \mathbf{r}_N) \quad (1.1)$$

where in the Hamiltonian one has the kinetic energy T_e, the potential energy V_{ne} describing the Coulomb interaction of the electrons with the nucleus of charge Ze and the electron-electron repulsive interaction V_{ee} (Fig. 1.1).

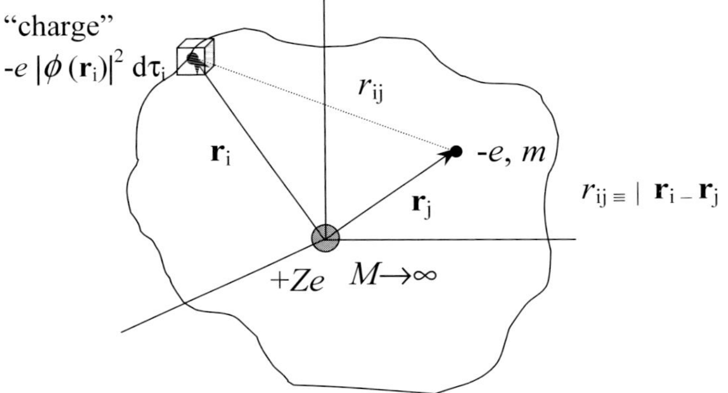

Fig. 1.1. Schematic view of multi-electrons atom. The nucleus is assumed as a point charge Ze, with mass M much larger than the mass m of the electron, of charge $-e$.

If the inter-electron interaction V_{ee} could be neglected, the total Hamiltonian would be $\mathcal{H} = \sum_i \mathcal{H}_i$, with $\mathcal{H}_i$ the one-electron Hamiltonian. Then $\psi(\mathbf{r}_1, \mathbf{r}_2, ...) = \prod_i \phi(\mathbf{r}_i)$, with $\phi(\mathbf{r}_i)$ the one-electron eigenfunction. V_{ee} does not allow one to separate the variables $\mathbf{r}_i$, in correspondence to the fact that the motion of a given electron does depend from the ones of the others. Furthermore V_{ee} is too large to be treated as a perturbation of $[T_e + V_{ne}]$. As we shall see (§2.2), even in the case of Helium atom, with only one pair of interacting electrons, the ground-state energy correction related to V_{ee} is about 30 percent of the energy of the unperturbed state correspondent to $V_{ee} = 0$.

The search for an approximate solution of Eq.1.1 can initiate by considering the form of the potential energy $V(\mathbf{r}_i)$, for a given electron, in the limiting cases of distances r_i from the nucleus much larger and much smaller than the average distance d of the other $(N-1)$ electrons:

$$r_i \gg d \qquad V(r_i) \simeq \frac{-e^2}{r_i}$$

$$r_i \ll d \qquad V(r_i) \simeq \frac{-Ze^2}{r_i} + \text{const.} \tag{1.2}$$

having taken into account that for neutral atoms $(N = Z)$ when $r_i \gg d$ the electrons screen $(Z-1)$ protons, while for $r_i \ll d$ $(N-1)$ electrons yield a constant effective potential, as expected for an average spherical charge distribution (Fig. 1.2). We shall discuss in detail the role of the screening cloud due to the inner electrons when dealing with alkali atoms (§2.1).

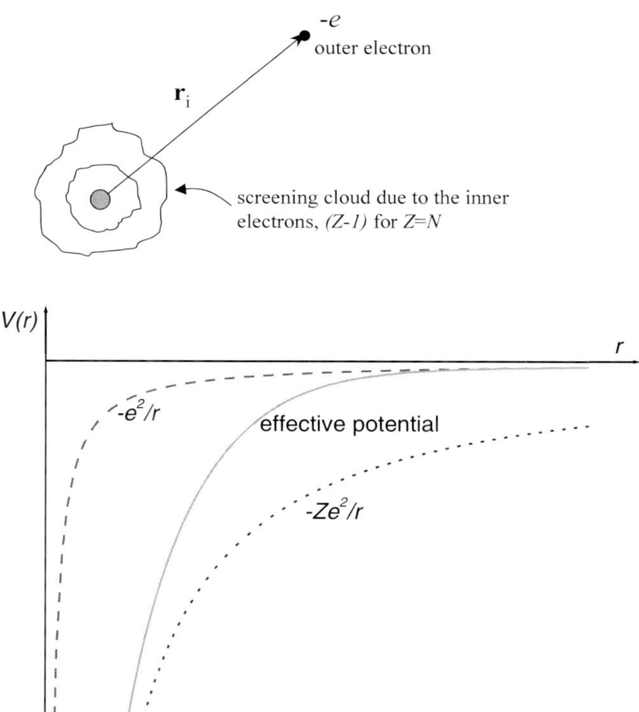

Fig. 1.2. Sketchy view of the electronic cloud screening the nuclear charge for an outer electron and correspondent forms of the potential energy in the limiting cases of large and small distances and of the effective central field potential energy (green solid line). Details on the role of the screening cloud shall be given in describing the alkali atoms (§2.1).

In the light of the form of the potential energy suggested by Eqs. 1.2 and neglecting correlation effects in the electronic positions, one deals with the **central field approximation**, first considered by **Hartree** and **Slater**. In this context any electron is moving in an **effective average field**, due to the nucleus and to the other electrons, which **depends only from the distance** $r \equiv |\mathbf{r}|$, with limiting expressions given by Eqs. 1.2.

Within this approximation Eq. 1.1 is rewritten

$$\sum_i \left[\frac{-\hbar^2}{2m} \nabla_i^2 + V(r_i) \right] \psi(\mathbf{r}_1, \mathbf{r}_2, ..., \mathbf{r}_N) = E\psi(\mathbf{r}_1, \mathbf{r}_2, ..., \mathbf{r}_N) \qquad (1.3)$$

implying

$$\psi(\mathbf{r}_1, \mathbf{r}_2, ..., \mathbf{r}_N) = \phi_a(\mathbf{r}_1)\phi_b(\mathbf{r}_2)...\phi_c(\mathbf{r}_i)...\phi_z(\mathbf{r}_N), \qquad (1.4)$$

where the one-electron eigenfunctions are solutions of the Equation

$$\{ \frac{-\hbar^2}{2m} \nabla_i^2 + V(r_i) \} \phi_a(\mathbf{r}_i) = E_i^a \phi_a(\mathbf{r}_i) \qquad (1.5)$$

in correspondence to a set of quantum numbers $a, b...$, and to one-electron eigenvalues $E_1^a, E_2^b...$. Moreover

$$E = \sum_i E_i^{a\cdots}. \qquad (1.6)$$

From the central character of $V(r_i)$, implying the commutation of $\mathcal{H}_i$ with the angular momentum operators, one deduces

$$\phi_a(\mathbf{r}_i) = R_{n(i)l(i)}(r_i) Y_{l(i)m(i)}(\theta_i, \varphi_i) \qquad (1.7)$$

where $Y_{l(i)m(i)}(\theta_i, \varphi_i)$ are the spherical harmonics and then the set of quantum numbers is $a \equiv n_i, l_i, m_i$.

Thus the one-electron states are labeled by the numbers $(n_1, l_1), (n_2, l_2)$ etc... or by the equivalent symbols $(1s), (2s), (2p)$ etc....

The spherical symmetry associated with $\sum_i V(r_i)$ implies that the total angular momentum $\mathbf{L} = \sum_i \mathbf{l}_i$ is a constant of motion. Then one can label the atomic states with quantum numbers $L = 0, 1, 2...$. $L(L + 1)\hbar^2$ is the square of the angular momentum of the whole atom, while the number M (the equivalent for the atom of the one-electron number m) characterizes the component $M\hbar$ of L along a given direction (usually indicated by z). It is noted that at this point we have no indication on how L and M result from the correspondent numbers l_i and m_i. The composition of the angular momenta will be discussed at Chapter 3. Anyway, since now we realize that the atomic states can be classified in the form S, P, D, F etc... in correspondence to the values $L = 0, 1, 2, 3$ etc....

1.2 Self-consistent construction of the effective potential

In the assumption that the one-electron wavefunctions $\phi_a(\mathbf{r}_i)$ have been found one can achieve a self-consistent construction of the effective potential energy $V(r_i)$. As it is known $-e|\phi_a(\mathbf{r})|^2 d\tau$ can be thought as the fraction of electronic charge in the volume element $d\tau$. Owing to the classical analogy, one can write the potential energy for a given j-th electron as [1] (see Fig. 11.1)

$$V(r_j) = -\frac{Ze^2}{r_j} + \sum_{i \neq j} \int \frac{e^2|\phi_a(\mathbf{r})|^2}{r_{ij}} d\tau_i \qquad (1.8)$$

This relationship between $V(r)$ and ϕ_a suggests that once a given $V(r)$ is assumed, Eq. 1.5 can be solved (by means of numerical methods) to obtain $\phi(\mathbf{r}_i)$ in the form 1.7. Then one can build up a new expression for $V(r_i)$ and iterate the procedure till the radial parts of the wavefunctions at the n-th step differ from the ones at the (n-1)th step in a negligible way. This is the physical content of the **self-consistent method** devised by **Hartree** to obtain the radial part of the one-electron eigenfunctions or, equivalently, the best **approximate** expression for $V(r_i)$. Here we only mention that a more appropriate procedure has to be carried out using eigenfunctions which include the spin variables and the dynamical equivalence (§1.3), with the **antisymmetry** requirement. Such a generalization of the Hartree method has been introduced by **Fock** and **Slater** and it is known as **Hartree-Fock** method. The appropriate many-electrons eigenfunctions have the **determinantal** form (see §2.3). A detailed derivation of the effective potential energy for the simplest case of two electrons on the basis of Eq. 1.8 is given in Prob. II.2.3.

The potential energy $V(r_i)$ can be conveniently described through an effective nuclear charge $Z_{eff}(r)$ by means of the relation

$$V(r) = -\frac{e^2}{r} Z_{eff}(r) \qquad (1.9)$$

(now the index i is dropped). The sketchy behavior of the effective nuclear charge is shown in Fig.1.3. The dependence on r at intermediate distance has to be derived, for instance, by means of the self-consistent method or by other numerical methods.

1.3 Degeneracy from dynamical equivalence

From Eqs. 1.3, 1.5 and 1.7 the N-electron wavefunction implies the assignment of a set of quantum numbers a_i to each i-th electron. This assignment

---

[1] Eq.1.8 can also be derived by applying the variational principle to the energy function constructed on the basis of the ϕ_a's with the complete Hamiltonian, for a variation $\delta\phi_a$ leaving the one-electron eigenfunction normalized.

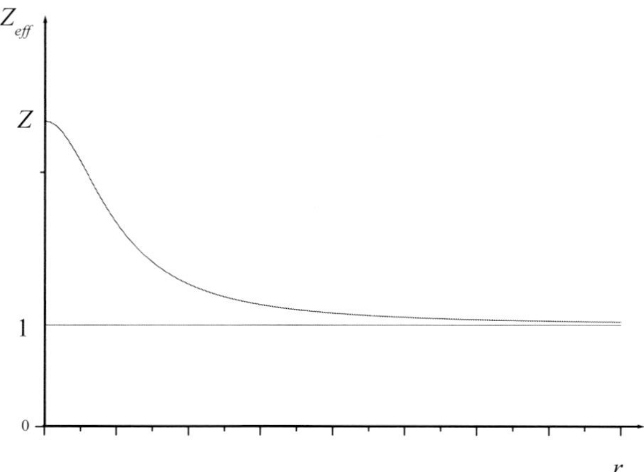

Fig. 1.3. Sketchy behavior of the effective nuclear charge acting on a given electron at the distance r from the nucleus of charge Ze, arising from the screening due to other electrons. The charge $(Z - N - 1)$ (1 for neutral atom with $Z = N$) is often called **residual charge** (for a quantitative estimate of $Z_{eff}(r)$ for Helium atom see Prob. II.2.3).

cannot be done in a unique way, since the electrons are **indistinguishable**, the Hamiltonian $\mathcal{H} = \sum_i \mathcal{H}_i$ being invariant upon exchange of the indexes (**exchange symmetry**). Therefore, for a state of the atom correspondent to a given eigenvalue, one has to write an eigenfunction combining with equal weights all the possible configurations, with the quantum numbers a_i variously assigned to different electrons. Therefore

$$\psi(\mathbf{r}_1, \mathbf{r}_2, ..., \mathbf{r}_N) = \sum_P P\left[\phi_{a_1}(\mathbf{r}_1)\phi_{a_2}(\mathbf{r}_2)...\phi_{a_n}(\mathbf{r}_N)\right] \qquad (1.10)$$

where P is an operator permuting electrons and quantum numbers.

It should be stressed that this remark on the role of the dynamical equivalence is incomplete and somewhat misleading. In fact we shall reformulate it after the introduction of a further quantum number, the spin number. Moreover, we will have to take into account the **Pauli principle**, that limits the acceptable wavefunctions obtained upon permutation to the ones changing sign (**antisymmetric**). This topic will be discussed after the analysis of Helium atom, with two electrons (§2.2). The eigenfunction in form of the **Slater determinant** (§2.3) does take into account the exchange degeneracy and the antisymmetry requirement.

We conclude these preliminary aspects observing that a proper quantum treatment, within a perturbative approach, at least should take into account the modifications to the central field approximation due to the Hamiltonian

$$\mathcal{H}_P = -\sum_i \frac{Ze^2}{r_i} + {\sum_{i \neq j}}' \frac{e^2}{r_{ij}} - \sum_i V(r_i) \, , \qquad (1.11)$$

resulting from the difference between Hamiltonian in 1.1 and the one in 1.3. This is the starting point of the **Slater theory for multiplets**.

1.4 Hydrogenic atoms: illustration of basic properties

The central field approximation allows one to reduce the Schrödinger equation to the form given by Eq. 1.3 and by Eq. 1.5. This latter suggests the opportunity to recall the basic properties for one-electron atoms, with Z protons at the nuclear site (Hydrogenic atoms). The Schrödinger equation is rewritten

$$\left[\frac{-\hbar^2}{2m} \nabla^2_{r,\theta,\varphi} - \frac{Ze^2}{r} \right] \phi_{n,l,m}(r,\theta,\varphi) = E_n \phi_{n,l,m}(r,\theta,\varphi) \qquad (1.12)$$

with $\phi_{n,l,m}$ of the form in Eq. 1.7. To abide by the description for the Hydrogen atom, one can substitute everywhere the proton charge $(+e)$ by Ze in the eigenvalues and in the wavefunctions. Then

$$E_n = -\frac{m(Ze)^2 e^2}{2\hbar^2} \frac{1}{n^2} = -Z^2 R_H hc \frac{1}{n^2} \qquad (1.13)$$

(with R_H Rydberg constant, given by $109,678$ cm^{-1}, correspondent to 13.598 eV). Y_{lm} are the spherical harmonics entering the wavefunction $\phi_{n,l,m}$ (see Eq. 1.7), reported in Tables I.4.1 up to $l = 3$.

The radial functions $R_{nl}(r)$ in Eq. 1.7 result from the solution of

$$\frac{d^2 R}{dr^2} + \frac{2}{r} \frac{dR}{dr} + \left[\frac{2m}{\hbar^2}(E + \frac{Ze^2}{r}) - \frac{l(l+1)}{r^2} \right] R = 0 \qquad (1.14)$$

or

$$\frac{-\hbar^2}{2mr^2} \frac{d}{dr} r^2 \frac{dR}{dr} + \left[\frac{l(l+1)\hbar^2}{2mr^2} - \frac{Ze^2}{r} \right] R = ER, \qquad (1.15)$$

namely a one-dimensional (1D) equation with an effective potential energy V_{eff} which includes the centrifugal term related to the non-inertial frame of reference of the radial axis. The shape of V_{eff} is shown in Fig.8.1. In comparison to the Hydrogen atom, Eq. 1.15 shows that in Hydrogenic atoms one has to rescale the distances by the factor Z. Instead of $a_0 = \hbar^2/me^2 = 0.529$ Å (radius of the first orbit in the Bohr atom, corresponding to an energy $-R_H hc = -e^2/2a_0$), the characteristic length becomes (a_0/Z).

Table I.4.1a. Normalized spherical harmonics, up to $l = 3$.

$s(l=0)$	$Y_{00} = \frac{1}{\sqrt{4\pi}}$
$p(l=1)$	$Y_{1-1} = \sqrt{\frac{3}{8\pi}}\frac{x-iy}{r} = \sqrt{\frac{3}{8\pi}}\sin\theta e^{-i\phi}$ $Y_{10} = \sqrt{\frac{3}{4\pi}}\frac{z}{r} = \sqrt{\frac{3}{4\pi}}\cos\theta$ $Y_{11} = -\sqrt{\frac{3}{8\pi}}\frac{x+iy}{r} = -\sqrt{\frac{3}{8\pi}}\sin\theta e^{i\phi}$
$d(l=2)$	$Y_{2-2} = \sqrt{\frac{15}{32\pi}}\frac{(x-iy)^2}{r^2} = \sqrt{\frac{15}{32\pi}}\sin^2\theta e^{-2i\phi}$ $Y_{2-1} = \sqrt{\frac{15}{8\pi}}\frac{z(x-iy)}{r^2} = \sqrt{\frac{15}{8\pi}}\sin\theta\cos\theta e^{-i\phi}$ $Y_{20} = \sqrt{\frac{5}{16\pi}}\frac{3z^2-r^2}{r^2} = \sqrt{\frac{5}{16\pi}}(3\cos^2\theta - 1)$ $Y_{21} = -\sqrt{\frac{15}{8\pi}}\frac{z(x+iy)}{r^2} = -\sqrt{\frac{15}{8\pi}}\sin\theta\cos\theta e^{i\phi}$ $Y_{22} = \sqrt{\frac{15}{32\pi}}\frac{(x+iy)^2}{r^2} = \sqrt{\frac{15}{32\pi}}\sin^2\theta e^{2i\phi}$
$f(l=3)$	$Y_{3-3} = \sqrt{\frac{35}{64\pi}}\frac{(x-iy)^3}{r^3} = \sqrt{\frac{35}{64\pi}}\sin^3\theta e^{-3i\phi}$ $Y_{3-2} = \sqrt{\frac{105}{32\pi}}\frac{z(x-iy)^2}{r^3} = \sqrt{\frac{105}{32\pi}}\sin^2\theta\cos\theta e^{-2i\phi}$ $Y_{3-1} = \sqrt{\frac{21}{64\pi}}\frac{(5z^2-r^2)(x-iy)}{r^3} = \sqrt{\frac{21}{64\pi}}(5\cos^2\theta - 1)\sin\theta e^{-i\phi}$ $Y_{30} = \sqrt{\frac{7}{16\pi}}\frac{(5z^2-3r^2)z}{r^3} = \sqrt{\frac{7}{16\pi}}(5\cos^2\theta - 3)\cos\theta$ $Y_{31} = -\sqrt{\frac{21}{64\pi}}\frac{(5z^2-r^2)(x+iy)}{r^3} = -\sqrt{\frac{21}{64\pi}}(5\cos^2\theta - 1)\sin\theta e^{i\phi}$ $Y_{32} = \sqrt{\frac{105}{32\pi}}\frac{z(x+iy)^2}{r^3} = \sqrt{\frac{105}{32\pi}}\sin^2\theta\cos\theta e^{2i\phi}$ $Y_{33} = -\sqrt{\frac{35}{64\pi}}\frac{(x+iy)^3}{r^3} = -\sqrt{\frac{35}{64\pi}}\sin^3\theta e^{3i\phi}$

Since $|\phi(r,\theta,\phi)|^2 d\tau$ corresponds to the probability to find the electron inside the volume element $d\tau = r^2\sin\theta dr d\theta d\phi$, from the form of the eigenfunctions the physical meaning of the spherical harmonics is grasped : $Y^*Y\sin\theta d\theta d\phi$ yields the probability that the vector $\mathbf{r}$, ideally following the electron in its motion, falls within the elemental solid angle $d\Omega$ around the direction defined by the polar angles θ and ϕ:

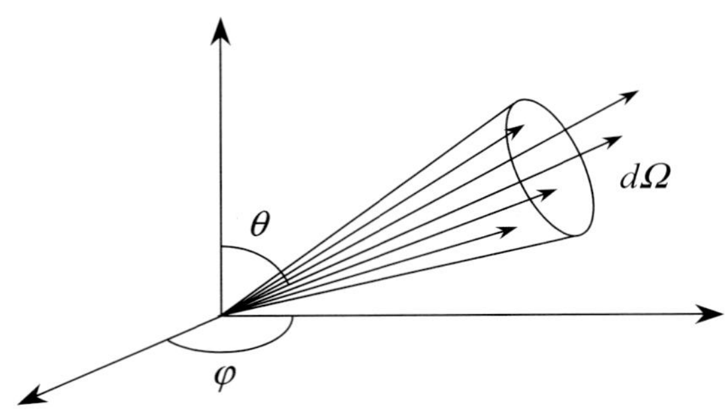

Table I.4.1b. Normalized spherical harmonics in the real form (see text).

$s(l=0)$	$Y_{00} = \frac{1}{\sqrt{4\pi}}$
$p(l=1)$	$Y_x = \sqrt{\frac{3}{4\pi}}\frac{x}{r} = \sqrt{\frac{3}{4\pi}}sin\theta cos\phi$ $Y_y = \sqrt{\frac{3}{4\pi}}\frac{y}{r} = \sqrt{\frac{3}{4\pi}}sin\theta sin\phi$ $Y_z = \sqrt{\frac{3}{4\pi}}\frac{z}{r} = \sqrt{\frac{3}{4\pi}}cos\theta$
$d(l=2)$	$Y_{z^2} = \sqrt{\frac{5}{16\pi}}\frac{3z^2-r^2}{r^2} = \sqrt{\frac{5}{16\pi}}(3cos^2\theta - 1)$ $Y_{zx} = \sqrt{\frac{15}{4\pi}}\frac{zx}{r^2} = \sqrt{\frac{15}{4\pi}}sin\theta cos\theta cos\phi$ $Y_{zy} = \sqrt{\frac{15}{4\pi}}\frac{zy}{r^2} = \sqrt{\frac{15}{4\pi}}sin\theta cos\theta sin\phi$ $Y_{x^2-y^2} = \sqrt{\frac{15}{16\pi}}\frac{x^2-y^2}{r^2} = \sqrt{\frac{15}{16\pi}}sin^2\theta cos2\phi$ $Y_{xy} = \sqrt{\frac{15}{4\pi}}\frac{xy}{r^2} = \sqrt{\frac{15}{16\pi}}sin^2\theta sin2\phi$
$f(l=3)$	$Y_{z^3} = \sqrt{\frac{7}{16\pi}}\frac{(5z^2-3r^2)z}{r^3} = \sqrt{\frac{7}{16\pi}}(5cos^2\theta - 3)cos\theta$ $Y_{z^2x} = \sqrt{\frac{21}{32\pi}}\frac{(5z^2-r^2)x}{r^3} = \sqrt{\frac{21}{32\pi}}(5cos^2\theta - 1)sin\theta cos\phi$ $Y_{z^2y} = \sqrt{\frac{21}{32\pi}}\frac{(5z^2-r^2)y}{r^3} = \sqrt{\frac{21}{32\pi}}(5cos^2\theta - 1)sin\theta sin\phi$ $Y_{z(x^2-y^2)} = \sqrt{\frac{105}{16\pi}}\frac{z(x^2-y^2)}{r^3} = \sqrt{\frac{105}{16\pi}}sin^2\theta cos\theta cos2\phi$ $Y_{zxy} = \sqrt{\frac{105}{4\pi}}\frac{zxy}{r^3} = \sqrt{\frac{105}{16\pi}}sin^2\theta cos\theta sin2\phi$ $Y_{x^2y} = \sqrt{\frac{35}{32\pi}}\frac{(3x^2y-y^3)}{r^3} = \sqrt{\frac{35}{32\pi}}sin^3\theta cos3\phi$ $Y_{y^2x} = \sqrt{\frac{35}{32\pi}}\frac{(x^3-3y^2x)}{r^3} = \sqrt{\frac{35}{32\pi}}sin^3\theta sin3\phi$

In the states labeled by the quantum numbers (n, l, m) the eigenvalue equations for the modulus square and for the z-component of the angular momentum are

$$\hat{l}^2\phi_{nlm} = R_{nl}(r)\hat{l}^2 Y_{lm}(\theta, \varphi) = R_{nl}(r)l(l+1)\hbar^2 Y_{lm}(\theta, \varphi);$$
$$\hat{l}_z\phi_{nlm} = R_{nl}(r)\hat{l}_z Y_{lm}(\theta, \varphi) = R_{nl}(r)\hat{l}_z \Theta_{lm}(\theta)e^{im\varphi} =$$
$$= R_{nl}(r)\Theta_{lm}(\theta)\hat{l}_z e^{im\varphi} = R_{nl}(r)Y_{lm}(\theta, \varphi)m\hbar \qquad (1.16)$$

Finally, it is noted that a given state of the Hydrogenic atom is Z^2 times more bound than the correspondent state in the Hydrogen atom because, on the average, the electron is $Z-$times closer to a nuclear charge increased by a factor Z.

The normalized wave functions for Hydrogenic atoms are reported in Table I.4.2. It is remarked that for $r \ll a_0/Z$ one has

$$(\phi_{nlm})_{r\rightarrow 0} \propto R_{nl} \propto r^l \qquad (1.17)$$

while for large distance

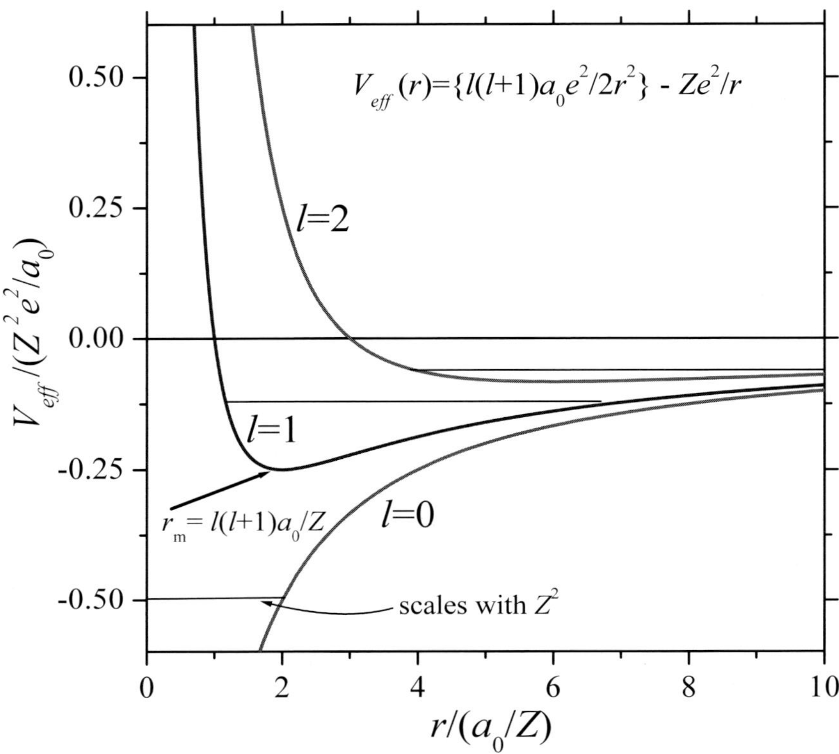

Fig. 1.4. Effective potential energy in the 1D Schrödinger equation for $R(r)$ (Eq. 1.15), for the lowest energy states. Horizontal lines indicate the eigenvalues for $n = 1, 2$ and 3, given by $-Z^2 R_H hc/n^2$ ($a_0 = \hbar^2/me^2$).

$$(\phi_{nlm})_{r\to\infty} \propto R_{nl} \propto e^{-\frac{r}{a_0}\frac{Z}{n}} \qquad (1.18)$$

From the wavefunctions relevant properties of the states, such as the radial probability density

$$P_{nl}(r) = \int d\varphi \int d\theta sin\theta r^2 |\phi_{nl}|^2 \qquad (1.19)$$

or the expectation values of any positional function $f(\mathbf{r})$

$$< f >_{nl} = \int |\phi_{nl}|^2 f(\mathbf{r}) d\tau \qquad (1.20)$$

can be derived. The radial probability densities for the $1s$, $2s$ and $2p$ states are depicted in Fig. 1.5.

Table I.4.2. Normalized eigenfunctions for Hydrogenic atoms, for $n = 1, 2$ and 3.

n	l	m	Eigenfunctions
1	0	0	$\phi_{100} = \frac{1}{\sqrt{\pi}}\left(\frac{Z}{a_0}\right)^{3/2} e^{-Zr/a_0}$
2	0	0	$\phi_{200} = \frac{1}{4\sqrt{2\pi}}\left(\frac{Z}{a_0}\right)^{3/2}\left(2 - \frac{Zr}{a_0}\right) e^{-Zr/2a_0}$
2	1	0	$\phi_{210} = \frac{1}{4\sqrt{2\pi}}\left(\frac{Z}{a_0}\right)^{3/2}\frac{Zr}{a_0} e^{-Zr/2a_0} \cos\theta$
2	1	± 1	$\phi_{21\pm 1} = \mp\frac{1}{8\sqrt{\pi}}\left(\frac{Z}{a_0}\right)^{3/2}\frac{Zr}{a_0} e^{-Zr/2a_0} \sin\theta e^{\pm i\varphi}$
3	0	0	$\phi_{300} = \frac{1}{81\sqrt{3\pi}}\left(\frac{Z}{a_0}\right)^{3/2}\left(27 - 18\frac{Zr}{a_0} + 2\frac{Z^2 r^2}{a_0^2}\right) e^{-Zr/3a_0}$
3	1	0	$\phi_{310} = \frac{\sqrt{2}}{81\sqrt{\pi}}\left(\frac{Z}{a_0}\right)^{3/2}\left(6 - \frac{Zr}{a_0}\right)\frac{Zr}{a_0} e^{-Zr/3a_0} \cos\theta$
3	1	± 1	$\phi_{31\pm 1} = \mp\frac{1}{81\sqrt{\pi}}\left(\frac{Z}{a_0}\right)^{3/2}\left(6 - \frac{Zr}{a_0}\right)\frac{Zr}{a_0} e^{-Zr/3a_0} \sin\theta e^{\pm i\varphi}$
3	2	0	$\phi_{320} = \frac{1}{81\sqrt{6\pi}}\left(\frac{Z}{a_0}\right)^{3/2}\left(\frac{Z^2 r^2}{a_0^2}\right) e^{-Zr/3a_0}(3\cos^2\theta - 1)$
3	2	± 1	$\phi_{32\pm 1} = \mp\frac{1}{81\sqrt{\pi}}\left(\frac{Z}{a_0}\right)^{3/2}\left(\frac{Z^2 r^2}{a_0^2}\right) e^{-Zr/3a_0} \sin\theta\cos\theta e^{\pm i\varphi}$
3	2	± 2	$\phi_{32\pm 2} = \frac{1}{162\sqrt{\pi}}\left(\frac{Z}{a_0}\right)^{3/2}\left(\frac{Z^2 r^2}{a_0^2}\right) e^{-Zr/3a_0} \sin^2\theta e^{\pm 2i\varphi}$

For spherical symmetry $P_{nl}(r)$ can be written as $4\pi r^2|\phi_{nl}|^2$. It should be remarked that for $Z = 1$ the maximum in P_{1s} occurs at $r = a_0$, corresponding to the radius of the first orbit in the Bohr model (see Problem I.4.4). For the states at $n = 2$ the correspondence of the maximum in $P_{nl}(r)$ with the radius of the Bohr orbit pertains to the $2p$ states.

The first excited state $(n = 2)$, corresponding to the eigenvalue $E_2 = -(Z^2 e^2/2a_0)(1/4)$, is the superposition of four degenerate states: $2s, 2p_1, 2p_0$ and $2p_{-1}$. To describe the $2p$ states, instead of the wavefunctions $\phi_{2p,m=\pm 1,0}$ (see Table I.4.2) one may use the linear combinations

$$\phi_{2px} = \frac{1}{\sqrt{2}}[\phi_{2p,1} + \phi_{2p,-1}] \propto \sin\theta\cos\varphi \propto x$$

$$\phi_{2py} = \frac{i}{\sqrt{2}}[\phi_{2p,1} - \phi_{2p,-1}] \propto \sin\theta\sin\varphi \propto y$$

$$\phi_{2pz} = \phi_{2p,0} \propto \cos\theta \propto z \tag{1.21}$$

From these expressions, also in the light of the $P_{2p}(r)$ depicted in Fig.1.5 and in view of the equivalence between the x, y and z directions, one can represent the **atomic orbitals** (the quantum equivalent of the classical orbits) in the form reported in Fig. 1.6.

The degeneracy in x, y, z is **necessary**, in view of the spherical symmetry of the potential. On the contrary the degeneracy in l, namely same energy for $s, p, d...$ states for a given n, is **accidental**, being the consequence of the particular, Coulombic form of the potential. We shall see that when the potential takes a different radial dependence because of $Z_{eff}(r)$ the degeneracy in l is removed (§2.1).

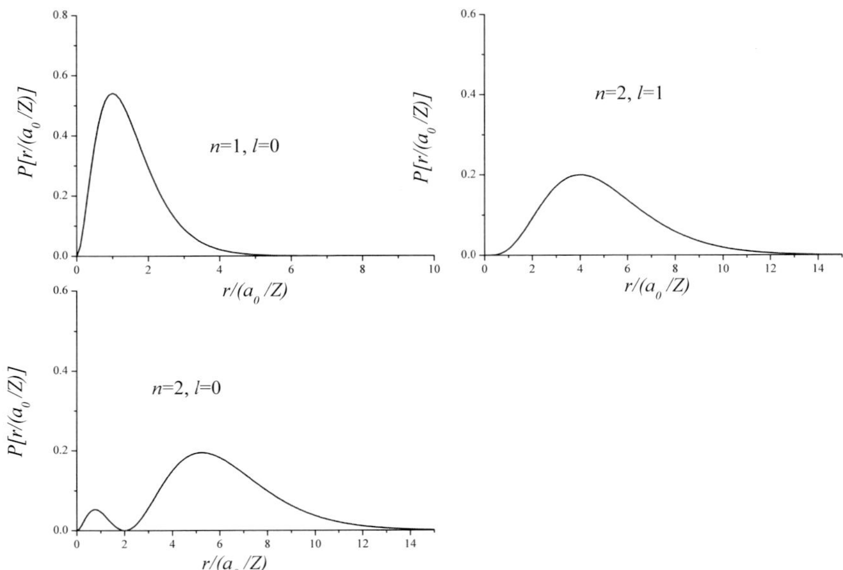

Fig. 1.5. Radial probability densities for $1s, 2s$ and $2p$ states in Hydrogenic atoms.

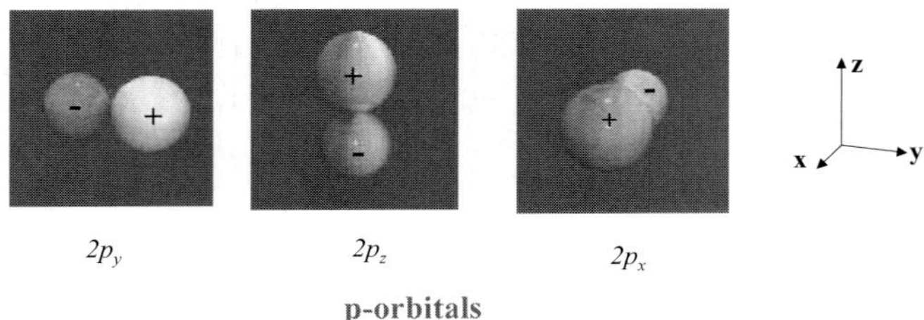

Fig. 1.6. Illustrative plots, for the $2p$ states of Hydrogenic atoms, of the atomic orbitals, defined as the shape of the surfaces where $|\phi_{nl}|^2 = constant$, meantime with probability of presence of the electron in the internal volume given by 0.9. It should be remarked that the sign $+$ or $-$, related to the sign of Y_{2p}, can actually be interchanged. However the **relationship** of the sign along the different directions is relevant, since it fixes the parity of the state under the operation of reversing the direction of the axes or, equivalently, of bringing $\mathbf{r}$ in $-\mathbf{r}$.

It is reminded that the difference between the $2p_{1,0,-1}$ and the $2p_{x,y,z}$ representation involves the eigenvalue for $\hat{l}_z$. The former are eigenfunctions of $\hat{l}_z$ while the latter are not, as shown for instance for ϕ_{2px}:

$$\hat{l}_z \phi_{2px} = -i\hbar \frac{\partial}{\partial \varphi} \phi_{2px} = -i\hbar \frac{\partial}{\partial \varphi} f(r) sin\theta cos\varphi = +i\hbar \phi_{2py} \qquad (1.22)$$

Obviously the difference is only in the description and no real modification occurs in regard of the measurements. This is inferred, for example, from the definition of ϕ_{2px} in terms of the basis of the eigenfunctions for $\hat{l}_z$ (see Eq. 1.21).

Finally in Fig. 1.7a the radial probability densities for the $n = 3$ states are plotted. The linear combinations of $3d$ states with different m's, leading to the most common representation, with the correspondent atomic orbitals are shown in Fig. 1.7b.

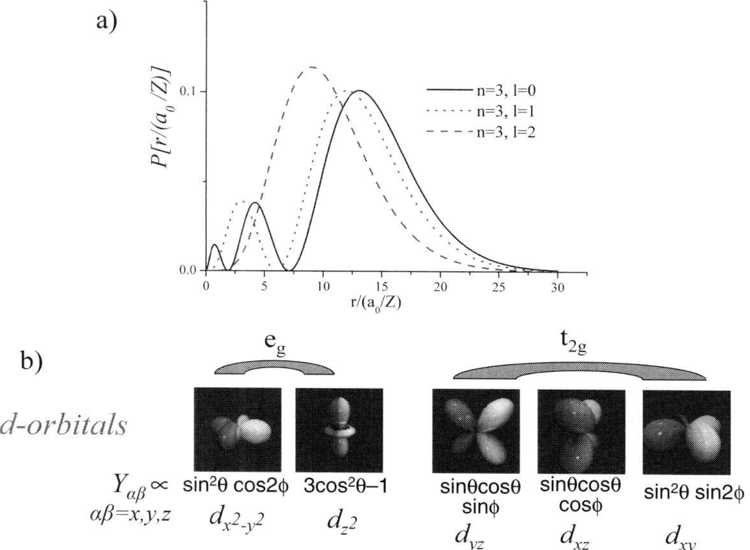

Fig. 1.7. Radial probability densities (a) for the $n = 3$ states of Hydrogenic atoms. In part b) of the Figure the angular distribution of the $3d$ atomic orbitals is reported. The d_{z^2} and $d_{x^2-y^2}$, grouped together are commonly called e_g levels, while the d_{xy}, d_{xz} and d_{yz} are called t_{2g} levels (we shall return to these aspects at §13.3)

Some expectation values of current use are reported in Table I.4.3

Table I.4.3 Expectation values of some quantities in Hydrogenic atoms

$$< r >_{nlm} \equiv \int \phi_{nlm}^*(\mathbf{r}) r \phi_{nlm}(\mathbf{r}) d\tau \equiv \int_0^\infty |R_{nl}|^2 r^3 dr =$$

$$= n^2 \frac{a_0}{Z}[1 + \frac{1}{2}(1 - \frac{l(l+1)}{n^2})] = \frac{a_0}{2Z}[3n^2 - l(l+1)]$$

$$< r^2 >_{nlm} = \frac{n^2}{2}(\frac{a_0}{Z})^2[5n^2 + 1 - 3l(l+1)]$$

$$< r^{-1} >_{nlm} = [n^2 \frac{a_0}{Z}]^{-1}$$

$$< r^{-2} >_{nlm} = \frac{Z^2}{a_0^2}[n^3(l + \frac{1}{2})]^{-1}$$

$$< V >_{nlm} = -\frac{Z^2 e^2}{a_0 n^2}$$

$$< T >_{nlm} = \frac{Z^2 e^2}{2 a_0 n^2}$$

$$< r^{-3} >_{nlm} = \frac{Z^3}{a_0^3 n^3 [l(l+1)(l+\frac{1}{2})]} \qquad (l \neq 0)$$

For $l = 0$ one has the divergence in the lower limit of the integral, since in $< r^{-3} >_{nlm} = \int \phi_{nl}^* \frac{1}{r^3} \phi_{nl} r^2 sin\theta dr d\theta d\varphi$
$\phi_{nl} \propto r^l$ for $r \to 0$ (see Eq. 1.17).

Problems I.4

Problem I.4.1 For two independent electrons, give a simple proof of Eqs. 1.4 and 1.6.

Solution:

From $\mathcal{H}_1 \phi_1 = E_1 \phi_1$ and $\mathcal{H}_2 \phi_2 = E_2 \phi_2$, by multiplying the first equation for ϕ_2 and the second for ϕ_1, recalling that $\mathcal{H}_{1,2}$ do not operate on $\phi_{2,1}$, respectively, one has $\mathcal{H}_1 \phi_1 \phi_2 = E_1 \phi_1 \phi_2$ and $\mathcal{H}_2 \phi_1 \phi_2 = E_2 \phi_1 \phi_2$. By summing

$$\mathcal{H}\phi = (\mathcal{H}_1 + \mathcal{H}_2)\phi_1 \phi_2 = (E_1 + E_2)\phi_1 \phi_2 = E\phi \ .$$

Problem I.4.2 One electron is in a state where the eigenvalue of the z-component of the angular momentum is $3\hbar$ while the square of the angular momentum is $12\hbar^2$. Evaluate the expectation value of the square of the x-component of the angular momentum.

Solution:

In $\hbar^2$ units, from $<\hat{l}_x^2> + <\hat{l}_y^2> = \hat{l}^2 - <\hat{l}_z^2>$, by taking into account that x and y directions are equivalent, one deduces $<\hat{l}_x^2> = (12-9)/2 = 1.5$. The expectation value of $\hat{l}_x$ is zero.

Problem I.4.3 Prove that the angular momentum operators $\hat{l}_z$ and $\hat{l}^2$ commute with the central field Hamiltonian and that a common set of eigenfunctions exists, so that Eq. 1.16 follows.

Solution:

In Cartesian coordinates, omitting $i\hbar$

$$\hat{l}_x \hat{l}_y - \hat{l}_y \hat{l}_x =$$

$$= (-y\frac{\partial}{\partial z} + z\frac{\partial}{\partial y})(-z\frac{\partial}{\partial x} + x\frac{\partial}{\partial z}) - (-z\frac{\partial}{\partial x} + x\frac{\partial}{\partial z})(-y\frac{\partial}{\partial z} + z\frac{\partial}{\partial y}) =$$

$$= y\frac{\partial}{\partial x} + yz\frac{\partial^2}{\partial z\partial x} - xy\frac{\partial^2}{\partial z^2} - z^2\frac{\partial^2}{\partial y\partial x} + xz\frac{\partial^2}{\partial y\partial z} -$$

$$-[zy\frac{\partial^2}{\partial z\partial x} - z^2\frac{\partial^2}{\partial x\partial y} - xy\frac{\partial^2}{\partial z^2} + x\frac{\partial}{\partial y} + xz\frac{\partial^2}{\partial z\partial y}] =$$

$$= (y\frac{\partial}{\partial x} - x\frac{\partial}{\partial y}) = \hat{l}_z$$

In analogous way the commutation rules for the components turn out

$$[\hat{l}_x, \hat{l}_y] = i\hbar\hat{l}_z, \quad [\hat{l}_z, \hat{l}_x] = i\hbar\hat{l}_y, \quad [\hat{l}_y, \hat{l}_z] = i\hbar\hat{l}_x.$$

In spherical polar coordinates, since

$$\hat{l}^2 = -\hbar^2\left[\frac{1}{sin\theta}\frac{\partial(sin\theta\frac{\partial}{\partial\theta})}{\partial\theta} + \frac{1}{sin^2\theta}\frac{\partial^2}{\partial\varphi^2}\right]$$

while $\hat{l}_z = -i\hbar\partial/\partial\varphi$ one finds $[\hat{l}^2, \hat{l}_z] = 0$.

For the central field Hamiltonian $\mathcal{H} = -\nabla^2 + V(r)$ and in Cartesian coordinates, for the kinetic energy

$$Tl_z = y\nabla^2\frac{\partial}{\partial x} - x\nabla^2\frac{\partial}{\partial y} = (y\frac{\partial}{\partial x} - x\frac{\partial}{\partial y})\nabla^2 = l_z T$$

while for the φ-independent potential energy $V(r)$ the commutation with $\hat{l}_z$ follows.

Now we prove that when an operator M commutes with the Hamiltonian a set of simultaneous eigenstates can be found, so that the two operators (one being $\mathcal{H}$, although the statement holds for any pair of commuting operators) describe observables with well defined values.

From $M\mathcal{H} - \mathcal{H}M = 0$ any matrix element involving the Hamiltonian eigenfunctions reads

$$< i|M\mathcal{H} - \mathcal{H}M|j > = < i|M\mathcal{H}|j > - < i|\mathcal{H}M|j > = 0.$$

From the multiplication rule

$$\sum_l < i|M|l > < l|\mathcal{H}|j > - \sum_k < i|\mathcal{H}|k > < k|M|j > = 0 \ .$$

$\mathcal{H}$ being diagonal one writes

$$< i|M|j > < j|\mathcal{H}|j > - < i|\mathcal{H}|i > < i|M|j > = 0,$$

namely $< i|M|j > (E_i - E_j) = 0$, that for $i \neq j$ proves the statement, when $E_j \neq E_i$ (for degenerate states the proof requires taking into account linear combinations of the eigenfunctions).

Problem I.4.4 In the Bohr model for the Hydrogen atom (with nuclear mass $M \to \infty$) the electron moves along circular orbits (**stationary states**) with no emission of electromagnetic radiation. The **Bohr-Sommerfeld condition** reads

$$\oint p_\theta d\theta = lh \qquad\qquad l = 1, 2, ...$$

p_θ being the moment conjugate to the polar angle in the plane of motion. Show that this quantum condition implies that the angular momentum is an integer multiple of $\hbar$ and derive the radius of the orbits and the correspondent energies of the atom.

Plot the energy levels in a scale of increasing energy and indicate the transitions allowed by the selection rule $\Delta l = \pm 1$, estimating numerically the wavelengths of the first lines in the **Balmer spectroscopic series** (transitions $n'' \to n'$, with $n' = 2$).

Compare the energy levels for H with the ones for He$^+$ and for Li^{2+}.

Finally consider the motion of the electron in three-dimensions and by applying the quantum condition to the polar angles, by means of vectorial arguments obtain the spatial quantization $\hat{l}_z = m\hbar$ for the z-component of the angular momentum.

Solution:
From the Lagrangian $\mathcal{L} = I(\partial\theta/\partial t)^2/2 + e^2/r$ one has $p_\theta = I\partial\theta/\partial t$, with I moment of inertia and $\partial\theta/\partial t = \omega = constant$. The quantum condition becomes

$$I\frac{\partial \theta}{\partial t} \oint d\theta = lh$$

so that $mr^2\omega 2\pi = lh$ and $mvr = l\hbar$.

From the latter equation and the equilibrium condition for the stationary orbits, where $mv^2/r = e^2/r^2$, the radii turn out

$$r_n = \frac{m^2 v^2 r^2}{me^2} = \frac{n^2\hbar^2}{me^2} = n^2 a_0$$

with $a_0 = \hbar^2/me^2 = 0.529$ Å.

The energy is

$$E = T + V = \frac{1}{2}mv^2 - \frac{e^2}{r} = -\frac{e^2}{2r}$$

(in agreement with the virial theorem, $<T>=<V>n/2$, with n exponent in $V \propto r^n$) and thus

$$E_n = -\frac{e^2}{2r_n} = -\frac{e^4 m}{2\hbar^2}\frac{1}{n^2}$$

as from Eq. 1.13, for $Z = 1$.

A pictorial view of the orbits is

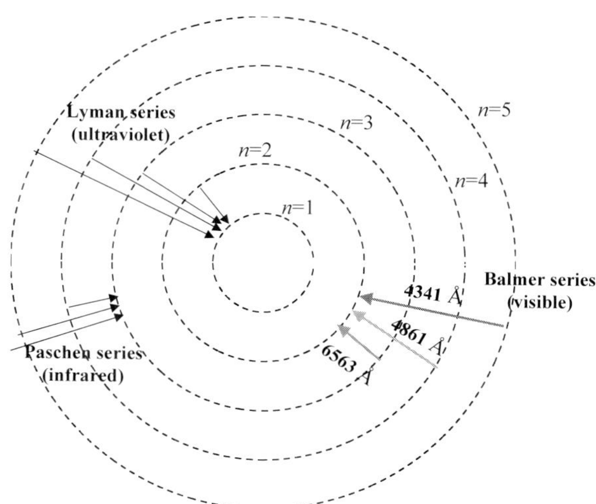

and the levels are

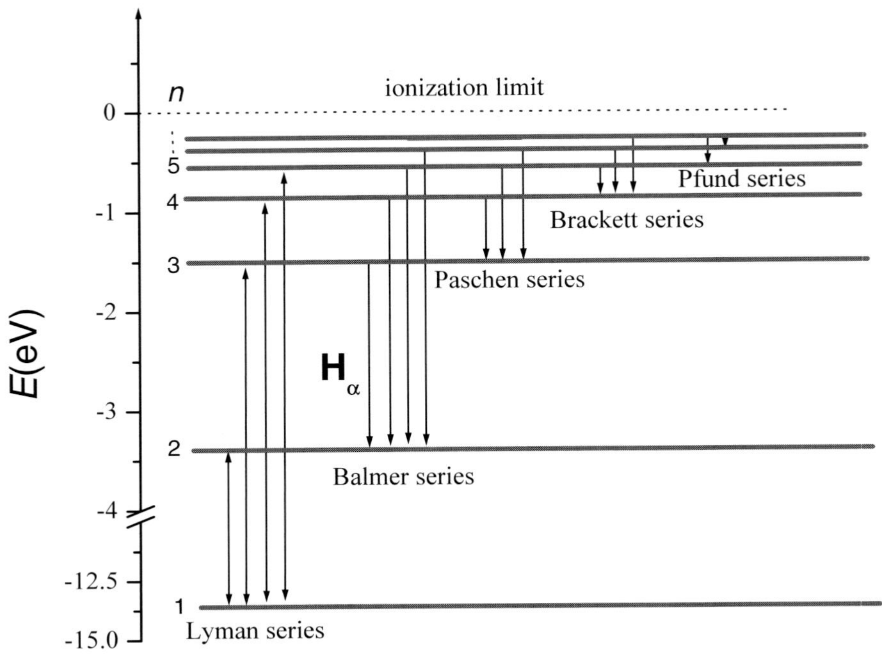

The Balmer series is shown below

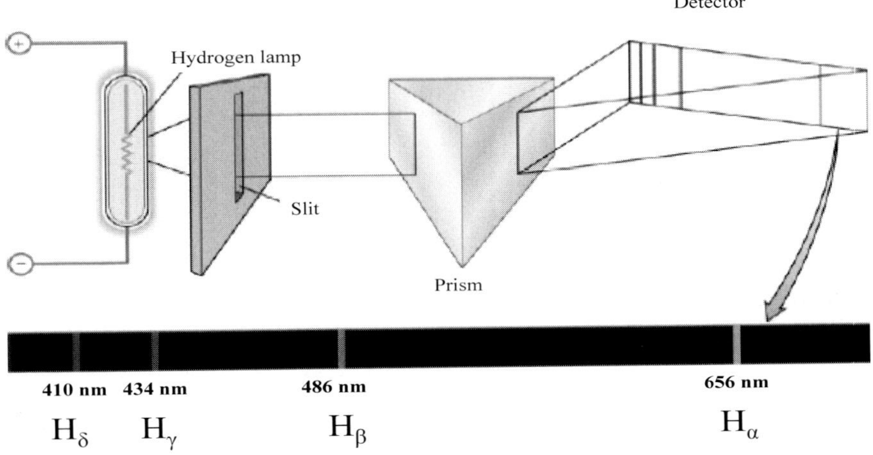

Comparison of the energy levels with the ones in He$^+$ and in Li^{2+}:

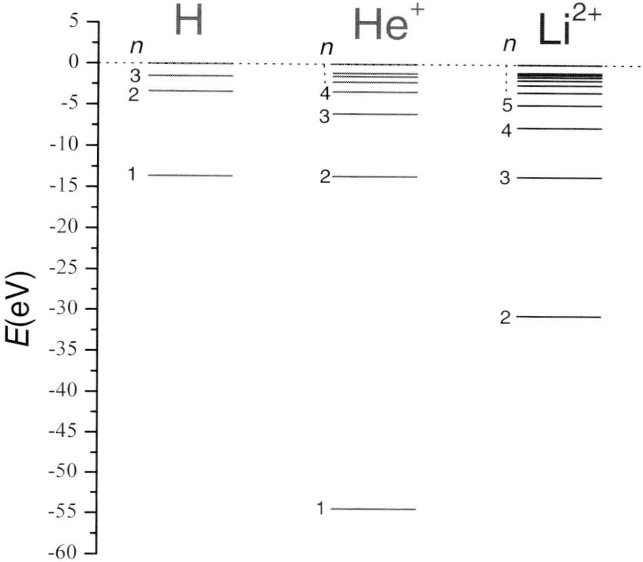

In the three-dimensional space

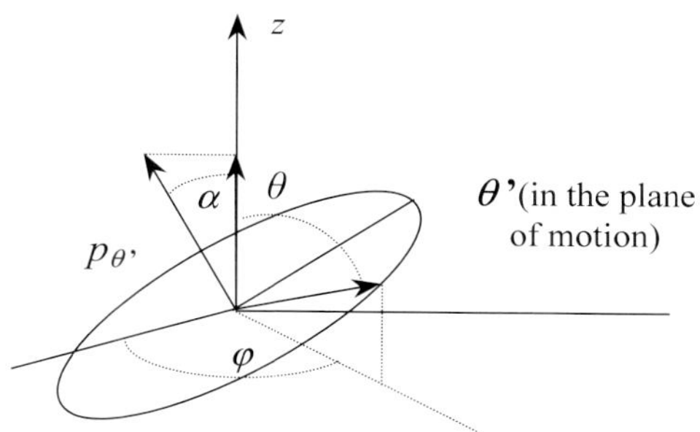

$$T = \frac{1}{2}[p_r \dot{r} + p_\theta \dot{\theta} + p_\varphi \dot{\varphi}]$$

and $p_{\theta'} d\theta' = p_\theta d\theta + p_\varphi d\varphi$ (since the energy is the same in the frame of reference (r, θ') and (r, θ, φ)). Thus

$$\oint p_\varphi d\varphi = mh$$

with m quantum number and p_φ constant, so that $p_\varphi = m\hbar$ and $p_{\theta'} = k\hbar$, while $\cos\alpha = m/k$, with $k = 1, 2, 3...$ and m varies from $-k$ to $+k$.

The pictorial view of the spatial quantization in terms of precession of the angular momentum for $l = 2$ (the "length" being $\sqrt{l(l+1)}\hbar$):

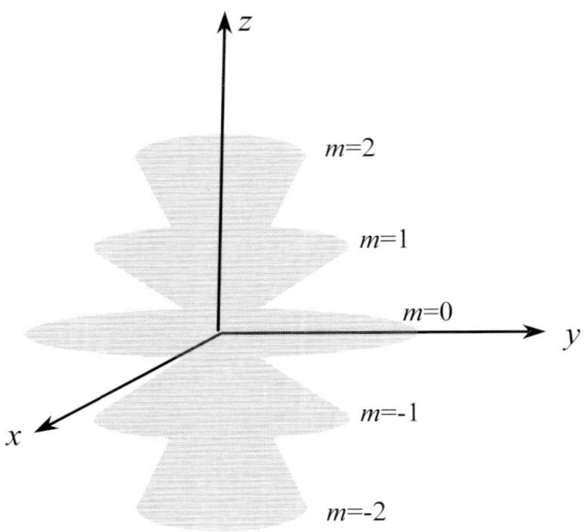

Problem I.4.5 In the first atomic model, due to **Thomson**, the atom was idealized as a uniform positive electric charge in a sphere, with point-charge electrons embedded in it. By referring to Hydrogen atom derive the motion of the electron and in the assumption that the radius of the sphere $R = 1$ Å estimate the frequency of the radiation expected in the classical description.

Solution:

The force at the distance r from the center of the sphere is

$$f(r) = -\frac{er^3}{R^3}\frac{e}{r^2}$$

and the electron motion is harmonic with angular frequency $\omega = \sqrt{e^2/mR^3}$. For $m = 9.09 \, 10^{-28}$ g, $e = 4.8 \times 10^{-10}$ u.e.s. and $R = 1$ Å the frequency turns out $\nu = 2.53 \times 10^{15}$ sec^{-1}. In the classical picture the emission is at the same frequency (and multiples) of the acceleration.

Problem I.4.6 In the assumption that the proton can be thought as a sphere with homogeneous charge distribution and radius $R = 10^{-13}$ cm, evaluate the shift in the ground state energy of the Hydrogen atom due to the finite size of the nucleus in the perturbative approach (Note that $R \ll a_0$). Repeat the calculation for uniform distribution onto the surface of the sphere.

Solution:

At the distance $r < R$ from the origin the potential energy is

$$V(r) = -\frac{e^2 r^3}{R^3 r} - \int_r^R \frac{e^2 4\pi r^2}{r 4\pi \frac{R^3}{3}} dr = -\frac{3}{2} e^2 \left[\frac{1}{R} - \frac{r^2}{3R^3} \right]$$

The difference with respect to the energy for point charge nucleus implies an energy shift given by

$$< 1s|V_{diff}|1s > = \frac{1}{\pi a_0^3} \int_0^R e^{-\frac{2r}{a_0}} \left[\frac{e^2}{r} + \frac{e^2 r^2}{2R^3} - \frac{3e^2}{2R} \right] 4\pi r^2 dr$$

and for $r < R \ll a_0$

$$< 1s|V_{diff}|1s > = -\frac{2e^2}{a_0^3} \left[R^2 - \frac{R^2}{5} - R^2 \right] = \frac{4}{5} \frac{e^2 R^2}{2a_0^3}$$

corresponding to about 3.9×10^{-9} eV.

For a uniform distribution onto the surface the perturbation Hamiltonian is $\mathcal{H}_P = +e^2/r - e^2/R$, for $0 \le r \le R$. The first order energy correction is

$$< 1s|\mathcal{H}_P|1s > = \frac{e^2}{\pi a_0^3} \int_0^R e^{-\frac{2r}{a_0}} [\frac{1}{r} - \frac{1}{R}] 4\pi r^2 dr =$$

$$= \frac{4e^2}{a_0^3} \int_0^R [r - \frac{r^2}{R}] dr = \frac{2e^2 R^2}{a_0^3} \frac{1}{3} \simeq 6.5 \times 10^{-9} \text{eV}$$

Problem I.4.7 For a Hydrogenic atom in the ground state evaluate the radius R of the sphere inside which the probability to find the electron is 0.9.
Solution:
From

$$\int_0^R 4\pi r^2 |\phi_{1s}|^2 dr = 0.9$$

with $\phi_{1s} = \sqrt{1/\pi}(Z/a_0)^{3/2} exp(-Zr/a_0)$, since

$$\int_0^R r^2 e^{-2Zr/a_0} dr = -e^{-2ZR/a_0} \left[\frac{R^2 a_0}{2Z} + \frac{R a_0^2}{2Z^2} + \frac{a_0^3}{4Z^3} \right] + \frac{a_0^3}{4Z^3}$$

a trial and error numerical estimate yields $R \simeq 2.66 a_0/Z$.

Problem I.4.8 In the assumption that the ground state of Hydrogenic atoms is described by an eigenfunction of the form $exp(-ar^2/2)$, derive the best approximate eigenvalue by means of variational procedure.

Solution:
The energy function is $E(a) = <\phi|\mathcal{H}|\phi> / <\phi|\phi>$, with

$$\mathcal{H} = -(\hbar^2/2m)[(d^2/dr^2) + (2/r)d/dr] - (Ze^2/r)$$

(see Eq. 1.14).
One has $<\phi|\phi> = 4\pi(1/4a)\sqrt{\pi/a}$, while

$$<\phi|\mathcal{H}|\phi> = 4\pi[(3\hbar^2/16m)\sqrt{\pi/a} - (Ze^2/2a)] \ .$$

Then

$$E(a) = \frac{3\hbar^2}{4m}a - 2Ze^2\sqrt{\frac{a}{\pi}}$$

From $dE/da = 0$ one has $a_{min}^{1/2} = 4mZe^2/3\hbar^2\sqrt{\pi}$ and $E_{min} = -4e^4Z^2m/3\pi\hbar^2 \simeq 0.849E_{1s}^H$ (for $Z = 1$).

Problem I.4.9 Prove that on the average the electronic charge distribution associated with $n = 2$ states in Hydrogenic atoms is spherically symmetric. Observe how this statement holds for multi-electrons atoms in the central field approximation.
  Solution:
  The charge distribution is controlled by

$$\frac{1}{4}|\phi_{2,0,0}|^2 + \frac{1}{4}[|\phi_{2,1,-1}|^2 + |\phi_{2,1,0}|^2 + |\phi_{2,1,1}|^2]$$

where the latter term (see Table I.4.1) is proportional to $[(1/2)sin^2\theta + cos^2\theta + (1/2)sin^2\theta] = 1$.
  In the central field approximation the statement holds, the wavefunctions being described in their angular dependence by spherical harmonics (This is a particular case of the **Unsold theorem** $\sum_{m=-l}^{m=+l} Y_{l,m}^* Y_{l,m} = (2l + 1)/4\pi$).

Problem I.4.10 On the basis of a perturbative approach evaluate the correction to the ground state energy of Hydrogenic atoms when the nuclear charge is increased from Z to $(Z + 1)$ ($\int_0^\infty x^n exp(-ax)dx = n!/a^{n+1}$).
  Solution:
  The exact result is $E_{Z+1} = -(e^2/2a_0)(Z + 1)^2$.
  The perturbative correction reads

$$E_{per}^{(1)} = -\int(\phi_Z^{1s})^* \frac{e^2}{r}(\phi_Z^{1s})d\tau = -\frac{4\pi e^2 Z^3}{\pi a_0^3}\int e^{-\frac{2Zr}{a_0}} rdr = -\frac{e^2 Z}{a_0}$$

In $(-e^2/2a_0)$ units the energy difference is $(2Z + 1)$ and for large Z this would practically coincide with $2Z$. It is noted that for the fractional correction goes as $1/Z$, since $E^0 \propto Z^2$.

1.5 Finite nuclear mass. Positron, Muonic and Rydberg atoms

To take into account the finite nuclear mass M in Hydrogenic atoms one can substitute the electron mass m with the reduced mass $\mu = Mm/(M+m)$. In fact this results from the very beginning, namely from the classical two-body Hamiltonian, the kinetic energy being

$$T = \frac{1}{2}\omega^2(Ma^2 + mb^2) = \frac{1}{2}\frac{Mm}{(M+m)}\omega^2 r^2 = \frac{1}{2}\mu\omega^2 r^2 \ ,$$

namely the one for a single mass μ rotating with angular velocity ω at the distance r:

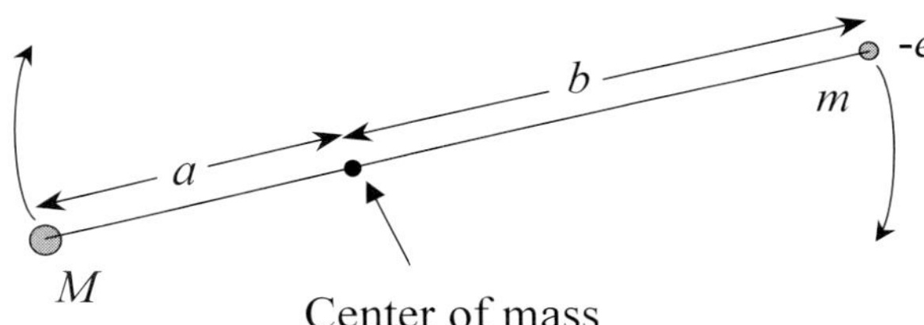

The potential energy does not change even though the nucleus is moving and therefore in order to account for the effects of finite nuclear mass, one simply substitutes m for μ in the eigenvalues and in the eigenfunctions. Then

$$E_n = -Z^2\frac{\mu e^4}{2\hbar^2}\frac{1}{n^2} = -\frac{e^2}{2a_0^*}\frac{Z^2}{n^2} \tag{1.23}$$

with $a_0^* = \hbar^2/\mu e^2$. In particular, the wavenumbers of the spectral lines (see Prob. I.4.4) are corrected according to

$$\bar{\nu} = Z^2 R_H \frac{1}{(1+\frac{m}{M})}\left(\frac{1}{n_f^2} - \frac{1}{n_i^2}\right) \tag{1.24}$$

where R_H is the Rydberg constant for the Hydrogen in the assumption of infinite nuclear mass (see Eq. 1.13).

The **Deuterium** has been discovered (1932) from slightly shifted weak spectroscopic lines (**isotopic shift**), related to the correction to the eigenvalues in Eq. 1.23, due to the different nuclear masses for H and D.

A two particle system where the correction due to finite "nuclear" mass is strongly marked is obviously the **positronium** i.e. the Hydrogen-like "atom" where the proton is substituted by the positron. The reduced mass in this case is $\mu = m/2$, implying strong corrections to the eigenvalues and to the correspondent spectral lines (and to other effects that we shall discuss in following Chapters).

In Hydrogenic atoms it is possible to substitute the electron with a negative muon. From high energy collisions of protons on a target, two neutrons and a negative pion are produced. The pion decays into an antineutrino and a negative muon, of charge $-e$ and mass about 206.8 times the electron mass. The muon decays into an electron and two neutrinos, with life-time $\tau \simeq 2.2\mu s$. Before the muon decays it can be trapped by atoms in "electron-like orbits", thus generating the so called **muonic atoms**.

Most of the results derived for Hydrogenic atoms can be transferred to muonic atoms by the substitution of the electron mass with the muon mass $m_\mu = 206.8m$. Thus the distances have to be rescaled by the same amount and the muonic atoms are very "small", the dimension being of the order or less of the nuclear size (see Fig. 1.8). It is obvious that in this condition the approximation of nuclear point charge and Coulomb potential must be abandoned.

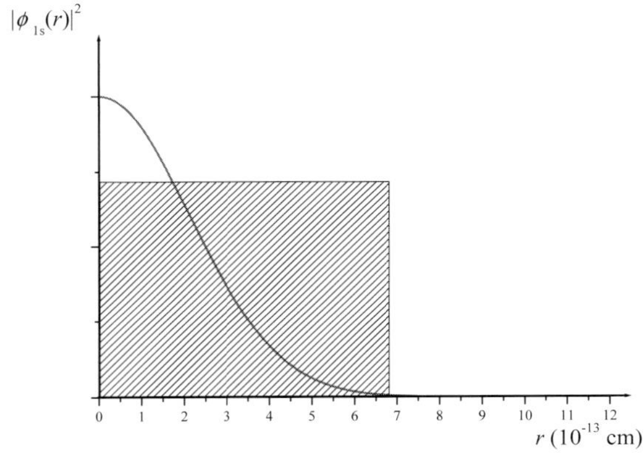

Fig. 1.8. Sketch of $|\phi_{1s}|^2$ for a muon in Pb ($Z = 82$) in the $1s$ state in the assumption of point charge nucleus (red line), in comparison with the charge distribution of the nucleus itself, of radius around 6 Fermi (dashed area).

However, qualitatively, in the muonic atoms the eigenvalues can still be obtained from the ones in Eq. 1.13 by multiplying for 206.8. Under this approximation the wavenumbers of the correspondent spectral lines become $\overline{\nu_\mu} = 206.8\,\overline{\nu_H}$ and the emission falls in the X-ray spectral range. The ionization potential is increased up to several MeV [2].

---

[2] It should be observed that dramatic effects in muonic atoms involve also other quantities or interactions, for instance the spin-orbit interaction and the hyperfine field (see §5.1)

Somewhat opposite to the muonic atoms are the "gigantic" **Rydberg atoms**, in which the electron, usually the one outside the inner shells (see the alkali atoms at §2.1) on the average is at very large distance from the nucleus. These atoms are found in interstellar spaces or can be produced in laboratory by irradiating atomic beams with lasers. The Rydberg atoms are therefore similar to Hydrogen atoms in excited states, the effective charge Z_{eff} (Fig. 1.3) being close to unit. Typically the quantum number n can reach several tens, hundreds in cosmic space. Since the expectation value of the distance (Table I.4.3) increases with n^2, the "dimension" of the Rydberg atoms can reach $10^3 - 10^4$ Å. In these states the life time is very long (we shall see in Appendix I.2 how the life time is related to the spontaneous emission of radiation) of the order of one second instead of the typical 10^{-8} s for inner levels in the Hydrogenic atoms. The eigenvalues scale with n^2 (Eq. 1.13) and become of the order of 10^{-2} eV. Thus the Rydberg atoms are easily ionized and highly polarizable, the electric polarizability increasing approximately with the seventh power of the quantum number n (see §4.2 and Problem F.IV.11).

Problems I.5

Problem I.5.1 In the Hydrogen atom the H_α line (see Prob. I.4.4) has a wavelength 6562.80 Å. In Deuterium the H_α line shifts to 6561.01 Å. Estimate the ratio of the proton to deuteron mass.

Solution:
From Eq. 1.24, $\lambda_D/\lambda_H = (1 + m/M_D)/(1 + m/M_H)$ and then

$$\frac{\Delta\lambda}{\lambda_H} = \frac{m(M_H - M_D)}{M_H M_D(1 + \frac{m}{M_H})} \simeq \frac{m(M_H - M_D)}{M_H M_D} = \frac{-1.79}{6562.8} \simeq \frac{\frac{M_H}{M_D} - 1}{1836} \;,$$

yielding $M_H/M_D \simeq 0.4992$, i.e. $M_D = 2.0032 M_H$.

Problem I.5.2 Show that in Rydberg atoms the frequency of the photon emitted from the transition between adjacent states at large quantum numbers n is close to the rotational frequency of the electron in the circular orbit of the Bohr atom (a particular case of the **correspondence principle**).

Solution
From Eq. 1.24, by neglecting the reduced mass correction, the transition frequency turns out

$$\nu = R_H c \frac{(n_i - n_f)(n_i + n_f)}{n_i^2 n_f^2}$$

which for $n_i, n_f \gg 1$ and $n_i - n_f = 1$ becomes $\nu \simeq 2R_Hc/n^3$.

The Bohr rotational frequency, (see Problem I.4.4) by taking into account the equilibrium condition $mv^2/r = e^2/r^2$, results

$$\nu_{rot} = \frac{mvr}{2\pi mr^2} = \frac{n\hbar m^2 e^4}{2\pi mn^4\hbar^4} = \frac{2R_Hc}{n^3} \ .$$

Problem I.5.3 By direct scaling arguments estimate the order of magnitude of the corrections in the wavefunction and in the eigenvalue for the ground state of Hydrogen when the electron is replaced by a negative muon.

Solution:

Since $\mu^{-1} = (m_P^{-1} + m_\mu^{-1})$, a_0 in the wavefunction is corrected by a factor $\simeq 186$. Since the eigenvalue depends linearly on the mass, the energy is larger than the one in Hydrogen atom by a factor $\simeq 186$. It is noted that these estimates neglect any modification in the potential energy. This is somewhat possible since $Z = 1$, while for heavy atoms (see Fig. 1.8) one should take into account the relevant modification in the potential energy. Similar considerations hold for **Protonium** (i.e. the "atom" with one positive and one negative proton), where only the states at small n are sizeably affected by the modified nuclear potential.

Problem I.5.4 By direct scaling arguments evaluate how the ground state energy, the wavelength of $(2p \to 1s)$ transition and the life time of the $2p$ state are modified from Hydrogen atom to Positronium (for the life time see Appendix I.3 and neglect the **annihilation process** related to the overlap of the wavefunctions in the $1s$ state).

Solution:

The reduced mass is about half of the one in Hydrogen. Therefore the eigenvalue for the ground state is 6.8 eV, instead of 13.6 eV. The transition frequency is at wavelength 2430 Å.

For the life-time, one has to observe from Appendix I.3 that the decay rate is proportional to the third power of the energy separation and to the second power of the dipole matrix element. Since the energy separation is one half while the length scale is twice, the decay rate is $1/2$ and the life time is increased by a factor 2, namely from 1.6 ns to 3.2 ns. One could remark that nuclear-size effects, which are relevant in high-resolution spectroscopy for Hydrogen (App. V.1), are absent for positronium.

Problem I.5.5 In experiments with radiation in cavity interacting with atoms, collimated beams of ^{85}Rb atoms in the $63p$ state are driven to the $61d$ state. On the basis of the classical analogy (see Problem I.5.2) estimate the frequency required for the transition, the "radius" of the atom (for $n = 63$) and the order of magnitude of the electric dipole matrix element.

Solution:

$\nu \simeq 2R_H c \Delta n / n^3 = 55.2$ GHz; $< r > \simeq n^2 a_0 = 2100.4$ Å; dipole matrix element $\delta \simeq e < r > = 1.009 \times 10^{-14}$ u.e.s. cm.

Problem I.5.6 In a Rydberg atom the outer electron is in the $n = 50$ state. Evaluate the electric field $\mathcal{E}$ required to ionize the atom (Hint: assume a potential energy of the form $V(r) = -e^2/r - er\mathcal{E}\cos\theta$ and disregard the possibility of quantum tunneling).

Solution:

From $dV/dr = 0$ the maximum in the potential energy is found at $r_m = \sqrt{e/\mathcal{E}}$, where $V(r_m) = -2e^{3/2}\sqrt{\mathcal{E}}$.

The energy of the Rydberg atom is approximately $E_n \simeq (-e^2/2a_0)(1/n^2)$ and equating it to $V(r_m)$ one obtains $(e^2/2a_0)(1/n^2) = 2e^{3/2}\sqrt{\mathcal{E}}$, i.e. $\mathcal{E} = e/16a_0^2 n^4$, corresponding to

$$\mathcal{E} \simeq 51\text{V/cm}$$

1.6 Orbital and spin magnetic moments and spin-orbit interaction

As we shall see in detail in Chapter 2, the spectral lines observed in moderate resolution (e.g. the yellow doublet resulting from the $3p \leftrightarrow 3s$ transition in the Na atom) indicate that also interactions of magnetic character have to be taken into account in dealing with the electronic structure of the atoms.

The magnetic moment associated with the orbital motion, somewhat corresponding to a current, can be derived from the Hamiltonian for an electron in a static magnetic field $\mathbf{H}$ along the z direction, with vector potential

$$\mathbf{A} = \frac{1}{2}\mathbf{H} \times \mathbf{r} \tag{1.25}$$

and scalar potential $\phi = 0$.

The one-electron Hamiltonian[3] is

$$\mathcal{H} = \frac{1}{2m}(\mathbf{p} + \frac{e}{c}\mathbf{A})^2 + V - e\phi \tag{1.26}$$

yielding, to the first order in $\mathbf{A}$, the operator

$$\mathcal{H} = \mathcal{H}_0 - i\frac{e\hbar}{mc}\mathbf{A}.\nabla \tag{1.27}$$

---

[3] This form of classical Hamiltonian associated with the force $\mathbf{F} = -e\mathcal{E} - e(\mathbf{v}/c)\times\mathbf{H}$ is required in order to have the kinetic energy expressed in terms of the generalized moment $\mathbf{p} = m\mathbf{v} - e\mathbf{A}/c$ (see the text by **Goldstein** quoted in the Preface) so that, in the quantum mechanical description, $\mathbf{p} = -i\hbar\nabla$.

where $\mathcal{H}_0$ is the Hamiltonian in the absence of magnetic or electric fields and where it has been taken into account that $\mathbf{A}$ and ∇ are commuting operators (**Lorentz gauge**). Therefore, in the light of Eq. 1.25 the Hamiltonian describing the effect of the magnetic field is

$$\mathcal{H}_{mag} = -i\frac{e\hbar}{2mc}\mathbf{H} \times \mathbf{r}.\nabla = \frac{e}{2mc}\mathbf{l}.\mathbf{H} \ . \tag{1.28}$$

Compared to the classical Hamiltonian $-\boldsymbol{\mu}.\mathbf{H}$ for a magnetic moment in a field, $\mathcal{H}_{mag}$ allows one to assign to the angular momentum $\mathbf{l}$ a magnetic moment operator given by

$$\boldsymbol{\mu}_l = -\frac{e}{2mc}\hbar\mathbf{l} = -\mu_B\mathbf{l} \tag{1.29}$$

where $\mu_B = e\hbar/2mc$ is called **Bohr magneton**, numerically 0.927×10^{-20} erg /Gauss. Equation 1.29 can be obtained even classically in the framework of the Bohr model for the Hydrogen atom (See Problem I.6.2).

Experimental evidences, such as spectral lines from atoms in magnetic field (see Chapter 4) as well the quantum electrodynamics developed by **Dirac**, indicate that an **intrinsic** angular momentum, the **spin s**, has to be assigned to the electron.

By extending the eigenvalue equations for the orbital angular momentum to spin, one writes

$$\mathbf{s}^2|\alpha > = s(s+1)\hbar^2|\alpha >, \qquad s_z|\alpha > = \frac{\hbar}{2}|\alpha >$$

$$\mathbf{s}^2|\beta > = s(s+1)\hbar^2|\beta >, \qquad s_z|\beta > = -\frac{\hbar}{2}|\beta > \tag{1.30}$$

$|\alpha >$ and $|\beta >$ being the spin eigenfunctions corresponding to quantum spin numbers $m_s = 1/2$ and $m_s = -1/2$, respectively, while $s = 1/2$.

As a first consequence of the spin, in the one-electron eigenfunction (**spin-orbital**) one has to include the spin variable, labeling the value of s_z. When the coupling between orbital and spin variables (the **spin-orbit interaction** that we shall estimate in the following) is weak, one can factorize the function in the form

$$\psi(r,\theta,\varphi,\mathbf{s}) = \phi(r,\theta,\varphi)\chi_{spin} \tag{1.31}$$

where χ_{spin} is $|\alpha >$ or $|\beta >$ depending on the value of the quantum number m_s.

To express the magnetic moment associated with $\mathbf{s}$ without resorting to quantum electrodynamics, one has to make an **ansatz** based on the experimental evidence. In partial analogy to Eq. 1.29 we write

$$\boldsymbol{\mu}_s = -2\mu_B\mathbf{s} \tag{1.32}$$

Due to the existence of elementary magnetic moments, an external magnetic field can be expected to remove the degeneracy in the z-component of the angular momenta. For instance for s_z, two sublevels are generated by the magnetic field, with energy separation $\Delta E = (e\hbar/mc)H$, a phenomenon that can be called **magnetic splitting** (Problem I.6.1).

Now we are going to derive the Hamiltonian describing the interaction between the orbital and the spin magnetic moments. This will be done in the semiclassical model first used by **Thomas** and **Frenkel**, assuming classical expressions for the electric and magnetic fields acting on the electron. By

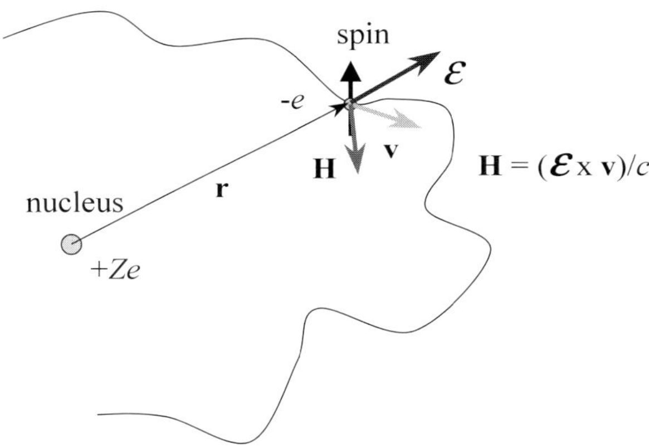

Fig. 1.9. Definition of the magnetic field **H** acting on the electron due to the relative motion of the nucleus of charge Ze, creating an electric field at the position **r**, in view of the relativistic transformation.

referring to Fig. 1.9, the electric field at the electron is $E = (1/er)(dV/dr)\mathbf{r}$ (where V is the central field energy). From the relativistic transformation and by adding a factor $1/2$ introduced by **Thomas** to account for the non-inertial motion, one has

$$\mathbf{H} = \frac{1}{2cer}\frac{dV}{dr}\mathbf{r}\times\mathbf{v}\ .\tag{1.33}$$

Thus the magnetic Hamiltonian becomes

$$\mathcal{H}_{spin-orbit} = -\boldsymbol{\mu}_s.\mathbf{H} = \frac{1}{2m^2c^2r}\frac{dV}{dr}(\mathbf{l.s}) \equiv \xi(r)\mathbf{l.s}\tag{1.34}$$

which can be viewed as an effective r-dependent magnetic field along **l** direction, acting on the spin magnetic moment when the electron is at the position **r**. It is noted that the function $\xi(r)$, of central character, is **essentially positive** and includes $\hbar^2$ from **l** and **s**.

An immediate physical interpretation of the Hamiltonian in Eq. 1.34 can be achieved by referring to Hydrogenic atoms, where

$$\xi(r) = \frac{Ze^2\hbar^2}{2m^2c^2r^3} \tag{1.35}$$

Then the energy associated with $\mathcal{H}_{spin-orbit}$ is of the order of

$$E_{SO} \simeq (Ze^2/2m^2c^2) < r^{-3} > n\hbar.(1/2)\hbar$$

and from Table I.4.3, where $< r^{-3} > \simeq Z^3/a_0^3 n^3 l^3$, one has

$$E_{SO} \simeq \frac{e^2\hbar^2 Z^4}{4m^2c^2 a_0^3 n^5} \tag{1.36}$$

displaying a strong dependence on the atomic number Z. For small Z the spin-orbit interaction turns out of the order of the correction related to the velocity dependence of the mass or to other relativistic terms, that have been neglected (see Problem F.I.15). Typical case is the Hydrogen atom, where the relativistic corrections of **Dirac** and **Lamb** are required in order to account for the detailed fine structure (see Appendix V.1).

From Eq. 1.36 one realizes that the effects of the spin-orbit interaction are strongly reduced for large quantum number n, as it is conceivable in view of the physical mechanism generating the effective magnetic field on the electron.

The energy corrections can easily be derived within the assumption that $\mathcal{H}_{spin-orbit}$ is sizeably weaker than $\mathcal{H}_0$, in Eq. 1.27. Then the perturbation theory can be applied to spin-orbital eigenfunctions, $\psi(r,\theta,\varphi,\mathbf{s}) = \phi(r,\theta,\varphi)\chi_{spin} \equiv \phi_{n,l,m,m_s}$, the operators $\hat{l}^2, \hat{s}^2, l_z, s_z$ being diagonal for the unperturbed system. Since the energy terms are often small in comparison to the energy separation between unperturbed states at different quantum numbers n and l , one can evaluate the energy corrections due to $\mathcal{H}_{spin-orbit}$ within the (nl) representation:

$$\Delta E_{SO} = \int R_{nl}^*(r)\xi(r)R_{nl}(r)r^2 dr \sum_{spin} \int \chi_{m_s'}^* Y_{lm'}^* (\mathbf{l}.\mathbf{s})\chi_{m_s} Y_{lm} sin\theta d\theta d\varphi \tag{1.37}$$

that can be written in the form [4]

$$(\Delta E_{SO})_{m',m_s',m,m_s} = \xi_{nl} < m'm_s'|\mathbf{l}.\mathbf{s}|mm_s > \ . \tag{1.38}$$

The spin orbit constant

$$\xi_{nl} = \int R_{nl}^*(r)\xi(r)R_{nl}(r)r^2 dr \tag{1.39}$$

---

[4] It could be remarked that the ϕ_{n,l,m,m_s} are not the proper eigenfunctions since $(\mathbf{l}.\mathbf{s})$ does not commute with l_z and s_z. However, when $(\mathbf{l}.\mathbf{s})$ is replaced by the linear combination of $\hat{j}^2, \hat{l}^2$ and $\hat{s}^2$ (see Eq. 1.41) and the eigenvalues are derived on the basis of the eigenfunctions of $\hat{j}^2$ and j_z, the appropriate ΔE_{SO} are obtained.

can be thought as a measure of the "average" magnetic field on the electron in the nl state. This average field is again along the direction of $\mathbf{l}$ and acting on $\boldsymbol{\mu}_s$ implies an interaction of the form $\mathcal{H}_{spin-orbit} \propto -\mathbf{h}_{eff}.\boldsymbol{\mu}_s$.

To evaluate the energy corrections due to the Hamiltonian $\xi_{nl}\mathbf{l}.\mathbf{s}$ instead of the formal diagonalization one can proceed with a first step of a more general approach (the so-called **vectorial model**) that we will describe in detail at Chapter 3. Let us define

$$\mathbf{j} = \mathbf{l} + \mathbf{s} \tag{1.40}$$

as the total, single-electron, angular momentum. For analogy with $\mathbf{l}$ and $\mathbf{s}$, $\mathbf{j}$ is specified by a quantum number j (integer or half-integer) and by the magnetic quantum number m_j taking the $(2j+1)$ values from $-j$ to $+j$, with the usual meaning in terms of quantization of the modulus and of the z-component of $\mathbf{j}$, respectively.

The operators $\mathbf{l}$ and $\mathbf{s}$ commute since they act on different variables, so that $\mathbf{l}.\mathbf{s}$ can be substituted by

$$\mathbf{l}.\mathbf{s} = \frac{1}{2}(\hat{j}^2 - \hat{l}^2 - \hat{s}^2) \tag{1.41}$$

involving only the modula, with eigenvalues $j(j+1), l(l+1)$ and $s(s+1)$ [4].

Therefore, for $l \neq 0$ one has the two cases, $j = l + 1/2$ and $j = l - 1/2$, that in a vectorial picture correspond to spin parallel and antiparallel to $\mathbf{l}$.

Then the energy corrections due to $\mathcal{H}_{spin-orbit}$ are

$$\Delta E_{SO} = \xi_{nl}\hbar^2 l/2,$$

for $j = l + 1/2$, and

$$\Delta E_{SO} = -\xi_{nl}\hbar^2(l+1)/2$$

for $j = (l - 1/2)$. Being ξ_{nl} **positive** the doublet sketched below is generated.

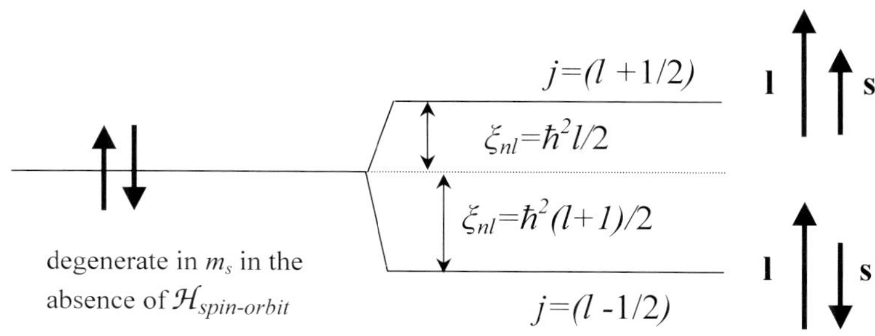

For s state only a **shift**, of relativistic origin, has to be associated with $\mathcal{H}_{spin-orbit}$ (see Problems I.6.3 and F.I.15).

Problems I.6

Problem I.6.1 Show that because of the spin magnetic moment, a magnetic field removes the degeneracy in m_s and two sublevels with energy separation $(e\hbar/mc)H$ are induced (**magnetic splitting**).

Solution:

From the Hamiltonian

$$\mathcal{H}_{mag} = -\boldsymbol{\mu}_s.\mathbf{H} = -(-2\mu_B s_z)H = \frac{e\hbar}{mc}H s_z$$

and the s_z eigenvalues $\pm 1/2$, one has

$$\Delta E = E_{1/2} - E_{-1/2} = \frac{e\hbar}{mc}H \equiv 2\mu_B H$$

with the splitting sketched below

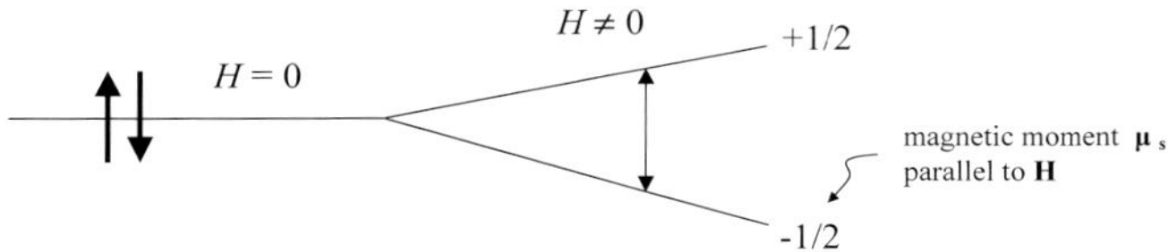

Problem I.6.2 By referring to the electron in the circular orbit of the Bohr model, derive the relationship between angular momentum and magnetic moment. By assigning to the electron the spin magnetic moment derive the correction to the energy levels due to spin-orbit interaction, comparing the results for $n = 2$ and $n = 3$ to the estimates in the Thomas-Frenkel approach (§1.6).

Solution:

The magnetic moment is $\boldsymbol{\mu} = (iA/c)\hat{n}$ with current $i = -e\nu_{rot}$ (see Problem I.5.2). A is the area of the orbit of radius r and $\hat{n}$ the normal. Thus $\boldsymbol{\mu} = -(e\nu\pi r^2/2\pi rc)\hat{n} \equiv -\mu_B \mathbf{l}$.

The magnetic field turns out

$$\mathbf{H} = -\frac{\boldsymbol{\mu}}{r^3} = \frac{e}{2cr^3}\mathbf{v} \times (-\mathbf{r})$$

Therefore the spin-orbit Hamiltonian is $\mathcal{H}_{spin-orbit} = -\boldsymbol{\mu}_s.\mathbf{H} = (e^2\hbar^2/2m^2c^2r^3)\mathbf{l}.\mathbf{s}$. For $r_n = n^2 a_0$ and Eq. 1.41 the energy correction is

$$E_{SO} = \frac{e^2\hbar^2}{2m^2c^2n^6a_0^3}\frac{1}{2}[j(j+1) - l(l+1) - s(s+1)]$$

By using for r_n^{-3} the expectation value

$$<r^{-3}> = \frac{1}{a_0^3 n^3 l(l+1)(l+\frac{1}{2})}$$

and indicating $e^2\hbar^2/4m^2c^2a_0^3 = 3.62 \times 10^{-4}$ eV with E_0, one writes

$$E_{SO} = E_0\frac{1}{n^3 l(l+1)(l+\frac{1}{2})}[j(j+1) - l(l+1) - s(s+1)]$$

and

$$
\begin{array}{ll}
n = 2, l = 1, j = 1/2 & E_{SO} = -\frac{E_0}{12} \\
n = 2, l = 1, j = 3/2 & E_{SO} = \frac{E_0}{24} \\
n = 3, l = 1, j = 1/2 & E_{SO} = -\frac{2E_0}{81} \\
n = 3, l = 1, j = 3/2 & E_{SO} = \frac{E_0}{81}
\end{array}
$$

Problem I.6.3 By referring to one-electron s states try to derive the correction to the unperturbed energy value due to $\mathcal{H}_{spin-orbit}$, making a remark on what has to be expected.

Solution:

$\mathcal{H}_{spin-orbit} = \xi_{nl}(\mathbf{l.s})$ with $\xi_{nl} \propto \int R_{nl}^*(r)\xi(r)R_{nl}(r)r^2 dr$.

Since $R_{nl}(r) \propto r^l$, for $l = 0$, ξ_{nl} diverges for $r \to 0$, while $\mathbf{l.s} = 0$.

The final result is an energy shift that cannot be derived along the procedure neglecting relativistic effects (see Problem F.I.15). A discussion of the fine and hyperfine structure in the Hydrogen atom, including the relativistic effects, is given in Appendix V.1.

Problem I.6.4 Evaluate the effective magnetic field that can be associated with the orbital motion of the optical electron in the Na atom, knowing that the transition $3p \to 3s$ yields a doublet with two lines at wavelenghts 5889.95 Å and 5895.92 Å.

Solution:

From the difference in the wavelengths the energy separation of the $3p$ levels turns out

$$|\Delta E| = \frac{hc|\Delta\lambda|}{\lambda^2} = 2.13 \times 10^{-3}\text{eV}$$

ΔE can be thought to result from an effective field $H = \Delta E/2\mu_B$ (see Prob. I.6.1). Thus, $H = 2.13 \times 10^{-3}/2 \times 5.79 \times 10^{-5}$ Tesla $= 18.4$ Tesla.

Problem I.6.5 The ratio (magnetic moment $\boldsymbol{\mu}$/angular momentum $\mathbf{L}$), often expressed as $(\boldsymbol{\mu}/\mu_B)/(\mathbf{L}/\hbar)$, is called **gyromagnetic ratio**. Assuming that the electron is a sphere of mass m and charge $-e$ homogeneously distributed onto the surface, rotating at constant angular velocity, show that the gyromagnetic ratio turns out $\gamma = \boldsymbol{\mu}/\mathbf{L} = -5e/6mc$.

Solution:
$m = (4\pi/3)\rho R^3$ while the angular momentum is
$L = \int_0^{2\pi} \int_0^{\pi} \int_0^{R} \rho\omega r^4 sin^3\theta d\theta d\varphi dr = (2/5)mR^2\omega$, ρ being the specific mass.

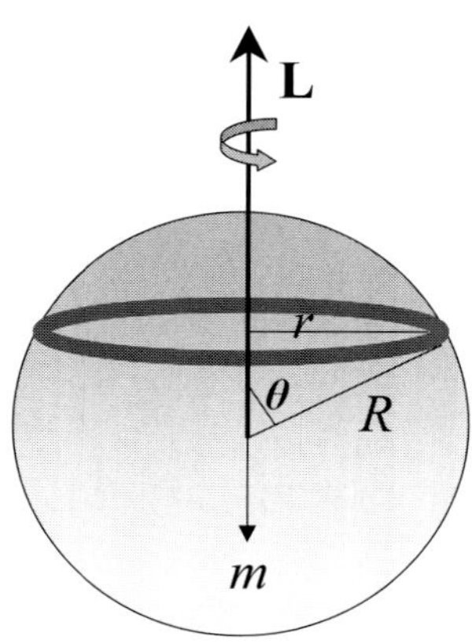

The surface charge density is $\sigma = -e/4\pi R^2$ and from
$\mu = Ai/c = (\pi R^4 \sigma\omega/c) \int sin^3\theta d\theta = -5eL/6mc$ one has $\gamma = -5e/6mc$.

Problem I.6.6 Express numerically the spin-orbit constant ξ_{nl} for the $3p, 3d$ and $4f$ states of the Hydrogen atom.

Solution:
From Eq. 1.35 and the expectation values of $<r^{-3}>$ (Table I.4.3)
$\xi_{3p} = 1.29 \times 10^{37}$ erg^{-1}sec^{-2} $\hbar^2 = 8.94 \times 10^{-6}$ eV,
$\xi_{3d} = 2.58 \times 10^{36}$ erg^{-1}sec^{-2} $\hbar^2 = 1.79 \times 10^{-6}$ eV,
$\xi_{4f} = 3.88 \times 10^{35}$ erg^{-1}sec^{-2} $\hbar^2 = 0.27 \times 10^{-6}$ eV.

Problem I.6.7 Show that when the spin-orbit interaction is taken into account the effective magnetic moment of an electron can be written
$\boldsymbol{\mu}_{\pm} = (-e/2mc)g_{\pm}(\mathbf{l} + \mathbf{s})$ with $g_{\pm} = 1 \pm [1/(2l + 1)]$.
Solution:

Here g is a particular case of the **Lande' g factor**, to be discussed at §3.2. $\pm$ means spin parallel or antiparallel to **l**.

For **s** $\parallel$ **l**

$$g_+ = 1 + \frac{(2l+1)(2l+3)+3-2l(2l+2)}{2(2l+1)(2l+3)} = 1 + \frac{1}{(2l+1)}$$

while for g_-, **s** antiparallel to **l**

$$g_- = 1 + \frac{(2l-1)(2l+1)+3-2l(2l+2)}{2(2l+1)(2l-1)} = 1 - \frac{1}{(2l+1)} \ .$$

1.7 Spectroscopic notation for multiplet states

In the light of spin-orbit interaction the one-electron states have to be labeled by quantum numbers n, l, j and m_j, with $s = 1/2$. Accordingly, a **fine structure** of the levels is induced, in form of **doublets**.

As we shall see in detail in Chapters 2 and 3, in the atom other couplings between $\mathbf{l}_i$ and $\mathbf{s}_i$ occur. At the moment we only state that the whole electronic structure of the atom can be described by the following quantum numbers:

L, taking possible values 0, 1, 2, 3...

S, taking possible values 0, 1/2, 1, 3/2, 2...

J, taking possible values 0, 1/2, 1, 3/2 , 2...

to be associated with the constants of motion

$\mathbf{L} = \sum_i \mathbf{l}_i$, the total angular momentum of orbital character,

$\mathbf{S} = \sum_i \mathbf{s}_i$, the total angular momentum of intrinsic character

and with the total (orbital and spin) angular momentum $\mathbf{J} = \mathbf{L} + \mathbf{S}$ or to $\mathbf{J} = \sum_i \mathbf{j}_i$.

It is customary to use the following notation for the multiplet state of the atom

$$^{2S+1}Letter_J$$

where *Letter* means S, P, D, F, etc... for $L = 0, 1, 2, 3$ etc..., $(2S+1)$ is the total number of the fine structure levels when $S < L$ $((2L+1)$ the analogous when $L < S)$.

The electronic configurations and the spectroscopic notations for the ground-state of the atoms are reported in the following pages.

Z	Element	Symbol	Configuration	Term
1	Hydrogen	H	$1s^1$	$^2S_{1/2}$
2	Helium	He	$1s^2$	$^1S_0$
3	Lithium	Li	$1s^2 2s^1$	$^2S_{1/2}$
4	Beryllium	Be	$1s^2 2s^2$	$^1S_0$
5	Boron	B	$1s^2 2s^2 2p^1$	$^2P_{1/2}$
6	Carbon	C	$1s^2 2s^2 2p^2$	$^3P_0$
7	Nitrogen	N	$1s^2 2s^2 2p^3$	$^4S_{3/2}$
8	Oxygen	O	$1s^2 2s^2 2p^4$	$^3P_2$
9	Fluorine	F	$1s^2 2s^2 2p^5$	$^2P_{3/2}$
10	Neon	Ne	$1s^2 2s^2 2p^6$	$^1S_0$
11	Sodium	Na	$[Ne]3s^1$	$^2S_{1/2}$
12	Magnesium	Mg	$[Ne]3s^2$	$^1S_0$
13	Aluminum	Al	$[Ne]3s^2 3p^1$	$^2P_{1/2}$
14	Silicon	Si	$[Ne]3s^2 3p^2$	$^3P_0$
15	Phosphorus	P	$[Ne]3s^2 3p^3$	$^4S_{3/2}$
16	Sulfur	S	$[Ne]3s^2 3p^4$	$^3P_2$
17	Chlorine	Cl	$[Ne]3s^2 3p^5$	$^2P_{3/2}$
18	Argon	Ar	$[Ne]3s^2 3p^6$	$^1S_0$
19	Potassium	K	$[Ar]4s^1$	$^2S_{1/2}$
20	Calcium	Ca	$[Ar]4s^2$	$^1S_0$
21	Scandium	Sc	$[Ar]3d^1 4s^2$	$^2D_{3/2}$
22	Titanium	Ti	$[Ar]3d^2 4s^2$	$^3F_2$
23	Vanadium	V	$[Ar]3d^3 4s^2$	$^4F_{3/2}$
24	Chromium	Cr	$[Ar]3d^5 4s^1$	$^7S_3$
25	Manganese	Mn	$[Ar]3d^5 4s^2$	$^6S_{5/2}$
26	Iron	Fe	$[Ar]3d^6 4s^2$	$^5D_4$
27	Cobalt	Co	$[Ar]3d^7 4s^2$	$^4F_{9/2}$
28	Nickel	Ni	$[Ar]3d^8 4s^2$	$^3F_4$
29	Copper	Cu	$[Ar]3d^{10} 4s^1$	$^2S_{1/2}$
30	Zinc	Zn	$[Ar]3d^{10} 4s^2$	$^1S_0$
31	Gallium	Ga	$[Ar]3d^{10} 4s^2 4p^1$	$^2P_{1/2}$
32	Germanium	Ge	$[Ar]3d^{10} 4s^2 4p^2$	$^3P_0$
33	Arsenic	As	$[Ar]3d^{10} 4s^2 4p^3$	$^4S_{3/2}$
34	Selenium	Se	$[Ar]3d^{10} 4s^2 4p^4$	$^3P_2$
35	Bromine	Br	$[Ar]3d^{10} 4s^2 4p^5$	$^2P_{3/2}$
36	Krypton	Kr	$[Ar]3d^{10} 4s^2 4p^6$	$^1S_0$

Z	Element	Symbol	Configuration	Term
37	Rubidium	Rb	$[\text{Kr}]5\text{s}^1$	$^2\text{S}_{1/2}$
38	Strontium	Sr	$[\text{Kr}]5\text{s}^2$	$^1\text{S}_0$
39	Yttrium	Y	$[\text{Kr}]4\text{d}^1 5\text{s}^2$	$^2\text{D}_{3/2}$
40	Zirconium	Zr	$[\text{Kr}]4\text{d}^2 5\text{s}^2$	$^3\text{F}_2$
41	Niobium	Nb	$[\text{Kr}]4\text{d}^4 5\text{s}^1$	$^6\text{D}_{1/2}$
42	Molybdenum	Mo	$[\text{Kr}]4\text{d}^5 5\text{s}^1$	$^7\text{S}_3$
43	Technetium	Tc	$[\text{Kr}]4\text{d}^5 5\text{s}^2$	$^6\text{S}_{5/2}$
44	Ruthenium	Ru	$[\text{Kr}]4\text{d}^7 5\text{s}^1$	$^5\text{F}_5$
45	Rhodium	Rh	$[\text{Kr}]4\text{d}^8 5\text{s}^1$	$^4\text{F}_{9/2}$
46	Palladium	Pd	$[\text{Kr}]4\text{d}^{10}$	$^1\text{S}_0$
47	Silver	Ag	$[\text{Kr}]4\text{d}^{10} 5\text{s}^1$	$^2\text{S}_{1/2}$
48	Cadmium	Cd	$[\text{Kr}]4\text{d}^{10} 5\text{s}^2$	$^1\text{S}_0$
49	Indium	In	$[\text{Kr}]4\text{d}^{10} 5\text{s}^2 5\text{p}^1$	$^2\text{P}_{1/2}$
50	Tin	Sn	$[\text{Kr}]4\text{d}^{10} 5\text{s}^2 5\text{p}^2$	$^3\text{P}_0$
51	Antimony	Sb	$[\text{Kr}]4\text{d}^{10} 5\text{s}^2 5\text{p}^3$	$^4\text{S}_{3/2}$
52	Tellurium	Te	$[\text{Kr}]4\text{d}^{10} 5\text{s}^2 5\text{p}^4$	$^3\text{P}_2$
53	Iodine	I	$[\text{Kr}]4\text{d}^{10} 5\text{s}^2 5\text{p}^5$	$^2\text{P}_{3/2}$
54	Xenon	Xe	$[\text{Kr}]4\text{d}^{10} 5\text{s}^2 5\text{p}^6$	$^1\text{S}_0$
55	Cesium	Cs	$[\text{Xe}]6\text{s}^1$	$^2\text{S}_{1/2}$
56	Barium	Ba	$[\text{Xe}]6\text{s}^2$	$^1\text{S}_0$
57	Lanthanum	La	$[\text{Xe}]5\text{d}^1 6\text{s}^2$	$^2\text{D}_{3/2}$
58	Cerium	Ce	$[\text{Xe}]4\text{f}^1 5\text{d}^1 6\text{s}^2$	$^1\text{G}_4$
59	Praseodymium	Pr	$[\text{Xe}]4\text{f}^3 6\text{s}^2$	$^4\text{I}_{9/2}$
60	Neodymium	Nd	$[\text{Xe}]4\text{f}^4 6\text{s}^2$	$^5\text{I}_4$
61	Promethium	Pm	$[\text{Xe}]4\text{f}^5 6\text{s}^2$	$^6\text{H}_{5/2}$
62	Samarium	Sm	$[\text{Xe}]4\text{f}^6 6\text{s}^2$	$^7\text{F}_0$
63	Europium	Eu	$[\text{Xe}]4\text{f}^7 6\text{s}^2$	$^8\text{S}_{7/2}$
64	Gadolinium	Gd	$[\text{Xe}]4\text{f}^7 5\text{d}^1 6\text{s}^2$	$^9\text{D}_2$
65	Terbium	Tb	$[\text{Xe}]4\text{f}^9 6\text{s}^2$	$^6\text{H}_{15/2}$
66	Dysprosium	Dy	$[\text{Xe}]4\text{f}^{10} 6\text{s}^2$	$^5\text{I}_8$
67	Holmium	Ho	$[\text{Xe}]4\text{f}^{11} 6\text{s}^2$	$^4\text{I}_{15/2}$
68	Erbium	Er	$[\text{Xe}]4\text{f}^{12} 6\text{s}^2$	$^3\text{H}_6$
69	Thulium	Tm	$[\text{Xe}]4\text{f}^{13} 6\text{s}^2$	$^2\text{F}_{7/2}$
70	Ytterbium	Yb	$[\text{Xe}]4\text{f}^{14} 6\text{s}^2$	$^1\text{S}_0$
71	Lutetium	Lu	$[\text{Xe}]4\text{f}^{14} 5\text{d}^1 6\text{s}^2$	$^2\text{D}_{3/2}$
72	Hafnium	Hf	$[\text{Xe}]4\text{f}^{14} 5\text{d}^2 6\text{s}^2$	$^3\text{F}_2$

Z	Element	Symbol	Configuration	Term
73	Tantalum	Ta	$[Xe]4f^{14}5d^36s^2$	$^4F_{3/2}$
74	Tungsten	W	$[Xe]4f^{14}5d^46s^2$	$^5D_0$
75	Rhenium	Re	$[Xe]4f^{14}5d^56s^2$	$^6S_{5/2}$
76	Osmium	Os	$[Xe]4f^{14}5d^66s^2$	$^5D_4$
77	Iridium	Ir	$[Xe]4f^{14}5d^76s^2$	$^4F_{9/2}$
78	Platinum	Pt	$[Xe]4f^{14}5d^96s^1$	$^3D_3$
79	Gold	Au	$[Xe]4f^{14}5d^{10}6s^1$	$^2S_{1/2}$
80	Mercury	Hg	$[Xe]4f^{14}5d^{10}6s^2$	$^1S_0$
81	Thallium	Tl	$[Xe]4f^{14}5d^{10}6s^26p^1$	$^2P_{1/2}$
82	Lead	Pb	$[Xe]4f^{14}5d^{10}6s^26p^2$	$^3P_0$
83	Bismuth	Bi	$[Xe]4f^{14}5d^{10}6s^26p^3$	$^4S_{3/2}$
84	Polonium	Po	$[Xe]4f^{14}5d^{10}6s^26p^4$	$^3P_2$
85	Astatine	At	$[Xe]4f^{14}5d^{10}6s^26p^5$	$^2P_{3/2}$
86	Radon	Rn	$[Xe]4f^{14}5d^{10}6s^26p^6$	$^1S_0$
87	Francium	Fr	$[Rn]7s^1$	$^2S_{1/2}$
88	Radium	Ra	$[Rn]7s^2$	$^1S_0$
89	Actinium	Ac	$[Rn]\,6d^17s^2$	$^2D_{3/2}$
90	Thorium	Th	$[Rn]\,6d^27s^2$	$^3F_2$
91	Protactinium	Pa	$[Rn]5f^26d^17s^2$	$^4K_{11/2}$
92	Uranium	U	$[Rn]5f^36d^17s^2$	$^5L_6$
93	Neptunium	Np	$[Rn]5f^46d^17s^2$	$^6L_{11/2}$
94	Plutonium	Pu	$[Rn]5f^67s^2$	$^7F_0$
95	Americium	Am	$[Rn]5f^77s^2$	$^8S_{7/2}$
96	Curium	Cm	$[Rn]5f^76d^17s^2$	$^9D_2$
97	Berkelium	Bk	$[Rn]5f^97s^2$	$^6H_{15/2}$
98	Californium	Cf	$[Rn]5f^{10}7s^2$	$^5I_8$
99	Einsteinium	Es	$[Rn]5f^{11}7s^2$	$^4I_{15/2}$
100	Fermium	Fm	$[Rn]5f^{12}7s^2$	$^3H_6$
101	Mendelevium	Md	$[Rn]5f^{13}7s^2$	$^2F_{7/2}$
102	Nobelium	No	$[Rn]5f^{14}7s^2$	$^1S_0$
103	Lawrencium	Lr	$[Rn]5f^{14}7s^27p^1$	$^2P_{1/2}$
104	Rutherfordium	Rf	$[Rn]\,5f^{14}6d^27s^2$	$^3F_2$
105	Dubnium	Db	$[Rn]\,5f^{14}6d^37s^2$	$^4F_{3/2}$
106	Seaborgium	Sg	$[Rn]\,5f^{14}6d^47s^2$	$^5D_0$
107	Bohrium	Bh	$[Rn]\,5f^{14}6d^57s^2$	$^6S_{5/2}$
108	Hassium	Hs	$[Rn]\,5f^{14}6d^67s^2$	$^5D_4$

Appendix I.1 Electromagnetic spectral ranges and useful numbers

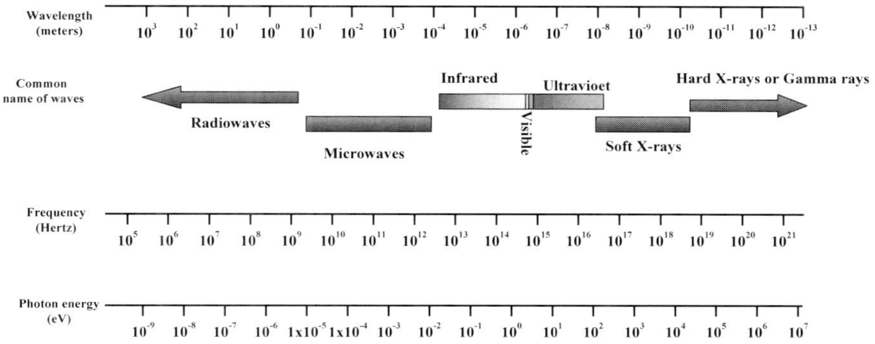

Useful numbers and fundamental constants (for magnetic quantities see App.IV.1)

Speed of light in vacuum	$c=2.99792 \times 10^{10}$ cm/s
Electron charge	$e=-1.60218 \times 10^{-19}$ Coulomb$=-4.8 \times 10^{-10}$ u.e.s.
Electron mass (at rest)	$m=9.10938 \times 10^{-28}$ g
Proton mass	$M=1.67262 \times 10^{-24}$ g
Neutron mass	$M_n=1.675 \times 10^{-24}$ g
Atomic mass unit (m(^{12}C)/12)	$u=1.661 \times 10^{-24}$ g
Planck constant	$h=6.62607 \times 10^{-27}$ erg.s$=4.1357 \times 10^{-15}$ eV.s
	$\hbar=(h/2\pi)=1.05457 \times 10^{-27}$ erg.s
Boltzmann constant	$k_B=1.38065 \times 10^{-16}$ erg/K
Stefan-Boltzmann constant (total emittance)	$\sigma=5.67 \times 10^{-5}$ erg/(s.cm^2.K^4)$= 5.67 \times 10^{-8}$ W/(m^2.K^4)
Bohr radius for atomic hydrogen (infinite nuclear mass)	$a_0=0.52918 \times 10^{-8}$ cm$= 0.52918$ Å
Rydberg constant (or Bohr energy e/2a$_0$) (infinite nuclear mass)	$R_H=109737$ cm$^{-1}= 13.606$ eV$=h.(3.29 \times 10^{15}$ Hz)
Bohr magneton	$\mu_B=e\hbar/2mc=0.9274 \times 10^{-20}$ erg/Gauss$= 0.9274 \times 10^{-23}$ A.m$^2= 0.9274 \times 10^{-23}$ J/Tesla (see App.IV.1)
Nuclear magneton	$M_N=\mu_B m/M= \mu_B/1836.15= =e\hbar/2Mc=5.0508 \times 10^{-24}$ erg/Gauss
Proton magnetic moment (maximum component)	$\mu_P=M_N g_N I=M_N(5.586)(1/2)=1.4106 \times 10^{-23}$ erg/Gauss
Neutron magnetic moment	$\mu_n=-1.91315\ M_N$
Avogadro number	$N_A=6.022 \times 10^{23}$ mol^{-1}
Electron volt	1 eV$= 1.602 \times 10^{-12}$ erg$= h.(2.418 \times 10^{14}$ Hz) 1 erg$= 6.242 \times 10^{11}$ eV
Gas constant	$R=N_A k_B= 8.31447 \times 10^7$ erg/(mol.K)
k$_B$T at room temperature	0.0259 eV$\approx 1/40$ eV
Fine structure constant	$\alpha=e^2/\hbar c=1/137.036$

Appendix I.2 Perturbation effects in two-levels system

We shall refer to a model system with two eigenstates, labeled $|1>$ and $|2>$, and correspondent eigenfunctions ϕ_1^0 and ϕ_2^0 forming a complete orthonormal basis. The model Hamiltonian is $\mathcal{H}_0$ and $\mathcal{H}_0\phi_m^0 = E_m\phi_m^0$, with $m = 1, 2$. In real systems the Hamiltonian $\mathcal{H}$ can differ from $\mathcal{H}_0$ owing to a small perturbation $\mathcal{H}_P$. Following a rapid transient (after turning on the perturbation) the stationary states are described by eigenfunctions that differ from the ones of the model system by a small amount, that can be written in terms of the unperturbed basis. This is equivalent to state that the eigenfunctions of the equation

$$\mathcal{H}\phi = E\phi \qquad\qquad (\text{A.I.2.1})$$

are

$$\phi = c_1\phi_1^0 + c_2\phi_2^0 \qquad\qquad (\text{A.I.2.2})$$

with c_1 and c_2 constants. By inserting ϕ in A.I.2.1 and multiplying by $<\phi_1^0|$ and by $<\phi_2^0|$ in turn, in the light of the orthonormality of the states, one derives for $c_{1,2}$

$$c_1(\mathcal{H}_{11} - E) + c_2\mathcal{H}_{12} = 0$$
$$c_1\mathcal{H}_{21} + c_2(\mathcal{H}_{22} - E) = 0$$

with $\mathcal{H}_{mn} =< m|\mathcal{H}|n >$. Non-trivial solutions imply

$$det \begin{pmatrix} \mathcal{H}_{11} - E & \mathcal{H}_{12} \\ \mathcal{H}_{21} & \mathcal{H}_{22} - E \end{pmatrix} = 0$$

and the eigenvalues turn out

$$E_{\mp} = \frac{1}{2}(\mathcal{H}_{11} + \mathcal{H}_{22}) \pm \frac{1}{2}\sqrt{(\mathcal{H}_{11} - \mathcal{H}_{22})^2 + 4\mathcal{H}_{12}\mathcal{H}_{21}} \qquad (\text{A.I.2.3})$$

When the diagonal elements of $\mathcal{H}_P$ are zero, A.I.2.3 reduces to

$$E_{\mp} = \frac{1}{2}(E_1 + E_2) \pm \frac{1}{2}\sqrt{(E_1 - E_2)^2 + 4\varepsilon^2} \qquad (\text{A.I.2.4})$$

where $\varepsilon^2 = | < 2|\mathcal{H}_P|1 > |^2$, $\mathcal{H}_P$ being Hermitian.

The perturbation effects strongly depend on the energy separation $\Delta E = E_2 - E_1$. For degenerate energy levels ($\Delta E = 0$) the largest shift of the levels occurs, given by 2ε. For perturbation much weaker than ΔE, Eq.A.I.2.4 can be expanded, to yield the second-order corrections

$$E_{\mp} = E_{2,1} \pm \frac{\varepsilon^2}{\Delta E} \qquad\qquad (\text{A.I.2.5})$$

The corrections to the unperturbed eigenvalues as a function of ΔE are illustrated below

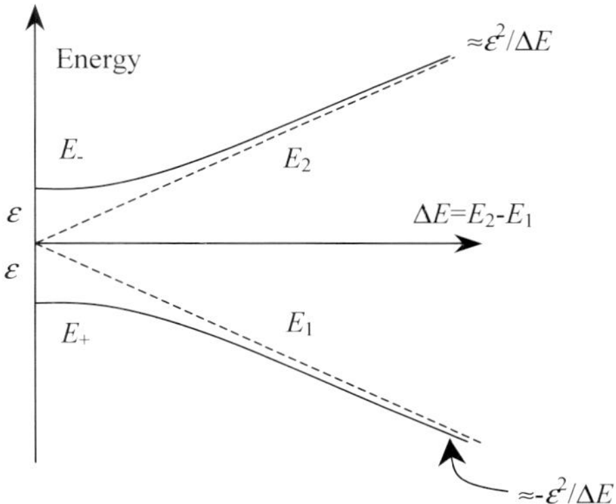

The eigenfunctions in the presence of $\mathcal{H}_P$ can be obtained by deriving the coefficients $c_{1,2}$ in A.I.2.2 in correspondence to $E = E_+$ and $E = E_-$. For widely separated unperturbed states one obtains

$$\phi_+ \simeq \phi_1^0 - \frac{\mathcal{H}_{12}}{\Delta E}\phi_2^0, \qquad \phi_- \simeq \phi_2^0 + \frac{\mathcal{H}_{12}}{\Delta E}\phi_1^0$$

while for degenerate eigenstates

$$\phi_+ = \frac{1}{\sqrt{2}}(\phi_1^0 + \frac{\mathcal{H}_{12}}{|\mathcal{H}_{12}|}\phi_2^0), \qquad \phi_- = \frac{1}{\sqrt{2}}(\phi_1^0 - \frac{\mathcal{H}_{12}}{|\mathcal{H}_{12}|}\phi_2^0) \qquad \text{(A.I.2.6)}$$

Now we turn to the **time evolution** of the system, by considering two cases. The first is the evolution of the system after a static, time-independent perturbation has been turned on, the second (to be discussed as Appendix I.3) when a periodic time-dependent perturbation is applied.

To deal with the time dependence one has to refer to the complete unperturbed eigenfunctions and to the time-dependent Schrodinger equation:

$$[\mathcal{H}_0 + \mathcal{H}_P(t)]\psi = i\hbar\frac{\partial\psi}{\partial t} \qquad \text{(A.I.2.7)}$$

The eigenfunction A.I.2.2 is now written with time dependent coefficients

$$\psi = c_1(t)\psi_1^0 + c_2(t)\psi_2^0 \qquad \text{(A.I.2.8)}$$

with $|c_1|^2 + |c_2|^2 = 1$. Let us assume the initial condition $c_1(t = 0) = 1$ and $c_2(t = 0) = 0$. The probability that at the time t after turning on the perturbation the system is found in the state $|2>$ is given by

$$P_2(t) = |c_2(t)|^2 \qquad \text{(A.I.2.9)}$$

The equation for $c_2(t)$ is obtained by inserting A.I.2.8 into A.I.2.7. Recalling that

$$\mathcal{H}_0 \psi^0_{1,2} = i\hbar \frac{\partial \psi^0_{1,2}}{\partial t}$$

one has

$$\mathcal{H}_P(c_1 \psi^0_1 + c_2 \psi^0_2) = i\hbar\left(\psi^0_1 \frac{dc_1}{dt} + \psi^0_2 \frac{dc_2}{dt}\right) \qquad \text{(A.I.2.10)}$$

By multiplying this equation by $(\psi^0_1)^*$, integrating over the spatial coordinates and by taking into account that $\psi^0_{1,2}(t) = \phi^0_{1,2} exp(-iE^o_{1,2}t/\hbar)$ one finds

$$c_1 < 1|\mathcal{H}_P|1 > + c_2 < 1|\mathcal{H}_P|2 > e^{-i\omega_{21}t} = i\hbar \frac{dc_1}{dt} \qquad \text{(A.I.2.11)}$$

where $\omega_{21} = (E^0_2 - E^0_1)/\hbar$; $< 1|\mathcal{H}_P|1 >= \mathcal{H}_{11} \equiv \int (\phi^0_1)^* \mathcal{H}_P \phi^0_1 d\tau$ and $< 1|\mathcal{H}_P|2 >= \mathcal{H}_{12} \equiv \int (\phi^0_1)^* \mathcal{H}_P \phi^0_2 d\tau$ are the matrix elements of the perturbation between the stationary states of the unperturbed system [5].
In analogous way, from A.I.2.10, multiplying by $(\psi^0_2)^*$ one derives

$$c_1 \mathcal{H}_{21} e^{+i\omega_{21}t} + c_2 \mathcal{H}_{22} = i\hbar \frac{dc_2}{dt} \qquad \text{(A.I.2.12)}$$

In order to illustrate these equations for $c_{1,2}$, let us refer to a perturbation which is constant in time, with no diagonal elements. Then $(\mathcal{H}_P)_{11} = (\mathcal{H}_P)_{22} = 0$ and $(\mathcal{H}_P)_{12} = \hbar\Gamma, (\mathcal{H}_P)_{21} = \hbar\Gamma^*$. Eqs. A.I.2.11 and A.I.2.12 become

$$\frac{dc_1}{dt} = -i\Gamma e^{-i\omega_{21}t} c_2 \qquad\qquad \frac{dc_2}{dt} = -i\Gamma^* e^{+i\omega_{21}t} c_1$$

By taking the derivative of the second and by using the first one, one has

$$\frac{d^2 c_2}{dt^2} = i\omega_{21} \frac{dc_2}{dt} - c_2 \Gamma^2$$

of general solution

$$c_2(t) = (Ae^{i\Omega t} + Be^{-i\Omega t})e^{\frac{i\omega_{21}t}{2}}$$

with $\Omega = (1/2)\sqrt{\omega^2_{21} + 4\Gamma^2}$. The constants A and B are obtained from the initial conditions already considered, yielding

$$c_2(t) = -\frac{i\Gamma}{\Omega} sin\Omega t e^{\frac{i\omega_{21}t}{2}}$$

---

[5] In the **Feynman** formulation the coefficients $c_i =< i|\psi(t) >$ are the **amplitudes** that the system is in state $|i >$ at the time t and one has $i\hbar(dc_i/dt) = \sum_j \mathcal{H}_{ij}(t)c_j(t)$, $\mathcal{H}_{ij}$ being the elements of the **matrix Hamiltonian**.

and therefore

$$P_2(t) = |c_2(t)|^2 = \frac{4\Gamma^2}{\omega_{21}^2 + 4\Gamma^2} sin^2 \frac{(\sqrt{\omega_{21}^2 + 4\Gamma^2})t}{2} \quad , \qquad (A.I.2.13)$$

known as **Rabi** equation. $P_1(t) = 1 - P_2(t)$.

It is worthy to illustrate the Rabi equation in the case of equivalent states, so that $E_1^0 = E_2^0$. We shall refer to such a situation in discussing the molecular Hydrogen ion H_2^+ where an electron is shared between two protons (§8.1). Then $\omega_{21} = 0$ and Eq. A.I.2.13 becomes

$$P_2(t) = sin^2 \Gamma t \qquad (A.I.2.14)$$

namely the system oscillates between the two states. After the time $t = \pi/2\Gamma$ the system is found in state $|2 >$, even though the perturbation is weak. For H_2^+ one can say that the electron is being exchanged between the two protons.

For widely separated states so that $\omega_{21}^2 \gg 4\Gamma^2$ Eq. A.I.2.13 yields

$$P_2(t) = (\frac{2\Gamma}{\omega_{21}})^2 sin^2 \frac{\omega_{21}t}{2} \qquad (A.I.2.15)$$

predicting fast oscillations but very small probability to find the system in state $|2 >$.

Pulse resonance techniques (see Chapter 6) can be thought as an application of the Rabi formula once that the two spin states (spin up and spin down in a magnetic field) are "forced to become degenerate" by the on-resonance irradiation at the separation frequency $(E_2^0 - E_1^0)/h$.

In the presence of a relaxation mechanism driving the system to the low-energy state, a term $-i\hbar\gamma$ (with γ the relaxation rate) should be included in the matrix element $\mathcal{H}_{22}$. In this case, from the solution of the equations for the coefficients $c_{1,2}(t)$ the probability $P_2(t)$ corrects Eq. A.I.2.14 for the Rabi oscillations with a damping effect. For strong damping the oscillator crosses to the overdamped regime: after an initial raise $P_2(t)$ decays to zero without any oscillation (see the book by **Budker, Kimball** and **De Mille** quoted in the preface). Some more detail on the relaxation mechanism for spins in a magnetic field will be given at Chapter 6.

Appendix I.3 Transition probabilities and selection rules

The phenomenological transition probabilities induced by electromagnetic radiation are defined in Problem F.I.1, where the **Einstein relations** are also derived. To illustrate the mechanism underlying the effect of the radiation one has to express the absorption probability W_{12} between two levels $|1 >$ and $|2 >$ in terms of the Hamiltonian describing the interaction of the radiation with the system. Here this description is carried out by resorting to the time-dependent perturbation theory.

The perturbation Hamiltonian $\mathcal{H}_P(t)$, already introduced in Appendix I.2, is then specified in the form

$$\mathcal{H}_P(t) = H_1 e^{i\omega t} \, , \tag{A.I.3.1}$$

appropriate to the electromagnetic (e.m.) radiation. In fact, from the one-electron Hamiltonian in fields (see Eq. 1.26)

$$\mathcal{H} = \frac{(\mathbf{p} + \frac{e\mathbf{A}}{c})^2}{2m} - e\varphi \tag{A.I.3.2}$$

($\mathbf{A}$ and φ vector and scalar potentials), recalling that $[\mathbf{p}, \mathbf{A}] = -i\hbar(\mathbf{\nabla}.\mathbf{A}) \propto div\mathbf{A} = 0$ in the Lorentz gauge and that for electromagnetic radiation $\mathbf{A}(\mathbf{r}, t) = \mathbf{A}_0 exp[i(\mathbf{k}.\mathbf{r} - \omega t)]$, the first order perturbation Hamiltonian turns out

$$\mathcal{H}_{rad} = -\frac{i\hbar e}{mc}\mathbf{A}.\mathbf{\nabla} \tag{A.I.3.3}$$

By expanding $\mathbf{A}(\mathbf{r}, t)$

$$\mathbf{A}(\mathbf{r}, t) = \mathbf{A}_0 e^{-i\omega t}[1 + i(\mathbf{k}.\mathbf{r}) + ...] \tag{A.I.3.4}$$

and limiting the attention to the site-independent term (**electric dipole approximation** or **long-wave length approximation**) one can show that[6]

$$\mathcal{H}_{rad} \propto \mathbf{A}_0.\mathbf{\nabla} \propto \mathbf{A}_0.\mathbf{r} \propto \mathbf{E}_0.\mathbf{r}\frac{c}{\omega_{21}}$$

Therefore H_1 in A.I.3.1 takes the form $H_1 = -e\mathbf{r}.\mathbf{E}_0$, with $\mathbf{E}_0$ amplitude of the e.m. field (**electric dipole mechanism of transition**).

Now we use the results obtained in Appendix A.I.2, again considering that $(\mathcal{H}_{rad})_{11} = (\mathcal{H}_{rad})_{22} = 0$ and $(\mathcal{H}_{rad})_{12} = (\mathcal{H}_{rad})_{21}^*$. The equations for the coefficients $c_{1,2}$ become

$$i\hbar\frac{dc_1}{dt} = c_2 e^{-i\omega_{21}t}cos\omega t < 1|x|2 > eE_0$$

$$i\hbar\frac{dc_2}{dt} = c_1 e^{+i\omega_{21}t}cos\omega t < 2|x|1 > eE_0 \tag{A.I.3.5}$$

for a given x-component of the operator $\mathbf{r}$. For Eqs. A.I.3.5 only approximate solutions are possible, essentially based on the perturbation condition

---

[6] It is recalled that $\mathbf{E} = -(1/c)\partial\mathbf{A}/\partial t$ and that the matrix element of the $\mathbf{\nabla}$ operator can be expressed in terms of the one for $\mathbf{r}$:

$$< 2|\mathbf{\nabla}|1 >= -m\omega_{21} < 2|\mathbf{r}|1 > /\hbar$$

$\mathcal{H}_P \ll \mathcal{H}_0$ (while they are solved exactly for $\omega = 0$, as seen in App.I.2). For ω around ω_{21} one finds

$$|c_2(t)|^2 = \frac{t^2}{\hbar^2} \frac{sin^2\left(\frac{(\omega-\omega_{21})t}{2}\right)}{(\omega-\omega_{21})^2 t^2} | < 2|\mathcal{H}_1|1 > |^2 \; . \qquad (\text{A.I.3.6})$$

$|c_2(t)|^2$ has the time dependence depicted below, with a maximum at $\omega = \omega_{21}$ proportional to t^2.

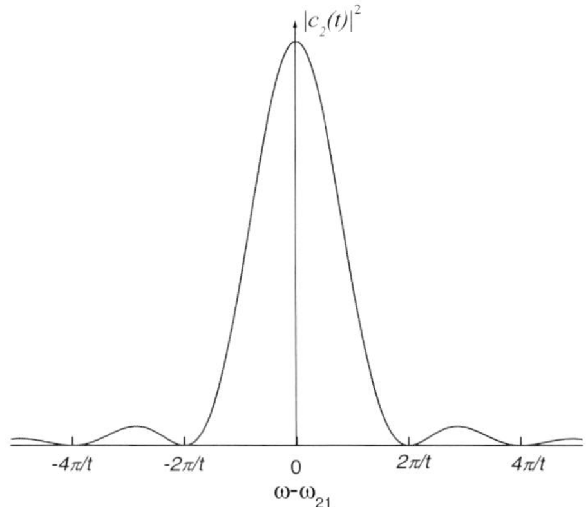

On increasing t the zeroes of the function tend to the origin while the maximum increases. Thus for $t \to \infty$ one has $|c_2(t)|^2 \propto \delta(\omega - \omega_{21})$, δ being the **Dirac delta function**. By taking into account the spread of the excited state due to the finite width (see Prob. F.I.1) or by resorting to the non-monocromatic character of the radiation, one writes

$$|c_2(t)|^2 \propto \frac{t^2}{\hbar^2} \int \rho(\omega) \frac{sin^2 \frac{(\omega-\omega_{21})t}{2}}{(\omega-\omega_{21})^2 \frac{t^2}{4}} d\omega \qquad (\text{A.I.3.7})$$

where the frequency distribution $\rho(\omega)$ of the radiation is a slowly varying function around ω_{21}. Then one can set $\rho(\omega) \simeq \rho(\omega_{21})$. The integration over ω yields $2\pi/t$, and thus the transition probability per unit time becomes

$$W_{12} = |c_2(t)|^2/t = \frac{2\pi}{\hbar^2} | < 2|\mathcal{H}_1|1 > |^2 \delta(\omega - \omega_{21})$$

For the electric dipole mechanism and linear polarization of the radiation along $\hat{\varepsilon}$ this Equation reads

$$W_{12} = \frac{2\pi}{\hbar^2} | < 2| - e\mathbf{r}.\hat{\varepsilon}|1 > |^2 E_0^2 \delta(\omega - \omega_{21}) \qquad (\text{A.I.3.8})$$

For random orientation of $\mathbf{r}$ with respect to the e.m. wave one has to average $cos^2\theta$ over θ, to obtain $1/3$. By introducing the energy density $\rho(\omega_{21})$ or $\rho(\nu_{21})$ ($\rho = < E^2 > /4\pi$) one finally obtains

$$W_{12} = \frac{2\pi}{3\hbar^2}\rho(\nu_{21})|\mathbf{R}_{21}|^2 \tag{A.I.3.9}$$

where $|\mathbf{R}_{21}|^2 = |<2|-ex|1>|^2 + |<2|-ey|1>|^2 + |<2|-ez|1>|^2$.

$\mathbf{R}_{21}$ represents an effective **quantum electric dipole associated with a pair of states**. The selection rules arise from the condition

$$\mathbf{R}_{21} \equiv <2|-e\mathbf{r}|1> \neq 0.$$

In the central field approximation the selection rules are

i) each electron makes a transition independently from the others ;

ii) neglecting the spin, the electric dipole transitions are possible when $\Delta l = \pm 1$ and $\Delta m = 0, \pm 1$ (according to parity arguments involving the spherical harmonics).

When the spin-orbit interaction is taken into account the selection rules are

$\Delta j = 0, \pm 1$ and $j = 0 \leftrightarrow j = 0$ transition not allowed ;

$\Delta m = 0, \pm 1$ and no transition from $m = 0 \leftrightarrow m = 0$, when $\Delta j = 0$ (in view of the conservation of the angular momentum).

The **magnetic dipole** transitions (mechanism associated with the term $(i\mathbf{k.r})$ in A.I.3.4) are controlled by the selection rules

$\Delta l = 0$ and $\Delta m = 0, \pm 1$

while for the transition driven by the **electric quadrupole** mechanism

$\Delta l = 0, \pm 2$ and $\Delta m = 0, \pm 1, \pm 2$ ($l = 0 \leftrightarrow l' = 0$ forbidden)

Further details on the selection rules will be given at §3.5. Here we remark that the transition probabilities associated with the magnetic dipole or with the electric quadrupole mechanisms are smaller than W_{12} in A.I.3.9 by a factor of the order of the square of the **fine structure constant** $\alpha = e^2/\hbar c \simeq 1/137$ (for e.m. radiation in the visible spectral range).

Problems F.I

Problem F.I.1 Refer to an ensemble of non-interacting atoms, each with two levels of energy E_1 (ground state) and E_2 (excited state). By applying the conditions of statistical equilibrium in a black-body radiation bath, derive the relationships among the probabilities of **spontaneous emission** A_{21}, of **stimulated emission** W_{21} and of **absorption** W_{12} (**Einstein relations**), in the assumption that the levels are non-degenerate. Repeat for degenerate levels, with statistical weights g_1 and g_2.

Then assume that at $t = 0$ all the atoms are in the ground state and derive the evolution of the statistical populations $N_1(t)$ and $N_2(t)$ as a function of the time t at which electromagnetic radiation at the transition frequency is turned on (consider the ground and the excited states non-degenerate). Comment the equilibrium situation in terms of the ratio W_{12}/A_{21}.

Briefly discuss some aspects of the Einstein relations in regards of the possible **maser** and **laser actions** and about the finite width of the spectral line (**natural broadening**), by comparing the result based on the **Heisenberg principle** with the classical description of emission from damped harmonic oscillator (**Lorentz** model).

Solution:

From the definition of transition probabilities,

the time dependence of the **statistical populations** are given by

$$\frac{dN_1}{dt} = -N_1 W_{12} + N_2 W_{21} + N_2 A_{21}$$

$$\frac{dN_2}{dt} = +N_1 W_{12} - N_2 W_{21} - N_2 A_{21}$$

The transition probabilities can be written in terms of the e.m. energy density at the transition frequency: $W_{12} = B_{12}\rho(\nu_{12})$, $W_{21} = B_{21}\rho(\nu_{12})$. B_{12} and B_{21} are the **absorption** and **emission coefficients**, respectively.

One can assume that the system attains the equilibrium at a given temperature T inside a cavity where the black-body radiation implies the energy density (see Problem F.I.2)

$$\rho(\nu_{12}) = \frac{8\pi h\nu_{12}^3}{c^3} \frac{1}{e^{\frac{h\nu_{12}}{k_B T}} - 1} \, .$$

At equilibrium $(dN_1/dt) = (dN_2/dt) = 0$. Then

$$\frac{N_1}{N_2} = \frac{W_{21} + A_{21}}{W_{12}} = \frac{\rho B_{21} + A_{21}}{\rho B_{12}}$$

while in accordance to Boltzmann statistics

$$\frac{N_1}{N_2} = e^{\frac{h\nu_{12}}{k_B T}}$$

These three equations are satisfied for

$$B_{21} = B_{12} \qquad \text{and} \qquad A_{21} = \frac{8\pi h\nu_{12}^3}{c^3} B_{21}$$

These **Einstein relations**, derived in equilibrium condition are assumed to hold also out of equilibrium.

For levels 1 and 2 with statistical weights g_1 and g_2 respectively, $N_1/N_2 = \frac{g_1}{g_2} e^{\frac{h\nu_{12}}{k_B T}}$ and from the equilibrium condition

$$A_{21} = \frac{8\pi h\nu_{12}^3}{c^3} \frac{g_1}{g_2} B_{12} \qquad \text{and} \qquad A_{21} = \frac{8\pi h\nu_{12}^3}{c^3} B_{21}$$

so that $g_1 B_{12} = g_2 B_{21}$.

Now the system in the presence of radiation at the transition frequency (with initial condition $N_1(t = 0) = N$ and $N_2(t = 0) = 0$) is considered. Since

$$\frac{dN_1}{dt} = -N_1 W_{12} + N_2 W_{21} + N_2 A_{21} \equiv -N_1 W + N_2(W + A) = -N_1(2W + A) + N$$

one derives

$$N_1(t) = \frac{N}{2W + A}\left(A + W + We^{-(2W+A)t}\right)$$

plotted below:

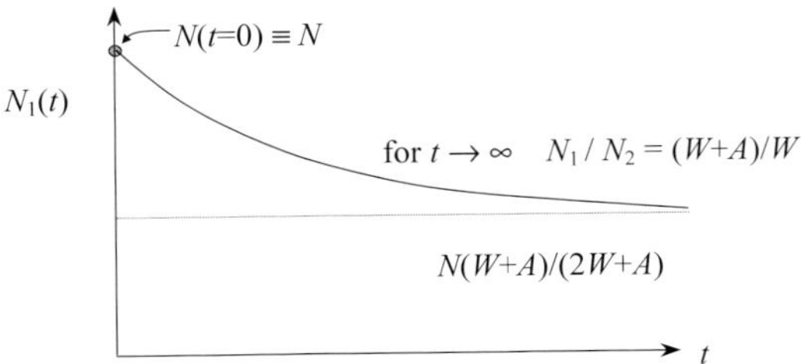

For $A \ll W \equiv W_{12} = W_{21}$ the saturation condition $N_1 = N_2 = N/2$ is achieved. It is noted that for $A \ll W$, by means of selective irradiation at the transition frequency the equilibrium condition implies a statistical temperature (describing N_1/N_2) different from the one of the thermostat. For $N_1 < N_2$ the **statistical temperature** would be negative (further discussion of these concepts is given at Chapter 6).

The condition of **negative temperature** (or **population inversion**) is a pre-condition for having radiation amplification in masers or in lasers. In the latter the **spontaneous emission** (i.e. A) acts as a disturbance, since the "signal" at the output is not driven by the signal at the input of the device.

Since $A_{21} \propto \nu_{12}^3$ the spontaneous emission can be negligible with respect to the **stimulated emission** $B_{12}\rho(\nu_{12})$ in the Microwave (MW) or in the Radiofrequency (RF) ranges, while it is usually rather strong in the visible range.

For the finite linewidth of a transition line the following is remarked. According to the uncertainty principle, because of the finite life-time τ of the excited state, of the order of A^{-1}, the uncertainty in the energy E_2 is $\Delta E \simeq A\hbar$ and then the linewidth is at least $\Delta\nu_{12} \simeq \tau^{-1}$. In the classical Lorentz description, the electromagnetic emission is related to a charge (the electron) in harmonic oscillation, with damping (**radiation damping**). The one-dimensional equation of motion of the charge can be written

$$m\frac{d^2x}{dt^2} + 2\Gamma m\frac{dx}{dt} + m\omega_0^2 x = 0$$

with solution $x(t) = x_0 exp(-\Gamma t)exp(-i\omega_0 t)$. The Fourier transform is $FT[x(t)] = 2x_0/[\Gamma - i(\omega - \omega_0)]$ implying an intensity of the emitted radiation proportional to

$$I(\omega, \Gamma) \propto |FT[x(t)]|^2 \propto \frac{\Gamma}{\Gamma^2 + (\omega - \omega_0)^2}$$

namely a Lorenztian curve, of width Γ:

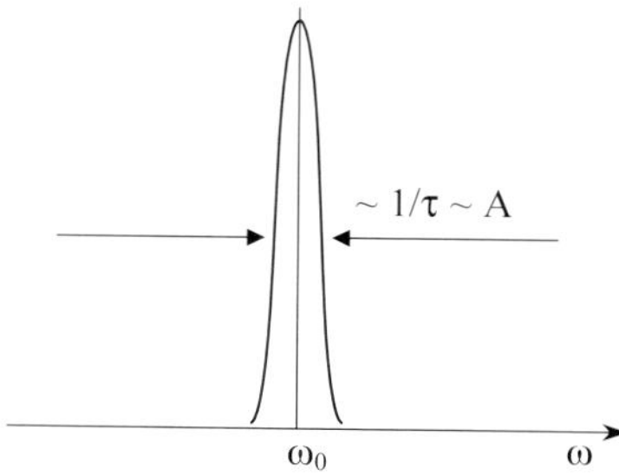

One can identify Γ with $\tau^{-1} \sim A$ and a certain equivalence of the classical description with the semi-classical theory of radiation is thus established.

Problem F.I.2 *(The **black-body** radiation).*

Black-body radiation is the one present in a cavity of a body (e.g. a hot metal) brought to a given temperature T. It is related to the emission of e.m. energy over a wide frequency range.

The energy density $u(\nu, T)$ per unit frequency range around ν can be measured from the radiation $\rho_S(\nu, T)$ coming out from a small hole of area S (the **black-body**), per unit time and unit area. Prove that $\rho_S(\nu, T) = u(\nu, T)c/4$.

The electromagnetic field inside the cavity can be considered as a set of harmonic oscillators (the **modes** of the radiation). From the **Planck**'s estimate of the **thermal statistical energy**, prove that the average number $< n >$ describing the degree of excitation of one oscillator is

$$< n >= 1/[exp(h\nu/k_B T) - 1] \ .$$

Then derive the **number of modes** $D(\omega)d\omega$ in the frequency range $d\omega$ around ω. (Note that $D(\omega)$ does not depend on the shape of the cavity).

By considering the photons as **bosonic** particles derive the **Planck distribution function**, the **Wien** law, the total energy in the cavity and the number of photons per unit volume.

Then consider the radiation as a thermodynamical system, imagine an expansion at constant energy and derive the exponent γ in the adiabatic transformation $TV^{\gamma-1} = const$. Evaluate how the entropy changes during the expansion.

Finally consider the e.m. radiation in the universe. During the expansion of the universe by a factor f each frequency is reduced by the factor $f^{1/3}$. Show that the Planck distribution function is retained along the expansion and derive the f-dependence of the temperature.

Solution:

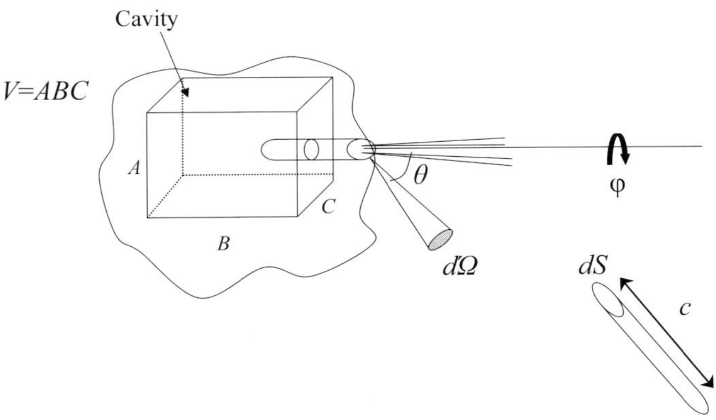

The energy emitted in $d\Omega$ from the element dS is

$$\rho_S dS d\nu = u(\nu, T)c \, cos\theta d\nu \frac{d\Omega}{4\pi} dS$$

Then

$$\rho_S(\nu, T) = \frac{u(\nu, T)c}{4\pi} \int_0^{2\pi} d\varphi \int_0^{\pi/2} cos\theta sin\theta d\theta = \frac{u(\nu, T)c}{4\pi}\frac{2\pi}{2} = \frac{u(\nu, T)c}{4}$$

In the **Planck** estimate the average energy per oscillator instead of being $< \varepsilon >= k_B T$ (as from the **equipartition principle** in the **Maxwell-Boltzmann** statistics) is evaluated according to

$$< \varepsilon >= \frac{\sum_{n=0}^{\infty} n\varepsilon_0 e^{-\frac{n\varepsilon_0}{k_B T}}}{\sum_{n=0}^{\infty} e^{-\frac{n\varepsilon_0}{k_B T}}}$$

where $\varepsilon_0 = h\nu$ is the quantum grain of energy for the oscillator at frequency ν. By defining $x = exp(-\varepsilon_0/k_B T)$ one writes

$$< \varepsilon >= \varepsilon_0 \frac{\sum_{n=0}^{\infty} nx^n}{\sum_{n=0}^{\infty} x^n}$$

and since $\sum_{n=0}^{\infty} x^n = 1/(1-x)$ for $x < 1$, while $\sum_{n=0}^{\infty} nx^n = xd(\sum_{n=0}^{\infty} x^n)/dx$, one obtains

$$< \varepsilon >= h\nu < n >, \qquad \text{with} \qquad < n >= \frac{1}{e^{\frac{h\nu}{k_B T}} - 1}$$

It is noted that for $k_B T \gg h\nu$, $< n > \rightarrow k_B T/h\nu$ and $< \varepsilon > \rightarrow k_B T$, the classical result for the average statistical energy of one-dimensional oscillator, i.e. for one of the modes of the e.m. field.

The number of modes having angular frequency between ω and $\omega + d\omega$ is conveniently evaluated by referring to the wavevector space and considering $k_x = (\pi/A)n_x$, $k_y = (\pi/B)n_y$ and $k_z = (\pi/C)n_z$ (see the sketch of the cavity). Since the e.m. waves must be zero at the boundaries, one must have an integer number of half-waves along A, B and C, i.e. $n_{x,y,z} = 1, 2, 3....$ By considering n as a continuous variable one has $dn_x = (A/\pi)dk_x$ and analogous expressions for the y and z directions. The number of $\mathbf{k}$ modes verifying the boundary conditions per unit volume of the reciprocal space, turns out

$$\frac{dn_x dn_y dn_z}{dk_x dk_y dk_z} \equiv D(\mathbf{k}) = \frac{ABC}{8\pi^3} = \frac{V}{8\pi^3} .$$

$D(\mathbf{k})$ is the **density of k-modes** or **density of k-states**[7].

---

[7] This concept will be used for the electronic states and for the vibrational states in crystals, Chapter 12 and Chapter 14. It is noted that the factor 8 is due to the fact that in this method of counting only positive components of the wave vectors have to be considered. For running waves, in the **Born-Von Karmann periodical conditions** (§12.4), the same number of excitations in the reciprocal space is obtained.

For photons the dispersion relation is $\omega = ck$ and the **number of modes** $D(\omega)$ in $d\omega$ can be estimated from the volume $d\mathbf{k}$ in the reciprocal space in between the two surfaces at constant frequency ω and $\omega + d\omega$:

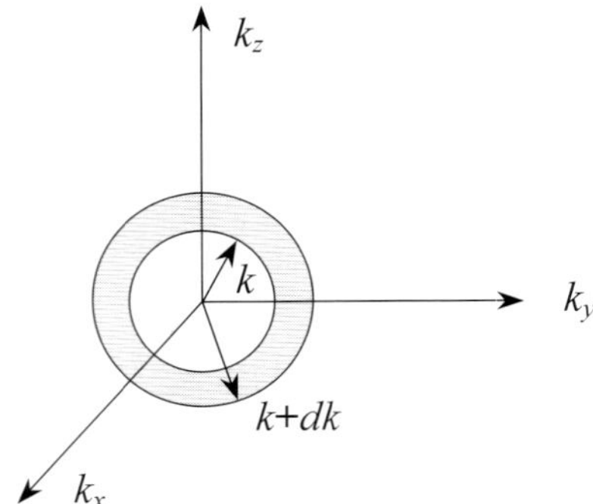

Then

$$D(\omega)d\omega = 2D(\mathbf{k})d\mathbf{k} = 2\frac{V}{8\pi^3}4\pi k^2 dk = \frac{V}{\pi^2 c^3}\omega^2 d\omega$$

and

$$D(\omega) = \frac{V}{\pi^2 c^3}\omega^2 \qquad \text{or} \qquad D(\nu) = \frac{8\pi V}{c^3}\nu^2$$

The factor 2 has been introduced to take into account the two polarization states of photons.

Photons are **bosonic** particles and therefore, by referring to the **Bose-Einstein statistical distribution function**

$$f_{BE} = 1/[exp(h\nu/k_B T) - 1]$$

one derives the **Planck distribution function** $\rho(\nu)$ (e.m. energy per unit volume in the unit frequency range) as follows.

The energy related to the number of photons dn_ν within $d\nu$ around the frequency ν is

$$dE(\nu) = h\nu dn_\nu \qquad \text{and} \qquad dn_\nu = f_{BE}D(\nu)d\nu = \frac{8\pi V \nu^2}{c^3}\frac{d\nu}{e^{\frac{h\nu}{k_B T}} - 1}$$

By definition $\rho(\nu)d\nu = dE(\nu)/V$ and then

$$\rho(\nu) = \frac{8\pi h\nu^3}{c^3}\frac{1}{e^{\frac{h\nu}{k_B T}} - 1}$$

The **Wien law** can be obtained by looking for the maximum in $\rho(\nu)$: $d\rho/d\nu = 0$ for $h\nu_{max}/k_B T \simeq 2.8214$, corresponding to $\nu_{max} \simeq T \times 5.88 \times 10^{10}$ Hz (for T in Kelvin). It can be remarked that $\lambda_{max} \neq c/\nu_{max}$. In fact $\lambda_{max} = (0.2898/T)$ cm.

The total energy per unit volume $U(T)$ is obtained by integrating over the frequency and taking into account the number of modes in $d\nu$ and the average energy per mode $h\nu < n >$:

$$U(T) = \int d\nu \frac{D(\nu)h\nu}{e^{\frac{h\nu}{k_B T}} - 1} = 3!\zeta(4)\frac{k_B^4 T^4}{\pi^2 c^3 \hbar^3}$$

where ζ is the Riemann zeta function, thus yielding

$$U(T) = \sigma T^4, \qquad \text{with} \qquad \sigma = 7.566 \times 10^{-15} \text{erg.cm}^{-3}\text{K}^{-4}$$

known as **Stefan-Boltzmann law**.

The density of photons is obtained by omitting in $U(T)$ the one-photon energy:

$$n_{tot}(T) = \int d\nu \frac{D(\nu)}{e^{\frac{h\nu}{k_B T}} - 1} = 2\zeta(3)\frac{k_B^3 T^3}{\pi^2 c^3 \hbar^3} = 20.29 \times T^3 \text{cm}^{-3}$$

For instance in the universe, with $T \simeq 2.73$ K, the number of photons per cubic centimeter turns out $n_{tot} \simeq 413$ cm^{-3}.

To derive the coefficient γ for expansion without exchange of energy, the radiation in the cavity is considered as a thermodynamical system of volume $V = ABC$ and temperature T (the temperature entering the energy distribution function). From $VU(T) = \sigma T^4 V = const$, one has $4VdT = -TdV$, i.e. $TV^{1/4} = const$ and therefore $\gamma = 5/4$.

During the expansion, since $N = n_{tot}V \propto T^3 V$, while $TV^{1/4} = const$, if the volume is increased by a factor f one has

$$T_{final} = T_{initial}\left(\frac{V_{initial}}{V_{initial}f}\right)^{\frac{1}{4}} = T_{initial}f^{-\frac{1}{4}}$$

and the number of photons becomes

$$N_{final} = N_{initial}f\left(\frac{T_{final}}{T_{initial}}\right)^3 = N_{initial}f^{\frac{1}{4}}$$

To evaluate the entropy the equation of state is required. The pressure of the radiation is obtained by considering the transfer of moment of the photons when they hit the surface and the well-known result $P = U/3$ is derived. For the entropy

$$dS = \frac{1}{T}d(UV) + \frac{PdV}{T} = \frac{V}{T}\frac{dU}{dT}dT + \frac{4U}{3T}dV$$

and since it has to be an exact differential $dU/dT = 4U/T$. Thus the equation of state is

$$PV = \frac{U_{tot}}{3}$$

where $U_{tot} = UV$ and then

$$dS = \frac{4U}{3T}dV + \frac{V}{T}4\sigma T^3 dT$$

From the condition of exact differential $S = 4U_{tot}/3T$. The decrease of T yields an increase of the entropy because the number of photons increases in the expansion.

It is noted that for $T \to 0$, S, P and U tend to zero.

For transformation at constant entropy, assumed reversible, one would have $dS = 0$ and then

$$\frac{4\sigma T^3}{3}dV + 4\sigma VT^2 dT = 0 \ ,$$

so that $TV^{1/3} = const.$

In the expansion of the universe by a factor f the **cosmological principle** (each galaxy is moving with respect to any other by a velocity proportional to the distance) implies that each frequency ν_i is shifted to $\nu_f = \nu_i/f^{1/3}$. As a consequence of the expansion the energy density $du(\nu_i, T_i)$ in a given frequency range $d\nu_i$ is decreased by a factor f because of the increase in the volume and by a factor $f^{1/3}$ because of the energy shift for each photon. Then

$$du_f = \frac{du_i}{f f^{\frac{1}{3}}} = \frac{8\pi h\nu_i^3}{c^3}\frac{d\nu_i}{e^{\frac{h\nu_i}{k_B T}} - 1}\frac{1}{f^{\frac{4}{3}}}$$

which can be rewritten in terms of the new frequency $\nu_f = \nu_i/f^{1/3}$

$$du_f = \frac{8\pi h\nu_f^3}{c^3}\frac{d\nu_f}{e^{\frac{h\nu_f f^{1/3}}{k_B T}} - 1} \ ,$$

namely the same existing before the expansion, provided that the temperature is scaled to $T/f^{1/3}$.

It is noted that since $U_{tot} = UV \propto T^4 V$ the entropy of the universe is constant during the expansion, while the energy decreases by a factor $f^{1/3}$. The number of photons is constant.

In the Figure

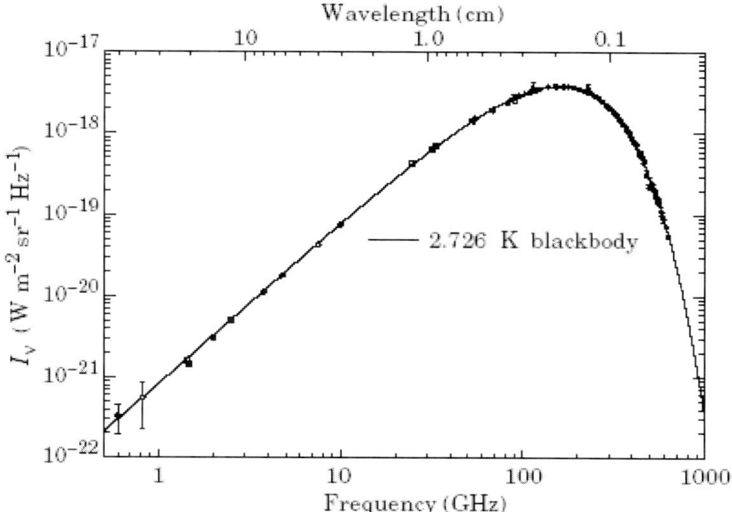

the Planck distribution function (solid line) for the cosmic background radiation, resulting from a series of experimental detections, is evidenced.

Problem F.I.3 Derive the life-time of the Hydrogen atom in the $2p$ state and the **natural broadening** of the line resulting from the transition to the ground state. By neglecting relativistic effects (see Problem F.I.15 and Appendix V.1) evaluate the energy split due to the spin-orbit interaction (§1.6) and the effective field of orbital origin acting on the electron in the $2p$ state.

Solution:

The life time (Prob. F.I.1) is $\tau = 1/A_{2p \to 1s}$, with $A_{2p \to 1s}$ the spontaneous emission transition probability. Then

$$A_{2p \to 1s} = \frac{32\pi^3 \nu^3}{3c^3 \hbar} \frac{1}{3} \sum_\alpha | < \phi_{2p\alpha} | - e\mathbf{r} | \phi_{1s} > |^2 \quad (\alpha \equiv 0, \pm 1)$$

From the evaluations of the matrix elements of the electric dipole components (See Eq. A.I.3.9 and Tables I.4.1a and I.4.1b) one obtains

$$A_{2p \to 1s} = (\frac{2}{3})^8 \frac{e^8}{c^3 a_0 \hbar^4} = 6.27 \times 10^8 \text{ s}^{-1} \ ,$$

or $\tau = 1.6 \times 10^{-9}$ s.

Then the natural line-width can be written $\Delta \nu = (2.54 \times 10^{-7}/2\pi)\nu_{2p\to 1s}$.

The energy split due to the spin-orbit interaction is $\Delta E = (3/2)\xi_{2p}$, with $\xi_{2p} = e^2\hbar^2/(2m^2c^2a_0^3 \times 24)$, so that $\Delta E = 4.53 \times 10^{-5}$ eV and therefore $H = (\Delta E/2\mu_B) \simeq 4$ kGauss (see also Problems II.1.2 and I.6.4).

Problem F.I.4 Show that the stimulated emission probability W_{21} due to thermal radiation is equivalent to the spontaneous emission probability A_{21} times the average number of photons (Prob. F.I.1).

Solution:

From $<n> = 1/[exp(h\nu/k_BT) - 1]$, while

$$\rho(\nu) = (8\pi h\nu^3/c^3)/[exp(h\nu/k_BT) - 1]$$

(Prob. F.I.2), and from the Einstein relation $B_{21}\rho(\nu) = A_{21} <n>$.

Problem F.I.5 By considering the sun as a source of black-body radiation at the temperature $T \simeq 6000$ K, evaluate the total power emitted in a bandwidth of 1 MHz around the wavelenght 3 cm (the diameter of the sun crown can be taken $2R = 10^6$ km).

Solution:

For $\lambda = 3$ cm the condition $h\nu \ll k_BT$ is verified. To each e.m. mode one can attribute an average energy $<\varepsilon> = k_BT$. The density of modes is $(8\pi/c^3)\nu^2$ and thus the energy in the bandwidth $\Delta\nu$ is $\Delta u = (8\pi/c^3)\nu^2\Delta\nu k_BT$. The power emitted per unit surface is $\rho_S = uc/4$ (see Prob. F.I.2) and therefore

$$\Delta P = \frac{8\pi}{c^3}\nu^2\Delta\nu k_BT\frac{c}{4}4\pi R^2 \simeq 1.8 \times 10^9 \text{Watt} .$$

Problem F.I.6 The energy flow from the sun arriving perpendicularly to the earth surface (neglecting atmospheric absorption) is $\Phi = 0.14$ Watt/cm^2. The distance from the earth to the sun is about 480 second-light. In the assumption that the sun can be considered as a black-body emitter, derive the temperature of the external crown.

Solution:

The flow scales with the square of the distances. Thus the power emitted per unit surface from the sun can be written $\Phi_{tot} = (d/R)^2\Phi$ (d average distance, R radius of the sun). Then $\Phi_{tot} = 8 \times 10^3$ Watt/cm^2 and since (Problem F.I.2) $\Phi = \sigma cT^4/4 = (5.67 \times 10^{-12} \times T^4)$ Watt/cm^2, one obtains $T_{Sun} \simeq 6129$ K.

Problem F.I.7 Because of the thermal motions of the atoms the shape of the emission line from a lamp is usually Gaussian. By referring to the yellow line emitted at about 5800 Å by Sodium atom, neglecting the lifetime broadening and assuming the Maxwellian distribution of the velocities, prove this statement. Estimate the order of magnitude of the broadening, for a temperature of the lamp of about 500 K.

Show that the shift due to the recoil of the atom when the photon is emitted is negligible in comparison to the motional broadening. Comment on the possibility of **resonance absorption** by atoms in the ground state. At which wavelength one could expect that the resonance absorption would hardly be achieved?

Solution:

Along the direction x of the motion the Doppler shift is

$$\lambda = \lambda_0(1 \pm \frac{v_x}{c})$$

The number of atoms $dn(v_x)$ moving with velocity between v_x and $v_x + dv_x$ is

$$dn(v_x) = N\sqrt{\frac{M}{2\pi k_B T}} e^{-\frac{M v_x^2}{2 k_B T}} dv_x$$

(N number of atoms with mass M). The number of atoms emitting in the range $d\lambda$ around λ is

$$dn(\lambda) = N\sqrt{\frac{Mc^2}{2\pi \lambda_0^2 k_B T}} e^{-\frac{Mc^2(\lambda-\lambda_0)^2}{2 k_B T \lambda_0^2}} d\lambda$$

The intensity $I(\lambda)$ in the emission spectrum is proportional to $dn(\lambda)$

$$I(\lambda) \propto \sqrt{\frac{1}{\pi\delta^2}} e^{-\frac{(\lambda-\lambda_0)^2}{\delta^2}}$$

with $\delta = \sqrt{2k_B T/M}(\lambda_0/c)$.

Numerically, for the Na yellow line one has a broadening of about 1700 MHz, in wave-numbers, $1/\delta \simeq 0.0576$ cm^{-1}.

The photon moment being $h\nu/c$, the recoil energy is $E_R = (h\nu/c)^2/2M \simeq 10^{-10}$ eV and the resonance absorption is not prevented. For wavelength in the range of the γ-rays the recoil energy would be larger than the Doppler broadening and without the **Mossbauer effect** (see §14.6) the resonance absorption would hardly be possible.

Problem F.I.8 X-ray emission can be obtained by removing an electron from inner states of atoms, with the subsequent transition of another electron from higher energy states to fill the vacancy. The X-Ray frequencies vary smoothly from element to element, increasing with the atomic number Z (see plot). Qualitatively justify the **Moseley law** $\lambda^{-1} \propto (Z - \sigma)^2$ (σ screening constant):

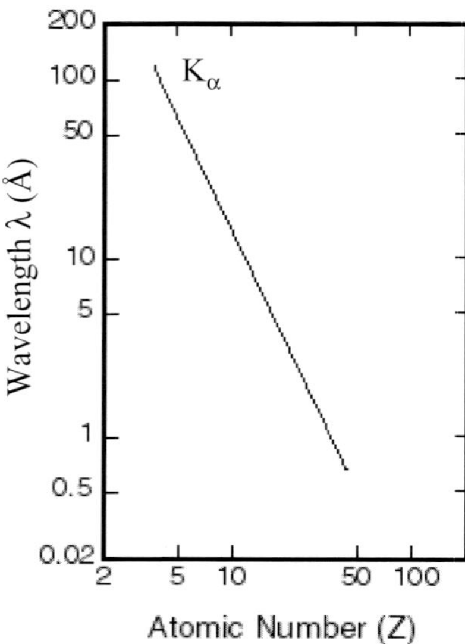

Fig. 1.10. Wavelength of the K_α line as a function of the atomic number.

Solution:
From the one-electron eigenvalues in central field with effective nuclear charge $(Z - \sigma)$ (σ reflecting the screening from other electrons, see §2.1 and §2.2), transitions between n_i and n_f imply the emission of a photon at energy

$$h\nu_{i \to f} = R_H hc (Z - \sigma)^2 \left(\frac{1}{n_f^2} - \frac{1}{n_i^2} \right)$$

The K-lines are attributed to the transitions to the final state $n_f = 1$. The K_α line corresponds to the longest wavelength ($n_i = 2$).

Problem F.I.9 Estimate the order of magnitude of the voltage in an X-ray generator with Fe anode yielding the emission of the K_α line and the wavelength of the correspondent photon.

Solution:
The energy of the K term is $E_K = +13.6(Z - \sigma)^2(3/4)$ eV. For $\sigma_K \simeq 2$ one would obtain for the voltage $V \simeq 5800$ Volts. The wavelength of the K_α line turns out around 1.8 Å.

Problem F.I.10 An electron is inside a sphere of radius $R_s = 1$ Å, with zero angular momentum. From the Schrödinger equation for the radial part of the wavefunction derive the lowest eigenvalue $E_{n=1}$ and the quantum pressure $P = -dE_{n=1}/dV$.

Solution:

The equation for $rR(r)$ reads

$$-\frac{\hbar^2}{2m}\frac{d}{dr^2}(rR) = E(rR)$$

(see Eq. 1.14). From the boundary condition $R(R_s) = 0$ one has $R \propto [sin(kr)]/kr$, with $k_n R_s = n\pi$ for $n = 1, 2, 3....$ Then

$$E_{n=1} = \frac{\hbar^2 k_{n=1}^2}{2m} = \frac{\pi^2 \hbar^2}{2mR_s^2}$$

and

$$P = \frac{\pi\hbar^2}{4mR_s^5}$$

For $R = 1$ Å one has $P = 9.6 \times 10^{12}$ dyne/cm^2. Compare this value with the one of the electron Fermi gas in a metal (§12.7).

Problem F.I.11 From the Boltzmann distribution of the molecular velocities in ideal gas, show that the number of molecules n_c that hit the unit surface of the container per second is given by $n < v > /4$ (n number of molecules per cm^3) with $< v >$ the average velocity. Then numerically estimate n_c for molecular Hydrogen at ambient temperature and pressure.

Solution:

From the statistical distribution of the velocities the number of molecules moving along a given direction x with velocity between v_x and $v_x + dv_x$ is

$$dn(v_x) = n\left(\frac{M}{2\pi k_B T}\right)^{1/2} e^{-Mv_x^2/2k_B T} dv_x$$

The molecules colliding against the unit surface in a second are

$$n_c = \int_0^\infty v_x dn(v_x) = n\left(\frac{M}{2\pi k_B T}\right)^{1/2}\left(-\frac{k_B T}{M}\right)\left[e^{-Mv_x^2/2k_B T}\right]_0^\infty = n\left(\frac{k_B T}{2\pi M}\right)^{1/2}$$

The average velocity is

$$< v > = \frac{1}{n}\int_0^\infty v\, dn(v) = \frac{1}{n}\int_0^\infty v\left[4\pi n\left(\frac{M}{2\pi k_B T}\right)^{3/2} v^2 e^{-Mv^2/2k_B T}\right] dv =$$

$$= \left(\frac{8k_B T}{\pi M}\right)^{1/2} = \frac{4n_c}{n}$$

Numerically, for H_2, $n_c = 1.22 \times 10^{24}$ molecules/(s.cm^2).

Problem F.I.12 Hydrogen atoms in the ground-state are irradiated at the resonance frequency $(E_{n=2} - E_{n=1})/h$, with e.m. radiation having the following polarization: a) linear; b) circular; c) unpolarized.

By considering only electric dipole transitions, discuss the polarization of the fluorescent radiation emitted when the atoms return to the ground-state.

Solution:

a) The only possible transition is to the $2p_z$ state, with z the polarization axis ($\Delta m = 0$). No radiation is re-emitted along z while it is emitted in the xy plane, with polarization of the electric field along z.

b) Only transitions to $2p_{\pm 1}$ state are possible ($\Delta m = \pm 1$). The fluorescent radiation when observed along the z direction is circularly polarized. By turning the observation axis from the z axis to the xy plane, the fluorescent radiation will progressively turn to the elliptical polarization, then to linearly polarized when the observation axis is in the xy plane.

c) Any transition $1s \rightarrow 2p_{\pm 1,0}$ is possible, with uniform distribution over all the solid angle. The atom will be brought in the superposition state and the fluorescent radiation will have random wave-vector orientation and no defined polarization state.

Problem F.I.13 An electron is moving along the x-axis under a potential energy $V(x) = (1/2)kx^2$, with $k = 5 \times 10^4$ dyne/cm. From the **Sommerfield quantization** (see Prob. I.4.4) obtain the amplitudes A of the motion in the lowest quantum states.

Solution:

From $x(t) = A sin[(\sqrt{k/m})t + \varphi]$ the quantum condition in terms of the period $T = 2\pi\sqrt{m/k}$ reads

$$\oint m\dot{x}dx = \int_0^T m\dot{x}\ddot{x}dt = A^2 k \int_0^T cos^2(\sqrt{\frac{k}{m}}t - \varphi)dt = A^2 k \frac{T}{2} = nh$$

Thus $A_0 = 0$ (the zero-point energy is not considered here), $A_1 = 1.47 \times 10^{-8}$cm, $A_2 = 2.07 \times 10^{-8}$cm.

Problem F.I.14 The emission of radiation from **intergalactic Hydrogen** occurs at a wavelength $\lambda' = 21$ cm (see §5.2). The galaxy, that can be idealized as a rigid disc with homogeneous distribution of Hydrogen, is rotating. Estimate the Doppler broadening $\Delta\nu_{rot}$ of the radiation, assuming a period of rotation of 10^8 years and a radius of the galaxy $R = 10$kps $= 3.091 \times 10^{18}$ cm. Prove that $\Delta\nu_{rot}$ is much larger than the broadening $\Delta\nu_T$ due to the thermal motion of the Hydrogen gas (assumed at a temperature $T = 100$ K) and larger than the shift due to the drift motion of the galaxy itself (at a speed of approximately $v_d = 10^7$ cm/s).

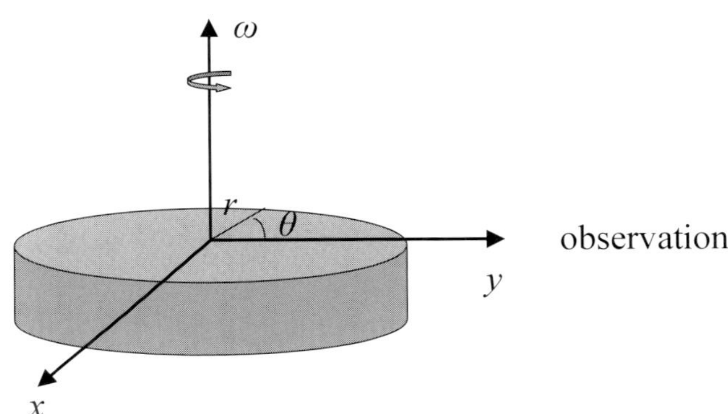

Solution:
The Doppler shift at large distance along the y direction is

$$\nu(r,\theta) \simeq \nu_0(1 + \frac{v}{c}cos\theta)$$

with $\nu_o = \lambda'/c$. The mean-square average frequency is

$$\bar{\nu_2} = \frac{1}{\pi R^2}\int_0^R rdr \int_0^{2\pi} \nu^2(r,\theta)d\theta = \nu_0^2 + \frac{\omega^2\nu_0^2}{c\pi R^2}\int_0^R r^3dr \int_0^{2\pi} cos^2\theta d\theta = \nu_0^2(1 + \frac{\omega^2 R^2}{4c^2})$$

Therefore

$$\Delta\nu_{rot} = \nu_0\frac{\omega R}{2c} = \nu_0\frac{\pi R}{cT} \simeq \nu_0.10^{-3}$$

From Problem F.I.7 one deduces the order of magnitude of the thermal broadening:

$$\Delta\nu_T = \frac{\nu_0}{\lambda'}\Delta\lambda = \frac{\nu_0}{\lambda'}\sqrt{\frac{2k_BT}{m_H}}\frac{\lambda'}{c} = \frac{\nu_0}{c}\sqrt{\frac{2k_BT}{m_H}} \simeq \nu_0(4\times 10^{-6})$$

For the drift associated with the linear motion of the galaxy one can approximately estimate the frequency shift of the order of $\Delta\nu_d = (v_d/c)\nu_0 \simeq 3.3\times 10^{-4}\nu_0$.

Problem F.I.15 In a description of the relativistic effects more detailed than the **Thomas-Frenkel** model (§1.6) to derive the one-electron spin-orbit Hamiltonian, the **Darwin term**

$$\mathcal{H}_D = \frac{\pi\hbar^2}{2m^2c^2} Ze^2 \delta(\mathbf{r}) \equiv \pi\alpha^2 \frac{Ze^2}{2a_0} a_0^3 \delta(\mathbf{r})$$

(with $\alpha = e^2/\hbar c = 1/137.036$ the **fine structure constant**) is found to be present.

Discuss the effects of $\mathcal{H}_D$ in Hydrogenic atoms, numerically comparing the corrections to the eigenvalues with the ones due to the spin-orbit Hamiltonian $\xi_{nl}\mathbf{l}.\mathbf{s}$.

Solution:

From

$$< \phi_{nl}|\mathcal{H}_D|\phi_{nl} > \equiv D \int \phi_{nl}^*(\mathbf{r})\delta(\mathbf{r})\phi_{nl}(\mathbf{r})d\mathbf{r} = D|\phi_{nl}(0)|^2 \ ,$$

with $D = \pi\alpha^2 Ze^2 a_0^2/2$, one sees that no effects due to $\mathcal{H}_D$ are present for non-s states (within the approximation of nuclear point-charge).

The shift for s states can be written (see Table I.4.2)

$$\Delta E_D = \frac{Z^2\alpha^2}{n}\left(\frac{e^2 Z^2}{2a_0 n^2}\right) \equiv -\frac{Z^2\alpha^2}{n} E_n^0$$

with $E_n^0 = -Z^2e^2/2a_0 n^2$ the unperturbed eigenvalues.

From

$$\xi_{nl} = (Ze^2/2m^2c^2) < r^{-3} >_{nlm}$$

and $< r^{-3} >_{nlm} = Z^3/[a_0^3 n^3 l(l + 1/2)(l + 1)]$ (see Table I.4.3), $(l \neq 0)$

$$\Delta E_{SO} = \frac{Z^2\alpha^2(-E_n^0)}{2nl(l + \frac{1}{2})(l + 1)}[j(j + 1) - l(l + 1) - 3/4] \ .$$

The relativistic corrections associated with the kinetic energy is

$$\Delta E_{kin} = -E_n^0 \frac{Z^2\alpha^2}{n^2}\left[\frac{3}{4} - \frac{n}{l + \frac{1}{2}}\right] \ .$$

From $\Delta E_D + \Delta E_{SO} + \Delta E_{kin}$ the eigenvalues of the **Dirac theory**, namely

$$E_{n,j} = -E_n^0 \frac{Z^2\alpha^2}{n^2}\left[\frac{3}{4} - \frac{n}{j + \frac{1}{2}}\right]$$

are obtained (see Appendix V.1).

2

Typical atoms

Topics

Effects on the outer electron from the inner core
Helium atom and the electron-electron interaction
Exchange interaction
Pauli principle and the antisymmetry requirement
Slater determinantal eigenfunctions

2.1 Alkali atoms

Li, Na, K, Rb, Cs and Fr are a particular group of atoms characterized by one electron (often called **optical** being the one involved in optical spectra) with expectation value of the distance from the nucleus $< r >$ considerably larger than the one of the remaining $(N-1)$ electrons, forming the internal "**core**". The alkali atoms are suited for analyzing the role of the core charge in modifying the Coulomb potential $(-Ze^2/r)$ pertaining to Hydrogenic atoms (§1.4), as well as to illustrate the effect of the spin-orbit interaction (§1.6).

From spectroscopy one deduces the diagram of the energy levels for Li atom reported in Fig. 2.1, in comparison to the one for Hydrogen.

In Fig. 2.2 the analogous level scheme for Na atom is shown, with the main electric-dipole transitions yielding the emission spectrum.

The quantum numbers for the energy levels in Fig. 2.1 are the ones pertaining to the outer electron. At first we shall neglect the fine structure related to the spin-orbit interaction, which causes the splitting in doublets of the states at $l \neq 0$, as indicated for Na in Fig. 2.2.

A summarizing collection of the energy levels for alkali atoms is reported in Fig. 2.3. It should be remarked that because of the different extent of

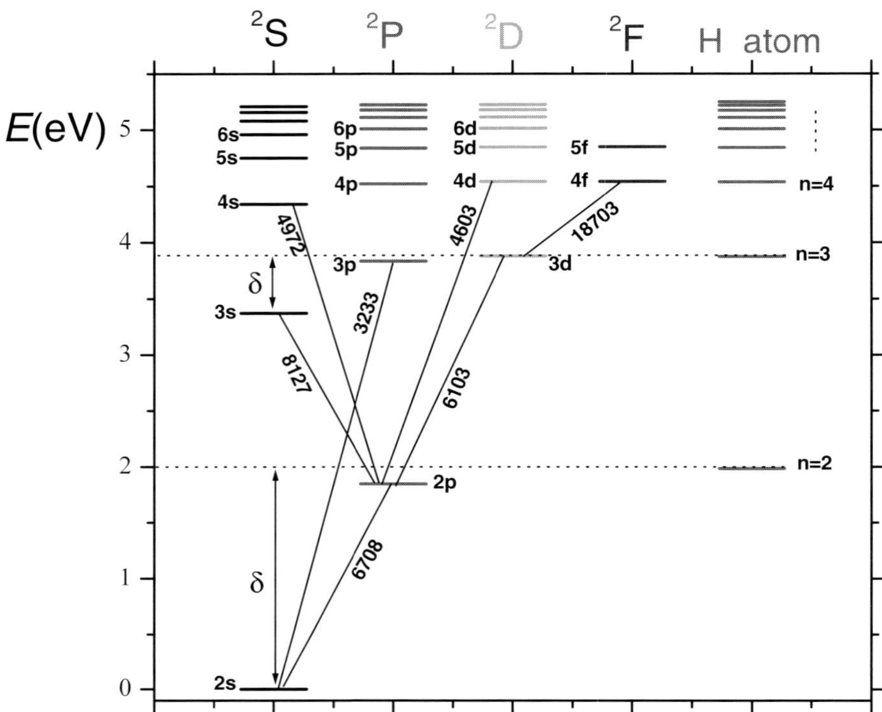

Fig. 2.1. Energy level diagram (**Grotrian diagram**) of Li atom, in term of the quantum numbers nl of the optical electron and comparison with the correspondent levels ($n > 1$) for H atom. The **quantum defect** δ (or **Rydberg** defect) indicated for $2s$ and $3s$ states, is a measure of the additional (negative) energy of the state in comparison to the correspondent state in Hydrogen. The wavelengths (in Å) for some transitions are reported.

penetration in the core (as explained in the following) an inversion of the order of the energy levels in terms of the quantum number n (namely $|E_n| > |E_{n-1}|$) can occur.

From the Grotrian diagrams one deduces the following:

i) the sequence of the energy levels is similar to the one for H, with more bound and no more l-degenerate states;

ii) the **quantum defect** δ for a given n-state (see Fig. 2.1) increases on decreasing the quantum number l;

iii) the ground state for Li is $2s$ ($3s$ for Na, etc...), with $L = l$ (and not the $1s$ state);

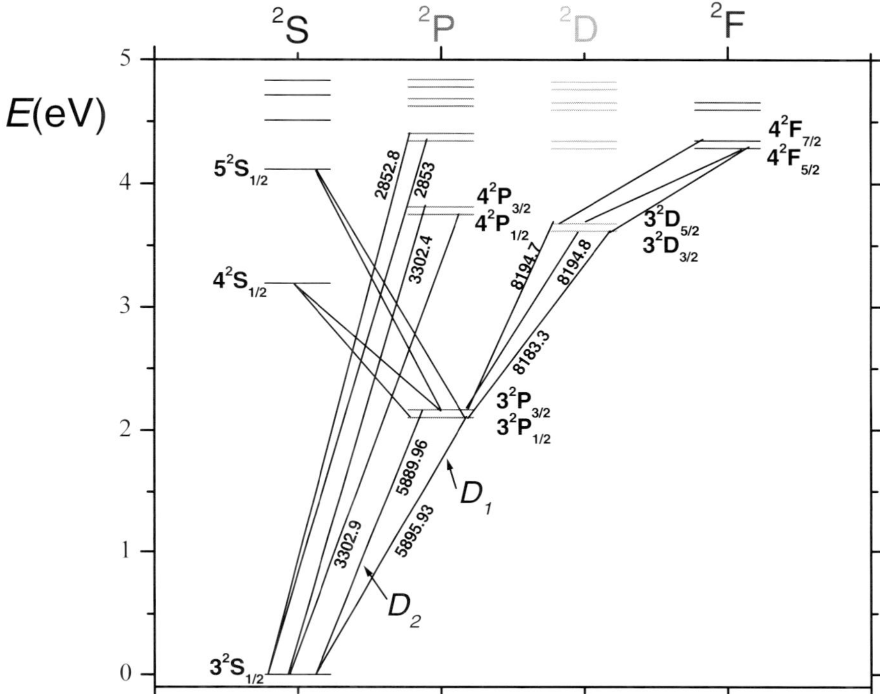

Fig. 2.2. Energy levels for Na atom with the electric dipole transitions ($\Delta l = \pm 1$) generating some spectral lines and correspondent wavelengths (in Å). The doublets related to spin-orbit interaction and resulting in states at different $j \equiv J$, are indicated (not in scale). The yellow emission line (a doublet) is due to the transition from the $^2P_{3/2}$ and $^2P_{1/2}$ states to the ground state $^2S_{1/2}$ with the optical electron in the $3s$ state.

iv) the transitions yielding the spectral lines obey the selection rule $\Delta l = \pm 1$.

These remarkable differences with respect to Hydrogen are related to an effective charge $Z_{eff}(r)$ for the optical electron (see §1.2) different from unit over a sizeable range of distance r from the nucleus.

In order to give a simple quantitative description of these effects we shall assume an *ad hoc* effective charge, of the form $Z_{eff} = (1 + b/r)$,

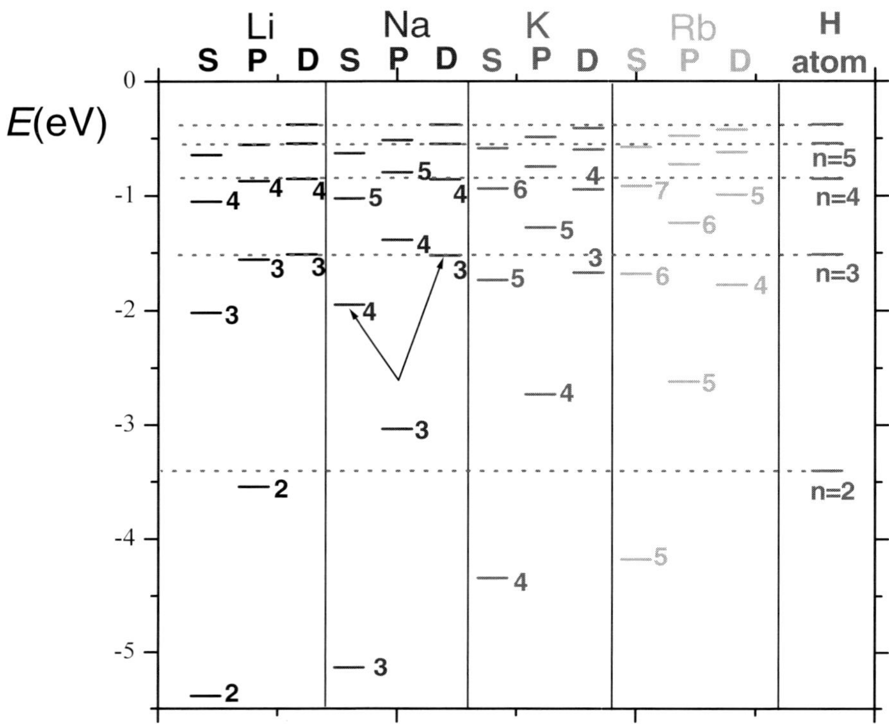

Fig. 2.3. Energy levels (neglecting the fine structure) for some alkali atoms, again compared with the states for Hydrogen at $n > 1$. The $4s$ state is more bound than the $3d$ state (see arrows), typical inversion of the order of the energies due to the extent of penetration of the s-electrons in the core, where the screening is not fully effective (see text and Fig. 2.6)

depicted in Fig. 2.4. The characteristic length b depends from the particular atom, it can be assumed constant over a large range of distance while for $r \to 0$ it must be such that $Z_{eff}(r) \to Z$.

As a consequence of that choice for $Z_{eff}(r)$ the radial part of the Schrodinger equation for the optical electron takes a form strictly similar to the one in Hydrogen (see §1.4):

$$\frac{d^2[rR(r)]}{dr^2} - \left[A - \frac{B}{r} + \frac{C}{r^2}\right]rR(r) = 0 \qquad (2.1)$$

where $B = 2/a_0$ and $C = l(l+1) - Bb$. It is remarked that for $b = 0$ the eigenvalues associated with Eq. 2.1 are $E_n = -R_H hc/n^2$ (Eq. 1.13, for $Z = 1$).

If an effective quantum number l^* such that

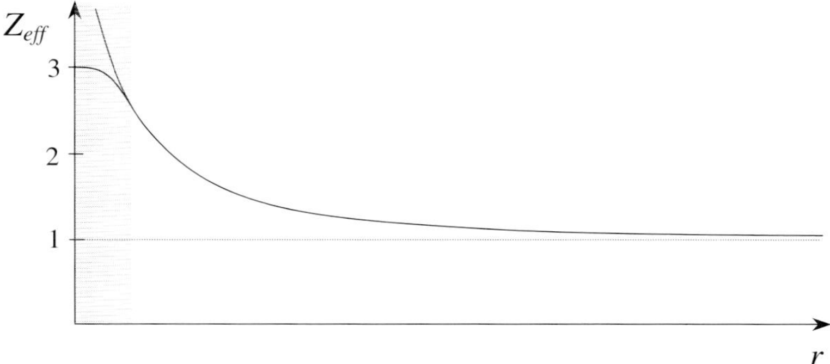

Fig. 2.4. Sketchy behavior of a plausible effective charge for the optical electron in Li atom. The dashed part of the Figure (not in scale) corresponds to the region of r not taken into account in the derivation of the energy levels. For Na, K, etc... atoms $Z_{eff}(r \to 0) \to Z$. A similar form of effective charge experimented by one electron because of the partial screening of the nuclear charge by the second electron is derived in Problem II.2.3 for He atom.

$$l^*(l^*+1) = C = l(l+1) - \frac{2me^2b}{\hbar^2} \equiv l(l+1) - Bb$$

is introduced, then in the light of the formal treatment for Hydrogen, from Eq. 2.1 one derives the eigenvalues

$$E_n = -\frac{R_H hc}{(n^*)^2}, \tag{2.2}$$

with n^* **not integer.** To evidence in these energy levels the numbers n and l pertaining to Hydrogen atom, we write $n^* = n - \delta l$, with $\delta l = l^* - l$, thus obtaining

$$E_{n,l} = -\frac{R_H hc}{(n - \delta l)^2} \ .$$

By neglecting the term in δl^2

$$E_{n,l} = -\frac{R_H hc}{(n - \frac{2b}{a_0(2l+1)})^2} \equiv -\frac{R_H hc}{(n - \delta_{n,l})^2} \ . \tag{2.3}$$

The eigenvalues are l-dependent, through a term that is atom-dependent (via b) and that decreases on increasing l, in agreement with the phenomenological findings.

The physical interpretation of the result described by Eq. 2.3 involves the amount of penetration of the optical electron within the core. In Fig. 2.5 it is shown that for $r \leq a_0$ the electron described by the $2s$ orbital has a radial probability of presence sizeably larger than the one for the $2p$ electron. This implies a reduced screening of the nuclear charge and then more bound state.

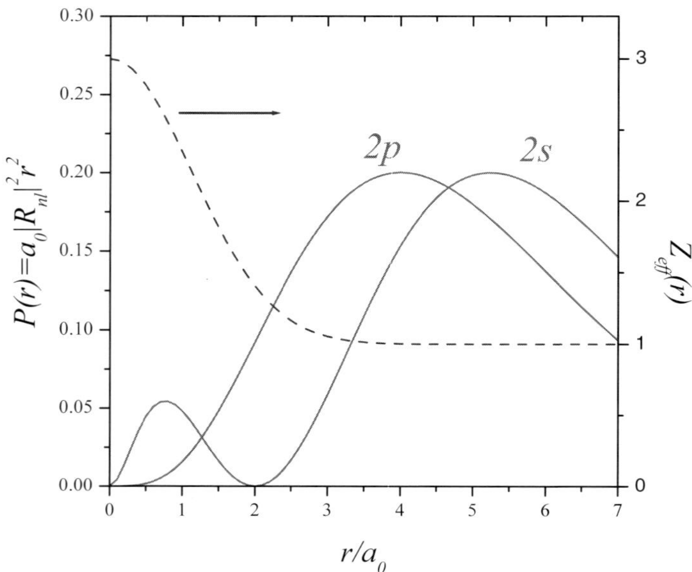

Fig. 2.5. Radial probability of presence for $2s$ and $2p$ electrons in Hydrogen and sketchy behavior of the effective charge for Li (see Fig. 2.4).

As a general rule one can state that the penetration within the core increases on decreasing l. In Fig. 2.6 it is shown how it is possible to have a more penetrating state for $n = 4$ rather than for $n = 3$, in spite of the fact that on the average the $3d$ electron is closer to the nucleus than the $4s$ electron. This effect is responsible of the inversion of the energy levels, with $|E_{4s}| > |E_{3d}|$, as already mentioned.

At the sake of illustration we give some quantum defects $\delta_{n,l}$ to be included in Eq. 2.3, for Na atom:

$\delta_{3s} = 1.373$	$\delta_{3p} = 0.883$	$\delta_{3d} = 0.01$
$\delta_{4s} = 1.357$	$\delta_{4p} = 0.867$	$\delta_{4d} = 0.011$
...	...	$\delta_{4f} \simeq 0$

These values for the quantum defects can be evaluated from the energy levels reported in Fig. 2.2 (see also Problem II.1.1).

Finally a comment on the selection rule $\Delta l = \pm 1$ is in order. This rule is consistent with the statement that each electron makes the transition indepen-

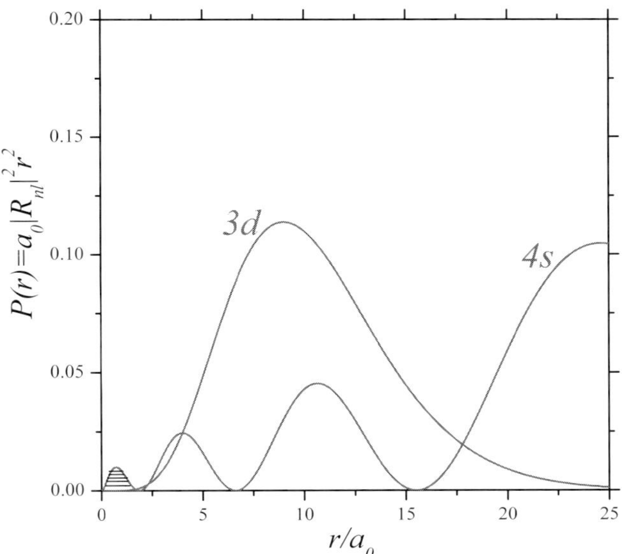

Fig. 2.6. Radial probability of presence for $3d$ and $4s$ electrons in Hydrogen. From the dashed area it is noted how the bump in $P(r)$ for $r \leq 2a_0$ grant a presence of the $4s$ electron in the vicinity of the nucleus larger than the one pertaining to the $3d$ state.

dently from the others, with the one-electron selection rule given in Appendix I.3. In fact, the total wavefunction for the alkali atom, within the central field approximation, can be written

$$\phi(\mathbf{r}_1, \mathbf{r}_2, ..., \mathbf{r}_N) = \phi_{core}\phi_{optical} \ .$$

The electric dipole matrix element associated to a given $1 \leftrightarrow 2$ transition becomes

$$\mathbf{R}_{1\leftrightarrow 2} = -e \int (\phi_{core}^{(2)})^*(\mathbf{r}_1, \mathbf{r}_2, ...)(\phi^{(2)}(\mathbf{r}_n))^*[\mathbf{r}_1 + \mathbf{r}_2 + ...\mathbf{r}_n + ... + \mathbf{r}_N]$$
$$\times \phi_{core}^{(1)}(\mathbf{r}_1, \mathbf{r}_2, ...)\phi^{(1)}(\mathbf{r}_n)d\tau_1 d\tau_2...d\tau_N$$

Because of the orthogonality conditions the above integral is different from zero in correspondence to a given term involving $\mathbf{r}_n$ only when $\phi_{core}^{(2)} = \phi_{core}^{(1)}$, while

$$\int (\phi^{(2)}(\mathbf{r}_n))^*[\mathbf{r}_n]\phi^{(1)}(\mathbf{r}_n)d\tau_n$$

yields the selection rule $(\Delta l)_n = \pm 1$ and $(\Delta m)_n = 0, \pm 1$.

Now we take into account the doublet structure of each of the states at $l \neq 0$ (see the illustrative diagram in Fig. 2.2). The doublets result from spin-orbit interaction, as discussed at §1.6. The splitting of the np states of the optical electron turns out

Li	Na	K	Rb	Cs	
2p	3p	4p	5p	6p	
0.337	17.2	57.7	238	554	cm^{-1}
0.042	2.1	7.2	29.5	68.7	meV

supporting the energy corrections derived in terms of the spin-orbit constant ξ_{nl} (see for instance Prob. II.1.2). It can be observed that because of the selection rule $\Delta j = 0, \pm 1$ ($0 \leftrightarrow 0$ forbidden) (see App. I.3) the spectral lines involving transitions between two non-S states in alkali atoms can display a fine structure in the form of three components (**compound doublets**).

Problems II.1

Problem II.1.1 The empirical values of the quantum defects $\delta_{n,l}$ (see Eq. 2.3) for the optical electron in the Na atom are

	Term	$n = 3$	$n = 4$	$n = 5$	$n = 6$
$l = 0$	s	1.373	1.357	1.352	1.349
$l = 1$	p	0.883	0.867	0.862	0.859
$l = 2$	d	0.010	0.011	0.013	0.011
$l = 3$	f	-	0.000	-0.001	-0.008

By neglecting spin-orbit fine structure, indicate how the main spectral series can be derived (see Fig. 2.2).

Solution:
The main spectral series are
principal (transitions from p to s terms), at wave numbers

$$\bar{\nu}_p = R_H \left[\frac{1}{[n_0 - \delta(n_0, 0)]^2} - \frac{1}{[n - \delta(n, 1)]^2} \right], \quad n \geq n_0, \ n_0 = 3;$$

sharp (transitions from s to p electron terms)

$$\bar{\nu}_s = R_H \left[\frac{1}{[n_0 - \delta(n_0, 1)]^2} - \frac{1}{[n - \delta(n, 0)]^2} \right], \quad n \geq n_0 + 1;$$

diffuse (transitions from d to p electron terms)

$$\bar{\nu}_d = R_H \left[\frac{1}{[n_0 - \delta(n_0, 1)]^2} - \frac{1}{[n - \delta(n, 2)]^2} \right], \quad n \geq n_0;$$

fundamental (transitions from f to d terms):

$$\bar{\nu}_f = R_H \left[\frac{1}{[n_0 - \delta(n_0, 2)]^2} - \frac{1}{[n - \delta(n, 3)]^2} \right], \quad n \geq n_0 + 1.$$

Problem II.1.2 The spin-orbit splitting of the $6^2 P_{1/2}$ and $6^2 P_{3/2}$ states in Cesium atom causes a separation of the correspondent spectral line (transition to the $^2 S_{1/2}$ ground-state) of 422 $\overset{\circ}{A}$, at wavelength around 8520 $\overset{\circ}{A}$. Evaluate the spin-orbit constant ξ_{6p} and the effective magnetic field acting on the electron in the $6p$ state.

Solution:

From $\lambda'' - \lambda' = \Delta\lambda = 422 \;\overset{\circ}{A}$ and $\nu d\lambda = -\lambda d\nu$ one writes

$$\Delta E = h\Delta\nu \simeq h \cdot \frac{c}{\lambda'^2} \cdot \Delta\lambda \simeq 0.07 \, \text{eV}.$$

From

$$\Delta E_{SO} = \frac{\xi_{6p}}{2} \{ j(j + 1) - l(l + 1) - s(s + 1) \}$$

one has

$$\Delta E = \frac{\xi_{6p}}{2} \left[\frac{15}{4} - \frac{3}{4} \right] = \frac{3}{2} \xi_{6p}$$

and then

$$\xi_{6p} = \frac{2}{3} \Delta E = 0.045 \, \text{eV} \ .$$

The field (operator, Eq. 1.33) is

$$\mathbf{H} = \frac{\hbar}{2emc} \frac{1}{r} \frac{dV}{dr} \mathbf{l}$$

with the spin-orbit hamiltonian

$$\mathcal{H}_{spin-orbit} = -\boldsymbol{\mu}_s \cdot \mathbf{H}_{nl} = \xi_{6p} \mathbf{l} \cdot \mathbf{s} \,.$$

Thus

$$|\mathbf{H}_{6p}| = \frac{0.045 \, \text{eV} \; |\mathbf{l}|}{2\mu_B} \simeq 5.6 \cdot 10^6 \, \text{Oe} \; = 560 \; \text{Tesla}$$

Problem II.1.3 In a maser ^{85}Rb atoms in the 63 $^2P_{3/2}$ state are driven to the transition at the 61 $^2D_{5/2}$ state. The quantum defects $\delta_{n,l}$ for the states are 2.64 and 1.34 respectively. Evaluate the transition frequency and compare it to the one deduced from the classical analogy for Rydberg atoms (§1.5). Estimate the isotopic shift for ^{87}Rb.

Solution:

From

$$E_{nl} = -R^* hc \frac{1}{n^{*2}}$$

where R^* is the Rydberg constant and $n^* = n - \delta_{n,l}$, the transition frequency turns out

$$\nu = -R^* c \left[\frac{1}{[n_i - \delta(n_i, l_i)]^2} - \frac{1}{[n_f - \delta(n_f, l_f)]^2} \right] \simeq 21.3\,\text{GHz}$$

The classical analogy (see Problem I.5.2) yields

$$\nu \approx -R^* c \frac{2\Delta n^*}{(\bar{n}^*)^3} = 3.29 \cdot 10^6\,\text{GHz} \cdot \frac{2}{62^3} = 27.6\,\text{GHz}.$$

The wavelengths are inversely proportional to the Rydberg constant:

$$\frac{\lambda_{87}}{\lambda_{85}} = \frac{R^*_{85}}{R^*_{87}} \approx \left(1 + \frac{1}{87 \cdot 1836} \right) \Big/ \left(1 + \frac{1}{85 \cdot 1836} \right) \approx 1 - 1.47 \cdot 10^{-7} \ .$$

Therefore the isotopic shift results $\Delta\nu \approx 3.16$ kHz or $\Delta\lambda \approx -20.6$ Å.

Problem II.1.4 By considering Li as a Hydrogenic atom estimate the ionization energy. Discuss the result in the light of the real value (5.39 eV) in terms of percent of penetration of the optical electron in the $(1s)^2$ core.

Solution:

By neglecting the core charge one would have $E_{2s} = -13.56\,Z^2/n^2 = -30.6$ eV, while for total screening (i.e. zero penetration and $Z = 1$) $E_{2s} = -13.56\,\text{eV}/4 = -3.4$ eV.

Then the effective charge experimented by the $2s$ electron can be considered $Z_{eff} \sim 1.27$, corresponding to about 15 % of penetration.

2.2 Helium atom

2.2.1 Generalities and ground state

The Helium atom represents a fruitful prototype to enlighten the effects due to the inter-electron interaction and then the arise of the central field potential $V(r)$, (see §1.1), the effects related to the exchange symmetry for indistinguishable electrons and to discuss the role of the spins and the **antisymmetry** of the total wavefunction.

First we shall start with the phenomenological examination of the energy levels diagram *vis-a-vis* to the one pertaining to Hydrogen atom (Fig. 2.7). A variety of comments is in order. It is noted that in He the state corresponding to the electronic configuration $(1s)(nl)$ when compared to the n state in Hydrogen shows the removal of the accidental degeneracy in l. This could be expected, being the analogous of the effect for the optical electron in alkali atoms (§2.1). A somewhat unexpected result is the occurrence of a **double** series of levels, in correspondence to the same electronic configuration $(1s)(nl)$. The first series includes the ground state, with first ionization energy 24.58 eV. It is labelled as the group of **parahelium** states and all the levels are **singlets** (classification $^1S, ^1P$, etc..., see §1.7). The second series has the lowest energy state at 19.82 eV above the ground-state and identifies the **orthohelium** states. These states are all **triplets**, namely characterized by a fine structure (detailed in the inset of the Figure for the 2^3P state). Each level has to be thought as the superposition of almost degenerate levels, the degeneracy being removed by the spin-orbit interaction (§1.6). The orthohelium states are classified $^3S, ^3P$, etc.... Among the levels of a given series the transition yielding the spectral lines correspond to the rule $\Delta L = \pm 1$, with an **almost** complete inhibition of the transitions from parahelium to orthohelium (i.e. almost no singlets$\leftrightarrow$ triplets transitions). Finally it can be remarked that while $(1s)^2$, at $S = 0$, is the ground state, the corresponding $(1s)^2$ triplet state is **absent** (as well as other states to be mentioned in the following).

In the assumption of infinite nuclear mass and by taking into account the Coulomb interaction only, the Schrodinger equation is

$$\left[-\frac{\hbar^2}{2m}(\nabla_1^2 + \nabla_2^2) - \frac{Ze^2}{r_1} - \frac{Ze^2}{r_2} + \frac{e^2}{r_{12}} \right] \phi(\mathbf{r}_1, \mathbf{r}_2) = E\phi(\mathbf{r}_1, \mathbf{r}_2) \qquad (2.4)$$

and it can be the starting point to explain the energy diagram. In Eq. 2.4 $Z = 2$ for the neutral atom.

Let us first assume that the inter-electron term e^2/r_{12} can be consider a perturbation of the hydrogenic-like Hamiltonian for two independent electrons (**independent electron approximation**). Then the unperturbed eigenfunction is

$$\phi_{n'l',n''l''}(\mathbf{r}_1, \mathbf{r}_2) = \phi_{n'l'}(\mathbf{r}_1)\phi_{n''l''}(\mathbf{r}_2) \qquad (2.5)$$

and

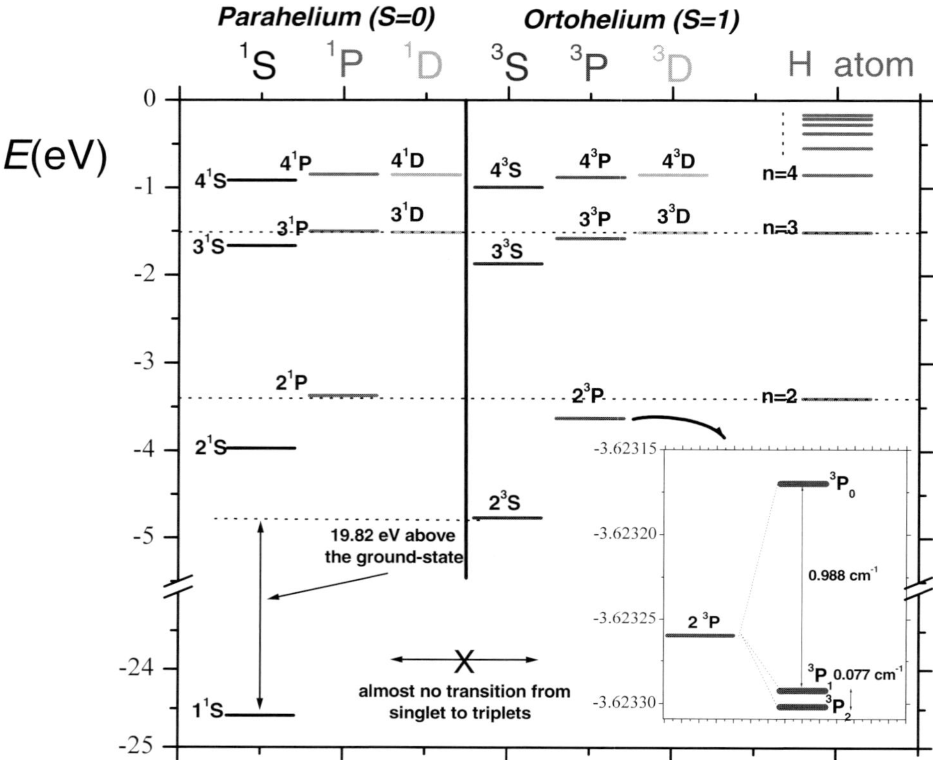

Fig. 2.7. Diagram of some energy levels for Helium atom and comparison with the correspondent levels for Hydrogen. The electron configuration of the states is $(1s)(nl)$. $E = 0$ corresponds to the first ionization threshold. The double-excited states (at weak transition probabilities and called **autoionizing states**) are unstable with respect to self-ionization (**Auger effect**) being at $E > 0$, within the continuum (Problem F.II.6). In the inset the fine structure of the $2\,^3P$ state is reported, to be compared with the separation, about 9000 cm^{-1}, between the $2\,^3S$ and the $2\,^3P$ states. Note that this fine structure does not follow the multiplet rules described at §3.2.

$$(E^0)_{n'l',n''l''} = Z^2 E^H_{n'l'} + Z^2 E^H_{n''l''} \tag{2.6}$$

E^H_{nl} being the eigenvalues for Hydrogen (degenerate in l).

For the ground state $(1s)^2$ one has

$$\phi_{1s,1s}(\mathbf{r}_1, \mathbf{r}_2) = \frac{Z^3}{\pi a_0^3} e^{-\frac{Z(r_1+r_2)}{a_0}} \tag{2.7}$$

and

$$E^0_{1s,1s} = 2Z^2 E^H_{1s} = -8\frac{e^2}{2a_0} \simeq -108.80 \text{ eV} \tag{2.8}$$

In this oversimplified picture the first ionization energy would be 54.4 eV, evidently far from the experimental datum (see Fig. 2.7). This discrepancy had to be expected since the effect of the electron-electron repulsion had not yet been evaluated.

At the first order in the perturbative approach the repulsion reads

$$E^{(1)}_{1s,1s} = \int \int \phi^*_{1s,1s}(\mathbf{r}_1,\mathbf{r}_2)\frac{e^2}{r_{12}}\phi_{1s,1s}(\mathbf{r}_1,\mathbf{r}_2)d\tau_1 d\tau_2 \equiv < 1s,1s|\frac{e^2}{r_{12}}|1s,1s > \equiv I_{1s,1s} \tag{2.9}$$

$I_{1s,1s}$ is called **Coulomb integral** in view of its classical counterpart, depicted in Fig. 2.8.

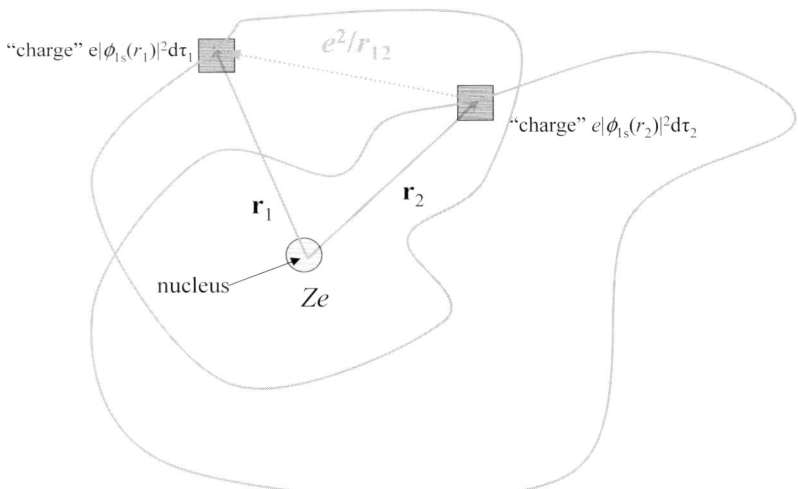

Fig. 2.8. Illustrative plot sketching the classical analogy of the first order perturbation term $< e^2/r_{12} >$ for the ground-state, in terms of electrostatic repulsion of two electronic clouds.

The estimate of the Coulomb integral can be carried out by expanding r_{12}^{-1} in term of the associated Legendre polynomials (see Problem II.2.1). For the particular case of $1s$ electrons, the Coulomb integral $I_{1s,1s}$ can be worked out in a straightforward way on the basis of the classical analogy for the electrostatic repulsion. The result is

$$I_{1s,1s} = \frac{5}{4}Z(-E^H_{1s}) = \frac{5}{8}\frac{e^2}{a_0}Z \tag{2.10}$$

The ground state energy corrected to the first order turns out

$$E_{1s,1s} = E^{(0)}_{1s,1s} + I_{1s,1s} = (2Z^2 - \frac{5}{4}Z)E^H_{1s} \simeq -74.8 \text{ eV} \qquad (2.11)$$

to be compared with the experimental value -78.62 eV.

The energy required to remove one electron is

$$[(2Z^2 - \frac{5}{4}Z) - Z^2].13.6\text{eV} \simeq 20.4 \text{ eV}$$

This estimate is not far from the value indicated in Fig. 2.7, in spite of the crudeness of the assumption for the unperturbed one-electron wavefunctions. An immediate refinement could be achieved by adjusting the hydrogen-like wave functions: in this way a good agreement with the experimental ionization energy would be obtained.

Another way to improve the description is to derive variationally an **effective nuclear charge** Z^*, which in indirect way takes into account the mutual screening of one electron by the other and the related correction in the wavefunctions. As shown in Problem II.2.2, this procedure yields $Z^* = Z - (5/16)$, implying for the ground state

$$E_{1s,1s} = 2(Z - \frac{5}{16})^2 . (\frac{-e^2}{2a_0}) = -77.5 \text{ eV}$$

One can remark how the perturbative approach, without modification of the eigenfunctions, is rather satisfactory, in spite of the relatively large value of the first order energy correction.

The ground state energy for He turns out about 94.6% of the "exact" one (numerically obtained via elaborate trial functions, see §3.4) with the first-order perturbative correction and 98% with the variationally-derived effective charge. The agreement is even better for atoms with $Z \geq 3$, as Li$^+$ or Be^{2+}. At variance the analogous procedure fails for H$^-$ (see Problem II.2.4).

2.2.2 Excited states and the exchange interaction

The perturbative approach used for the ground state could be naively attempted for the excited states with an electron on a given nl state. For a trial wavefunction of the form

$$\phi(\mathbf{r}_1, \mathbf{r}_2) = \phi_{1s}(\mathbf{r}_1)\phi_{nl}(\mathbf{r}_2) \qquad (2.12)$$

the energy

$$E_{1s,nl} = E^0_{1s,nl} + \; <1s,nl|\frac{e^2}{r_{12}}|1s,nl>$$

would not account for the experimental data, numerically falling approximately in the middle of the singlet and triplet $(1s, nl)$ energy levels. The striking discrepancy is evidently the impossibility to infer two energy levels in correspondence to the same electronic configuration from the wavefunction in Eq. 2.12. The obvious inadequacy of the tentative wavefunction is that it disregards the **exchange symmetry** (discussed at §1.3). At variance with Eq. 2.12 one has to write the functions

$$\phi_{1s,nl}^{sym}(\mathbf{r}_1,\mathbf{r}_2) = \frac{1}{\sqrt{2}}\left[\phi_{1s}(\mathbf{r}_1)\phi_{nl}(\mathbf{r}_2) + \phi_{nl}(\mathbf{r}_1)\phi_{1s}(\mathbf{r}_2)\right] \tag{2.13}$$

$$\phi_{1s,nl}^{ant}(\mathbf{r}_1,\mathbf{r}_2) = \frac{1}{\sqrt{2}}\left[\phi_{1s}(\mathbf{r}_1)\phi_{nl}(\mathbf{r}_2) - \phi_{nl}(\mathbf{r}_1)\phi_{1s}(\mathbf{r}_2)\right] \tag{2.14}$$

granting indistinguishable electrons, the same weights being attributed to the configurations $1s(1)nl(2)$ and $1s(2)nl(1)$. The labels **sym** and **ant** correspond to the **symmetrical** and **antisymmetrical character** of the wavefunctions upon exchange of the electrons.

On the basis of the functions 2.13 and 2.14, along the same perturbative procedure used for the ground state, instead of Eq.2.11 one obtains

$$E_+^{sym} = Z^2 E_{1s}^H + Z^2 E_{nl}^H + I_{1s,nl} + K_{1s,nl} \tag{2.15}$$

and

$$E_-^{ant} = Z^2 E_{1s}^H + Z^2 E_{nl}^H + I_{1s,nl} - K_{1s,nl} \tag{2.16}$$

where

$$K_{1s,nl} = \int\int \phi_{1s}^*(\mathbf{r}_1)\phi_{nl}^*(\mathbf{r}_2)\frac{e^2}{r_{12}}\phi_{1s}(\mathbf{r}_2)\phi_{nl}(\mathbf{r}_1)d\tau_1 d\tau_2 \tag{2.17}$$

is the **exchange integral, essentially positive** and without any classical interpretation, at variance with the Coulomb integral $I_{1s,nl}$. Thus double series of levels is justified by the quantum effect of **exchange symmetry**.

The wavefunctions 2.13 and 2.14 are not complete, spin variables having not yet been considered. In view of the weakness of the spin-orbit interaction, as already stated (§1.6), one can factorize the spatial and spin parts. Then, again by taking into account indistinguishable electrons, the spin functions are

$$\alpha(1)\alpha(2), \quad \beta(1)\beta(2), \quad \frac{1}{\sqrt{2}}[\alpha(1)\beta(2) + \alpha(2)\beta(1)] \quad \text{for } S = 1$$

$$\frac{1}{\sqrt{2}}[\alpha(1)\beta(2) - \alpha(2)\beta(1)] \quad \text{for } S = 0 \tag{2.18}$$

The first group can be labelled $\chi_{S=1}^{sym}$ and it includes the three eigenfunctions corresponding to $S = 1$. The fourth eigenfunction is the one pertaining to $S = 0$. $\chi_{S=0}^{ant}$ is antisymmetrical upon the exchange of the electrons, while $\chi_{S=1}^{sym}$ are symmetrical.

Therefore the complete eigenfunctions describing the excited states of the Helium atom are of the form $\phi_{tot} = \phi_{1s,nl}\,\chi_S$ and in principle in this way one would obtain 8 spin-orbitals. However, from the comparison with the experimental findings (such as the spectral lines from which the diagram in Fig. 2.7 is derived) one is lead to conclude that only four states are actually found in reality. These states are the ones for which the total (spatial and spin) wavefunctions are **antisymmetrical** upon the exchange of the two electrons.

This requirement of antisymmetry is also known as **Pauli principle** and we shall see that it corresponds to require that the electrons differ at least in one of the four quantum numbers n, l, m and m_s. For instance, the lack of the triplet $(1s)^2$ is evidently related to the fact that in this hypothetical state the two electrons would have the same quantum numbers, meantime having a wavefunction of symmetric character $\phi_{tot} = \phi_{1s}\phi_{1s}\chi_{S=1}^{sym}$. Thus $\phi^{sym}\chi_{S=0}^{ant}$ describes the singlet states, while $\phi^{ant}\chi_{S=1}^{sym}$ describes the triplet states.

Accordingly, one can give the following pictorial description

when	S	χ	$\phi(\mathbf{r})$	ϕ_{tot}	Energy
↑↓	0	ant	sym	ant	E_+
↑↑	1	sym	ant	ant	E_-

In other words, because of the exchange symmetry a kind of relationship, arising from electron-electron repulsion, between the mutual "direction" of the spin momenta and the energy correction does occur. For "parallel spins" one has $E_- < E_+$, the repulsion is decreased as the electron should move, on the average, more apart.

The dependence of the energy from the spin orientation can be related to an exchange **pseudo-spin interaction**, in other words to an Hamiltonian operator of the form [1]

$$H = -2K\mathbf{s}_1.\mathbf{s}_2 \tag{2.19}$$

In fact if we extend the vectorial picture to spin operators (in a way analogous to the definition of the $\mathbf{j}$ angular momentum for the electron (see §1.6)) and write

$$\mathbf{S} = \mathbf{s}_1 + \mathbf{s}_2, \tag{2.20}$$

by "**squaring**" this sum one deduces $\mathbf{s}_1.\mathbf{s}_2 = (1/2)[\mathbf{S}^2 - \mathbf{s}_1{}^2 - \mathbf{s}_2{}^2]$. Thus, from the Heisenberg Hamiltonian 2.19 the two energy values

$$E' = -2K(1/2)[S(S+1) - 2(1/2)(1+1/2)] = -K/2$$

for $S = 1$ and

$$E'' = 3K/2$$

---

[1] This Hamiltonian, known as **Heisenberg Hamiltonian**, is often assumed as starting point for quantum magnetism in bulk matter. Below a given temperature, in a three-dimensional array of atoms, this Hamiltonian implies a spontaneous ordered state, with magnetic moments cooperatively aligned along a common direction (see §4.4 for a comment).

for $S = 0$ are obtained. In other words, from the Hamiltonian 2.19, for a given $1snl$ configuration, the singlet and the triplet states with energy separation and classification consistent with Eq. 2.15 and 2.16, are deduced.

Now it is possible to justify the weak singlet$\leftrightarrow$triplet transition probability indicated by the optical spectra. The electric dipole transition element connecting parahelium to ortohelium states can be written

$$\mathbf{R}_{S=0 \leftrightarrow S=1} \propto < \chi_{ant} | \chi_{sym} > \int \int \phi^*_{sym} [\mathbf{r}_1 + \mathbf{r}_2] \phi_{ant} d\tau_1 d\tau_2 \ . \qquad (2.21)$$

This matrix element is zero, both for the orthogonality of the spin states and because the function in the integral changes sign upon exchange of the indexes 1 and 2, then requiring zero as physically acceptable result. Thus one understands why orthohelium cannot be converted to parahelium and *vice-versa*. This selection rule would seem to prevent any transitions (including the ones related to magnetic dipole or electric quadrupole mechanisms) and then do not admit any violation. The weak singlet-triplet transitions actually observed in the spectrum are related to the non-total validity of the factorization in the form $\phi_{tot} = \phi(\mathbf{r}_1, \mathbf{r}_2)\chi_{spin}$. The spin-orbit interaction, by coupling spin and positional variables, partially invalidates that form of the wavefunctions. This consideration is supported by looking at the transitions in an atom similar to Helium, with two electrons outside the core. Calcium has the ground state electronic configuration $(1s)^2...(4s)^2$ and the diagram of the energy levels is strictly similar to the one in Fig. 2.7. At variance with Helium, because of the increased strength of the spin-orbit interaction, the lines related to $S = 0 \leftrightarrow S = 1$ transitions are very strong. Analogous case is Hg atom (see Fig. 3.9).

Problems II.2

Problem II.2.1 Evaluate the Coulomb integral for the ground state of the Helium atom.

Solution:

In the expectation value

$$< \frac{1}{r_{12}} >= \frac{Z^6}{\pi^2} \int e^{-2Z(r_1+r_2)} \frac{1}{r_{12}} d\mathbf{r}_1 d\mathbf{r}_2.$$

$1/r_{12}$ is expanded in Legendre polynomials

$$\frac{1}{r_{12}} = \frac{1}{r_1} \sum_{l=0}^{\infty} \left(\frac{r_2}{r_1}\right)^l P_l(\cos\theta), \qquad r_1 > r_2$$

$$= \frac{1}{r_2} \sum_{l=0}^{\infty} \left(\frac{r_1}{r_2}\right)^l P_l(\cos\theta), \qquad r_1 < r_2$$

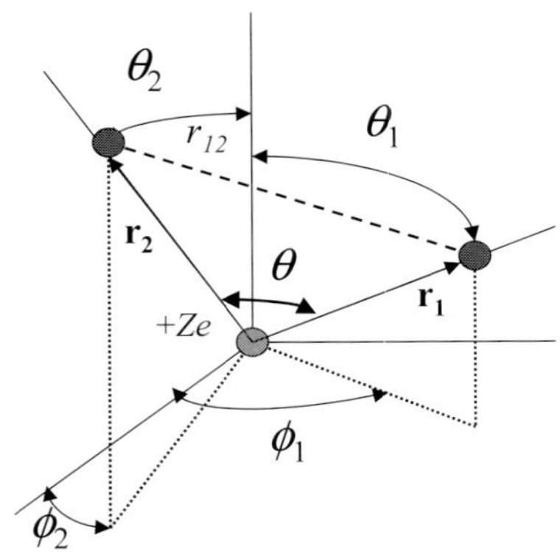

where θ is the angle between the vectors $\mathbf{r}_1$ and $\mathbf{r}_2$ and

$$\cos\theta = \cos\theta_1 \cos\theta_2 + \sin\theta_1 \sin\theta_2 \cos(\phi_1 - \phi_2)$$

In compact form

$$\frac{1}{r_{12}} = \sum_{l=0}^{\infty} \frac{(r_<)^l}{(r_>)^{l+1}} P_l(\cos\theta)$$

where $r_<$ is the smallest and $r_>$ the largest between r_1 and r_2. From the **addition theorem** one writes

$$\frac{1}{r_{12}} = \sum_{l=0}^{\infty} \sum_{m=-l}^{+l} \frac{4\pi}{(2l+1)} \frac{(r_<)^l}{(r_>)^{l+1}} Y_{lm}^*(\theta_1,\phi_1) Y_{lm}(\theta_2,\phi_2).$$

The function $exp[-2Z(r_1 + r_2)]$ is spherically symmetric and $Y_{00} = (4\pi)^{-\frac{1}{2}}$. By integrating over the polar angles one has

$$I'_{1s,1s} = \frac{Z^6}{\pi^2} \sum_{l=0}^{\infty} \sum_{m=-l}^{+l} \frac{(4\pi)^2}{(2l+1)} \int_0^{\infty} dr_1 r_1^2 \int_0^{\infty} dr_2 r_2^2 e^{-2Z(r_1+r_2)} \frac{(r_<)^l}{(r_>)^{l+1}}$$
$$\cdot \delta_{l,0}\delta_{m,0}.$$

All terms in the sum vanish, except the one for $l = m = 0$. Then

$$I'_{1s,1s} = 16Z^6 \int_0^{\infty} dr_1 r_1^2 \int_0^{\infty} dr_2 r_2^2 e^{-2Z(r_1+r_2)} \frac{1}{r_>}$$
$$= 16Z^6 \int_0^{\infty} dr_1 r_1^2 e^{-2Zr_1} \left[\frac{1}{r_1} \int_0^{r_1} dr_2 r_2^2 e^{-2Zr_2} \right.$$
$$\left. + \int_{r_1}^{\infty} dr_2 r_2 e^{-2Zr_2} \right] = \frac{5}{8} Z$$

and including a_0 and e in the complete form of the ϕ's $I_{1s,1s} = \frac{5}{4}Ze^2/2a_0$.

For spherically symmetric wavefunctions one can evaluate the Coulomb integral from the classical electrostatic energy:

$$I_{1s,1s} = \frac{Ze^2}{32\pi^2 a_0} \int \frac{e^{-\rho_1} e^{-\rho_2}}{\rho_{12}} d\tau_1 d\tau_2$$

where

$$\rho_{1,2} = \frac{2Zr_{1,2}}{a_0}, \qquad \rho_{12} = \frac{2Zr_{12}}{a_0}$$

and

$$d\tau_{1,2} = \rho_{1,2}^2 \sin\theta_{1,2}\, d\rho_{1,2}\, d\theta_{1,2}\, d\phi_{1,2}.$$

The electric potential from the shell $d\rho_1$ at ρ_1 is

$$d\Phi(r) = 4\pi \rho_1^2 e^{-\rho_1} d\rho_1 \frac{1}{\rho_1} \text{ for } r < \rho_1,$$

$$4\pi \rho_1^2 e^{-\rho_1} d\rho_1 \frac{1}{r} \text{ for } r > \rho_1.$$

Then the total potential turns out

$$\Phi(r) = \frac{4\pi}{r} \int_0^r e^{-\rho_1} \rho_1^2 d\rho_1 + 4\pi \int_r^\infty e^{-\rho_1} \rho_1 d\rho_1 = \frac{4\pi}{r}\{2 - e^{-r}(r+2)\}$$

and therefore

$$I_{1s,1s} = \frac{Ze^2}{32\pi^2 a_0} \int \Phi(\rho_2) e^{-\rho_2} d\tau_2 = \frac{Ze^2}{2a_0} \int_0^\infty [2 - e^{-\rho_2}(\rho_2 + 2)]e^{-\rho_2} \rho_2^2 d\rho_2 = \frac{Ze^2}{2a_0} \frac{5}{4}.$$

Problem II.2.2 By resorting to the variational principle, evaluate the effective nuclear charge Z^* for the ground state of the Helium atom.

   Solution:
   The energy functional is

$$E[\phi] = \frac{< \phi|H|\phi >}{< \phi|\phi >}$$

where

$$\phi(r_1, r_2) = \frac{Z^{*3}}{\pi} e^{-Z^*(r_1 + r_2)}$$

with Z^* variational parameter (a_0 is omitted).

   Then

$$E[\phi] = \left\langle \phi \left| T_1 + T_2 - \frac{Z}{r_1} - \frac{Z}{r_2} + \frac{1}{r_{12}} \right| \phi \right\rangle$$

and

$$\langle \phi | T_1 | \phi \rangle = \langle \psi_{1s}^{Z^*} | T_1 | \psi_{1s}^{Z^*} \rangle = \frac{1}{2} Z^{*2},$$

$$\langle \phi | T_2 | \phi \rangle = \langle \phi | T_1 | \phi \rangle,$$

while

$$\left\langle \phi \left| \frac{1}{r_1} \right| \phi \right\rangle = \left\langle \psi_{1s}^{Z^*} \left| \frac{1}{r_1} \right| \psi_{1s}^{Z^*} \right\rangle = Z^* = \left\langle \phi \left| \frac{1}{r_2} \right| \phi \right\rangle$$

Since

$$\left\langle \phi \left| \frac{1}{r_{12}} \right| \phi \right\rangle = \frac{5}{8} Z^*$$

one has

$$E[\phi] \equiv E(Z^*) = Z^{*2} - 2ZZ^* + \frac{5}{8} Z^* \ .$$

From

$$\frac{\partial E}{\partial Z^*} = 2Z^* - 2Z + \frac{5}{8} = 0$$

one deduces $Z^* = Z - 5/16$.

Problem II.2.3 In the light of the interpretation of the Coulomb integral in terms of repulsion between two spherically symmetric charge distributions, evaluate the effective potential energy for a given electron in the ground state of He atom and the effective charge $Z_{eff}(r)$.

Solution:

The electric potential due to a spherical shell of radius R (thickness dR and density $-e\rho(R)$) at distance r from the center of the sphere is

$$-\frac{1}{4\pi e} d\phi(r) = R^2 \rho(R) \frac{dR}{R} \quad \text{for } r \leq R,$$

$$R^2 \rho(R) \frac{dR}{r} \quad \text{for } r \geq R.$$

By integrating over R and taking into account that

$$\varrho(r) \equiv |\psi_{1s}(r)|^2 = \left(\frac{Z}{a_0} \right)^3 \frac{e^{-\frac{2Zr}{a_0}}}{\pi},$$

one has

$$-\frac{\phi(r)}{4\pi e} = \frac{1}{\pi} \left(\frac{Z}{a_0} \right)^3 \left[\frac{1}{r} \int_0^r dR R^2 e^{-\frac{2ZR}{a_0}} + \int_r^\infty dR R e^{-\frac{2ZR}{a_0}} \right]$$

$$= \frac{1}{4\pi} \left(\frac{Z}{a_0} \right) \left[\frac{1}{u} \int_0^u dx x^2 e^{-x} + \int_u^\infty dx x e^{-x} \right]$$

$$= \frac{1}{4\pi} \left(\frac{Z}{a_0} \right) \frac{1}{u} [2 - e^{-u}(u + 2)],$$

where $u = \frac{2Zr}{a_0}$. Therefore

$$\phi(r) = -\frac{e}{r}\left[1 - e^{-\frac{2Zr}{a_0}}\left(\frac{Zr}{a_0} + 1\right)\right]$$

and from

$$-\frac{Z_{eff}(r)e^2}{r} = -\frac{Ze^2}{r} - e\phi(r)$$

for $Z = 2$ one finds

$$Z_{eff}(r) = 1 + e^{-\frac{4r}{a_0}}\left(1 + \frac{2r}{a_0}\right)$$

plotted below.

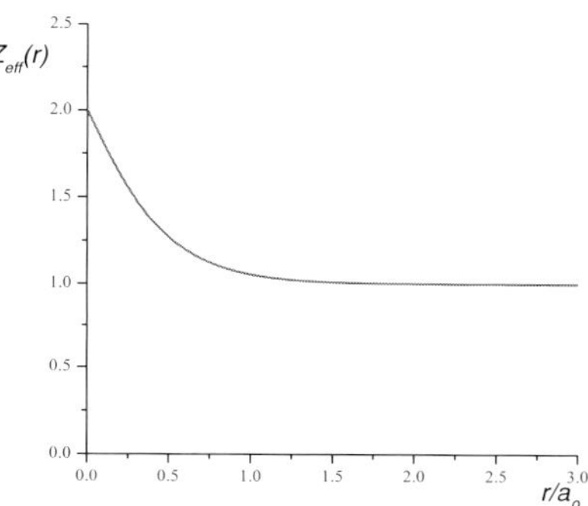

Problem II.2.4 The **electron affinity** (energy gain when an electron is acquired) for Hydrogen atom is 0.76 eV. Try to derive this result in the framework of a perturbative approach for the ground state of H^-, as well as by considering a reduced nuclear charge Z^*.

Comment the results in the light of the almost-exact value which, at variance, is obtained only by means of a variational procedure with elaborate trial wavefunctions.

Solution:

For H^-, by resorting to the results for He and setting $Z = 1$, in the perturbative approach one would obtain

$$E'_{H^-} = -2Z^2 R_H hc + \frac{5}{4}Z R_H hc = -\frac{3}{4} R_H hc$$

to be compared with $-R_H hc$ for H. With the variational effective charge $Z_{eff} = (1 - \frac{5}{16})$

$$E''_{H^-} = -2(0.473) R_H hc$$

again less bound than the ground-state for neutral Hydrogen.

Only more elaborate calculations yield the correct value, the reason being that for small Z the perturbation is too large with respect to the unperturbed energy. By repeating the estimate for $Z = 3$ (Li$^+$), for $Z = 4$ (Be^{2+}) and for $Z = 5$ (B^{3+}) a convergence is noted towards the "exact" values of the ground state energy (namely 198.1 eV, 371.7 eV and 606.8 eV, respectively) obtained from the variational procedure with elaborate trial functions. It should be remarked that the real experimental eigenvalues cannot be derived simply on the basis of the Hamiltonian in Eq. 2.4 which does not include the finite nuclear mass, the relativistic and the radiative terms (see for the Hydrogen atom the recall in Appendix V.1).

2.3 Pauli principle, determinantal eigenfunctions and superselection rule

In the light of the analysis of the properties of the electronic states in Helium atom, one can state the **Pauli principle: the total wavefunction** (spatial and spin) of electrons, particles at half integer spin, **must be antisymmetrical** upon exchange of two particles. This statement is equivalent to the one inhibiting a given set of the four quantum numbers $(nlmm_s)$ to more than one electron. For instance, this could be realized by considering an hypothetical triplet ground state $(1s)^2$ for orthohelium, for which the wavefunction would be $\phi_{1s}(\mathbf{r}_1)\phi_{1s}(\mathbf{r}_2)\alpha(1)\alpha(2)$ (or $\beta(1)\beta(2)$ or $(1/\sqrt{2})[\alpha(1)\beta(2)+\alpha(2)\beta(1)]$), and the quantum numbers n, l, m, m_s would be the same for both electrons. At variance, one only finds the singlet ground state, for which $m_s = \pm 1/2$.

From the specific case of Helium now we go back to the general properties of multi-electron atoms (see §1.1 and §1.3). Because of the exchange degeneracy and of the requirement of antisymmetrical wavefunction, the total eigenfunction, instead of Eq. 1.10 must be written

$$\varphi_{tot} = \frac{1}{\sqrt{N!}} \sum_P P(-1)^P \varphi_\alpha(1)\varphi_\beta(2)...\varphi_\nu(N) \tag{2.22}$$

where $\alpha, \beta, ...$ here indicate the group of quantum numbers $(nlmm_s)$ and the numbers $1, 2, 3, ...N$ include spatial and spin variables. P is an operator exchanging the electron i with the electron j and the wavefunction changes (does not change) sign according to an odd (even) number of permutations. The sum includes all possible permutations.

A total eigenfunction complying with exchange degeneracy and antisymmetry is the **determinantal wavefunction** devised by Slater [2]

$$\varphi_{tot} = \frac{1}{\sqrt{N!}} \begin{pmatrix} \varphi_\alpha(1) & \varphi_\alpha(2) & ... & \varphi_\alpha(N) \\ \varphi_\beta(1) & \varphi_\beta(2) & ... & \varphi_\beta(N) \\ ... & ... & ... & ... \\ \varphi_\nu(1) & \varphi_\nu(2) & ... & \varphi_\nu(N) \end{pmatrix}$$

---

[2] This form is the basis for the multiplet theory in the perturbation approach dealing with operators r_i^{-1} and r_{ij}^{-1} (see §3.4).

accounting for all the possible index permutations with change of sign when two columns are exchanged. On the other hand the determinant goes to zero when two groups of quantum numbers (and then two rows) are the same.

Now it can be proved that no transition, by any mechanism, is possible between globally antisymmetric and symmetric states (in the assumption that they exist), sometimes known as **superselection rule**. In fact such a transition would be controlled by matrix elements of the form

$$\mathbf{R}_{ANT\leftrightarrow SYM} \propto \int \phi^*_{SYM}[\mathbf{O}_1 + \mathbf{O}_2 + ...]\phi_{ANT}d\tau_{gen} \qquad (2.23)$$

that must be zero in order to avoid the unacceptable result of having a change of sign upon exchange of indexes, since the integrand is globally antisymmetric.

In the light of what has been learned from the analysis of alkali atoms and of Helium atom, now we can move to a useful description of multi-electrons atoms which allows us to derive the structure of the eigenvalues and their classification in terms of proper quantum numbers (The **vectorial model**, Chapter 3). Other typical atoms, such as N, C and transition metals (Fe, Co, *etc...*) shall be discussed in that framework.

Problems F.II

Problem F.II.1 By means of the perturbation approach for independent electrons derive the energy levels for the first excited states of Helium atom, in terms of Coulomb and exchange integrals, writing the eigenfunctions and plotting the energy diagram.

Solution:

The first excited $1s2l$ states are

$$
\begin{aligned}
u_1 &= 1s(1)2s(2) & u_5 &= 1s(1)2p_y(2) \\
u_2 &= 1s(2)2s(1) & u_6 &= 1s(2)2p_y(1) \\
u_3 &= 1s(1)2p_x(2) & u_7 &= 1s(1)2p_z(2) \\
u_4 &= 1s(2)2p_x(1) & u_8 &= 1s(2)2p_z(1)
\end{aligned}
$$

From the unperturbed Hamiltonian without the electron-electron interaction, by setting $\hbar = 2m = e = 1$, one finds

$$\mathcal{H}^\circ u_i(1,2) = -4(1 + \frac{1}{4})u_i(1,2) = -5u_i(1,2)$$

The secular equation involves the integrals

$$I_s = \left\langle 1s(1)2s(2) \left| \frac{1}{r_{12}} \right| 1s(1)2s(2) \right\rangle$$

$$I_p = \left\langle 1s(1)2p(2) \left| \frac{1}{r_{12}} \right| 1s(1)2p(2) \right\rangle$$

$$K_s = \left\langle 1s(1)2s(2) \left| \frac{1}{r_{12}} \right| 1s(2)2s(1) \right\rangle$$

$$K_p = \left\langle 1s(1)2p(2) \left| \frac{1}{r_{12}} \right| 1s(2)2p(1) \right\rangle$$

(p here represents p_x, p_y or p_z) and it reads

$$\begin{vmatrix} I_s - E' & K_s & 0 & 0 & 0 \\ K_s & I_s - E' & & & \\ & I_p - E' & K_p & 0 & 0 \\ 0 & K_p & I_p - E' & & \\ & & I_p - E' & K_p & 0 \\ 0 & 0 & K_p & I_p - E' & \\ & & & I_p - E' & K_p \\ 0 & 0 & 0 & K_p & I_p - E' \end{vmatrix} = 0$$

From the first block $E' = I_s \pm K_s$, with the associated eigenfunctions

$$\phi_{1,2} = \frac{1}{\sqrt{2}}[1s(1)2s(2) \pm 1s(2)2s(1)] \ .$$

From the second block $E'' = I_p \pm K_p$, with eigenfunctions

$$\phi_{3,4} = \frac{1}{\sqrt{2}}[1s(1)2p_x(2) \pm 1s(2)2p_x(1)]$$

and the analogous for y and z. Thus the following diagram is derived (I and $K > 0$ and $I_p > I_s$).

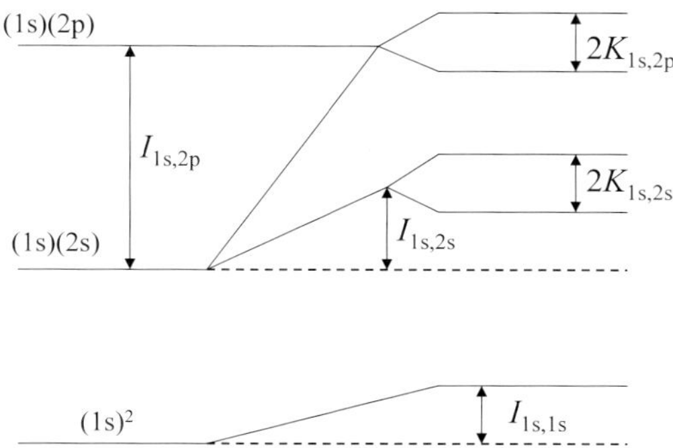

Problem F.II.2 For the optical electron in Li atom consider the **hybrid orbital**

$$\Phi = (1 + \lambda^2)^{-\frac{1}{2}}[\phi_{2s} + \lambda\phi_{2p_z}]$$

ϕ_{2s} and ϕ_{2p_z} being normalized hydrogen-like wavefunctions, with effective nuclear charge Z. Find the pseudo-dipole moment $\mu = e\langle z \rangle$ and the value of λ yielding the maximum of μ (relevant connections for situations where hybrid orbitals are actually induced are to be found at §4.2 and §9.2).

Solution:

The pseudo-dipole moment turns out

$$\mu = e \int \Phi^* z \Phi d\tau =$$

$$= \frac{e}{(1 + \lambda^2)} \left[\int \phi_{2s}^2(r) z d\tau + \lambda^2 \int \phi_{2p_z}^2(r) z d\tau + 2\lambda \int \phi_{2s}(r)\phi_{2p_z}(r) z d\tau \right] ,$$

where the first two integrals are 0. From Table I.4.2 (setting for simplicity $e = a_0 = 1$)

$$\mu = \frac{2\lambda}{1 + \lambda^2} \left(\frac{Z}{2}\right)^3 \frac{1}{4\pi} \int_0^{2\pi} d\phi \int_0^\pi \cos^2\theta \sin\theta d\theta \int_0^\infty Z r^4 (Z r - 2) e^{-Z r} dr =$$

$$= \frac{\lambda}{1 + \lambda^2} \frac{Z^3}{12} \left[\frac{Z^2 5!}{Z^6} - \frac{2Z 4!}{Z^5} \right] = \frac{\lambda}{1 + \lambda^2} \frac{6}{Z} ,$$

i.e. $\mu = (6 e a_0 / Z)\lambda/(1 + \lambda^2)$ in complete form.

From

$$\frac{d\mu}{d\lambda} = \frac{6}{Z}\frac{(1+\lambda^2 - 2\lambda^2)}{(1+\lambda^2)^2} = \frac{6}{Z}\frac{(1-\lambda^2)}{(1+\lambda^2)^2} = 0$$

the maximum is found for $\lambda = 1$, as it could be expected.

Problem F.II.3 Prove that the two-particles spin-orbital

$$\psi_{ANT} = \frac{1}{\sqrt{2}}\{\alpha(1)\beta(2)[\phi_a(1)\phi_b(2)] - \alpha(2)\beta(1)[\phi_a(2)\phi_b(1)]\}.$$

represents an eigenstate for the z-component of the total spin at zero eigenvalue. Then evaluate the expectation value of $\mathbf{S}^2$.

Solution:

From

$$S_1^z\psi_{ANT} = \frac{1}{2}\frac{1}{\sqrt{2}}\{\alpha(1)\beta(2)[\phi_a(1)\phi_b(2)] + \alpha(2)\beta(1)[\phi_a(2)\phi_b(1)]\} = \frac{1}{2}\psi_{SYM}.$$

and

$$S_2^z\psi_{ANT} = -\frac{1}{2}\psi_{SYM}$$

Thus

$$S^z\psi_{ANT} = (S_1^z + S_2^z)\psi_{ANT} = 0$$

Since

$$S_1^z S_2^z\psi_{ANT} = -\frac{1}{4}\psi_{ANT}$$

while

$$S_1^x S_2^x\psi_{ANT} = \frac{1}{4}\frac{1}{\sqrt{2}}\{\beta(1)\alpha(2)[\phi_a(1)\phi_b(2)] - \alpha(1)\beta(2)[\phi_a(1)\phi_b(2)]\} \equiv \psi'_{ANT}$$

and

$$< \psi_{ANT}|\psi'_{ANT} > \equiv -|\int \phi_a^*(\mathbf{r})\phi_b(\mathbf{r})d\tau|^2 \equiv -\mathcal{A}$$

(with the same result for the y component). By taking into account that $(\mathbf{S})^2 = (\mathbf{S}_1)^2 + (\mathbf{S}_2)^2 + 2\mathbf{S}_1.\mathbf{S}_2$, then

$$< \psi_{ANT}|(S_1^{x,y} + S_2^{x,y})^2|\psi_{ANT} > = \frac{1}{2}\{1 - \mathcal{A}\}$$

and

$$< \psi_{ANT}|(\mathbf{S})^2|\psi_{ANT} > = 1 - \mathcal{A} \ .$$

Problem F.II.4 At Chapter 5 it will be shown that between one electron and one proton an hyperfine interaction of the form $A\mathbf{I}.\mathbf{S}\delta(\mathbf{r})$ occurs, where $\mathbf{I}$ is the nuclear spin (**Fermi contact interaction**). An analogous term, i.e. $\mathcal{H}_p = A\mathbf{s}_1.\mathbf{s}_2\delta(\mathbf{r}_{12})$ (with $\mathbf{r}_{12} \equiv \mathbf{r}_1 - \mathbf{r}_2$) describes a relativistic interaction between the two electrons in the Helium atom. In this case A turns out $A = -(8\pi/3)(e\hbar/mc)^2$. Discuss the first-order perturbation effect of $\mathcal{H}_p$ on the lowest energy states of orthohelium and parahelium, showing that only a small shift of the ground-state level of the latter occurs (return to Prob. F.I.15 for similarities) .

Solution:

For orthohelium the lowest energy states is described by the spin-orbital

$$\phi_{tot}(1,2) = \chi_{sym}^{S=1}\left[\phi_{1s}(\mathbf{r}_1)\phi_{2s}(\mathbf{r}_2) - \phi_{2s}(\mathbf{r}_1)\phi_{1s}(\mathbf{r}_2)\right]$$

The evaluation of the expectation value of $\mathcal{H}_p$ yields zero since two electrons at parallel spin cannot have the same spatial coordinates. For the ground state of parahelium since $\mathbf{s}_1.\mathbf{s}_2 = -3/4$ (see Eq. 2.20), by using hydrogenic wave functions $\phi_{1s}(\mathbf{r}_1)$ and $\phi_{1s}(\mathbf{r}_2)$ one estimates

$$< 1s, 1s|\mathcal{H}_p|1s, 1s >= -\frac{3A}{4}\frac{Z^6}{\pi^2 a_0^6}\int\int e^{-2\frac{Z(r_1+r_2)}{a_0}}\delta(\mathbf{r}_{12})d\tau_1 d\tau_2 =$$

$$= -\frac{3A}{4}\frac{Z^6}{\pi^2 a_0^6}4\pi\int_0^\infty e^{-4Z\frac{r}{a_0}}r^2 dr = \frac{3}{32}\left(\frac{e\hbar}{mc}\right)^2\frac{Z^3}{a_0^3} \simeq 10^{-3}\text{eV},$$

a small shift compared to -78.62 eV.

Problem F.II.5 The spin-orbit constant ξ_{2p} for the $2p$ electron in Lithium turns out $\xi_{2p} = 0.34\text{cm}^{-1}$. Evaluate the magnetic field causing the first crossing between $P_{3/2}$ and $P_{1/2}$ levels, in the assumption that the field does not correct the structure of the doublet.

Solution:

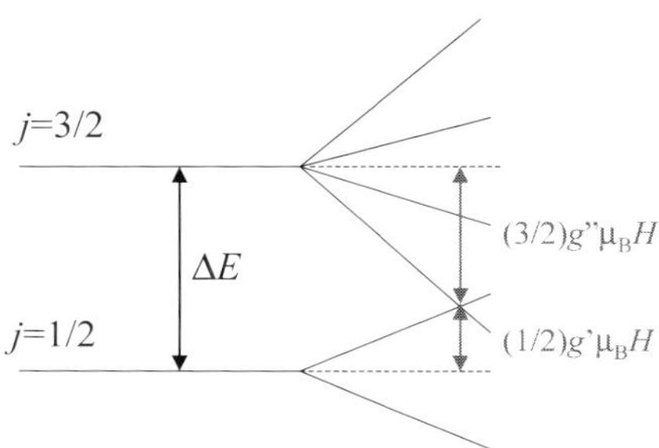

The first crossing takes place when

$$\left(\frac{3}{2}g" + \frac{1}{2}g'\right)\mu_B H = \Delta E$$

Since $g' = 2/3$ and $g" = 4/3$ the crossing occurs for $H = \Delta E/(7\mu_B/3)$.
The correction associated with the spin-orbit interaction is

$$\frac{\xi_{2p}}{2}[j(j+1) - l(l+1) - s(s+1)]$$

Then $\Delta E = (3/2)\xi_{2p}$ and $H \simeq 4370$ Oe.
When the **weak field** condition is released and the perturbation theory
for the full Hamiltonian $\xi_{nl}\mathbf{l.s} + \mu_B \mathbf{H.(l + 2s)}$ is applied, the crossing is found
at a slightly different field. Try to estimate it after having read Chapter 4 (or
see Prob. 1.1.20 in the book by **Balzarotti, Cini** and **Fanfoni** quoted in the
Preface).

Problem F.II.6 Refer to the double-excited electron state $2s4p$ of the Helium atom. In the assumption that the $2s$ electron in practice is not screened
by the $4p$ electron, which in turn feels just the residual charge $Z(r) \simeq 1$
(see §2.1), evaluate the wavelength of the radiation required to promote the
transition from the ground state to that double-excited state. After the **au-
toionization** of the atom, and decay to the ground-state of He^+, one electron
is ejected. Estimate the velocity of this electron.
Solution:
$E(2s, 4p) = -14.5$ eV, then $\lambda = c/\nu = 192$ Å and $v = 3.75 \times 10^8$ cm/s.

The shell vectorial model

Topics

"Aufbau" of the electronic structure and closed shells
Coupling of angular momenta (LS and jj schemes)
Rules for the ground state
Low energy states of C and N atoms
Effective magnetic moments and gyromagnetic ratio
Approximate form of the radial wavefunctions
Hartree-Fock-Slater theory for multiplets
Selection rules

3.1 Introductory aspects

By resorting to the principles of quantum mechanics and after having dealt
with specific atoms, one can now proceed to the description of the electronic
structure in generic multi-electron atoms. We shall see that the sequence of
electron states accounts for the microscopic origin of the periodic Table of the
elements.

First the one-electron states, described by orbitals of the form
$\phi_{nlm} = R_{nl}Y_{lm}$, have to be placed in the proper energy scale (**diagram**).
Then the atom can be thought to result from the progressive accommodation
of the electrons on the various levels, with the related eigenfunctions. This
build-up principle (called **aufbau** from the German) has to be carried out by
taking into account the **Pauli principle** (§2.3). Therefore a limited number
of electrons can be accommodated on a given level and each electron has
associated one (and only one) spin-orbital eigenfunction, differing in one or
more of the quantum numbers n, l, m, m_s from the others.

The maximum number of electrons characterized by a given value of n is

$$\sum_{l=0}^{n-1} 2(2l+1) = 2n + 4\sum_{l=0}^{n-1} l = 2n + 4\frac{n(n-1)}{2} \qquad (3.1)$$

When this maximum number is attained one has a **closed shell**. A closed sub-shell, often called **nl shell**, occurs when a given nl state (which defines the energy in the absence of spin-orbit and exchange interactions) accommodates $2(2l+1)$ electrons, in correspondence to the degeneracy in the z-component of the orbital momentum and of the spin degeneracy.

A complete sub-shell (or shell) implies electron charge distribution at **spherical symmetry**[1] and the quantum numbers L (for the total orbital momentum) and S (total spin) are zero, obviously implying $J = 0$ and spectral notation (see §1.7) $^1S_0$.

For the electrons outside the closed nl shells one has to take into account the spin-orbit interaction yielding $\mathbf{j} = (\mathbf{l} + \mathbf{s})$ and the electron-electron interaction leading to the Coulomb and exchange integrals, as it has been discussed for alkali atoms (§2.1) and for Helium atom (§2.2). As a consequence, a variety of "couplings" is possible and a complex distribution of the energy levels occurs, the detailed structure depending on the relative strengths of the couplings. For instance, the sequence of levels seen for Helium (§2.2), with spin-orbit terms much weaker than the Coulomb and exchange integrals, can be considerably modified on increasing the atomic number, when the spin-orbit interaction is stronger than the inter-electron effects.

In order to take into account the various couplings and to derive the qualitative sequence of the eigenvalues (with the proper classification in terms of good quantum numbers corresponding to constants of motion) one can abide by the so-called **vectorial model**. Initiated by **Heisenberg** and by **Dirac**, this model leads to the structure of the energy levels and to their classification in agreement with more elaborate theories for the multiplets, although it does not provide the quantitative estimate of the energy separation of the levels.

In the vectorial model the angular momenta and the associated magnetic moments are thought as classical vectors, as seen in the *ad hoc* definition of $\mathbf{J}$ and of $\mathbf{L}$ and $\mathbf{S}$ at §1.6, 1.7 and 2.2.2. Furthermore, somewhat classical equations of motion are used (for instance the precessional motion is often recalled). Moreover constraints are taken into account in the couplings, so that the final results do have characteristics in agreement with the quantum conditions. For example, the angular momenta of two p electrons are coupled and pictorially sketched as shown in Fig. 3.1

The interactions are written in the form

$$a) \quad a_{ik}\mathbf{l}_i.\mathbf{s}_k \qquad b) \quad b_{ik}\mathbf{l}_i.\mathbf{l}_k \qquad c) \quad c_{ik}\mathbf{s}_i.\mathbf{s}_k \qquad (3.2)$$

where $a)$ can be considered a generalization of the spin-orbit interaction ($a_{ii} > 0$, as proved at §1.6); $b)$ is the analogous for the orbital couplings, while $c)$ is the

---

[1] The rule $\sum_{m=-l}^{+l} Y_{l,m}^{*}(\theta, \varphi) Y_{l,m}(\theta, \varphi) = (2l+1)/4\pi$ is known as the **Unsold theorem** (See Problem I.4.9 for a particular case).

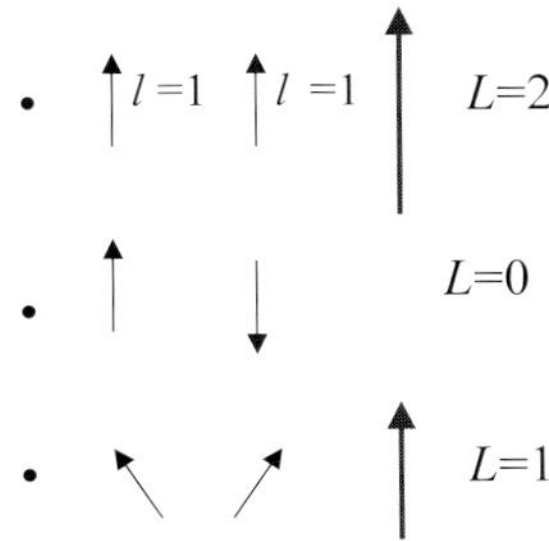

Fig. 3.1. Illustrative coupling of the angular momenta for two p electrons to yield the $L = 0, L = 1$ and $L = 2$ states. It is noted that the effective "lengths" of the "vectors" must be considered $\sqrt{l(l+1)}$ and $\sqrt{L(L+1)}$.

extension of the exchange interaction discussed in Helium ($c_{jk} = -2K < 0$). In these coupling forms the constants a, b and c have usually the dimensions of energy, the angular momenta thus being in $\hbar$ units.

On the basis of Eqs. 3.2 the energy levels are derived by coupling the electrons outside the closed shells and the states are classified in terms of good quantum numbers. The values of a, b and c are left to be estimated on the basis of the experimental findings, for instance from the levels resulting from the optical spectra.

In spite of these simplifying assumptions the many-body character of the problem prevents suitable solutions when a, b and c are of the same order of magnitude. Two limiting cases have to be considered:

i) "small" atoms (nuclear charge Z not too large) so that the spin-orbit interaction is smaller than other coupling terms and the condition $a \ll c$ can be assumed. This assumption leads to the so-called **LS scheme**;

ii) "heavy" atoms at large Z, where the strong spin-orbit interaction implies $a \gg c$ (**jj scheme**).

3.2 Coupling of angular momenta

3.2.1 LS coupling model

Within this scheme one couples $\mathbf{s}_i$ to obtain $\mathbf{S}$ and $\mathbf{l}_i$ for $\mathbf{L}$ (in a way to account for the quantum prescriptions). For

$$\mathbf{S} = \sum_i \mathbf{s}_i \quad \text{and} \quad \mathbf{L} = \sum_i \mathbf{l}_i \tag{3.3}$$

the total spin number $S = 0, 1/2, 1, 3/2, ...$ and the total orbital momentum number $L = 0, 1, 2, ...$ are defined. Then the spin orbit interaction is taken into account with an Hamiltonian of the form

$$\mathcal{H}_{SO} = \xi_{LS}\,\mathbf{L}.\mathbf{S}\,, \tag{3.4}$$

an extension of the Hamiltonian derived at §1.6 (see §3.2.2 in order to understand that the precessions of $\mathbf{l}_i$ yield an average orbital momentum along $\mathbf{L}$, while the average spin momentum is along $\mathbf{S}$: then Eq. 3.4 follows).

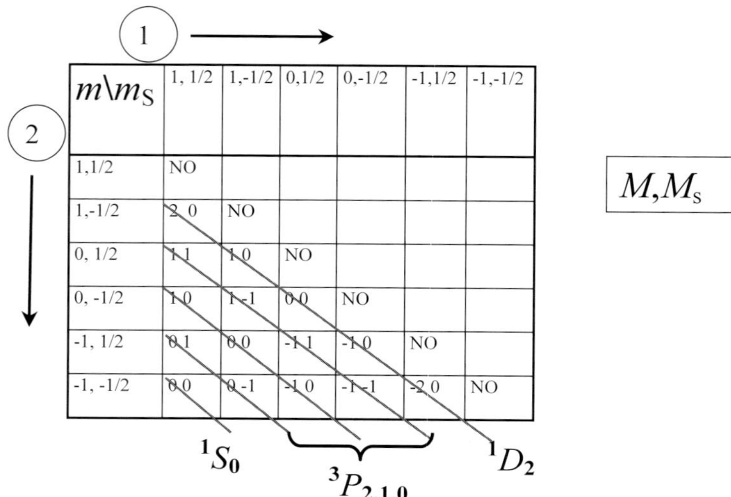

$m\backslash m_S$	1, 1/2	1,-1/2	0,1/2	0,-1/2	-1,1/2	-1,-1/2
1,1/2	NO					
1,-1/2	2 0	NO				
0, 1/2	1 1	1 0	NO			
0, -1/2	1 0	1 -1	0 0	NO		
-1, 1/2	0 1	0 0	-1 1	-1 0	NO	
-1, -1/2	0 0	0 -1	-1 0	-1 -1	-2 0	NO

$${}^1S_0 \qquad {}^3P_{2,1,0} \qquad {}^1D_2$$

Table III.2.1 Derivation of the electronic states compatible with the Pauli principle for two equivalent p electrons. It is noted that the group ${}^3D$ cannot exists since states with $M = 2$ and $M_s = 1$ are not found. The values $M_s = 0$ and with M running from -2 to $+2$ are present and they correspond to ${}^1D$ states at $S = 0$, implying $J = 2$. The states at $M = 1$ and $M_s = 1$ are all found and then the multiplet $S = 1$ and $L = 1$ does exist, implying the values $J = 2, 1, 0$. Finally the last case corresponds to the singlet state at $S = 0$ and $L = 0$. The total number of original states is 36 (corresponding to $2 \times 2 \times (2l_1 + 1) \times (2l_2 + 1)$) and only 15 of them are allowed. Six states are eliminated because they violate Pauli principle. Of the remaining 30 states, only half are distinguishable.

When the electrons to be coupled are **equivalent**, namely with the same quantum numbers n and l, one has to reject the coupling configurations that would invalidate the Pauli principle. In other words, one has to take into account the antisymmetry requirement for the total wavefunction and this corresponds to the problem of the **Clebsch-Gordan coefficients**. A simple method to rule out unacceptable states is shown in Table III.2.1 for two np electrons. All the possible values for m and m_s are summed up to give M and M_s. Then the states along the diagonal are disregarded, since they correspond to four equal quantum numbers. The states above the diagonal are also to be left out, since they correspond to the exchange of equivalent electrons, the exchange degeneracy being taken into account by the spin-spin interaction.

Finally the electronic states compatible with the values of M and M_s are found by inspection. This method corresponds to a brute-force counting of the states, as it is shown in the Problems for the low-energy electronic states in C and in N atoms (Problems III.2.1 and III.2.2).

When the electrons are **inequivalent** (differing in n or in l) no restrictions to the possible sums has to be considered (see Problem III.2.3).

Once that L and S are found and the structure of the levels expected from the couplings 3.2 is derived, then in the **LS scheme** one defines

$$\mathbf{J} = \mathbf{L} + \mathbf{S} \ , \tag{3.5}$$

characterized by the quantum number J. The spin-orbit interaction is taken into account according to Eq. 3.4 in order to derive the multiplets. Pictorially

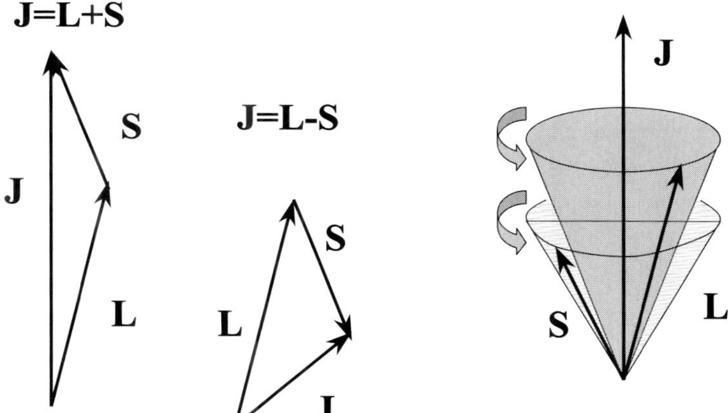

with coupling energy $E_{SO} = \xi_{LS}|\mathbf{L}|.|\mathbf{S}|cos\theta$ (θ angle between $\mathbf{L}$ and $\mathbf{S}$).

It is reminded that according to the classical equation of motion, a magnetic moment $\boldsymbol{\mu}_L \propto -\mathbf{L}$ in magnetic field precesses with angular frequency $\omega_L = \gamma H$, with γ the gyromagnetic ratio given by $\gamma = \boldsymbol{\mu}_L/\mathbf{L}$ (Problem III.2.4). In terms of $\mathbf{L}$ and $\mathbf{S}$ and of the related torque of modulus $-\partial E_{SO}/\partial \theta$, a precession of each of them around the direction of $\mathbf{J}$ has to be expected. To show this one writes

$$\frac{d\mathbf{L}}{dt} = \xi_{LS}\mathbf{S} \times \mathbf{L} \tag{3.6}$$

$$\frac{d\mathbf{S}}{dt} = \xi_{LS}\mathbf{L} \times \mathbf{S}. \tag{3.7}$$

and since $\mathbf{S} \times \mathbf{S} = \mathbf{L} \times \mathbf{L} = 0$

$$\frac{d\mathbf{L}}{dt} = \xi_{LS}\mathbf{J} \times \mathbf{L}$$

$$\frac{d\mathbf{S}}{dt} = \xi_{LS}\mathbf{J} \times \mathbf{S},$$

implying the precessional motions of $\mathbf{L}$ and $\mathbf{S}$ around an effective magnetic field along the direction of $\mathbf{J}$, the angular frequency being proportional to ξ_{LS} (see Problem III.2.4).

Therefore J and M_J are good quantum numbers while L_z and S_z are no longer constant of motion (z is here an arbitrary direction). Then the energies of the multiplet are derived by adding the corrections due to the spin-orbit Hamiltonian (in the form 3.4) to the energy $E^0(L,S)$ resulting from the couplings between $\mathbf{s}_i$ and between $\mathbf{l}_i$ (see examples in subsequent Figures). From the definition of J (Eq. 3.5), again by the usual "**squaring**", one obtains

$$E(L, S, J) = E^0(L, S) + \frac{1}{2}\xi_{LS}\left[J(J+1) - L(L+1) - S(S+1)\right] \qquad (3.8)$$

An empirical rule for ξ_{LS} is $\xi_{LS} \simeq \pm\xi_{nl}/2$, with the sign $+$ when the number of the electrons in the sub-shell in less then half of the maximun number that can be accommodated and $-$ in the opposite case (according to §1.6 $\xi_{nl} = a_{11}$ in Eq. 3.1). For sign $+$ the multiplet is called **regular**, namely the state at lowest energy is the one corresponding to $\mathbf{J}$ **minimum** (pictorially with $\mathbf{L}$ and $\mathbf{S}$ antiparallel). For sign $-$ the multiplet is **inverted**, the state at lowest energy being the one with maximum value for J (i.e. $\mathbf{L}$ and $\mathbf{S}$ parallel).

For regular multiplets one immediately derives the **interval rule**, giving the energy separation between the states at J and $(J+1)$. From Eq. 3.8

$$\Delta_{J,J+1} = (J+1)\xi_{LS} \qquad (3.9)$$

implying, for example for $L = 2$ and $S = 1$, the structure of the levels shown in Fig. 3.2. This rule can be used as a test to check the validity of the **LS coupling** scheme. It is noted that the "center of gravity" of the levels, namely the mean perturbation of all the states of a given term, is not affected by the spin-orbit interaction. In fact

$$< \Delta(E - E^0) > = \sum_{J=|L-S|}^{L+S} \frac{\xi_{LS}}{2}(2J+1)\left[J(J+1) - L(L+1) - S(S+1)\right] = 0$$

$$(3.10)$$

(see Figs. 3.2, 3.4 and Problem F.III.5).

When more than two electrons are involved in the coupling, the procedure outlined above has to be applied by combining the third electron with the results of coupling the first two and so on. Examples (Problems III.2.2) will clarify how to deal with more than two electrons.

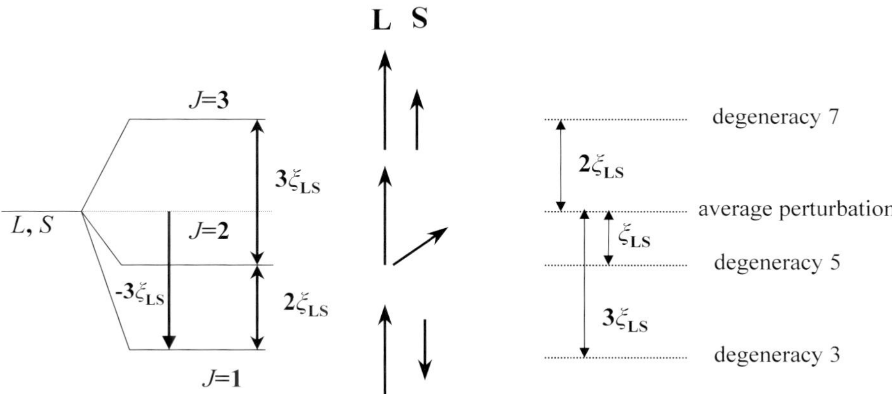

Fig. 3.2. Illustration of the interval rule for the multiplet arising from the $L = 2$ and $S = 1$ state.

3.2.2 The effective magnetic moment

At §1.6 the effect of an external magnetic field on one single electron has been considered. The quantum description for multi-electrons atom shall be given at Chapter 4. Here we derive the atomic magnetic moment that effectively interacts with the external field in the framework of the vectorial model and of the **LS scheme**.

The magnetic field, acting on $\boldsymbol{\mu}_L$ and $\boldsymbol{\mu}_S$, induces torques on **L** and on **S** while they are coupled by the spin-orbit interaction. A general solution for the motions of the momenta and for the energy corrections in the presence of the field can hardly be obtained. Rigorous results are derived in the limiting cases of **strong** and of **weak** magnetic field, namely for situations such that $\boldsymbol{\mu}_{L,S}.\mathbf{H} \gg \xi_{LS}$ and $\boldsymbol{\mu}_{L,S}.\mathbf{H} \ll \xi_{LS}$, respectively. Let us first discuss the case of weak magnetic field (Fig. 3.3)

In view of the meaning of **L.J** and of **S.J**, the angles between **L** and **J** and **S** and **J** can be written

$$cos\widehat{LJ} = \frac{L(L+1) + J(J+1) - S(S+1)}{2\sqrt{L(L+1)}\sqrt{J(J+1)}} \tag{3.11}$$

$$cos\widehat{SJ} = \frac{S(S+1) + J(J+1) - L(L+1)}{2\sqrt{S(S+1)}\sqrt{J(J+1)}} \tag{3.12}$$

Then the magnetic moment along the **J** direction, after averaging out the transverse components of **L** and **S** (due to fast precession induced by spin

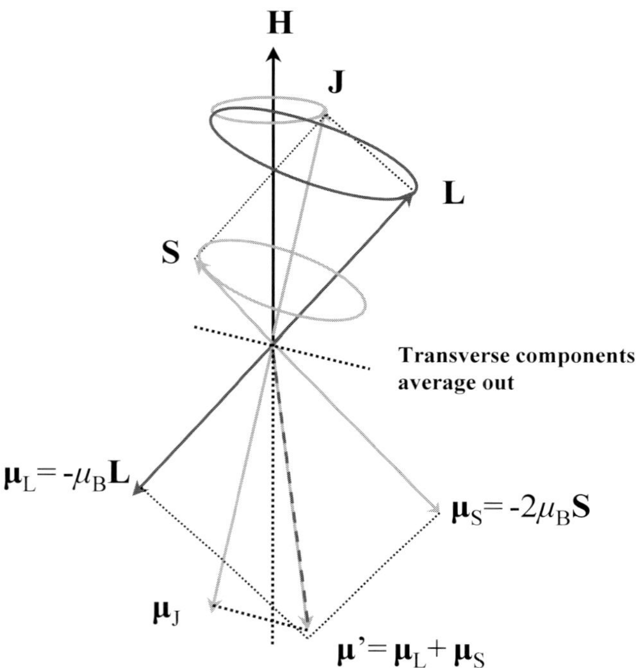

Fig. 3.3. Vectorial description of angular and magnetic moments in magnetic field, within the **LS** model. The interaction with the field is weak in comparison to the spin-orbit interaction and fast precessions of **L** and **S** around **J** occur, controlled by ξ_{LS}. Only the "result" of the precessional motion can effectively interact with the field: the precession of **J** at the Larmor frequency $\omega_L = \gamma H$ is induced. ω_L is much smaller than the precessional frequency of **L** and **S** around **J** (see Problem III.2.4).

orbit interaction) is [2]

$$\boldsymbol{\mu}_J = -2\mu_B \mathbf{S} cos\widehat{SJ} - \mu_B \mathbf{L} cos\widehat{LJ} \ .$$

Therefore the effective magnetic moment turns out

$$\boldsymbol{\mu}_J = -\mu_B g \mathbf{J} \tag{3.13}$$

where g, called the **Landé factor**, is

$$g = 1 + \frac{J(J+1) + S(S+1) - L(L+1)}{2J(J+1)} \tag{3.14}$$

---

[2] The formal proof is based on the **Wigner-Eckart theorem** (see §4.3)

Hence the energy corrections associated with the magnetic Hamiltonian are $\Delta E = -\boldsymbol{\mu}_J.\mathbf{H} = -\mu_J^z H = g\mu_B H M_J$. Thus the magnetic field removes the degeneracy in M_J and the energy levels, in weak magnetic field, turn out

$$E(L, S, J, M_J) = E^0(L, S, J) + \mu_B g H M_J. \tag{3.15}$$

In the opposite limit when the magnetic field is strong enough that the Hamiltonians $\boldsymbol{\mu}_L.\mathbf{H}$ and $\boldsymbol{\mu}_S.\mathbf{H}$ prevail over the spin-orbit interaction, one can first disregard this latter and the energy levels are derived in terms of the quantum magnetic numbers M and M_S. Vectorially this corresponds to the decoupling of the orbital and spin momenta and to their independent quantization along the axis of the magnetic field, around which they precess at high angular frequency. The magnetic moment is the sum of the independent components and therefore the energy correction is written

$$\Delta E = -[\mu_L^z H + \mu_S^z H] = \mu_B M H + 2\mu_B M_s H \tag{3.16}$$

The spin-orbit interaction can be taken into account subsequently, as perturbation of the states labelled by the quantum magnetic numbers M and M_S. This will be described at Chapter 4 as the so-called **Paschen-Back regime**.

3.2.3 Illustrative examples and the Hund rules for the ground state

In the framework of the **LS scheme**, by taking into account the signs of the coupling constants for the spin-orbit interaction (§1.6) and for the spin-spin interaction (§2.2.2), one can figure out simple rules to predict the configuration pertaining to the ground state of the atom. This is an important step for the description of the magnetic properties of matter. The rules, first empirically devised by **Hund**, are the following:

i) **maximize** the quantum number **S**. The reason for this is related to the sign of the exchange integral, since in the spin-spin coupling c_{12} plays the role of $-2K$, as already observed;

ii) **maximize L**, in a way compatible with Pauli principle;

iii) **minimize J for regular multiplets** while **maximize** it for **inverted multiplets**. This rule follows from the sign of ξ_{nl} and then of ξ_{LS} (see Eq. 3.8).

As illustrative examples let us consider one atom of the transition elements, with incomplete $3d$ shell (Fe) and one of the rare earth group, with incomplete $4f$ shell (Sm). The electronic configuration of iron is $(1s)^2(2s)^2(2p)^6(3s)^2(3p)^6$ $(3d)^6(4s)^2$. Maximization of S implies the spin vectorial coupling in the form $\uparrow\uparrow\uparrow\uparrow\uparrow\downarrow$ yielding $S = 2$. The coupling of five of the six orbital momenta must be zero, since the m numbers must be all different (from -2 to +2) in order to

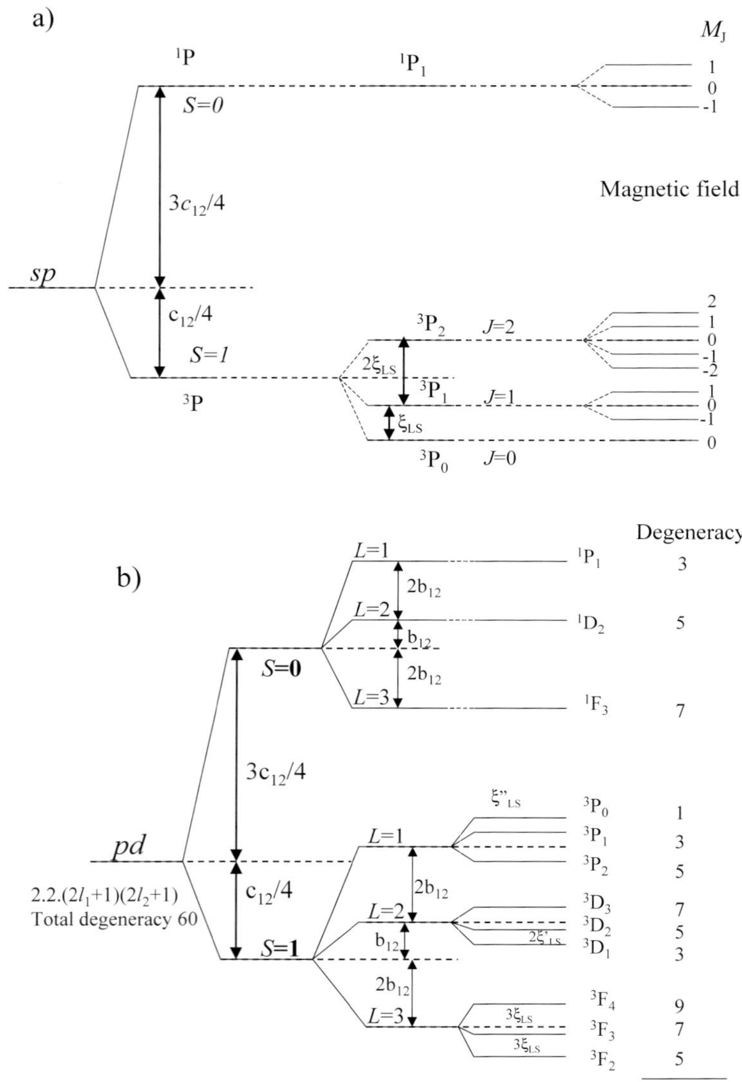

Fig. 3.4. Diagram of the energy levels and labeling of the electronic states within the **LS scheme** for: a) one s and one p electron; b) one p and one d electron (outside closed shells). For case a) it is shown how a magnetic field removes all the degeneracies, while in case b) the number of degenerate states are indicated on the right (ξ''_{LS} has been taken negative).

preserve Pauli principle. Then for the sixth electron we take the maximum, namely $L = 2$. The multiplet is inverted, because the maximum number of electrons that can be accommodated in the $3d$ sub-shell is 10. Then $J = 4$. Thus the ground state for iron is $^5D_4$.

Samarium has the electronic configuration ending with $(4f)^6(6s)^2$. Maximizing S yields $S = 3$. To complete half of the shell (that would give $L = 0$) one electron is missing. Then by taking the maximum possible value one has $L = 3$. The multiplet is regular and therefore the ground state is the one with $J = 0$, namely $^7F_0$. Other ground states are derived in Problem III.2.5.

In Table III.2.2 the ground state of some $4f$ magnetic ions often involved in paramagnetic crystals, with their effective magnetic moment $|\boldsymbol{\mu}| = g\sqrt{J(J+1)}$ are reported. As illustrative examples of the structure and classification of the energy levels in the **LS scheme** according to the prescriptions described above, in Fig. 3.4 the cases of atoms with one s and one p electron and with one p and one d electron outside the closed shells are shown.

Ion	Shell	S	L	J	Atomic Configuration	$|\boldsymbol{\mu}|$ (in Bohr magneton)
Ce^{3+}	$4f^1$	1/2	3	5/2	$^2F_{5/2}$	2.54
Pr^{3+}	$4f^2$	1	5	4	$^3H_4$	3.58
Nd^{3+}	$4f^3$	3/2	6	9/2	$^4I_{9/2}$	3.62
Pm^{3+}	$4f^4$	2	6	4	$^5I_4$	2.68
Sm^{3+}	$4f^5$	5/2	5	5/2	$^6H_{5/2}$	0.85
Eu^{3+}	$4f^6$	3	3	0	$^7F_0$	0
Gd^{3+}	$4f^7$	7/2	0	7/2	$^8S_{7/2}$	7.94
Tb^{3+}	$4f^8$	3	3	6	$^7F_6$	9.72
Dy^{3+}	$4f^9$	5/2	5	15/2	$^6H_{15/2}$	10.65
Ho^{3+}	$4f^{10}$	2	6	8	$^5I_8$	10.61
Er^{3+}	$4f^{11}$	3/2	6	15/2	$^4I_{15/2}$	9.58
Tm^{3+}	$4f^{12}$	1	5	6	$^3H_6$	7.56
Yb^{3+}	$4f^{13}$	1/2	3	7/2	$^2F_{7/2}$	4.54
Lu^{3+}	$4f^{14}$	0	0	0	$^1S_0$	0

Table III.2.2 Ground state of some magnetic ions of the $4f$ sub-shell, according to Hund's rules, and correspondent values of the effective magnetic moments. It should be remarked that these data refer to free ions, while the magnetic properties can change when the crystalline electric field is acting (see §13.3).

In Fig. 3.5 the energy levels of the p^2 configuration are reported and the transitions to the sp configuration (see Fig. 3.4a), driven by electric dipole mechanism, are indicated (see §3.5).

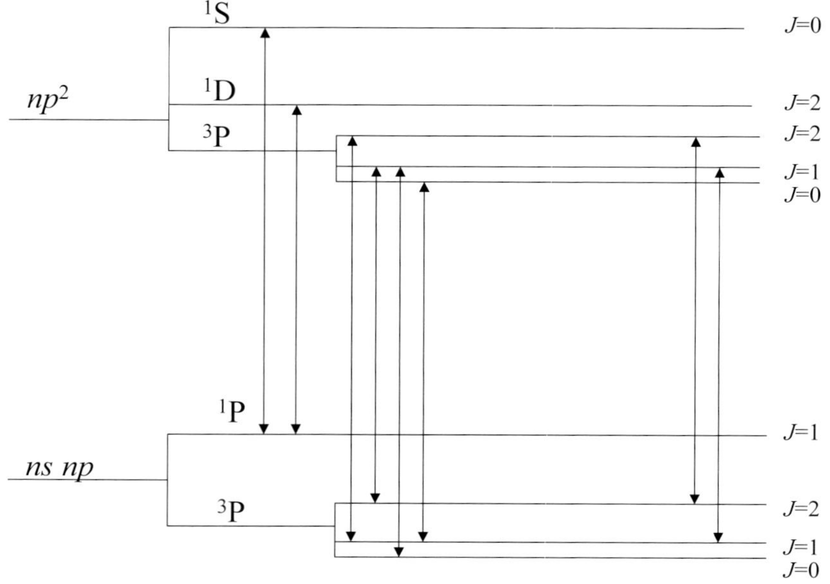

Fig. 3.5. Multiplet structure in the LS scheme for the $nsnp$ and the np^2 configurations and transitions allowed by the electric dipole mechanism (see §3.5).

Problems III.2

Problem III.2.1 Derive, with proper labeling, the first low-energy states of the **carbon** atom (ground state configuration $(1s)^2 \, (2s)^2 \, (2p)^2$) by taking into account the inter-electronic interactions, first disregarding the spin-orbit interaction.

Solution:

The method to rule out unacceptable states for **equivalent** $2p$ electrons is shown in Table III.2.1. Equivalently, by indicating $|m = 1, m_s = \frac{1}{2} >\equiv a$, $|0, -\frac{1}{2} >\equiv d$, $|1, -\frac{1}{2} >\equiv b$, $|-1, \frac{1}{2} >\equiv e$, $|0, \frac{1}{2} >\equiv c$, $|-1, -\frac{1}{2} >\equiv f$ one has the possibilities listed below:

	M_L	M_S	
ab	2	0	♯
ac	1	1	●
ad	1	0	♯
ae	0	1	●
af	0	0	◇
bc	1	0	●
bd	1	-1	●
be	0	0	●
bf	0	-1	●
cd	0	0	♯
ce	-1	1	●
cf	-1	0	♯
de	-1	0	●
df	-1	-1	●
ef	-2	0	♯

♯ terms correspond to $L = 2$ and $S = 0$, ● to $L = 1$ and $S = 1$, while ◇ to $L = 0$ and $S = 0$ (see Table III.2.1).

The first low-energy states are $^1S_0$, $^1D_2$ and $^3P_{0,1,2}$, according to the vectorial picture and to the Hund rules:

$$\uparrow\downarrow \ \uparrow\downarrow \qquad ^1S_0 \qquad L = 0 \ S = 0$$

$$\uparrow\uparrow \ \uparrow\downarrow \qquad ^1D_2 \qquad L = 2 \ S = 0$$

$$\diagdown\diagup\uparrow\uparrow \quad ^3P_{0,1,2} \quad L = 1 \ S = 1$$

The correspondent energy diagram, including the experimentally detected splitting of the lowest energy ^3P state due to spin-orbit interaction is

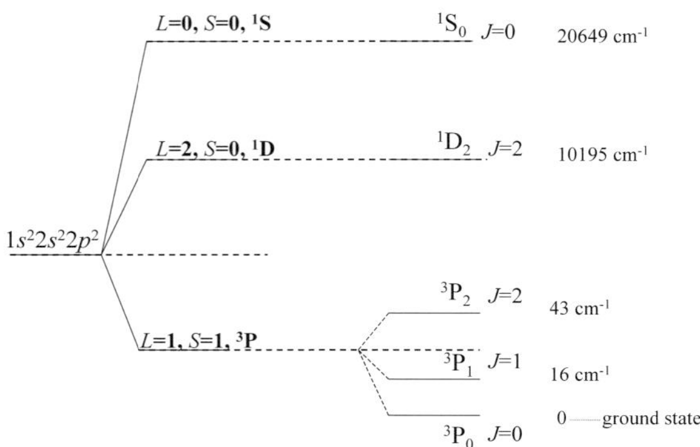

An extended energy levels diagram of the atom (with spin-orbit splitting not detailed) is

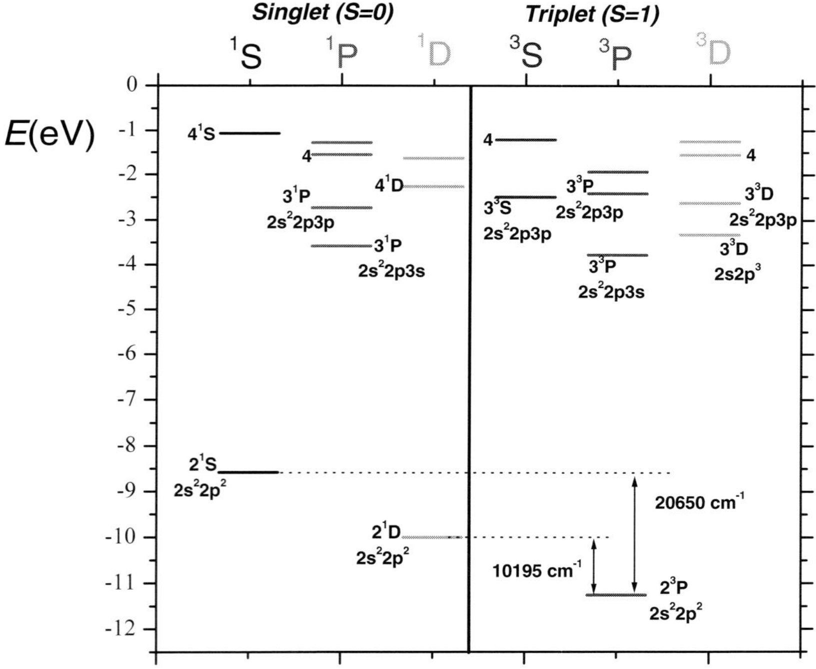

Problem III.2.2 Derive the first low-energy states for the N atom (electronic configuration $(1s)^2 (2s)^2 (2p)^3$) by taking into account the inter-electron couplings, labelling the states with the good quantum numbers. By assuming a spin-spin interaction of the form $\sum'_{i,j} A\mathbf{s}_i.\mathbf{s}_j$ evaluate the shift of the ground state.

Solution:

According to the notation used in Problem III.2.1, the possible one-electron states are $a\ b\ c\ d\ e\ f$.

The complete states, in agreement with the Pauli principle, are

		M_L	M_S				M_L	M_S
♯	abc	2	1/2			bcd	1	-1/2
♯	abd	2	-1/2			bce	0	1/2
♯	abe	1	1/2			bcf	0	-1/2
♯	abf	1	-1/2			bde	0	-1/2
	acd	1	1/2		♯	cde	-1	1/2
	ace	0	3/2		♯	cdf	-1	-1/2
	acf	0	1/2		♯	def	-2	-1/2
♯	ade	0	1/2			bef	-1	-1/2
♯	adf	0	-1/2		♯	cef	-2	1/2
	aef	-1	1/2			bdf	0	-3/2

(♯ terms corresponding to $L = 2$, $S = 1/2$, i.e $^2D_{5/2,3/2}$, etc...).

Thus the three low-energy states are

$$^2P_{\frac{3}{2},\frac{1}{2}} \qquad ^2D_{\frac{5}{2},\frac{3}{2}} \qquad ^4S_{\frac{3}{2}}$$

correspondent to the vectorial picture

$\searrow\nearrow \rightarrow \uparrow\downarrow\uparrow$	$^2P_{\frac{3}{2},\frac{1}{2}}$	$L = 1\ S = \frac{1}{2}$
$\uparrow\uparrow \rightarrow\ \uparrow\downarrow\uparrow$	$^2D_{\frac{5}{2},\frac{3}{2}}$	$L = 2\ S = \frac{1}{2}$
$\uparrow\downarrow \leftarrow\ \uparrow\uparrow\uparrow$	$^4S_{\frac{3}{2}}$	$L = 0\ S = \frac{3}{2}$

The diagram is

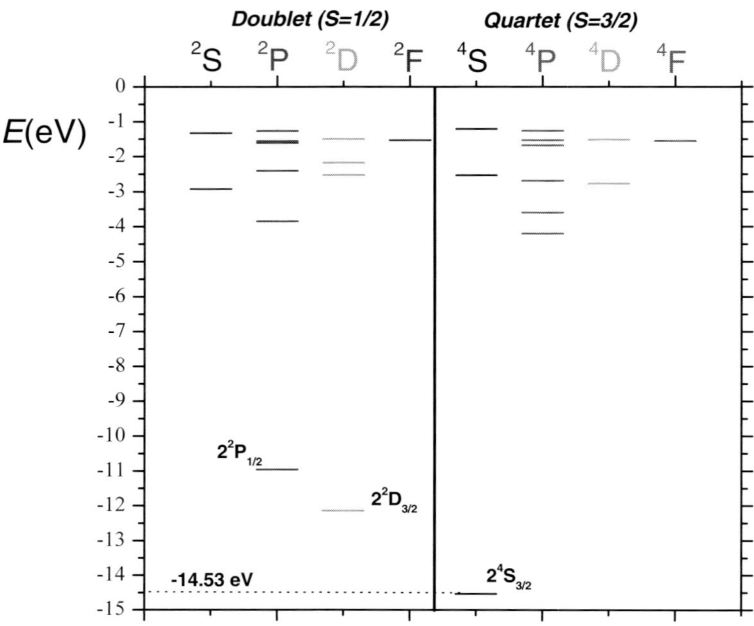

The shift of the ground state due to the spin-spin interaction is $3A/4$. In fact

$$\mathbf{S}^2 = \mathbf{s}_1^2 + \mathbf{s}_2^2 + \mathbf{s}_3^2 + 2[\mathbf{s}_1.\mathbf{s}_2 + \mathbf{s}_2.\mathbf{s}_3 + \mathbf{s}_1.\mathbf{s}_3]$$

and then

$$[\mathbf{s}_1.\mathbf{s}_2 + \mathbf{s}_2.\mathbf{s}_3 + \mathbf{s}_1.\mathbf{s}_3] = \frac{\mathbf{S}^2 - \mathbf{s}_1^2 - \mathbf{s}_2^2 - \mathbf{s}_3^2}{2} = \frac{3}{4}$$

The same structure and classification of the electronic states hold for **phosphorous** atom, in view of the same configuration $s^2 p^3$ outside the closed shells. On increasing the atomic number along the V group of the periodic Table, the increase in the spin-orbit interaction can be expected to invalidate the **LS scheme** (see §3.3). However, for three electrons in the p sub-shell, since ξ_{LS} is almost zero (see Eq. 3.8), the $^2P_{1/2}$ and $^2P_{3/2}$ states, for instance, have approximately the same energy ($\Delta E_{SO} \simeq 3.1$ meV).

Problem III.2.3 Reformulate the vectorial coupling for two inequivalent p electrons in the **LS** scheme, indicating the states that would not occur for equivalent electrons.

Solution:

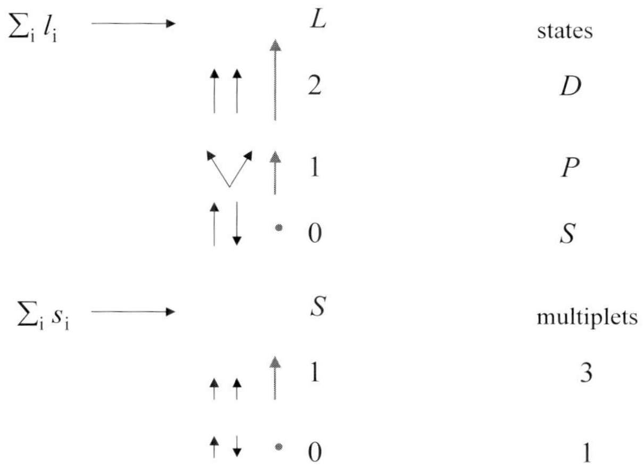

$$\sum_i l_i \longrightarrow \qquad L \qquad\qquad \text{states}$$

$$\uparrow\uparrow \quad\Big|\quad 2 \qquad\qquad D$$

$$\vee \quad\uparrow\quad 1 \qquad\qquad P$$

$$\uparrow\downarrow \quad\bullet\quad 0 \qquad\qquad S$$

$$\sum_i s_i \longrightarrow \qquad S \qquad\qquad \text{multiplets}$$

$$\uparrow\uparrow \quad\uparrow\quad 1 \qquad\qquad 3$$

$$\uparrow\downarrow \quad\bullet\quad 0 \qquad\qquad 1$$

$$\mathbf{J} = \mathbf{L} + \mathbf{S} \qquad D \qquad J = 3, 2, 1 \ \text{ for } \ S = 1$$

$$J = 2 \ \text{ for } \ S = 0$$

$$P \qquad J = 2, 1, 0 \ \text{ for } \ S = 1$$

$$J = 1 \ \text{ for } \ S = 0$$

$$S \qquad J \equiv S = 1, 0 \ \ .$$

For equivalent electrons only $^3P_{2,1,0}$, $^1D_2$ and $^1S_0$ are present (see Table III.2.1).

Problem III.2.4 By referring to a magnetic moment $\boldsymbol{\mu}_L$ in magnetic field $\mathbf{H}$, derive the precessional motion of $\mathbf{L}$ with the **Larmor frequency** $\boldsymbol{\omega}_L = \gamma \mathbf{H}$, where γ is the **gyromagnetic ratio** (see Problem I.6.5).

Solution:

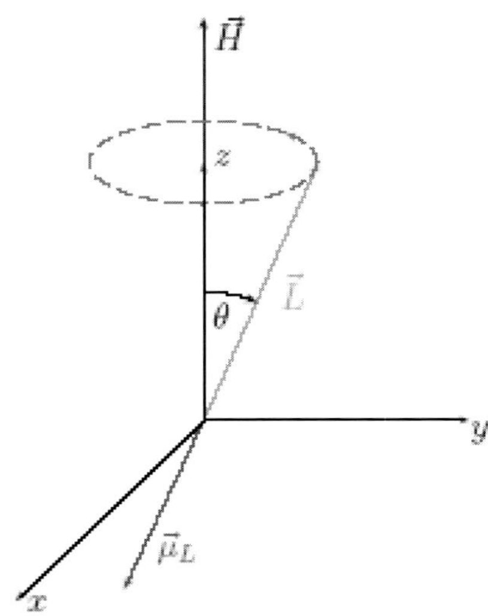

The equation of motion is

$$\frac{d\mathbf{L}}{dt} = \boldsymbol{\mu}_L \times \mathbf{H} = -\gamma \mathbf{L} \times \mathbf{H}$$

i.e.

$$\frac{dL_z}{dt} = 0 \qquad \Rightarrow \qquad L_z \equiv L \cos\theta = const$$

$$\dot{L}_x = -\gamma L_y H \qquad \dot{L}_y = \gamma L_x H$$

Then

$$\frac{d^2 L_x}{dt^2} = -\gamma^2 H^2 L_x$$

(and analogous for L_y), implying coherent rotation of the components in the $x - y$ plane with $\omega_L = eH/2mc$.

The frequency of the precessional motion can be obtained by writing (see Figure)

$$|\Delta \mathbf{L}| = L \sin\theta \, \omega_L \, \Delta t$$

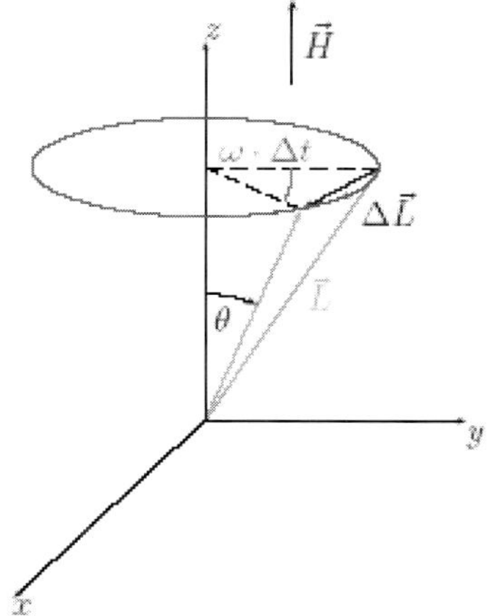

so that

$$\omega_L = \frac{|\Delta \mathbf{L}|}{\Delta t} \frac{1}{L \sin \theta} = \frac{\mu_L H \sin \theta}{L \sin \theta} = \gamma H$$

Problem III.2.5 Derive the ground states for Fe^{++}, V^{+++}, Co, As, La, Yb^{+++} and Eu^{++}, in the framework of the **LS** coupling scheme (a similar Problem is F.III.3).

Solution:

The ion Fe^{++} has six $3d$ electrons. According to the Pauli principle and the Hund rules Then $S = 2$ $L = 2$ $J = L + S = 4$ $\implies$ state $^5D_4$;

m spin

$$\begin{array}{cc} 2 & \uparrow \ \downarrow \\ 1 & \uparrow \\ 0 & \uparrow \\ -1 & \uparrow \\ -2 & \uparrow \end{array}$$

V^{+++} has incomplete $3d$ shell (2 electrons):

$$m \quad \text{spin}$$

$$
\begin{array}{cc}
2 & \uparrow \\
1 & \uparrow \\
0 & \\
-1 & \\
-2 &
\end{array}
$$

Then $S = 1 \quad L = 3 \quad J = L - S = 2 \implies$ state $^3F_2$.

Similarly

Co $\quad (3d)^7 (4s)^2 \quad S = \frac{3}{2} \quad L = 3 \quad J = \frac{9}{2} \implies$ state $^4F_{\frac{9}{2}}$;

As $\quad (3d)^{10} (4s)^2 (4p)^3 \quad S = \frac{3}{2} \quad L = 0 \quad J = \frac{3}{2} \implies$ state $^4S_{\frac{3}{2}}$;

La $\quad (5d)^1 (6s)^2 \quad S = \frac{1}{2} \quad L = 2 \quad J = \frac{3}{2} \implies$ state $^2D_{\frac{3}{2}}$;

Yb^{+++} $\quad (4f)^{13} \quad S = \frac{1}{2} \quad L = 3 \quad J = L + S = \frac{7}{2} \implies$ state $^2F_{\frac{7}{2}}$;

Eu^{++} $\quad (4f)^7 \quad S = \frac{7}{2} \quad L = 0 \quad J = S = \frac{7}{2} \implies$ state $^8S_{\frac{7}{2}}$.

(see Table III.2.2)

3.3 jj coupling scheme

The experimental findings indicate that the interval rule (Eq. 3.9), characteristic of the **LS** scheme, no longer holds for heavy atoms. This can be expected in view of the increase of the spin-orbit interaction upon increasing Z, thus invalidating the condition $a \ll c$ at the basis of the **LS** coupling. In the opposite limit of $a \gg c$ one first has to couple the single-electron orbital and spin momenta to define $\mathbf{j}$ and then construct the total momentum $\mathbf{J}$:

$$\mathbf{j}_i = \mathbf{l}_i + \mathbf{s}_i \qquad \text{with good quantum numbers } j_i \text{ and } (m_j)_i \qquad (3.17)$$

and

$$\mathbf{J} = \sum_i \mathbf{j}_i \qquad \text{with good quantum numbers } J \text{ and } M_j \qquad (3.18)$$

The final state is characterized by l, s, j of each electron and by J and M_J of the whole atom. The vectorial picture is shown in Fig. 3.6. j_1 and j_2 are half integer while J is always integer. To label the states, the individual j_i's are usually written between parentheses while J is written as subscript.

In a way analogous to the couplings in Eqs. 3.3 and 3.5, $\mathbf{j}_1 . \mathbf{j}_2$ leads to

$$\mathbf{j}_1 . \mathbf{j}_2 = \frac{J(J+1) - j_1(j_1+1) - j_2(j_2+1)}{2} \qquad (3.19)$$

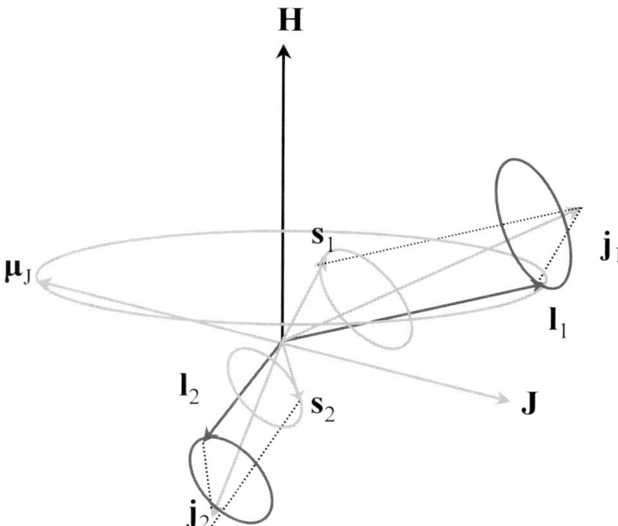

Fig. 3.6. Vectorial sketch of the **jj** coupling and of the precessional motions for two electrons, leading to the total **J** and $\boldsymbol{\mu}_J$ precessing around the external magnetic field.

The structure of the levels and their labelling is evidently different from the one derived within the **LS** scheme, as it appears from the example for one s and one p electron in Fig. 3.7 (to be compared with Fig. 3.4a). In Fig. 3.8 the comparison of the **LS** and **jj** schemes for two equivalent p electrons is shown. The jj coupling for two inequivalent p electrons is indicated below

j_1	j_2	J	Notation	Degeneracy
3/2	3/2	3,2,1,0	$(3/2,3/2)_{3,2,1,0}$	16
3/2	1/2	2,1	$(3/2,1/2)_{2,1}$	8
1/2	1/2	1,0	$(1/2,1/2)_{1,0}$	4
1/2	3/2	2,1	$(1/2,3/2)_{2,1}$	8

Total number of states 36

For equivalent p electrons the following cases are excluded

j_1	j_2	J
3/2	3/2	3
3/2	3/2	1
1/2	1/2	1

number of states excluded 13

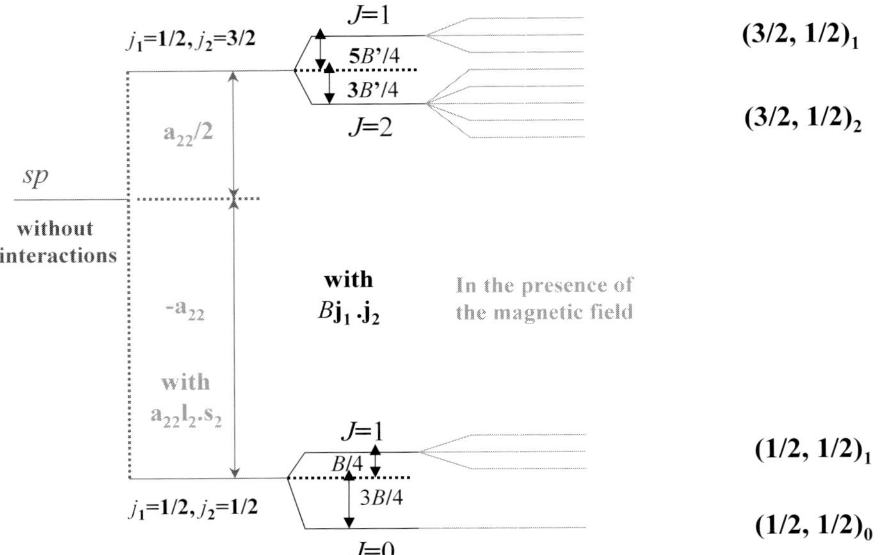

Fig. 3.7. jj coupling for s and p electrons. It is noted that for the state $j_1 = 1/2$ and $j_2 = 3/2$ the energy constant B' describing the coupling is equal and of opposite sign of the one (B) for the $j_1 = 1/2$ and $j_2 = 1/2$ state (this is proved in Problem III.3.1).

The first case implies parallel orbital momenta as well as parallel spins. The third case corresponds to $l_1 = l_2 = 0$ and parallel spins. The middle term is not pictorially evident (it is the analogous of the ^1P states at Table III.2.1) and corresponds to a level for which no additional distinguishable states are available.

The states allowed for equivalent p electrons are listed below, where the M_J degeneracy can be removed by a magnetic field:

j_1	j_2	J	Spectroscopic notation	Degeneracy
3/2	3/2	2,0	$(\frac{3}{2}, \frac{3}{2})_{2,0}$	6
3/2	1/2	2,1	$(\frac{3}{2}, \frac{1}{2})_{2,1}$	8
1/2	1/2	0	$(\frac{1}{2}, \frac{1}{2})_0$	1
–	–	–	–	Total 15

see Fig. 3.8

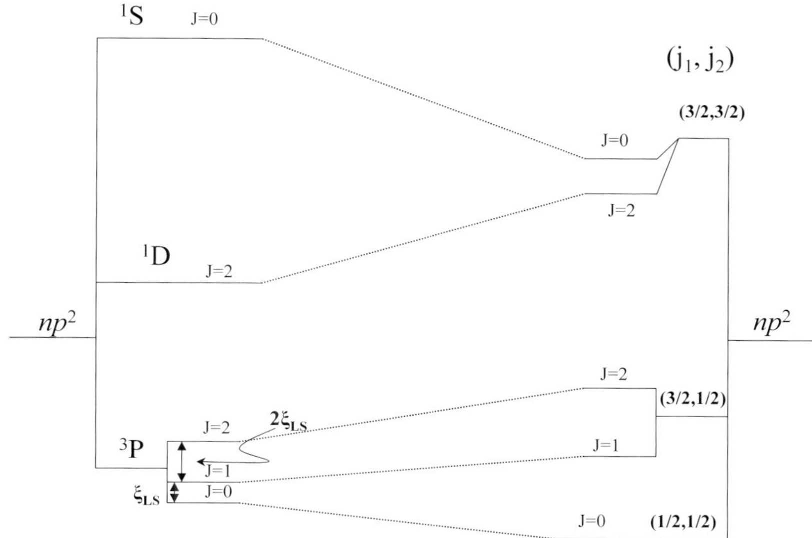

Fig. 3.8. Comparison of the structure and classification of the levels for two equivalent p-electrons in the **LS** (left) and **jj** (right) coupling schemes.

It is noted that the state $(\frac{3}{2}, \frac{1}{2})_{2,1}$ is indistinguishable from the $(\frac{1}{2}, \frac{3}{2})_{2,1}$ and this accounts for the other 8 states missing with respect to the original 36 states.

An example of heavy atom where a coupling intermediate between the **LS** and the **jj** schemes is Mercury. The energy diagram (simplified) is shown in Fig. 3.9.

Besides the violation of the interval rule one should remark that the strongest lines in the spectral emission of a mercury lamp originate from the intercombination of the $^1S_0$ and $^3P_1$ states. At the sake of illustration, since the line at 2537 Å would be forbidden in the **LS** scheme (because of the orthogonality of singlet and triplet states), one realizes the breakdown of **LS** coupling.

In very heavy atoms pure **jj** coupling does occur. The tendency from **LS** to **jj** coupling scheme is shown schematically in Fig. 3.10 for the sequence C, Si, Ge, Sn, Pb, in terms of the (sp) outer electrons configuration.

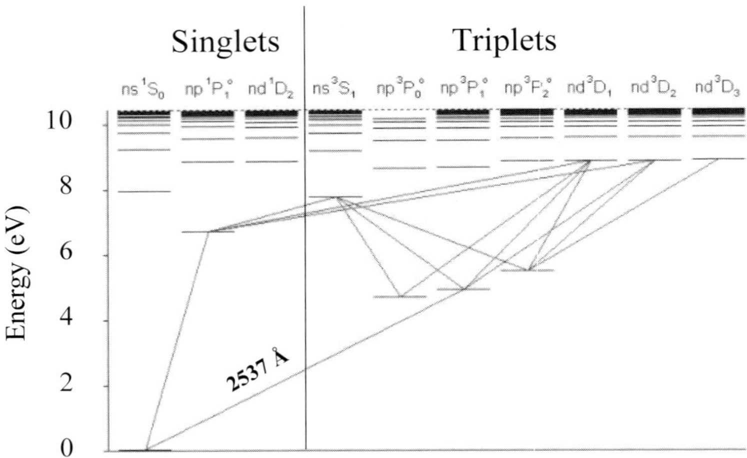

Fig. 3.9. Energy diagram for Hg, emphasizing the strength of the intercombination lines between singlets and triplet states (at variance with Fig. 2.7). In the triplet $6^3D_3 - 6^3D_2 - 6^3D_1$ the experimental measure of the separation 6^3D_3 - 6^3D_2 is 35 cm^{-1}, while the separation for 6^3D_2 - 6^3D_1 is 60 cm^{-1}. The ratio of the intervals turns out 0.58, whereas in the **LS** scheme one would have 1.5 (Eq. 3.9).

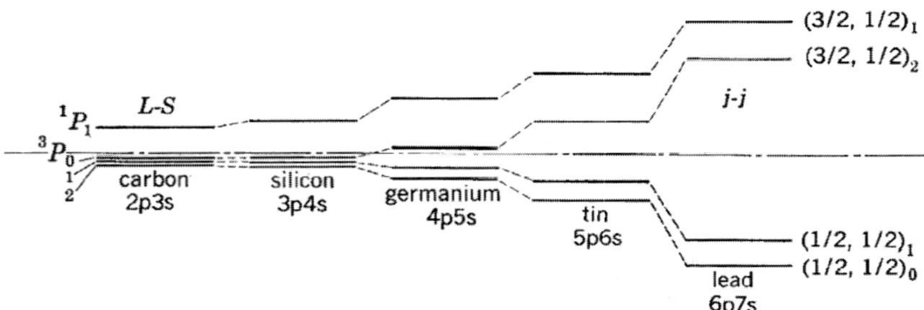

Fig. 3.10. Schematic view of the progressive changeover from **LS** scheme towards **jj** scheme on increasing the atomic number for the two electrons energy levels. It should be remarked that the **LS** scheme is often used to label the states eventhough their structure is rather close to the one pertaining to the **jj** coupling scheme.

Problems III.3

Problem III.3.1 Prove that for the **jj** coupling of one s and one p electrons in the state at $j_1 = 1/2$ and $j_2 = 3/2$ the fine structure constant B' is equal to $-B$ (see Fig. 3.7).

Solution:

The couplings are

$$a_{11}\mathbf{l}_1 \cdot \mathbf{s}_1 + a_{22}\mathbf{l}_2 \cdot \mathbf{s}_2 + B(j_1, j_2)\mathbf{j}_1 \cdot \mathbf{j}_2 \quad \text{with} \quad a_{11} \text{ and } a_{22} > 0.$$

For s electron $l_1 = 0$ $s_1 = \frac{1}{2}$ $j_1 = \frac{1}{2}$

For p electron $l_2 = 1$ $s_2 = \frac{1}{2}$ $j_2 = \frac{3}{2}\,\frac{1}{2}$
corresponding to the configuration

$$
\begin{array}{ccc}
j_1 & j_2 & J \\
\frac{1}{2} & \frac{3}{2} & 2, 1 \\
\frac{1}{2} & \frac{1}{2} & 1, 0
\end{array}
$$

with $B\left(\frac{1}{2}, \frac{3}{2}\right) \equiv B'$ (a) and $B\left(\frac{1}{2}, \frac{1}{2}\right) \equiv B$ (b).

For case (a)

$$B\mathbf{j}_1 \cdot \mathbf{j}_2 = B'\mathbf{s}_1 \cdot (\mathbf{l}_2 + \mathbf{s}_2) = \underbrace{B'\mathbf{l}_2 \cdot \mathbf{s}_1}_{\text{negligible}} + B'\mathbf{s}_1 \cdot \mathbf{s}_2$$

while for case (b)

$$B\mathbf{j}_1 \cdot \mathbf{j}_2 = B\mathbf{s}_1 \cdot (\mathbf{l}_2 - \mathbf{s}_2) = \underbrace{B\mathbf{l}_2 \cdot \mathbf{s}_1}_{\text{negligible}} - B\mathbf{s}_1 \cdot \mathbf{s}_2$$

Thus $B \equiv -c_{12} > 0$ and $B' = -B$.

Problem III.3.2 For an electron in the $lsjm_j$ state, express the expectation values of s_z, l_z, l_z^2 and l_x^2 (z is an arbitrary direction and x is perpendicular to z).

Solution:

By using arguments strictly similar to the ones at §2.2.2 (see Eq. 3.12) and taking into account that because of the spin-orbit precession $\mathbf{s}$ must be projected along $\mathbf{j}$:

$$\mathbf{s}_j = |\mathbf{s}|cos(\widehat{\mathbf{s}\mathbf{j}}) \text{ with}$$

$$cos(\widehat{\mathbf{s}\mathbf{j}}) = \left[s(s+1) + j(j+1) - l(l+1)\right]/2\sqrt{s(s+1)}\sqrt{j(j+1)}.$$

Then $< s_z > = \mathbf{s}_j\, m_j/|j| = m_j A,$

with $A = [s(s+1) + j(j+1) - l(l+1)]/2j(j+1)$

$$< l_z >=< j_z > - < s_z >= m_j(1 - A)$$

$$< l_z^2 >=< (j_z - s_z)^2 >=< j_z^2 > + < s_z^2 > -2 < j_z s_z >=$$

$$= m_j^2 + \frac{1}{4} - 2m_j < s_z >= m_j^2 + \frac{1}{4} - 2m_j^2 A$$

Since $< l^2 >=< l_z^2 > +2 < l_x^2 > (< l_x^2 >=< l_y^2 >)$ then

$$< l_x^2 >= \frac{1}{2}\left[l(l+1) - \frac{1}{4} - m_j^2(1 - 2A) \right]$$

The same result for $< s_z >$ is obtained from the Wigner-Eckart theorem (Eq. 4.25): $< l, s, j, m_j|\mathbf{s_z}|l, s, j, m_j >=< |(\mathbf{s} \cdot \mathbf{j})j_z| > /j(j + 1) =$

$$= m_j < |\mathbf{s} \cdot \mathbf{j}| > /j(j + 1) = m_j < |\mathbf{j}^2 - \mathbf{l}^2 + \mathbf{s}^2| > /2j(j + 1)$$

3.4 Quantum theory for multiplets. Slater radial wavefunctions

From the perturbative Hamiltonian reported in Eq. 1.11 and on the basis of the Slater determinantal eigenfunctions $D(1, 2, 3, ...)$ described at §2.3, one can develop a quantum treatment at the aim of deriving the multiplet structure discussed in the framework of the vectorial model. The perturbation theory for degenerate states has to be used. A particular form of this approach is described in Problem II.1 for the $1s2l$ states of Helium. At §2.2.2 a similar treatment was practically given, without involving *a priori* the degenerate eigenfunctions corresponding to a specific electronic configuration.

In general the direct solution of the secular equation is complicated and the matrix elements include operators of the form r_i^{-1} and r_{ij}^{-1} and the spin-orbit term. Again two limiting cases of predominance of the spin-spin or of the spin-orbit interaction have to be used in order to fix the quantum numbers labelling the unperturbed states associated with the zero-order degenerate eigenfunctions. The eigenvalues are obtained in terms of generalized Coulomb and exchange integrals. First we shall limit ourselves to a schematic illustration of the results of the Slater theory for the electronic configuration $(np)^2$, to be compared with the results obtained at §3.2.3 in the framework of the vectorial model.

For two non-equivalent p electrons (say $2p$ and $3p$) the Slater multiplet theory yields the following eigenvalues in the **LS** scheme, $I_{0,2}$ and $K_{0,2}$ being Coulomb and exchange integrals for different one-electron states:

a) $E(^3D) = E_0 + I_0 + \frac{I_2}{25} - K_0 - \frac{K_2}{25}$

b) $E(^3P) = E_0 + I_0 - \frac{5I_2}{25} + K_0 - \frac{5K_2}{25}$

c) $E(^3S) = E_0 + I_0 + \frac{10I_2}{25} - K_0 - \frac{10K_2}{25}$

d) $E(^1D) = E_0 + I_0 + \frac{I_2}{25} + K_0 + \frac{K_2}{25}$

e) $E(^1P) = E_0 + I_0 - \frac{5I_2}{25} - K_0 + \frac{5K_2}{25}$

f) $E(^1S) = E_0 + I_0 + \frac{10I_2}{25} + K_0 + \frac{10K_2}{25}$

(the indexes 0 and 2 result from the expansion of $1/r_{12}$ in terms of **Legendre polynomials**). For equivalent $2p$ electrons only states b), d) and f) occur, with energies (Fig. 3.11)

$$E(^3P) = E_0 + I_0 - \frac{5I_2}{25}$$

$$E(^1D) = E_0 + I_0 + \frac{I_2}{25}$$

$$E(^1S) = E_0 + I_0 + \frac{10I_2}{25}$$

(the exchange integral formally coincides with the Coulomb integral here).

The quantitative estimate of the energy levels cannot be given unless numerical computation of I and K in terms of one-electron eigenfunctions is carried out.

Approximate analytical expressions for the radial parts of the one-electron eigenfunction can be obtained as follows.

An effective potential energy of the form

$$V(r) = \frac{-(Z - \sigma)e^2}{r} + \frac{n^*(n^* - 1)\hbar^2}{2mr^2} \tag{3.20}$$

is assumed, with σ and n^* parameters to be determined. This form is strictly similar to the one for Hydrogenic atoms, with a screened Coulomb term and a centrifugal term (see §1.4). Thus the associated eigenfunctions are

$$\phi_{nlm}(r, \theta, \varphi) = NY_{l,m}(\theta, \varphi)r^{n^* - 1}e^{-\frac{(Z-\sigma)r}{n^* a_0}} \tag{3.21}$$

with N normalization factor.

The eigenvalues are similar to the ones at §1.4 and depend on σ and n^*. Then $E(\sigma, n^*)$ is minimized to find the best approximate values for σ and n^* and the radial part of the eigenfunctions is derived.

Empirical rules to assign the proper values to σ and n^* are the following. For quantum number n one has the correspondence

$n = 1$, 2, 3, 4, 5, and 6

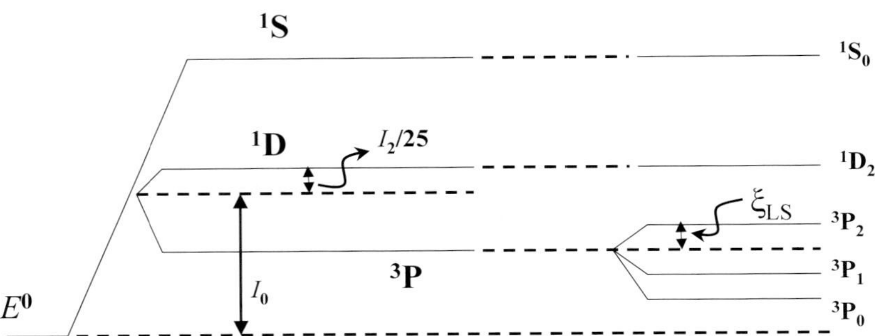

Fig. 3.11. Schematic diagram for equivalent p^2 electron configuration as derived in the Slater theory, in terms of Coulomb and exchange generalized integrals. The comparison with the results of the vectorial model (see Problem III.2.1) clarifies that the same structure and classification of the levels is obtained. Quantitative estimates require the knowledge of the radial parts of the one-electron eigenfunctions.

$n^* = 1, 2, 3, 3.7, 4$, and 4.2
while Table III.4.1 gives the rules to derive $(Z - \sigma)$.

	H							**He**
1s	1							1.6875
	Li	**Be**	**B**	**C**	**N**	**O**	**F**	**Ne**
1s	2.6906	3.6848	4.6795	5.6727	6.6651	7.6579	8.6501	9.6421
2s	1.2762	1.9120	2.5762	3.2166	3.8474	4.4916	5.1276	5.7584
2p			2.4214	3.1358	3.8340	4.4532	5.1000	5.7584
	Na	**Mg**	**Al**	**Si**	**P**	**S**	**Cl**	**Ar**
1s	10.6259	11.6089	12.5910	13.5754	14.5578	15.5409	16.5239	17.5075
2s	6.5714	7.3920	8.2136	9.0200	9.8250	10.6288	11.4304	12.2304
2p	6.8018	7.8258	8.9634	9.9450	10.9612	11.9770	12.9932	14.0082
3s	2.5074	3.3075	4.1172	4.9032	5.6418	6.3669	7.0683	7.7568
3p			4.0656	4.2852	4.8864	5.4819	6.1161	6.7641

Table III.4.1 The **Clementi-Raimondi** values for $Z - \sigma$ (ground states). It can be noted that for He atom, since $n^* = n = 1$ the value of $Z - \sigma$ must coincide with Z^* variationally derived at Prob. II.2.2.

The best atomic orbitals are actually obtained by the numerical solutions along the lines devised by **Hartree** with the improvement by **Fock** and

Slater to include the electron exchange interaction. The so-called **Hartree-Fock** equations for the one-electron eigenfunctions can be derived, by means of a rather lengthy procedure, applying the variational principle to the energy function, for a variation that leaves the determinantal Slater eigenfunctions normalized. The Hartree-Fock equation for the orbital $\phi_\alpha(\mathbf{r}_i)$ of the i-th electron can be written in the form

$$\{\mathcal{H}_i + \sum_\beta [2I_\beta - K_\beta]\}\phi_\alpha(\mathbf{r}_i) = E_\alpha^i \phi_\alpha(\mathbf{r}_i) \tag{3.22}$$

$\mathcal{H}_i$ is the one-electron **core Hamiltonian** ($T_i - Z^* e^2/r_i$, with $Z^* \equiv Z$ if no screening effects are considered), while I_β and K_β are the Coulomb and exchange operators that generalize the correspondent terms derived at §2.3 for He:

$$I_\beta \phi_\alpha(\mathbf{r}_i) = \left[\phi_\beta^*(\mathbf{r}_j) \frac{e^2}{r_{ij}} \phi_\beta(\mathbf{r}_j) d\tau_j \right] \phi_\alpha(\mathbf{r}_i)$$

$$K_\beta \phi_\alpha(\mathbf{r}_i) = \left[\phi_\beta^*(\mathbf{r}_j) \frac{e^2}{r_{ij}} \phi_\alpha(\mathbf{r}_j) d\tau_j \right] \phi_\beta(\mathbf{r}_i) \tag{3.23}$$

E_α in Eq. 3.22 is the one-electron energy. After an iterative numerical procedure, once the best self-consistent ϕ's are obtained, by multiplying both sides of Eq. 3.22 by $\phi_\alpha^*(\mathbf{r}_i)$ and integrating, one obtains for the i-th electron

$$E_\alpha = E_\alpha^o + \sum_\beta (2I_{\alpha\beta} - K_{\alpha\beta}) \tag{3.24}$$

with $E_\alpha^o \equiv < \alpha|\mathcal{H}_i|\alpha >$ and $I_{\alpha\beta}$ and $K_{\alpha\beta}$ are the Coulomb and exchange integrals, respectively (with $I_{\beta\beta} \equiv K_{\beta\beta}$). A sum over all the energies E_α would count all the interelectron interactions twice. Thus, by taking into account that each orbital in a closed shell configuration is double occupied, the total energy of the atom is written

$$E_T = 2 \sum_\alpha E_\alpha - \sum_{\alpha,\beta} (2I_{\alpha\beta} - K_{\alpha\beta}) \tag{3.25}$$

Although the eigenvalues obtained along the procedure outlined above are generally very close to the experimental data for the ground-state (for light atoms within 0.1 percent) still one could remark that any approach based on the model of independent electrons necessarily does not entirely account for the **correlation effects**.

Suppose that an electron is removed and that the other electrons do not readjust their configurations. Then the one-electron energy E_α corresponds to the energy required to remove a given electron from its orbital. This is the physical content of the **Koopmans theorem**, which identifies $|E_\alpha|$ with the ionization energy. Its validity rests on the assumption that the orbitals of the ion do not differ sizeably from the ones of the atom from which the electron has been removed.

The Hartree-Fock procedure outlined here for multi-electron atoms is widely used also for molecules and crystals, by taking advantage of the fast computers available nowadays which allow one to manipulate the Hartree-Fock equations. When the spherical symmetry of the central field approximation has to be abandoned numerical solutions along Hartree-Fock approach are anyway hard to be carried out. Thus particular manipulations of the equations have been devised, as the widely used **Roothaan's** one. Alternative methods are based on the **density functional theory** (DFT), implemented by the **local density approximation** (LDA). Correlation and relativistic effects are to be taken into account when detailed calculations are aimed, particularly for heavy atoms. Chapter 9 of the book by **Atkins** and **Friedman** (quoted in the preface) adequately deals with the basic aspects of the computational derivation of the electronic structure.

Finally we mention that for atoms with a rather high number of electrons and when dealing in particular with the radial distribution function of the electron charge in the ground-state (and therefore to the expectation values), the semiclassical method devised by **Thomas** and **Fermi** can be used. This approach is based on the statistical properties of the so-called Fermi gas of independent non-interacting particles obeying to Pauli principle, that we shall encounter in a model of solid suited to describe the metals (§12.7.1). The Thomas-Fermi approach is often used as a first step in the self-consistent numerical procedure that leads to the Hartree-Fock equations.

3.5 Selection rules

Here the selection rules that control the transitions among the electronic levels in the **LS** and in the **jj** coupling schemes are recalled. Their formal derivation (the extension of the treatment in Appendix I.3) requires the use of the **Wigner-Eckart theorem** and of the properties of the **Clebsch-Gordan coefficients**. We will give the rules for electric dipole, magnetic dipole and electric quadrupole transition mechanisms, again in the assumption that one electron at a time makes the transition. This is the process having the strongest probability with respect to the one involving two electrons at the same time, that would imply the breakdown of the factorization of the total wavefunction, at variance to what has been assumed, for instance, at §2.1.

A) **Electric dipole transition**

LS *coupling*
$\Delta L = 0, \pm 1$ and $\Delta S = 0$, non rigorous ($L = 0 \rightarrow L' = 0$ forbidden)

$\Delta J = 0, \pm 1$, transition $0 \leftrightarrow 0$ forbidden [3];
$\Delta M_J = 0, \pm 1$ for $\Delta J = 0$ the transition $M_J = 0 \rightarrow M'_J = 0$ forbidden [3].

For the electron making the transition one has $\Delta l = \pm 1$, according to parity arguments (see App.I.3).

jj *coupling*

For the atom as a whole

$\Delta J = 0, \pm 1$, transition $0 \leftrightarrow 0$ forbidden [3];
$\Delta M_J = 0, \pm 1$ for $\Delta J = 0$ the transition $M_J = 0 \rightarrow M'_J = 0$ is forbidden [3].

For the electron making the transition $\Delta l = \pm 1$, $\Delta j = 0, \pm 1$.

B) **Magnetic dipole transitions**

$\Delta J = 0, \pm 1$ and $\Delta M_J = 0, \pm 1$ (general validity)

LS *scheme*
$\Delta S = 0$, $\Delta L = 0$, $\Delta M = \pm 1$

C) **Electric quadrupole mechanism**

$\Delta J = 0, \pm 1, \pm 2$ (general validity),

LS *scheme*
$\Delta L = 0, \pm 1, \pm 2$
$\Delta S = 0$

Fig. 3.5 shows an example of the selection rules given in A).

Problems F.III

Problem F.III.1 A beam of Ag atoms (in the ground state $5\,^2S_{1/2}$) flows with speed $v = 10^4$ cm/s, for a length $l_1 = 5$ cm, in a region of inhomogeneous magnetic field, with $dH/dz = 10^4$ Gauss/cm. After the exit from this region the beam is propagating freely for a length $l_2 = 10$ cm and then collected on a screen, where a separation of about 0.6 cm between the split beam is observed (**Stern-Gerlach** experiment). From these data obtain the magnetic moment of Ag atom .

Solution:

---

[3] Rules of general validity in both schemes.

In the first path l_1 the acceleration is

$$a = \frac{F}{M_{Ag}} = \frac{\mu_z}{M_{Ag}} \frac{dH}{dz}$$

and the divergence of the atomic beam along z turns out

$$d' = \frac{1}{2} a \left(\frac{l_1}{v} \right)^2$$

In the second path l_2, with $v_z = a l_1 / v$ and then $d'' = a l_1 l_2 / v^2$.

The splitting of the two beams with different z-component of the magnetic moment $(S = J = 1/2)$ turns out

$$d = 2(d' + d'') = \frac{a}{v^2} (l_1^2 + 2 l_1 l_2)$$

Then

$$\mu_z = \frac{M_{Ag} v^2 d}{\frac{dH}{dz} (l_1^2 + 2 l_1 l_2)} \simeq 0.93 \cdot 10^{-20} \frac{\text{erg}}{\text{Gauss}}.$$

Problem F.III.2 In the **LS** coupling scheme, derive the electronic states for the configurations $(ns, n's)$ (i), $(ns, n'p)$ (ii), $(nd)^2$ (iii) and $(np)^3$ (iv). Then schematize the correlation diagram to the correspondent states in the **jj** scheme, for the nd^2 and for the np^3 configurations.

Solution:

i)$S = 1$ $L = 0$ $^3S_1$
  $S = 0$ $L = 0$ $^1S_0$

ii) $^1P_1$,$^3P_{0,1,2}$

iii)

m_l	2	2	1	1	0	0	-1	-1	-2	-2
m_s	$\frac{1}{2}$	$-\frac{1}{2}$	$\frac{1}{2}$	$-\frac{1}{2}$	$\frac{1}{2}$	$-\frac{1}{2}$	$\frac{1}{2}$	$-\frac{1}{2}$	$\frac{1}{2}$	$-\frac{1}{2}$
$2\ \ \frac{1}{2}$										
$2\ -\frac{1}{2}$	4,0									
$1\ \ \frac{1}{2}$	3,1	3,0								
$1\ -\frac{1}{2}$	3,0	3,-1	2,0							
$0\ \ \frac{1}{2}$	2,1	2,0	1,1	1,0						
$0\ -\frac{1}{2}$	2,0	2,-1	1,0	1,-1	0,0					
$-1\ \ \frac{1}{2}$	1,1	1,0	0,1	0,0	-1,1	-1,0				
$-1\ -\frac{1}{2}$	1,0	1,-1	0,0	0,-1	-1,0	-1,-1	-2,0			
$-2\ \ \frac{1}{2}$	0,1	0,0	-1,1	-1,0	-2,1	-2,0	-3,1	-3,0		
$-2\ -\frac{1}{2}$	0,0	0,-1	-1,0	-1,-1	-2,0	-2,-1	-3,0	-3,-1	-4,0	
$m_l\ \ m_s$										

then $^1S_0$, $^3P_{2,1,0}$, $^1D_2$, $^3F_{4,3,2}$, $^1G_4$

The total number of states is $\begin{pmatrix} 10 \\ 2 \end{pmatrix} = 45$

iv) $^4S_{\frac{3}{2}}$, $^2P_{\frac{1}{2},\frac{3}{2}}$, $^2D_{\frac{3}{2},\frac{5}{2}}$

The total number of states is $\begin{pmatrix} 6 \\ 3 \end{pmatrix} = 20$

The correlation between the two schemes is given below for the p^3 and d^2 configurations:

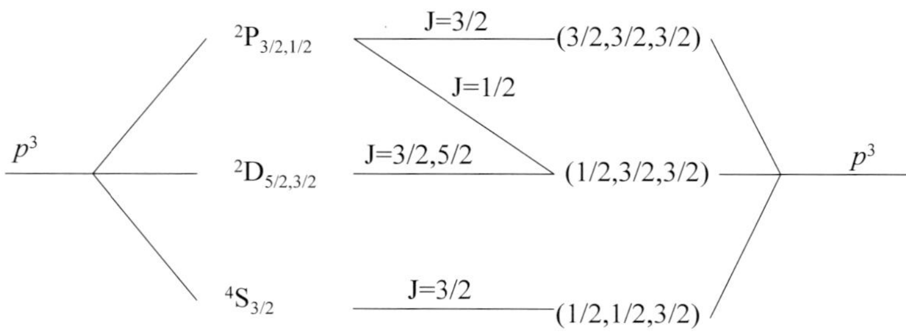

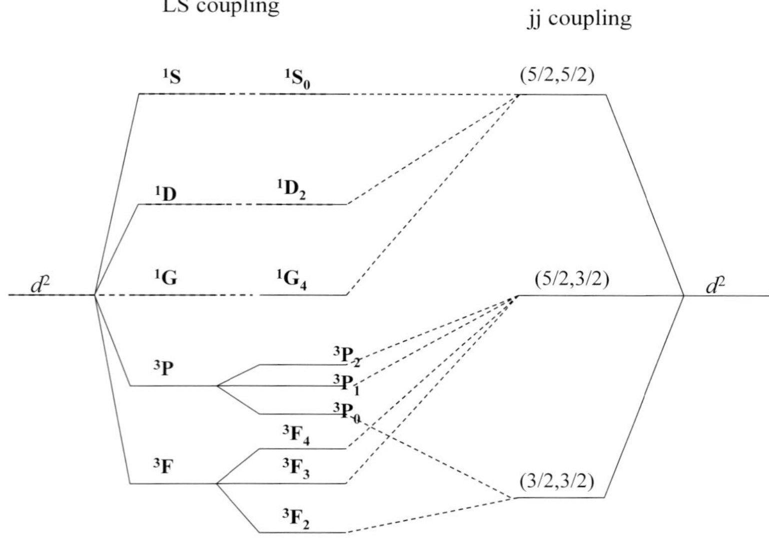

Problem F.III.3 By resorting to the Hund rules derive the effective magnetic moments for Dy^{+++}, Cr^{+++} and Fe^{+++} (See Table III.2.2).

Solution:

The ion Dy^{+++} has incomplete $4f$ shell (9 electrons).

According to the Pauli principle and the Hund rules

m spin

$$
\begin{array}{rll}
3 & \uparrow & \downarrow \\
2 & \uparrow & \downarrow \\
1 & \uparrow & \quad S=\tfrac{5}{2} \quad L=5 \quad J=L+S=\tfrac{15}{2} \quad \Longrightarrow \quad \text{state} \quad {}^{6}H_{\frac{15}{2}} \\
0 & \uparrow & \\
-1 & \uparrow & \\
-2 & \uparrow & \\
-3 & \uparrow & \\
\end{array}
$$

The Landé factor is

$$
g = 1 + \frac{J(J+1)+S(S+1)-L(L+1)}{2J(J+1)} = \frac{4}{3} = 1.33
$$

thus

$$
p = \frac{\mu}{\mu_B} = g\sqrt{J(J+1)} = 1.33 \cdot \sqrt{\frac{15}{2}\left(\frac{15}{2}+1\right)} = 10.65
$$

In similar way

$$
Cr^{+++} \quad (3d)^3 \quad S=\tfrac{3}{2} \quad L=3 \quad J=|L-S|=\tfrac{3}{2} \quad \Longrightarrow \quad \text{state} \quad {}^{4}F_{\frac{3}{2}};
$$

$$
g = 0.4 \text{ and } p = 0.77
$$

$$
Fe^{+++} \quad (3d)^5 \quad S=\tfrac{5}{2} \quad L=0 \quad J\equiv S=\tfrac{5}{2} \quad \Longrightarrow \quad \text{state} \quad {}^{6}S_{\frac{5}{2}};
$$

$$
g = 2 \text{ and } p = 5.92
$$

Problem F.III.4 Derive the multiplets for the ${}^{3}F$ and the ${}^{3}D$ states and sketch the transitions allowed by the electric dipole mechanism.

Solution:

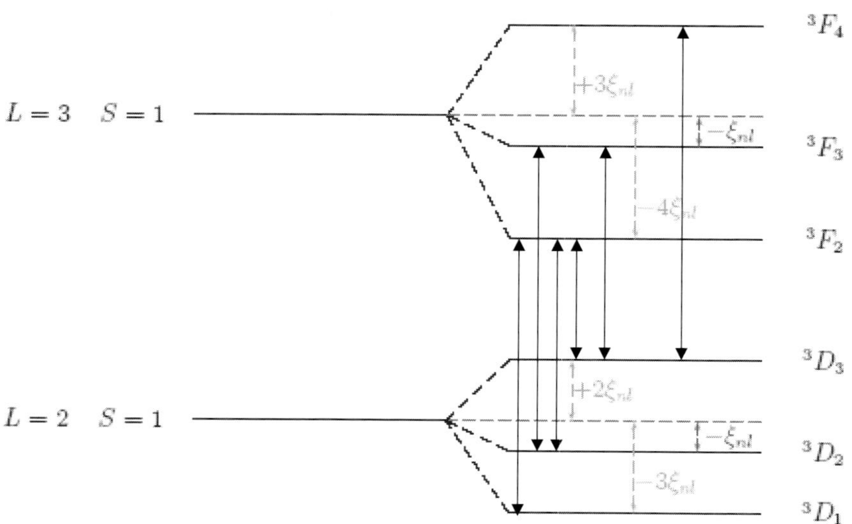

Problem F.III.5 When accelerated protons collide on ^{19}F nuclei an excited state of ^{20}Ne is induced and transition to the ground state yields γ **emission**. The emission spectrum, as a function of the energy of colliding protons, displays a line centered at 873.5 keV, with full width at half intensity of 4.8 keV. Derive the life time of the excited state of ^{20}Ne. Comment about the difference with the emission spectrum of ^{57}Fe, where the transition to the ground state from the first excited state yields a γ-photon at 14.4 keV, with life time 10^{-7} s.

By referring to ^{57}Fe, considering that the transition is due to a proton and assuming as radius of the nucleus of 10^{-12} cm, by means of order of magnitude estimates discuss the transition mechanism (**electric dipole, electric quadrupole, magnetic dipole**) driving the γ transition at 14.4 keV in ^{57}Fe.

Solution:

From

$$\tau \simeq \frac{\hbar}{\Delta E} \simeq \frac{1.05 \cdot 10^{-27}\ erg\,sec}{4.8 \cdot 10^3 \cdot 1.6 \cdot 10^{-12}\ erg} = 1.37 \cdot 10^{-19}\ s.$$

for ^{20}Ne, while for ^{57}Fe

$$\Delta E \simeq \frac{\hbar}{\tau} = 1.05 \cdot 10^{-20}\ erg \simeq 6.6 \cdot 10^{-12} \text{keV}$$

The transition mechanism driving the γ transition at 14.4 keV in ^{57}Fe is discussed as follows:

a) for electric dipole transition the spontaneous emission probability (see App.I.3) is

$$A_{21}^{E} = \frac{32\pi^3(E_2 - E_1)^3}{3c^3\hbar\, h^3}|<2|e\mathbf{R}|1>|^2 \simeq \frac{992 \cdot (14.4 \cdot 10^3 \cdot 1.6 \cdot 10^{-12}\, erg)^3}{8.1 \cdot 10^{31} \cdot 306.79 \cdot 10^{-108}}(e\cdot R_N)^2$$

$$\simeq 1.1 \times 10^{11}\ s^{-1}.$$

Then one would expect

$$\tau^E \sim (A_{21}^E)^{-1} \sim 10^{-11}\ s$$

b) for the electric quadrupole mechanism

$$\frac{A_{21}^E}{A_{21}^Q} \simeq \left(\frac{\lambda}{R_N}\right)^2 \frac{1}{4\pi^2} \simeq \left(\frac{6.6 \cdot 10^{-27} \cdot 3 \cdot 10^{10}}{10^{-12} \cdot 14.4 \cdot 10^3 \cdot 1.6 \cdot 10^{-12}}\right)^2 \frac{1}{4\pi^2}.$$

Thus

$$\tau^Q \sim \tau^E \cdot 1.9 \cdot 10^6 \simeq 1.9 \cdot 10^{-5}\ s$$

c) for the magnetic dipole mechanism

$$\frac{A_{21}^E}{A_{21}^M} \sim \left[\frac{eR_N}{0.09 \cdot \mu_N}\right]^2 \simeq \left(\frac{4.8 \cdot 10^{-10} \cdot 10^{-12}}{5 \cdot 10^{-24}}\right)^2 \simeq 4100$$

$$\tau_M \sim \tau^E \cdot 4100 \sim 4 \cdot 10^{-7}\ s.$$

On the basis of the experimental value it may be concluded that the transition is due to magnetic dipole mechanism.

Problem F.III.6 Estimate the order of magnitude of the ionization energy of ^{92}U in the case that Pauli principle should not operate (assume that the screened charge is $Z/2$) and compare it with the actual ionization energy (4 eV).

Solution:

From

$$E = -\frac{\mu Z^2 e^4}{2\hbar^2 n^2} = -\frac{Z^2}{n^2} \cdot 13.6\,\mathrm{eV}$$

and for $n = 1$ and $Z = 46$, the ionization energy would be

$$|E| = (46)^2 \cdot 13.6\,eV \simeq 3 \cdot 10^4\,\mathrm{eV}.$$

Problem F.III.7 The structure of the electronic states in the Oxygen atom can be derived in a way similar to the one for Carbon (Problem III.2.1) since the electronic configuration $(1s)^2 (2s)^2 (2p)^4$ has two "holes" in the $2p$ shell, somewhat equivalent to the $2p$ two electrons. Discuss the electronic term structure for oxygen along these lines.

Solution:

From Table III.2.1 taking into account that for $(2p)^6$ one would have $M_L = 0$ and $M_S = 0$ the term ^3P, ^1D, ^1S are found. Since one has four electrons the spin-orbit constant changes sign, the multiplet is inverted and the ground state is $^3P_2$ instead of $^3P_0$ (see Prob. III.2.1).

4

Atoms in electric and magnetic fields

Topics

Electric polarizability of the atom
Linear and quadratic field dependences of the atomic energies
Energy levels in strong and weak magnetic fields
Atomic paramagnetism and diamagnetism
Paramagnetism in the presence of mean field interactions

4.1 Introductory aspects

The analysis of the effects of magnetic or electric fields on atoms favors deeper understanding of the quantum properties of matter. Furthermore, electric or magnetic fields are tools currently used in several experimental studies.

In classical physics the prototype atom is often considered as an electron rotating on circular orbit around the fixed nucleus. In the presence of electric and magnetic fields (see Fig. 4.1), the equation of motion for the electron becomes

$$m\frac{d^2\mathbf{r}}{dt^2} = -\frac{e^2\mathbf{r}}{r^3} - e\boldsymbol{\mathcal{E}} - \frac{e}{c}\left(\frac{d\mathbf{r}}{dt} \times \mathbf{H}\right) \tag{4.1}$$

For a static magnetic field $\mathbf{H}$ only (then the external electric field $\boldsymbol{\mathcal{E}} = 0$) from Eq. 4.1 it is found that the Lorentz force induces a precessional motion of the charge around z, with angular frequency (see Problem IV.2.1)

$$\omega = \pm\sqrt{(\frac{eH}{2mc})^2 + \frac{e^2}{mr^3}} + \frac{eH}{2mc} \simeq \omega_L + \sqrt{\frac{e^2}{mr^3}} \tag{4.2}$$

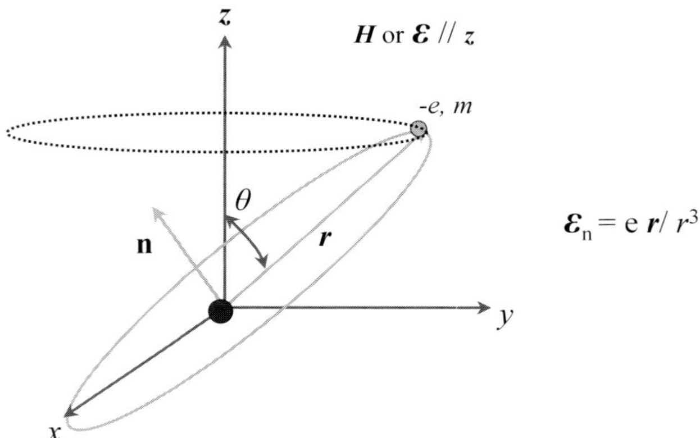

Fig. 4.1. Variables used to account for the effects of electric or magnetic field in the classical atom (Eq. 4.1).

To give orders of magnitude, the orbital frequency in the plane of motion is $\omega_0 = e/\sqrt{mr^3} \sim 10^{16}$ rad s^{-1} while the Larmor frequency $\omega_L = eH/2mc$ is around 10^{11} rad s^{-1}, for field $H = 10^4$ Oe (1 Tesla).

The current related to the orbital motion corresponds to the magnetic moment $\boldsymbol{\mu}' = \mu_B \mathbf{n}$ (see Problem I.6.2): its alignment along the field, contrasted by thermal excitation, implies the temperature dependent **paramagnetism**. The effective z component of the magnetic moment is expected of the order of $(\boldsymbol{\mu}')_z \sim \mu_B(\mu_B H/k_B T)$ (formal description will be given at §4.4). Therefore the paramagnetic susceptibility $\chi_{para} = N(\mu'_z)/H$, for a number $N = 10^{22}$ of atoms per cubic cm, is of the order of $\chi_{para} \simeq N\mu_B^2/k_B T \sim 6 \times 10^{-5}$ (for $T \simeq 100$ K).

The current related to the precessional motion of the orbit is $i = (-e\omega_L/2\pi) = -e^2 H/4\pi mc$, along a ring of area $A = \pi(r\sin\theta)^2$ (see Fig. 4.1). The associated magnetic moment is

$$(\mu")_z = \frac{iA}{c} = -\frac{e^2 H}{4\pi mc^2}\pi r^2 sin^2\theta\,,$$

yielding a **diamagnetic susceptibility** $\chi_{dia} = N\mu"/H \simeq -e^2 Nr^2/4mc^2$, as order of magnitude around -10^{-6} (again for $N = 10^{22}$ atoms per unit volume).

On the ground of qualitative arguments the effect of an electric field $\mathcal{E} \parallel \hat{z}$ can be understood by referring to the displacement δz of the orbit along the field direction: the component of the Coulomb force $e^2\delta z/r^3$ equilibrates the force $e\mathcal{E}$ (see the sketch below).

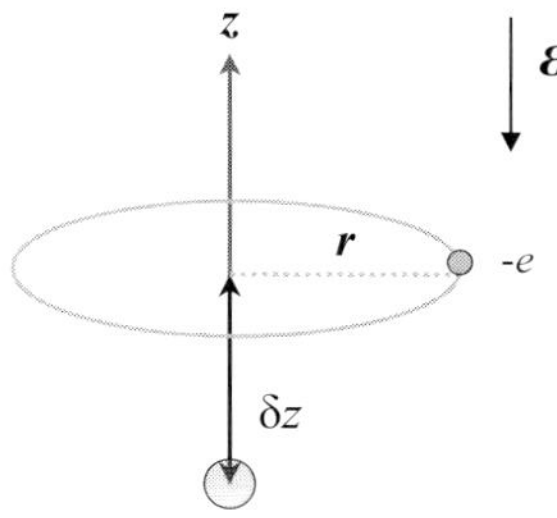

Then the dipole moment turns out $e\delta z = \mathcal{E}r^3$ and an **atomic polarizability** given by $\alpha \sim e\delta z/\mathcal{E} \sim r^3 \sim 10^{-24}$ cm^3 can be predicted.

In the quantum mechanical description the electric and magnetic forces imply the one-electron Hamiltonian (see Eq. 1.26)

$$\mathcal{H} = \frac{1}{2m}\left(\mathbf{p} + \frac{e}{c}\mathbf{A}\right)^2 + V(r) + 2\mu_B\mathbf{s}.rot\mathbf{A} - e\varphi =$$

$$= \frac{p^2}{2m} + V(r) + \underbrace{2\mu_B\mathbf{s}.rot\mathbf{A} - e\varphi + \frac{e}{2mc}\left[(\mathbf{p}.\mathbf{A}) + (\mathbf{A}.\mathbf{p})\right] + \frac{e^2A^2}{2mc^2}}_{\mathcal{H}_P} \equiv \mathcal{H}_0 + \mathcal{H}_P$$

$$(4.3)$$

Here the magnetic term related to the spin moment (Eq. 1.32) has been added, while $V(r)$ in $\mathcal{H}_0$ is the central field potential energy. φ and $\mathbf{A}$ in Eq. 4.3 are the scalar and vector potentials describing the perturbation applied to the atom.

For static and homogeneous electric field

$$\mathbf{A} = 0, \qquad \text{and} \qquad \varphi = -\int_0^z \mathcal{E}dz = -z\mathcal{E} \qquad (4.4)$$

while for static and homogeneous magnetic field

$$\varphi = 0, \qquad \text{and} \qquad \mathbf{A} = \frac{1}{2}\mathbf{H} \times \mathbf{r} \qquad (4.5)$$

The corrections to the energy levels can be evaluated on the basis of the eigenfunctions of the zero-field Hamiltonian $\mathcal{H}_0$. In multi-electrons atoms this perturbative approach is generally hard to carry out, in view of the inter-electron couplings (as it can be realized by recalling the description in the framework of the vectorial model (Chapter 3)). In the following we shall describe the basic aspects of the effects due to the fields by deriving the corrections to the atomic energy levels in some simplifying conditions.

4.2 Stark effect and atomic polarizability

Stark effect is usually called the modification to the energy levels in the presence of the Hamiltonian $\mathcal{H}_P = \sum_i e z_i \mathcal{E}$ (first studied in the Hydrogen atom also by **Lo Surdo**). In the perturbative approach energy corrections linear in the field in general are not expected, the matrix elements of the form $\int \phi^*(\mathbf{r}_i) z_i \phi(\mathbf{r}_i) d\tau_i$ being zero.

The second order correction can be put in the form

$$\Delta E^{(2)} = -\frac{\alpha}{2}\mathcal{E}^2 \tag{4.6}$$

where, in the light of the classical analogy for the electric dipole [1]

$$\mu_e = -\frac{\partial \Delta E}{\partial \mathcal{E}} \quad, \tag{4.7}$$

α defines the **atomic polarizability**. In fact, one can attribute to the atom an **induced electric dipole moment** $\mu_e = \alpha \mathcal{E}$. The polarizability depends in a complicated way from the atomic state, in terms of the quantum numbers J and M_J: $\alpha = \alpha(J, M_J)$.

Let us first evaluate the atomic polarizability α_{1s} for Hydrogen in the ground state. Instead of carrying out the awkward sum of the second order matrix elements we shall rather estimate the limits within which α_{1s} falls. From Eqs. 4.3 and 4.4 one has

$$\Delta E^{(2)} = -\sum_{n>1} \frac{|<1s|\mathcal{H}_P|nlm>|^2}{E_n - E_1} = -\frac{1}{2}\alpha_{1s}\mathcal{E}^2 \quad . \tag{4.8}$$

$|<1s|\mathcal{H}_P|nlm>|^2$ is always positive and E_n increases on increasing n. Therefore one can set the limits of variability of $\alpha_{1s}/2$:

$$-\frac{e^2}{E_1}\sum|<1s|z|nlm>|^2 < \frac{\alpha_{1s}}{2} < \frac{e^2}{(E_2 - E_1)}\sum|<1s|z|nlm>|^2. \tag{4.9}$$

(note that the state $n = 1$ can be included in the sum, since $<1s|z|1s> = 0$). On the other hand

$$\sum <1s|z|nlm><nlm|z|1s> = <1s|z^2|1s> = \frac{1}{\pi a_0^3}\int \frac{4\pi}{3}r^4 e^{-2r/a_0}dr = a_0^2 \quad .$$

---

[1] Note that the field-related energy is $\Delta E = -\int_0^{\mathcal{E}} \mu_e d\mathcal{E}'$, so that for $\mu_e = \alpha \mathcal{E}'$ Eq. 4.6 follows.

From $E_1 = -e^2/2a_0$ while $(E_2 - E_1) = 3e^2/8a_0$, one deduces

$$4a_0^3 < \alpha_{1s} < \frac{16}{3}a_0^3 \,.$$

It is recalled that the "brute-force" second order perturbative calculation yields $\alpha_{1s} = 4.66a_0^3$. Thus the electric polarizability turns out of the order of magnitude of the "size" of the atom to the third power, as expected from the qualitative argument at §4.1.

An approximate estimate of the polarizability of the ground state of the Hydrogen atom can also be obtained by means of variational procedures, on the basis of a trial function involving the mixture of the $1s$ and the $2p_z$ states:

$$\phi_{var} = c_1\phi_{1s} + c_2\phi_{2p_z}. \tag{4.10}$$

This form could be expected on the ground of physical arguments, as sketched below in terms of atomic orbitals:

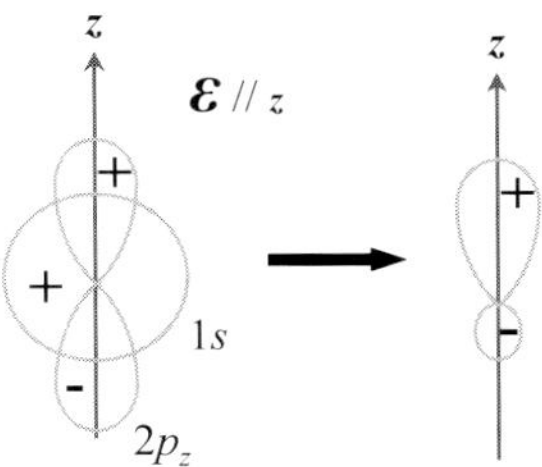

(see Problem F.II.2).

The energy function is

$$E(c_1, c_2) = \frac{\int \phi_{var}^* \mathcal{H}\phi_{var}\, d\tau}{\int \phi_{var}^* \phi_{var}\, d\tau} \tag{4.11}$$

where $\mathcal{H}$ is the total Hamiltonian, while

$$\mathcal{H}_{11} \equiv\, <1s|\mathcal{H}|1s>\,, \mathcal{H}_{22} \equiv\, <2p_z|\mathcal{H}|2p_z>\,, \mathcal{H}_{12} \equiv\, <1s|\mathcal{H}|2p_z>,$$
$$S_{12} \equiv\, <1s|2p_z>=\, 0\,, S_{11} = S_{22} = 1 \tag{4.12}$$

From $\partial E/\partial c_{1,2} = 0$

$$c_1(\mathcal{H}_{11} - E) + c_2\mathcal{H}_{12} = 0$$
$$c_1\mathcal{H}_{12} + c_2(\mathcal{H}_{22} - E) = 0 \,, \tag{4.13}$$

with secular equation

$$\begin{pmatrix} \mathcal{H}_{11} - E & \mathcal{H}_{12} \\ \mathcal{H}_{21} & \mathcal{H}_{22} - E \end{pmatrix} = 0 \tag{4.14}$$

Since $\mathcal{H}_{11} = E_{1s}^0$, $\mathcal{H}_{22} = E_{1s}^0/4$, while from Table I.4.2 for $Z = 1$
$\mathcal{H}_{12} = <1s|\mathcal{H}_0|2p_z> + <1s|z|2p_z> e\mathcal{E} = e\mathcal{E}2^8 a_0/3^5 \sqrt{2} \equiv A$, Eq. 4.14
becomes

$$\begin{pmatrix} E_{1s}^0 - E & A \\ A & \frac{E_{1s}^0}{4} - E \end{pmatrix} = 0 \ , \tag{4.15}$$

of roots

$$E_{\pm} = \frac{5}{8} E_{1s}^0 \pm \frac{1}{2} \sqrt{\frac{9(E_{1s}^0)^2}{16} \left(1 + \frac{64 A^2}{9(E_{1s}^0)^2}\right)} \tag{4.16}$$

By taking into account that $A \ll E_{1s}^0$, from $(1+x)^{1/2} \simeq 1 + x/2$ the lowest
energy level turns out

$$E = E_{1s}^0 + \frac{4}{3} \frac{A^2}{E_{1s}^0} \equiv E_{1s}^0 - 2.96 \frac{a_0^3}{2} \mathcal{E}^2 \ , $$

corresponding to the polarizability $\alpha_{1s} = 2.96 a_0^3$.

In the particular case of **accidental degeneracy** (see §1.4) Stark effect
linear in the field occurs. Let us consider the $n=2$ states of Hydrogen atom.
The zero-order wavefunction is

$$\phi_l = c_1^{(l)} \phi_{2s} + c_2^{(l)} \phi_{2p_1} + c_3^{(l)} \phi_{2p_0} + c_4^{(l)} \phi_{2p_{-1}} \tag{4.17}$$

and the corrected eigenvalues are obtained from

$$\begin{pmatrix} <2s| - ez\mathcal{E}|2s> - E & \dots & \dots \\ \dots & <2p_1| - ez\mathcal{E}|2p_1> - E & \dots \\ \dots & \dots & \dots \\ \dots & \dots & \dots \end{pmatrix} = 0 \tag{4.18}$$

Again recalling the selection rules for the z-component of the electric dipole
(App.I.3), this determinant is reduced to

$$\begin{pmatrix} -E & 0 & B & 0 \\ 0 & -E & 0 & 0 \\ B & 0 & -E & 0 \\ 0 & 0 & 0 & -E \end{pmatrix} = 0 \tag{4.19}$$

where $B = -3a_0 e\mathcal{E}$.

From the roots $R_{1,2} = 0$ and $R_{3,4} = \pm B$ the structure of the $n = 2$ levels
in the presence of the field is deduced in the form depicted in Fig. 4.2.

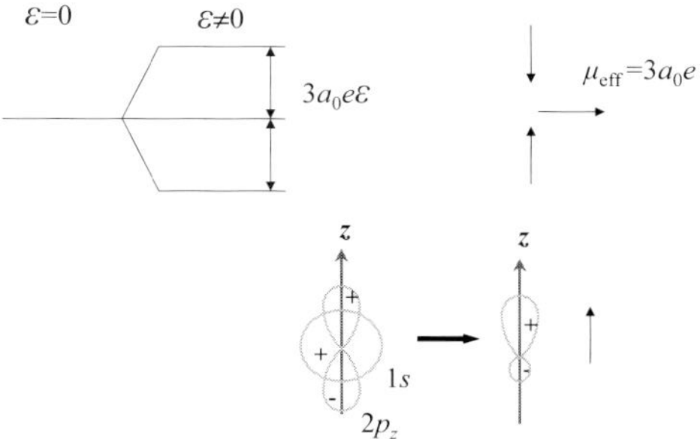

Fig. 4.2. Effect of the electric field on the $n = 2$ states of Hydrogen atom, illustrating how in the presence of accidental degeneracy a kind of **pseudo-orientational polarizability** arises, with energy correction linear in the field $\mathcal{E}$.

The first-order Stark effect is observed in Hydrogen and in **F-centers** in crystals (where a vacancy of positive ion traps an electron and causes an effective potential of Coulombic character which yields the accidental degeneracy).

Finally in Fig. 4.3 the experimental observation of the Stark effect on the $D_{1,2}$ doublet of Na atom (see Fig. 2.2) is depicted. It is noted that the degeneracy in $\pm M_J \equiv \pm m_j$ is not removed, the energy correction being independent from the versus of the field.

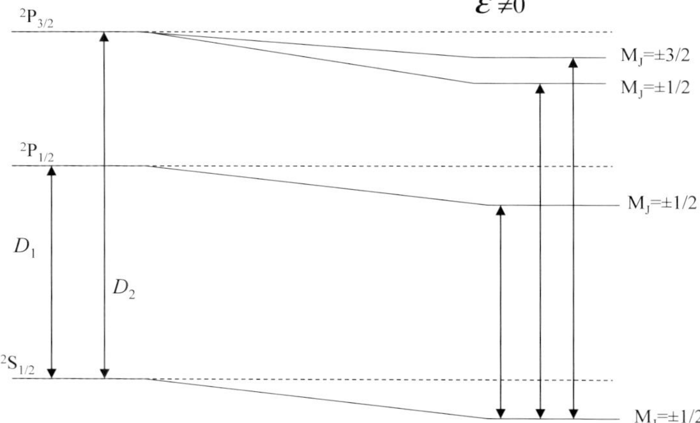

Fig. 4.3. Ground state and first excited states of Na atom upon application of electric field and modification of the D doublet. The energy shift of the ground state is 40.56 kHz/(kV/cm)^2, corresponding to an electric polarizability $\alpha = 24.11 \cdot 10^{-24}$ cm^3. The shifts of the P states are about twice larger.

Problems IV.2

Problem IV.2.1 Show how the classical equation for the electron in orbit around the nucleus in the presence of static and homogeneous magnetic field implies the precessional motion of the orbit with the Larmor frequency.

Solution:

From the force

$$\mathbf{F} = -e^2 \frac{\mathbf{r}}{r^3} - \frac{e}{c}(\mathbf{v} \times \mathbf{H})$$

for $\mathbf{H}$ along z

$$m\frac{dv_x}{dt} + \frac{e}{c}Hv_y + \frac{e^2 x}{r^3} = 0$$

$$m\frac{dv_y}{dt} - \frac{e}{c}Hv_x + \frac{e^2 y}{r^3} = 0.$$

By transforming to polar coordinates

$$x(t) = r\cos\omega t \quad \Longrightarrow \quad v_x(t) = -r\omega\sin\theta$$

$$y(t) = r\sin\omega t \quad \Longrightarrow \quad v_y(t) = r\omega\cos\theta$$

where $\omega = d\theta/dt$ (see Fig. 4.1), one writes

$$\frac{dv_x}{dt} = -r\omega^2\cos\theta; \quad \frac{dv_y}{dt} = -r\omega^2\sin\theta.$$

By substitution into the equations of motion

$$\omega^2 - \left(\frac{eH}{mc}\right)\omega - \frac{e^2}{mr^3} = 0 \ ,$$

yielding

$$\omega = \frac{eH}{2mc} \pm \sqrt{\left(\frac{eH}{2mc}\right)^2 + \frac{e^2}{mr^3}}$$

Since, from order of magnitude estimates (see §4.1)

$$\left(\frac{eH}{2mc}\right)^2 \ll \frac{e^2}{mr^3} \ ,$$

Eq. 4.2 follows.

Problem IV.2.2 By extending the procedure given at §4.2 for the Hydrogen states at $n = 2$ it can be shown that the linear Stark effect yields correction energy of the form $\Delta E = (\frac{3ea_0}{2})\mathcal{E}n(n_1 - n_2)$, with n_1 and n_2 running from zero to $(n\text{-}1)$. By neglecting the second order effect, evaluate the partition function and then the induced dipole moment, in the high temperature limit.

Solution:

When the quantum numbers $n_{1,2}$ run from 0 to $n-1$ the partition function turns out

$$Z = \sum_{n_1,n_2=0}^{n-1} e^{-n(n_1-n_2)x} = \left(\frac{1-e^{-n^2x}}{1-e^{-nx}}\right)\left(\frac{1-e^{n^2x}}{1-e^{nx}}\right) = \left(\frac{\sinh \frac{1}{2}n^2x}{\sinh \frac{1}{2}nx}\right)^2$$

where $x = \beta\lambda\mathcal{E}$ ($\beta = 1/k_BT$ and $\lambda = 3ea_0/2$).

The average thermal energy is

$$U = -\frac{\partial}{\partial\beta}logZ = -n\mathcal{E}\lambda\left[n\coth\left(\frac{1}{2}n^2x\right) - \coth\left(\frac{1}{2}nx\right)\right].$$

At low temperatures ($k_BT \ll ea_0\mathcal{E}$)

$$U \simeq -n(n-1)\lambda\mathcal{E} + \mathcal{O}(e^{-nx}),$$

while at high temperatures

$$U \simeq -\frac{1}{6}n^2(n^2-1)\mathcal{E}^2\lambda^2\beta,$$

corresponding to the induced dipole moment $< \mu_z > \simeq \frac{3}{4}n^2(n^2-1)e^2a_0^2\mathcal{E}\beta$.

Problem IV.2.3 In the classical model for the atom and for the electromagnetic radiation source (**Thomson** and **Lorentz** models) the electron was thought as an harmonic oscillator, oscillating around the center of a sphere of uniform positive charge (see Problem I.4.5). Show that the electric polarizability was $\alpha = e^2/k$, with effective elastic constant $k = 4\pi\rho e/3$, ρ being the (uniform) positive charge density.

By resorting to the second-order perturbative derivation of the polarizability for the quantum oscillator show that the same result is obtained and that it is actually the exact result.

Solution:

The restoring force is $F = -(4\pi x^3\rho/3)e/x^2$ and then $k = 4\pi\rho e/3$, corresponding for the electron to an oscillating frequency $\nu_0 = (1/2\pi)\sqrt{k/m} \simeq 2.53\times10^{15}$ s^{-1}. From $e\mathcal{E} = F = kx$ and dipole moment $ex = e^2\mathcal{E}/k$, $\alpha = e^2/k$ follows.

From Eq. 4.8, with perturbation Hamiltonian $\mathcal{H}_p = -e\mathcal{E}z$ and quantum oscillator ground and excited states

$$\alpha = 2e^2\sum_{exc\neq f}\frac{|<exc|z|f>|^2}{E_{exc}^0 - E_f^0}.$$

From the matrix elements $< v|z|v - 1 >= \sqrt{v\hbar/2m\omega}$ (according to the properties of Hermite polynomials) only the first excited state $|exc >$ has to be taken into account. Then

$$\alpha = \frac{2e^2}{\hbar^2} \frac{| < f + 1|z|f > |^2}{\omega_0} = \frac{2e^2}{\hbar^2} \frac{\hbar}{2m\omega_0^2} = \frac{e^2}{k}$$

The proof that this is the **exact result** is achieved by rewriting the Hamiltonian of the linear oscillator to include the electric energy $ez\mathcal{E}$ and observing that a shift of the eigenvalues by $-(e\mathcal{E})^2/2k$ occurs (see the analogous Problem X.5.6 for the vibrational motion of molecules, where it is also shown that α does not depend from the state $|v >$ of the oscillator).

4.3 Hamiltonian in magnetic field

From Eqs. 4.3 and 4.5, by including now the spin-orbit interaction, the perturbation of the central field Hamiltonian for multi-electron atoms is written

$$\mathcal{H}_P^{(1)} = \mu_B H \sum_i l_z^i + 2\mu_B H \sum_i s_z^i + \sum_i \xi_{nl}^i \mathbf{l}_i . \mathbf{s}_i . \tag{4.20}$$

The term

$$\mathcal{H}_P^{(2)} = \sum_i \frac{e^2 A_i^2}{2mc^2} \tag{4.21}$$

has been left out: it shall be taken into account in discussing the diamagnetism (§4.5). In writing Eq. 4.20 we have used the interaction in the form $-\boldsymbol{\mu}_{l,s}.\mathbf{H}$, as it has been proved possible at §1.6. The magnetic field is considered static, homogeneous and applied along the z-direction.

One could emphasize that in the hypothetical absence of the spin Eq. 4.20 would reduce to

$$\mathcal{H}_P^{(1)} = \mu_B H L_z \tag{4.22}$$

implying corrections to the energy levels in the form $\Delta E = \mu_B M H$. Therefore, in the light of the selection rule $\Delta M = 0, \pm 1$ (see §3.5), one realizes that for a given emission line the magnetic field should induce a triplet, characteristic of the so-called **normal Zeeman effect** (this terminology being due to the fact that for such a triplet an explanation in terms of classical Lorentz oscillators appeared possible, see Problem IV.3.1). The experimental observation that the effect of the magnetic field on the spectral lines is more complex, as shown in the following, can be considered stringent evidence for the existence of the spin. The real **Zeeman effect** (at first erroneously considered as **"anomalous"**) in general does not consists in a triplet (see the case of the Na doublet in the following). The triplet actually can occur, in principle, in the presence of very strong field (**Paschen-Back effect**), as we shall see at §4.3.2.

4.3.1 Zeeman regime

In order to derive the energy of the atom from the Hamiltonian 4.20 one has to consider the relative magnitude of the terms $\mu_B H$ (magnetic field energy) and ξ_{nl} (spin-orbit energy). In the **weak field regime**, for $\mu_B H \ll \xi_{nl}$ and in the **LS** coupling scheme, the Hamiltonian is considered in the form

$$\mathcal{H}_P^{(1)} = \mu_B \mathbf{H} \cdot (\mathbf{L} + 2\mathbf{S}) \tag{4.23}$$

and acting as a perturbation on the states $|E_0, J, M_J >$ resulting from the central field Hamiltonian, with the coupling $\sum_i \mathbf{l}_i$ and $\sum_i \mathbf{s}_i$ and the spin-orbit interaction in the form $\xi_{LS}\mathbf{L}.\mathbf{S}$.

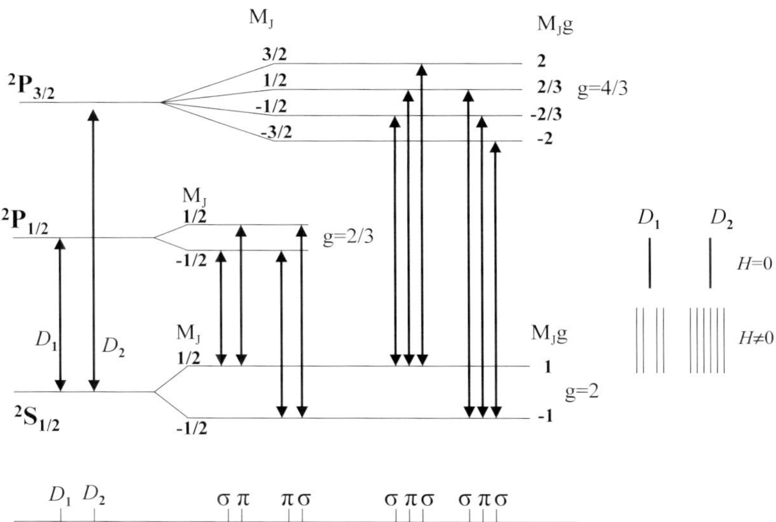

Fig. 4.4. Structure of the $^2S_{1/2}$ ground-state and of the ^2P doublet of Na atom in a magnetic field and transitions allowed by the electric dipole selection rules $\Delta S = 0$, $\Delta J = 0, \pm 1$ and $\Delta M_J = 0, \pm 1$. The D_1 line splits into four components, the D_2 line into six. Similar structure of the levels hold for the other alkali atoms. On increasing the magnetic field strength the structure of the lines, here shown for the **weak field regime**, progressively changes towards a central π line and two σ^+ and σ^- doublets (see Problem IV.3.5). π lines correspond to $\Delta M_J = 0$, while σ lines to $\Delta M_J = \pm 1$.

The operator $(\mathbf{L} + 2\mathbf{S})$ has to be projected along $\mathbf{J}$ by using **Wigner-Eckart** theorem

$$< E_0, J, M_J' |L_z + 2S_z|E_0, J, M_J >= g < E_0, J, M_J'|J_z|E_0, J, M_J >= gM_J\delta_{M_J',M_J} , \tag{4.24}$$

the constant g being obtained from the component of $(\mathbf{L} + 2\mathbf{S})$ along $\mathbf{J}$:

$$g = < E_0, J, L, S| \frac{(\mathbf{L} + 2\mathbf{S}).\mathbf{J}}{J^2} |E_0, J, L, S > =$$

$$= < E_0, J, L, S| \frac{(\mathbf{L} + \mathbf{S}).\mathbf{J} + \mathbf{S}.\mathbf{J}}{J^2} |E_0, J, L, S > =$$

$$= 1 + \frac{J(J+1) + S(S+1) - L(L+1)}{2J(J+1)} \tag{4.25}$$

This result is in close agreement with the deduction of the **Lande' factor** within the vectorial coupling model (§3.2.2). Then the energy corrections are given by

$$\Delta E = \mu_B H g M_J \, , \tag{4.26}$$

the result that one would anticipate by assigning to the atom a magnetic moment $\boldsymbol{\mu}_J = -\mu_B g \mathbf{J}$ and by writing the perturbation Hamiltonian as $\mathcal{H}_P = -\boldsymbol{\mu}_J.\mathbf{H}$.

As a consequence of Eqs. 4.25 and 4.26, in general the structure of the atomic levels in the magnetic field, in the Zeeman regime, is more complicated than the one for $S = 0$. The spectral lines are modified in a form considerably different from a triplet. At the sake of illustration, the case of the Na doublet D_1 and D_2 is schematically reported in Fig. 4.4. By taking into account the selection rules $\Delta M_J = 0, \pm 1$, for Na coinciding with the ones for single electron (see §2.1), also the polarization of the emission lines is justified.

4.3.2 Paschen-Back regime

When the strength of the magnetic field is increased the structure of the spectral lines predicted within the **LS** coupling model and weak field condition is progressively altered and in the limit of very strong field the condition of a triplet (as one would expect for $S = 0$) is restored. This crossover is related to the fact that for $\mu_B H \gg \xi_{LS}$ the effect of the magnetic perturbation has to be evaluated for unperturbed states characterized by quantum numbers M and M_S pertaining to L_z and S_z, while the spin-orbit interaction can be taken into account only as a subsequent perturbation. This is the so-called **Paschen-Back, or strong field, regime.**

From the field-related Hamiltonian in Eq. 4.23, in a way similar to the derivation within the vectorial model (see §3.2), the energy correction turns out

$$\Delta E = \mu_B H (M + 2M_S). \tag{4.27}$$

From the selection rules $\Delta M = 0, \pm 1$ and $\Delta M_S = 0$ (the spin-orbit interaction being absent at this point) one sees that the frequency $\nu_{12}^{(0)}$ of a given line related to the transition $|2 > \rightarrow |1 >$ in zero-field condition, is modified by the field in

$$\nu_{12}^{(H)} = \nu_{12}^{(0)} + \frac{\mu_B H}{h} \left[(M^2 - M^1) + 2(M_S^2 - M_S^1) \right] \tag{4.28}$$

implying the triplet, with two lines symmetrically shifted by $(e/4\pi mc)H$.

Then the spin-orbit interaction can be taken into account, yielding

$$\Delta E' = \xi_{LS} < |L_x S_x + L_y S_y + L_z S_z| >= \xi_{LS} < |L_z S_z| >= \xi_{LS} M M_S \quad (4.29)$$

and causing a certain structure of the triplet (see Problems IV.3.2 and IV.3.5).

Finally we mention that the effect of magnetic fields in the **jj** coupling scheme can be described by operating directly on the single-electron **j** moment and considering the relationship between the magnetic energy and the inter-electron coupling leading to total **J**. Again one has to use the Wigner-Eckart theorem and the results anticipated in the framework of the vectorial model (§3.3) are derived.

Problems IV.3

Problem IV.3.1 By taking into account the Larmor precession (Problem IV.2.1), the classical picture of the Lorentz radiation in magnetic field implies a triplet for observation perpendicular to the field and a doublet for longitudinal observation.

Discuss the polarization of the radiation in terms of the selection rules for the quantum magnetic number.

Solution:
The sketch of the experimental observation for classical oscillator in a magnetic field is given below:

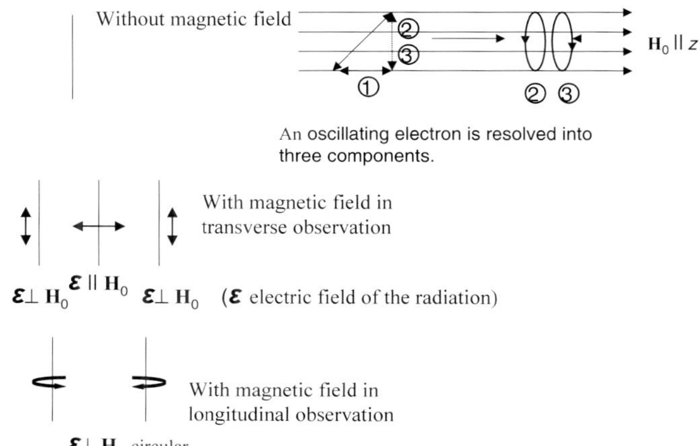

The frequency shift $\delta\omega$ can be calculated as follows (see Problem IV.2.1). For homogeneous magnetic field $\mathbf{H}_0$ along the z direction

$$m\,\ddot{x} + m\omega_0^2 x + \frac{e}{c}\,\dot{y}\,H_0 = 0,$$

$$m\,\ddot{y} + m\omega_0^2 y - \frac{e}{c}\,\dot{x}\,H_0 = 0,$$

$$m\,\ddot{z} + m\omega_0^2 z = 0,$$

The frequency of the electron oscillating in the z direction (see sketch above) remains unchanged. To solve the equations for x and y we substitute $u = x + iy$ and $v = x - iy$ and to find

$$u = u_0 e^{i\left(\omega_0 + \frac{eH_0}{2mc}\right)t} \qquad \text{and} \qquad v = v_0 e^{i\left(\omega_0 - \frac{eH_0}{2mc}\right)t}.$$

namely the equations for left-hand and right-hand circular motions at frequencies $\omega_0 \pm \delta\omega$, with $\delta\omega = \frac{eH_0}{2mc}$. The oscillators 2 and 3 in the sketch above have to emit or absorb radiation at frequency $(\omega_0 \pm \delta\omega)$, **circularly polarized** when detected along $\mathbf{H}_0$.

Oscillator 1 is along the field and therefore the intensity of the radiation is zero along that direction. If the radiation from the oscillators 2 and 3 is observed along the perpendicular direction is **linearly polarized**.

The polarizations of the Zeeman components have their quantum correspondence in the $\Delta M_J = 0$ and $\Delta M_J = \pm 1$ transitions. These rules are used in the so-called **optical pumping**: the exciting light is polarized in a way to allow one to populate selectively individual Zeeman levels, thus inducing a given spin orientation (somewhat equivalent to the magnetic resonance, see Chapter 6).

Problem IV.3.2 Illustrate the Paschen-Back regime for the $2P \longleftrightarrow 2S$ transition in Lithium atom, by taking into account *a posteriori* the spin-orbit interaction. Sketch the levels structure and the resulting transitions, with the correspondent polarizations.

Solution:
The degenerate block

$$\mu_B H < nlm'm_s'|l_z + 2s_z|nlmm_s > = \mu_B H(m + 2m_s)$$

is diagonal. The degeneracy is not completely removed. For the non-degenerate levels the spin-orbit interaction yields the correction

$$\xi_{nl} < mm_s|l_z s_z|mm_s > = \hbar^2 \xi_{nl} mm_s.$$

For the degenerate levels one has to diagonalize the corresponding block. It is noted that the terms $l_+ s_-$ and $l_- s_+$ have elements among the degenerate states equal to zero (the perturbation does not connect the degenerate states with $m = 1$, $m_s = -1/2$ and $m = -1$, $m_s = 1/2$). So the degeneracy is not removed.

The levels structure and the transitions are sketched below:

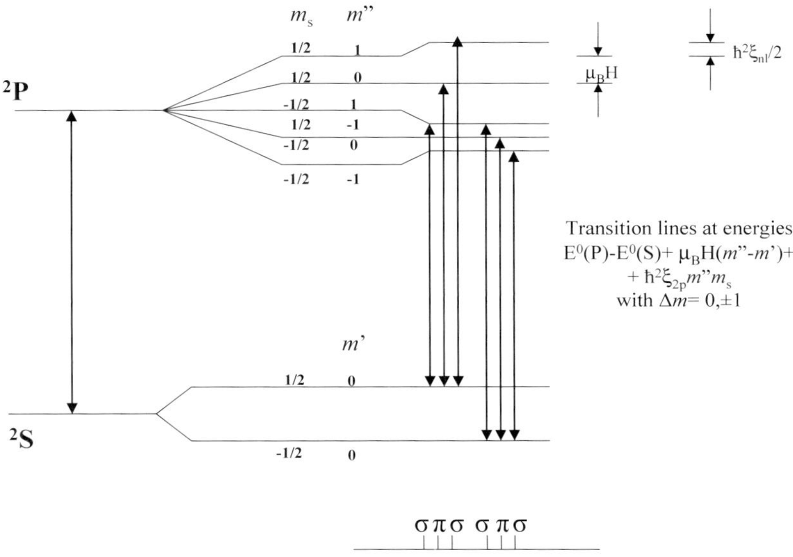

Problem IV.3.3 Evaluate the shift of a spectral line at $\lambda = 1894.6$ Å due to the transition from the $^1P_1$ to the $^1S_0$ state when a magnetic field of 1 Tesla is applied.

Solution:

The magnetic field gives rise to the triplet $(\Delta M_J = 0, \pm 1)$ and the separation between the components is

$$\Delta\nu = \frac{E_1 - E_0}{h} = \frac{g\mu_B H}{h} \simeq 1.4 \cdot 10^{10} \, Hz$$

From $\Delta\lambda \simeq -(\lambda_0/\nu_0)\Delta\nu = -(\lambda_0^2/c)\Delta\nu$ one has $\Delta\lambda \simeq 0.02$ Å.

Problem IV.3.4 Show that in positronium atom in the lowest energy state no Zeeman effect occurs (the magnetic moment of the positron is $\boldsymbol{\mu}_p = \mu_B g \mathbf{S}^p$).

Solution:

The Hamiltonian is

$$\mathcal{H} = -(\boldsymbol{\mu}_e + \boldsymbol{\mu}_p) \cdot \mathbf{H} = a(S_z^e - S_z^p) \, ,$$

with $a = \mu_B g H$.

From the energy correction

$$E = a < \phi | S_z^e - S_z^p | \phi >$$

since in the singlet state the spin eigenfunction is antisymmetric and the operator $(S_z^e - S_z^p)$ is antisymmetric, the matrix element must be zero. A more formal proof can be obtained by applying the operator $(S_z^e - S_z^p)$ on the four spin eigenfunctions $\alpha_p \alpha_e$, $\beta_p \beta_e$, etc... for the two particles.

Problem IV.3.5 From the Paschen-Back structure of the $D_{1,2}$ doublet of Sodium atom imagine to decrease the magnetic field until the Zeeman weak field regime is reached. Classify the states and connect the levels in the two regimes.

Solution:

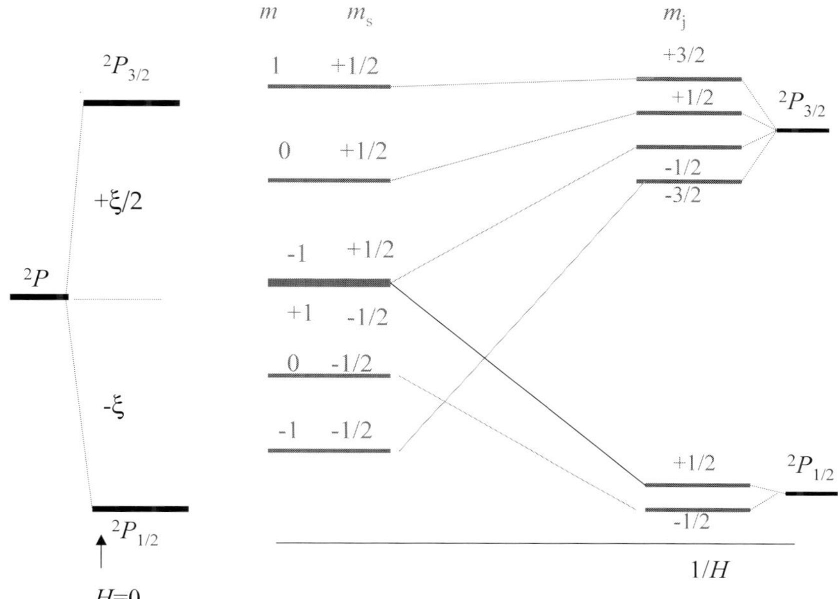

Paschen-Back regime

$$\Delta E = \mu_B H(m + 2m_s)$$

with $E_0(n, l, m, m_s)$

Zeeman regime

$$\Delta E = g\mu_B H m_j$$

with $E_0(n, l, j, m_j)$

Upon increasing the field (from right to left in the Figure) the D_1 and D_2 lines (Fig. 4.4) modify their structures as schematically shown below:

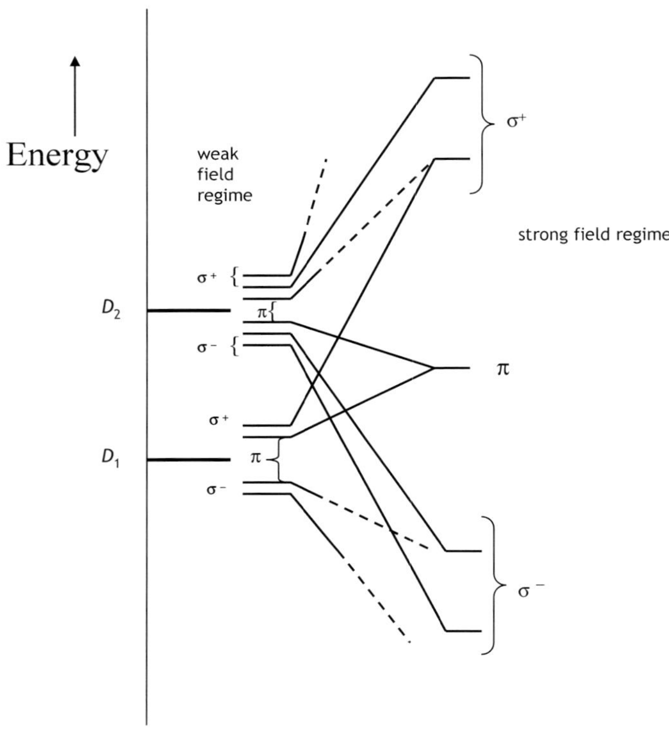

4.4 Paramagnetism of non-interacting atoms and mean field interaction

From the energy corrections induced by a magnetic field (Eq. 4.26) in the weak field regime and in the light of the classical analogy, one can attribute to the atom a magnetic moment $\boldsymbol{\mu}_J = -\mu_B g \mathbf{J}$, with $\mathbf{J}$ the total angular momentum. This statement, already used in the vectorial description at §3.2, is at the basis of the theory for the magnetic properties of matter.

As illustrative example we shall show how the magnetic properties of an assembly of atoms can be derived by referring to the statistical distribution on the levels, when the thermal equilibrium at a given temperature T is achieved. The atoms will first be considered as non-interacting (the only weak interactions occurring with the other degrees of freedom of the thermal reservoir, so that statistical equilibrium can actually be attained).

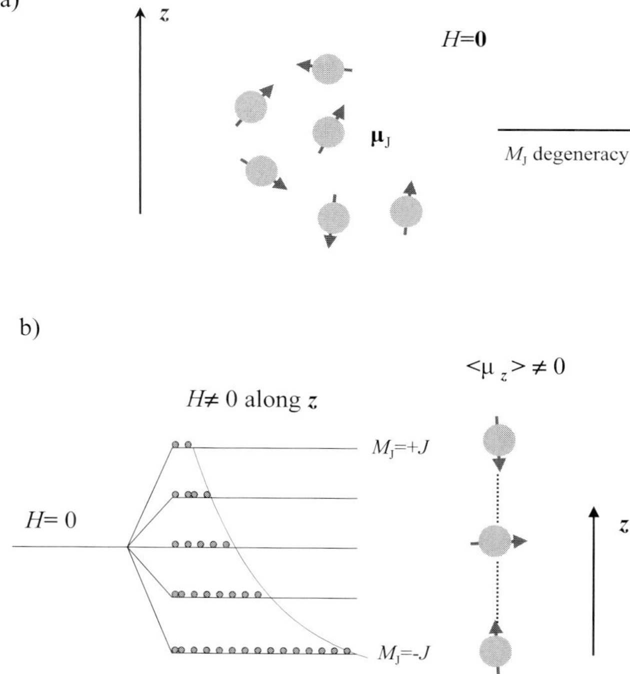

Fig. 4.5. Pictorial sketch of non-interacting atomic magnetic moments in the absence (a) and in the presence (b) of the field. The field removes the degeneracy in M_J and after some time (of the order of T_1) the statistical distribution yields an excess population on the low energy levels so that an effective component of the magnetic moment along the field is induced.

In the absence of field, degeneracy in the magnetic quantum number M_J occurs, pictorially corresponding to equiprobable orientations of the magnetic moments with respect to a given z-direction, as sketched in Fig. 4.5. When the field is switched on, in a characteristic time usually called **spin-lattice relaxation time** T_1 (for some detail on this process see Chapter 6), statistical equilibrium is achieved, with the populations on the magnetic levels as depicted in Fig. 4.5b and with an average (statistical) expectation value of the magnetic moment along the field $< \mu_z > \neq 0$.

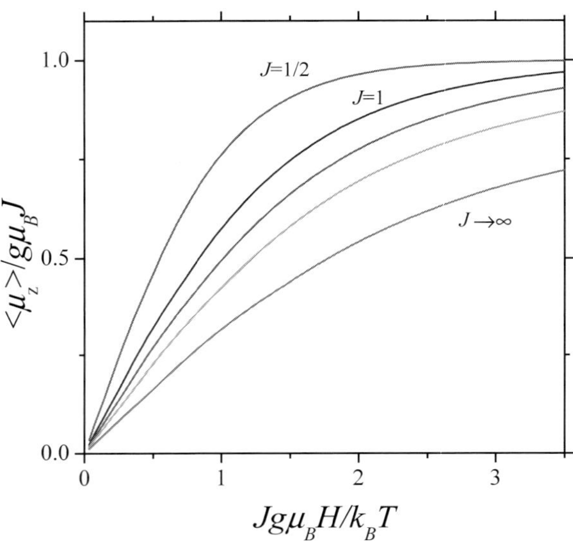

Fig. 4.6. Normalized value of the effective magnetic moment along the field direction as a function of the dimensionless variable $(Jg\mu_B H/k_B T)$, according to Eq. 4.32, for different J's.

$< \mu_z >$ is written

$$< \mu_z >= -g\mu_B \frac{\sum_{M_J} M_J e^{-xM_J}}{\sum_{M_J} e^{-xM_J}} \tag{4.30}$$

where $x = g\mu_B H/k_B T$. For $x \ll 1$ one has

$$< \mu_z > \simeq -g\mu_B \frac{\sum_{M_J} M_J(1 - xM_J)}{\sum_{M_J}(1 - xM_J)}$$

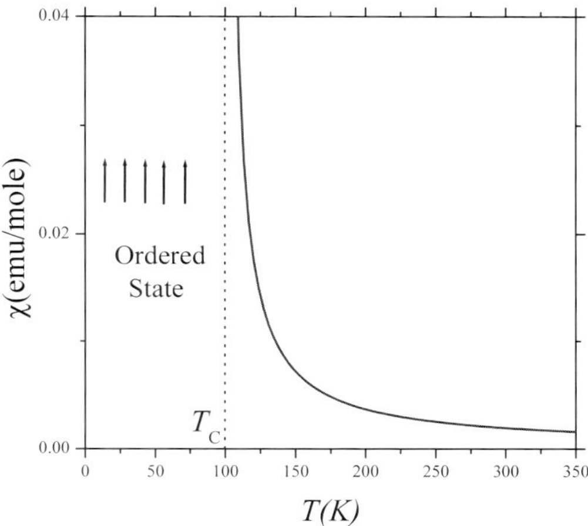

Fig. 4.7. Sketchy behavior of the temperature dependence of the paramagnetic susceptibility in presence of interactions among the magnetic moments. The state below T_c corresponds to spontaneous ordering of the magnetic moments along a given direction as a consequence of a cooperative process, typical of **phase transitions** in **many-body systems**, driven by the interaction among the components.

and since $\sum_{M_J} M_J^2 = J(J+1)(2J+1)/3$,

$$< \mu_z >= g\mu_B x \frac{J(J+1)(2J+1)}{3(2J+1)} = \frac{\mu_J^2 H}{3k_B T} \tag{4.31}$$

with

$$|\boldsymbol{\mu}_J| = g\mu_B \sqrt{J(J+1)} \ .$$

The volume paramagnetic susceptibility is $\chi = N\chi_a$, with N number of atoms per unit volume and χ_a atomic susceptibility, given by $\chi_a = \mu_J^2/3k_B T$, according to Eq. 4.31. Thus the quantum derivation of the Curie law has been obtained.

Without the approximation of low field (or high temperature), Eq. 4.30 gives

$$< \mu_z >= g\mu_B J \left[\frac{2J+1}{2J} coth \frac{(2J+1)x}{2} - \frac{1}{2J} coth \frac{x}{2} \right] \ , \tag{4.32}$$

the function depicted in Fig. 4.6 and known as **Brillouin function**. For $J \to \infty$, the Brillouin function becomes the **Langevin function**, while for $J = 1/2$ it reduces to $tanh(x/2)$.

The saturation magnetization $M_{sat} = N < \mu_z >_{T \to 0}$ corresponds to the situation where all the atoms are found on the lowest energy level of Fig. 4.5b and $(< \mu_z >)_{T \to 0} = g\mu_B J$.

According to Eq. 4.31 on decreasing temperature the paramagnetic susceptibility (in evanescent field) diverges as $1/T$. However, when the temperature is approaching zero so that the condition $x \ll 1$ no longer holds, partial saturation is achieved and χ reaches a maximum and then decreases on cooling. In practice this can happen only in strong fields (of the order of several Tesla) and at low temperature.

The assumption of ideal paramagnet in practice corresponds to the assumption that the local magnetic field is the one externally applied (apart from the diamagnetic correction, see §4.5). This condition does not hold when some type of interaction among the atomic magnetic moments is active. In this case the susceptibility can diverge at finite temperature, as sketched in Fig. 4.7.

A simple method to deal with the interactions is the **mean field approximation**, namely to assume that the local field is the external one $\mathbf{H}_{ext}$ plus a second contribution, related to the interactions, proportional to the magnetization:

$$\mathbf{H} = \mathbf{H}_{ext} + \lambda \mathbf{M}$$

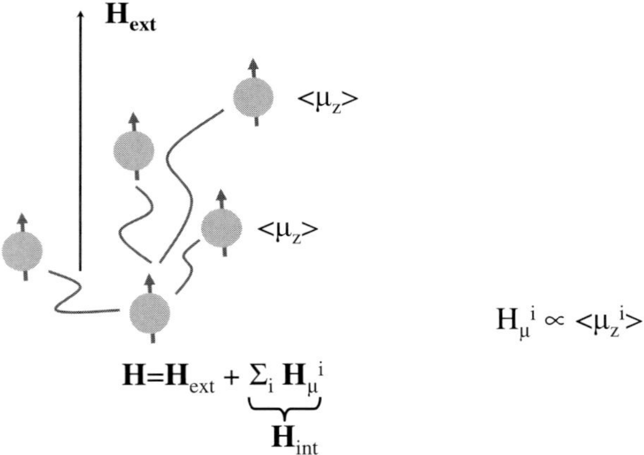

Then the magnetization reads

$$M = N\left[\chi_0(\mathbf{H}_{ext} + \mathbf{H}_{int})\right] \tag{4.33}$$

and the susceptibility turns out

$$\chi = \frac{\chi_0}{1 - \lambda\chi_0} \tag{4.34}$$

where χ_0 is the **bare susceptibility** of the ideal paramagnet, the one without interactions. Eq. 4.34 is a particular case of a more general equation, for any system in the presence of many-body interactions (in the framework of the **linear response theory** and the so-called **random phase approximation**).

By taking into account Eq. 4.31, Eq. 4.34 can be rewritten in the form

$$\chi = \frac{N\mu_J^2}{3k_B(T - T_c)}, \quad \text{where} \quad T_c = \frac{N\mu_J^2\lambda}{3k_B} \tag{4.35}$$

For $T \to T_c^+$ one has the divergence of the magnetic response and a **phase transition** to an ordered state, with spontaneous magnetization in zero field, is induced. Typical transition is the one from the paramagnetic to the ferromagnetic state and it can be expected to occur when the thermal energy k_BT is of the order of the interaction energy.

It is noted that the values of T_c's in most ferromagnets (as high as $T_c = 1044$ K, for instance for Fe bcc), indicate that the transition is driven by interactions much stronger than the dipolar one. This latter, in fact, for an interatomic distance d of the order of 1 Å, would imply $T_c \sim \mu_J^2/d^3k_B$, of the order of a few degrees K. Instead the interaction leading to the ordered states (ferromagnetic or antiferromagnetic, depending on the sign of λ in Eq. 4.34) is the one related to the **exchange integral**, as mentioned at §2.2 (for details see Appendix XIII.1).

4.5 Atomic diamagnetism

The magnetic Hamiltonian (Eq. 4.3) also implies the one-electron term (see Eq. 4.21 and 4.5)

$$\mathcal{H}_P^{(2)} = \frac{e^2 A^2}{2mc^2}$$

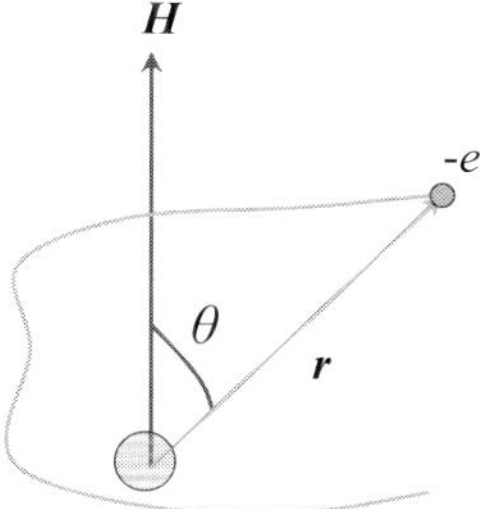

with $\mathbf{A} = (1/2)\mathbf{H} \times \mathbf{r} = (1/2)Hr\sin\theta$, the term usually neglected in comparison with the one linear in the field and leading to paramagnetism. Instead $\mathcal{H}_P^{(2)}$ is responsible of the atomic **diamagnetism**.

Let us refer to atoms in the ground state where $\mu_L = \mu_S = 0$. The effect of $\mathcal{H}_P^{(2)}$ can be evaluated in the form of perturbation for states having L, S, M and M_S as good quantum numbers, the spin-orbit interaction being absent. Thus, from first-order perturbation theory the energy correction due to $\mathcal{H}_P^{(2)}$ is

$$\Delta E = \frac{e^2 H^2}{8mc^2} \sum_i < |r_i^2 sin^2\theta_i| > \tag{4.36}$$

where the sum is over all the electrons. By resorting to $\mu = -(\partial E/\partial H)$, Eq. 4.36 implies an atomic magnetic moment **linear** in the field and in the **opposite direction**. Therefore the diamagnetic susceptibility is written

$$\chi_{dia} = -N\frac{e^2}{4mc^2} \sum_i \frac{2}{3} < r_i^2 > \tag{4.37}$$

(N number of atoms per unit volume), the assumption of isotropy having been made, so that $< x^2 >=< y^2 >= 1/3 < r^2 >$. In the Table below the molar diamagnetic susceptibilities for inert-gas atoms (to a good approximation the same values apply in condensed matter) are reported:

	He	Ne	Ar	Kr	Xe
$\chi_{dia}(cm^3/mole)(\times 10^{-6})$	-2.36	-8.47	-24.6	-36.2	-55.2
Z	2	10	18	36	54

When the perturbation effects from the magnetic Hamiltonian are extended up to the second order, a mixture of states is induced and a further energy correction is obtained, **quadratic in the field** and causing a **decrease** of the energy. Thus, even in atoms where in the ground state no paramagnetic moment is present, a **positive** paramagnetic-like susceptibility (**Van Vleck paramagnetism**) of the form

$$\chi_{vv} = 2N\mu_B^2 \sum_{n\neq 0} \frac{| < \phi_0|(L_z + 2S_z)|\phi_n > |^2}{E_n^0 - E_0^0} \tag{4.38}$$

is found. For a quantitative estimate the electronic wavefunctions ϕ_0 of the ground and of the excited states ϕ_n are required. The Van-Vleck susceptibility is usually temperature-independent and small with respect to Curie susceptibility.

Problems IV.5

Problem IV.5.1 Evaluate the molar diamagnetic susceptibility of Helium in the ground state, by assuming Hydrogen-like wavefunctions with the effective nuclear charge derived in the variational procedure (Problem II.2.2). Estimate the variation of the atom energy when a magnetic field of 1 Tesla is applied.

Solution:

From Eq. 4.37

$$\chi_{dia} = -\frac{Ne^2}{6mc^2}[< r_1^2 > + < r_2^2 >]$$

and in hydrogenic atoms (Table I.4.3) $< r^2 > = 3 \left(\frac{a_0}{Z}\right)^2$. For effective charge $Z^* = Z - \frac{5}{16} = \frac{27}{16}$

$$\chi_{dia} \simeq -0.6 \cdot 10^{-6} \frac{emu}{mole}.$$

The energy variation is $\Delta E = (e^2 H^2/12mc^2)[< r_1^2 > + < r_2^2 >] \simeq 10^{-10}$ eV, very small compared to the ground state energy.

Problem IV.5.2 In a diamagnetic crystal Fe^{3+} paramagnetic ions are included, with density $d = 10^{21}$ ions/cm^3. By neglecting interactions among the ions and the diamagnetic contribution, derive the magnetization at $T = 300\,K$, in a magnetic field $H = 1000$ Oe. Then estimate the magnetic contribution to the specific heat (per unit volume).

Solution:

From Problem F.III.3 for Fe^{3+} in the ground state the effective magnetic moment is $\mu = p\mu_B$ with $p = g\sqrt{J(J+1)} = \sqrt{35}$.

From Eq. 4.31

$$M = d\frac{\mu^2 H}{3k_B T} = 0.0242 \; \mathrm{erg\, cm^{-3}\, Oe^{-1}}.$$

The energy density is $E = -\mathbf{M} \cdot \mathbf{H}$ and for $\mu_B H \ll k_B T$ the specific heat is

$$C_V = \left(\frac{\partial E}{\partial T}\right)_V = d\frac{\mu^2 H^2}{3k_B T^2} = 0.0808 \; \mathrm{erg\, K^{-1}\, cm^{-3}}.$$

Problem IV.5.3 For non-interacting spins in external magnetic field, in the assumption of high temperature, derive the Curie susceptibility from the density matrix for the expectation value of the magnetization.

Solution:

The density matrix is

$$\rho = \frac{1}{Z} exp(-\frac{\mathcal{H}_{Zeeman}}{k_B T})$$

with Z the partition function, then

$$\overline{< M_z >} = \frac{1}{Z} Tr \left\{ M_z exp(-\frac{\mathcal{H}_{Zeeman}}{k_B T}) \right\} \simeq \frac{1}{Z} Tr \left\{ M_z \left(1 - \frac{\mathcal{H}_{Zeeman}}{k_B T} \right) \right\}$$

with

$$M_z = -g\mu_B S_z \qquad \mathcal{H}_{Zeeman} = S_z g\mu_B H \ .$$

Since

$$Tr \ S_z^2 = \frac{1}{3}(S+1)S(2S+1) \ \ \text{and} \ \ Z \simeq 2S+1$$

one obtains

$$\chi = \frac{S(S+1)g^2\mu_B^2}{3k_B T}$$

(as in Eq. 4.31 for $\mathbf{S} \equiv \mathbf{J}$).

Appendix IV.1 Electromagnetic units and Gauss system

Throughout this book we are using the CGS system of units that when involving the electromagnetic quantities is known as the **Gauss system**. This system corresponds to have assumed for the dielectric constant ε_0 and for the magnetic permeability μ_0 of the vacuum the dimensionless values $\varepsilon_0 = \mu_0 = 1$, while the velocity of light in the vacuum is necessarily given by $c = 3 \times 10^{10}$ cm/s.

As it is known, the most common units in practical procedures (such as Volt, Ampere, Coulomb, Ohm and Faraday) are better incorporated in the MKS system of units (and in the international SI). These systems of units are derived when in the Coulomb equation instead of assuming as arbitrary constant $k = 1$, one sets $k = 1/4\pi\varepsilon_0$, with $\varepsilon_0 = 8.85 \times 10^{-12}$ Coulomb^2/Nm^2, as electrical permeability of the vacuum. In the SI system the magnetic field $\mathbf{B}$, defined through the Lorentz force

$$\mathbf{F} = q\mathbf{E} + q\mathbf{v} \times \mathbf{B}$$

is measured in Weber/m^2 or **Tesla**.

The auxiliary field $\mathbf{H}$ is related to the current due to the free charges by the equation $H = nI$, corresponding to the field in a long solenoid with n turns per meter, for a current of I Amperes. The unit of H is evidently Ampere/m. Thus in the vacuum one has $\mathbf{B} = \mu_0\mathbf{H}$, with $\mu_0 = 4\pi 10^{-7}$ N/Ampere$^2 \equiv 4\pi 10^{-7}$ Henry/m.

In the matter the magnetic field is given by

$$\mathbf{B} = \mu_0(\mathbf{H} + \mathbf{M})$$

where $\mathbf{M}$ is the magnetic moment per unit volume.

The SI system is possibly more convenient in engineering and for some technical aspects but it is not suited in physics of matter. In fact, the Maxwell equations in the vacuum are symmetric in the magnetic and electric fields only when $\mathbf{H}$ is used, while $\mathbf{B}$ and not $\mathbf{H}$ is the field involved in the matter. The SI system does not display in a straightforward way the electromagnetic symmetry. In condensed matter physics the Gauss system should be preferred.

Thus within this system the electric and magnetic fields have the same dimensions, the Lorentz force is

$$\mathbf{F} = q\mathbf{E} + \frac{q}{c}\mathbf{v} \times \mathbf{B}, \tag{A.IV.1}$$

$\mathbf{B}$ is related to $\mathbf{H}$ by

$$\mathbf{B} = \mathbf{H} + 4\pi\mathbf{M} = \mu\mathbf{H} \qquad \text{with} \qquad \mu = 1 + 4\pi\chi. \tag{A.IV.2}$$

$\mathbf{M} = \chi\mathbf{H}$ defines the dimensionless **magnetic susceptibility** χ. For $\mu_B H \ll k_B T$, often called **evanescent field condition**, χ is field independent. As already mentioned μ_0 and ε_0 are equal to unit, dimensionless.

The practical units can still be used, just by resorting to the appropriate conversion factors, such as

1 volt $\rightarrow$	$\frac{1}{299.8}$ statvolt or erg/esu (esu electrostatic unit)
1 ampere	2.998×10^9 esu/sec
1 Amp/m	$4\pi \times 10^{-3}$ Oersted (see below)
1 ohm	1.139×10^{-12} sec/cm
1 farad	0.899×10^{12} cm
1 henry	1.113×10^{-12} sec^2/cm
1 Tesla	10^4 Gauss
1 Weber	10^8 Gauss/cm^2

The Bohr magneton, which is **not an SI unit**, is often indicated as $\mu_B = 9.274 \times 10^{-24}$ Joule/Tesla, equivalent to our definition $\mu_B = 0.9274 \times 10^{-20}$ erg/Gauss. The gyromagnetic ratio is measured in the Gauss system in (rad/s.Oe) and in the SI system in (rad.m/Amp.s).

Unfortunately, some source of confusion is still present when using the Gauss system. According to Eq. A.IV.2, $\mathbf{B}$ and $\mathbf{H}$ have the same dimensions and are related to the currents (measured in esu/s) in the very same way. In spite of that, while B is measured in **Gauss**, without serious reason the unit of **H** is called **Oersted**. Furthermore, there are two ways to describe electromagnetism in the framework of the CGS system. One with electrostatic units **(esu)** and the other with electromagnetic units **(emu)**. The latter is usually preferred in magnetism. Thus the magnetic moment is measured in the **emu unit**, which is nothing else than a volume and therefore cm^3. The magnetic susceptibility (per unit volume) is dimensionless and often indicated as emu/cm^3. The symmetric Gauss-Hertz-Lorentz system (commonly known as Gauss system) corresponds to a mixing of the **esu** and of the **emu** systems, having assumed both $\varepsilon_0 = 1$ and $\mu_0 = 1$.

Here we do not have the aim to set the final word on the *vexata quaestio* of the most convenient system of units. Further details can be achieved from the books by **Purcell** and by **Blundell**, quoted in the foreword.

A Table is given below for the magnetic quantities in the Gauss system and in the SI system, with the conversion factors.

Quantity	Symbol	Gauss	SI	Conversion factor*
Magnetic Induction	**B**	G$\equiv Gauss$	T	10^{-4}
Magnetic field intensity	**H**	Oe	A m^{-1}	$10^3/4\pi$
Magnetization	**M**	erg/(G cm^3)	A m^{-1}	10^3
Magnetic moment	$\boldsymbol{\mu}$	erg/G($\equiv$ emu)	J/T($\equiv$ Am^2)	10^{-3}
Specific magnetization	σ	emu/g	A m^2/kg	1
Magnetic flux	ϕ	Mx (maxwell)	Wb (Weber)	10^{-8}
Magnetic energy density	E	erg/cm^3	J/m^3	10^{-1}
Demagnetizing factor	N_d	-	-	$1/4\pi$
Susceptibility (per unit volume)	χ	-	-	4π
Mass susceptibility	χ_g	erg/(G g Oe)	m^3/kg	$4\pi \times 10^{-3}$
Molar susceptibility	χ_{mol}	emu/(mol Oe)	m^3/mol	$4\pi \times 10^{-6}$
Magnetic permeability	μ	G/Oe	H/m	$4\pi \times 10^{-7}$
Vacuum permeability	μ_0	G/Oe	H/m	$4\pi \times 10^{-7}$
Anisotropy constant	K	erg/cm^3	J/m^3	10^{-1}
Gyromagnetic ratio	γ	rad/(s Oe)	rad m/(A s)	$4\pi \times 10^{-3}$

*To obtain the values of the quantities in the SI, the corresponding Gauss values should be multiplied by the conversion factor.

Finally a mention to the **atomic units** (a.u.), frequently used, is in order. In this system of units one sets $e = \hbar = m = 1$. Thus the Bohr radius for atomic Hydrogen (infinite nuclear mass) becomes $a_0 = 1$, the ground state energy becomes $E_{n=1} = -1/2$ a.u., the a.u. for velocity is $v_0 = \alpha c$ with $\alpha \simeq 1/137$ the **fine structure constant**, so that the speed of light is $c \simeq 137$ a.u.. Less practical are the a.u.'s for other quantities. For instance the a.u. for the magnetic field corresponds to 2.35×10^5 Tesla and the one for the electric field to 5.13×10^9 V/cm.

Problems F.IV

Problem F.IV.1 The magnetization curves for crystals containing paramagnetic ions Gd^{3+}, Fe^{3+} and Cr^{3+} display the saturation (for about $H/T = 20$ kGauss/K) about at the values $7\mu_B$, $5\mu_B$, and $3\mu_B$ (per ion), respectively. From the susceptibility measurements at $T = 300$ K for evanescent magnetic field one evaluates the magnetic moments $7.9\mu_B$, $5.9\mu_B$, and $3.8\mu_B$, respectively. Comment on the differences. Then obtain the theoretical values of the magnetic moments for those ions and prove that **quenching of the orbital momenta** occurs (see Problem F.IV.2).

Solution:
The susceptibility $\chi = Ng^2 J(J+1)\mu_B^2/3k_B T$ involves an effective magnetic moment $\mu_{eff} = g\mu_B\sqrt{J(J+1)}$ different from $< \mu_z >_{max} = g\mu_B J$ obtained from the saturation magnetization, related to the component of $\mathbf{J}$ along the direction of the field.

For Gd^{3+}, electronic configuration $(4f)^7$, one has $S = \frac{7}{2}$, $L = 0$, $J = \frac{7}{2}$ and $g = 2$.
Then $\mu_{eff} = g\mu_B\sqrt{S(S+1)} \simeq 7.9\,\mu_B$, while $< \mu_z >_{max} \simeq 2\mu_B 7/2 = 7\mu_B$, in satisfactory agreement with the data.

For Fe^{3+} (see Problem IV.5.2) $J = 5/2$ and $g = 2$ and then $\mu_{eff} = 5.92\,\mu_B$ and $< \mu_z >_{max} \simeq 5\,\mu_B$.

For Cr^{3+}, electronic configuration $(3d)^3$, $S = 3/2$, $L = 3$, $J = 3/2$ and $g = 2/5 = 0.4$. For unquenched $\mathbf{L}$ one would have $\mu_{eff} = (2/5)\mu_B\sqrt{15/4} = 0.77\,\mu_B$, while for $L = 0$, $\mu_{eff} = 2\mu_B\sqrt{15/4} \simeq 3.87\,\mu_B$.

Problem F.IV.2 By referring to the expectation value of l_z in $2p$ and $3d$ atomic states, in the assumption that the degeneracy is removed by crystal field, justify the quenching of the orbital momenta.

Solution:
When the degeneracy is removed the wavefunction $\phi_{2px,y,z}$ **are real**. Then, since

$$< l_z >= -i\hbar \int \phi^* \frac{\partial}{\partial \varphi}\phi d\tau$$

cannot be imaginary, one must have $l_z = 0$. Analogous consideration holds for $3d$ states. Details on the role of the crystal field in quenching the angular momenta are given at §13.3 and at Problem F.XIII.3.

Problem F.IV.3 In Hydrogen, the lines resulting from the transitions $^2P_{3/2} \longrightarrow \, ^2S_{1/2}$ and $^2P_{1/2} \longrightarrow \, ^2S_{1/2}$ occur at $(1210 - 3.54 \cdot 10^{-3})$ Å and $(1210 + 1.77 \cdot 10^{-3})$ Å, respectively. Evaluate the effect of a magnetic field of 500 Gauss, by estimating the shifts in the wavelengths of these lines, in the **weak field** regime.

Solution:

The relationship between the splitting of lines and the applied field is found from

$$dE_2 - dE_1 = -\frac{hc}{\lambda^2}\,d\lambda \ ,$$

namely

$$d\lambda = -\frac{\lambda^2}{hc}\,(dE_2 - dE_1) = -\frac{(1210\,\text{Å})^2}{12.4 \cdot 10^3\,\text{eV} \cdot \text{Å}}\,(dE_2 - dE_1) = \left(-118\,\frac{\text{Å}}{\text{eV}}\right)(dE_2 - dE_1)$$

The values for dE_2 and dE_1 are given in Table below. There are 10 transitions that satisfy the electric dipole selection rule $\Delta M_J = \pm 1, 0$. The deviation of each of these lines from $\lambda_0 = 1210\,$Å is also given.

$d\lambda_0$ Å $\cdot 10^{-3}$	Transition	dE_2 eV $\cdot 10^{-5}$	dE_1 eV $\cdot 10^{-5}$	$d\lambda$ Å $\cdot 10^{-3}$	$d\lambda_T = d\lambda_0 + d\lambda$ Å $\cdot 10^{-3}$
-3.54	a	+0.579	+0.289	-0.342	-3.88
-3.54	b	+0.193	+0.289	+0.114	-3.43
-3.54	c	+0.193	-0.289	-0.570	-4.11
-3.54	d	-0.193	+0.289	+0.570	-2.97
-3.54	e	-0.193	-0.289	-0.114	-3.65
-3.54	f	-0.579	-0.289	+0.342	-3.20
1.77	g	+0.096	+0.289	+0.228	+2.00
1.77	h	+0.096	-0.289	-0.456	+1.31
1.77	i	-0.096	+0.289	+0.456	+2.23
1.77	j	-0.096	-0.289	-0.228	+1.54

Problem F.IV.4 Refer to the H_α line in Hydrogen (see Problem I.4.4). Report the splitting of the s and p levels and the structure of the correspondent line when a magnetic field of 45000 Gauss is applied. By taking into account that the separation between two adjacent lines is $6.29 \cdot 10^{10}$ Hz and by ignoring the fine structure, evaluate the specific electronic charge (e/m). Compare the estimate of (e/m) with the one obtained from the observation that a field of 30000 Gauss induces the splitting of the spectral line in Ca atom at 4226 Å in a triplet with separation 0.25 Å (do not consider in this case the detailed structure of the energy levels).

Solution:

In the Paschen-Back regime the energy correction is

$$\Delta E_{m,m_s} = \mu_B H_0 (m + 2m_s) \ ,$$

with electric dipole selection rules. One can observe three lines with splitting $\Delta \bar\nu = 2.098 \, cm^{-1}$. Then from

$$\Delta E = \mu_B H_0 = \frac{e\hbar}{2m} H_0 = h\Delta\nu$$

one has

$$\frac{|e|}{m} = \frac{4\pi \Delta\nu}{H_0} = 5.28 \times 10^{17} \frac{u.e.s.}{g} \ .$$

For Ca, from $\Delta E \simeq e\hbar H / 2mc = -hc\Delta\lambda/\lambda^2$ again

$$\frac{|e|}{m} = \frac{4\pi c^2}{H} \frac{\Delta\lambda}{\lambda^2} \simeq 5.2 \cdot 10^{17} \frac{u.e.s.}{g} \ .$$

Problem F.IV.5 Two particles at spin $s = 1/2$ and magnetic moments $a\boldsymbol{\sigma}_1$ and $b\boldsymbol{\sigma}_2$, with $\boldsymbol{\sigma}_{1,2}$ spin operators, interact through the Heisenberg exchange Hamiltonian (see §2.2). Derive eigenstates and eigenvalues in the presence of magnetic field.

Solution:

From the Hamiltonian

$$\mathcal{H} = -\frac{(a+b)}{2}(\sigma_{1z} + \sigma_{2z})H - \frac{(a-b)}{2}(\sigma_{1z} - \sigma_{2z})H + K\boldsymbol{\sigma}_1 \cdot \boldsymbol{\sigma}_2.$$

one writes

$$\mathcal{H} = -(a+b)HS_z + \frac{K}{2}(4S(S+1) - 6) - \frac{(a-b)}{2}(\sigma_{1z} - \sigma_{2z})H.$$

The first two terms are diagonal (in the representation in which the total spin is diagonal). Both the triplet and the singlet states have definite parity for exchange of particles (even and odd respectively). Thus the only non-zero matrix element of the last term (which is odd for exchange) is the one connecting singlet and triplet states. One finds

$$(\sigma_{1z} - \sigma_{2z})\left(\frac{a(1)b(2) - a(2)b(1)}{\sqrt{2}}\right) = 2\left(\frac{a(1)b(2) + a(2)b(1)}{\sqrt{2}}\right).$$

the only non-zero matrix element being

$$< 10|\mathcal{H}|00 >= -(a-b)H.$$

Therefore $S = 1$, $S_z = \pm 1$ classify the eigenstates, with eigenvalues

$$E_\pm = \mp(a + b)H + K,$$

For the states with $S_z = 0$, the Hamiltonian can be represented by the matrix

$$\mathcal{H}(S_z = 0) = \begin{pmatrix} K & -(a-b)H \\ -(a-b)H & -3K \end{pmatrix},$$

where the out of diagonal elements involve the triplet-singlet mixture. From the secular equation the eigenvalues turn out

$$E_\pm(S_z = 0) = -K \pm \sqrt{4K^2 + (a-b)^2 H^2}.$$

Problem F.IV.6 The saturation magnetization (per unit volume) of Iron (Fe^{2+}) is often reported to be $1.7 \cdot 10^6$ A/m. Derive the magnetic moment per atom and compare it to the theoretical estimate (density of iron $7.87\,\text{g/cm}^3$).
  Solution:
  From
$$M_{sat} = 1.7 \cdot 10^3 \text{ erg Gauss}^{-1} \text{ cm}^{-3}$$
and $n_{atom} = 0.85 \cdot 10^{23}$ cm^{-3}, one derives

$$\mu_a \simeq 2 \cdot 10^{-20} \text{ erg Gauss}^{-1}$$

or equivalently $\mu_a = 2.2\mu_B$. For Fe^{2+} ($S = 2$, $L = 2$, $J = 4$ and $g = \frac{3}{2}$) one would expect $\mu = g\mu_B J = 6\mu_B$. For quenched orbital momentum $\mu = 2S\mu_B = 4\mu_B$ (see Problem F.IV.2 and §13.3).

Problem F.IV.7 A bulb containing Hg vapor is irradiated by radiation propagating along the x axis and linearly polarized along z, along which a constant magnetic field is applied. When the wavelength of the radiation is $2537\,\text{Å}$, absorption and meantime re-emission of light along the y direction, with the same polarization, is detected. When a RF coil winding the bulb along the y direction is excited at the frequency 200 MHz one notes re-emission of light also along the z direction, light having about the same wavelength and circular polarization. Explain such a phenomenology and estimate the strength of the field.
  Solution:
Since spin-orbit interaction is very strong the weak-field regime holds (see §3.3 and Fig. 3.9). The electric dipole selection rule $\Delta M_J = 0$ requires linearly polarized radiation. In the absence of radio frequency excitation, π radiation is re-emitted again, observed along y. Along the z direction the radiation is not observed.

The radio-frequency induces magnetic dipole transitions at $\Delta M_J = \pm 1$ among the Zeeman levels. The re-emission of light in such a way is about at the same wavelength λ. In fact

$$\Delta \lambda \simeq \frac{\lambda^2}{c}\, \Delta \nu = 4.29 \cdot 10^{-4}\, \overset{\circ}{A} \ .$$

On the other hand, since among the levels involved in the emission $\Delta M_J = \pm 1$, one has circular σ polarization and so the radiation along z can be observed.

From the resonance condition

$$\nu = \frac{\Delta E}{h} = \frac{g\mu_B H}{h}$$

with $g = 3/2$, one deduces

$$H = \frac{h\nu}{g\mu_B} = 95.26\,\mathrm{Oe}.$$

The levels (in the **LS** scheme and in the weak field regime) for the Hg $^1S_0$ and $^3P_1$ states (see Fig. 3.9) and the transitions are sketched below:

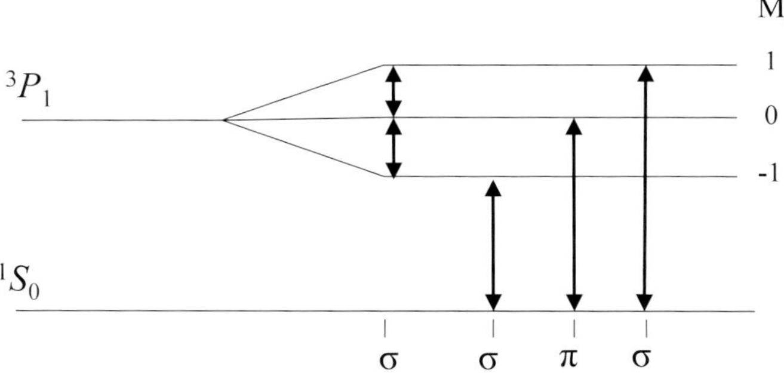

Problem F.IV.8 Consider a paramagnetic crystal, with non-interacting magnetic ions at $J = 1/2$. Evaluate the fluctuations $< \Delta M^2 >$ of the magnetization and show that it is related to the susceptibility $\chi = \partial < M >/\partial H$ by the relation $\chi = < \Delta M^2 >/k_B T$ (particular case of the **fluctuation-dissipation theorem**).

Solution:

The density matrix is $\rho = (1/Z)e^{\beta H M_z}$ ($\beta \equiv 1/k_B T$) and the partition function $Z = Tr\left[e^{\beta H M_z}\right]$. The magnetization can be written

$$< M_z >= \frac{1}{H}\frac{\partial}{\partial \beta} \log Z.$$

Then

$$\chi \equiv \frac{\partial < M_z >}{\partial H} = \beta \left[\frac{Tr(e^{\beta H M_z} M_z^2)}{Tr(e^{\beta H M_z})} - \left[\frac{Tr(e^{\beta H M_z} M_z)}{Tr(e^{\beta H M_z})}\right]^2 \right] = \beta < \Delta M_z^2 > .$$

Without involving the density matrix, from the single-ion fluctuations

$$< \Delta \mu_z^2 >=< \mu_z^2 > -(< \mu_z >)^2$$

with $< \mu_z >$ statistical average of $\boldsymbol{\mu} = -g\mu_B \mathbf{J}$, since

$$< \mu_z^2 >= \frac{1}{4}g^2 \mu_B^2$$

and (see §4.4)

$$(< \mu_z >)^2 = \frac{1}{4}g^2\mu_B^2 \tanh^2\left(\frac{1}{2}\frac{g\mu_B H}{k_B T}\right),$$

by taking into account that

$$\chi = \frac{\partial M}{\partial H} = \frac{1}{4}g^2\mu_B^2 \frac{N}{k_B T}\cosh^{-2}\left(\frac{1}{2}\frac{g\mu_B H}{k_B T}\right)$$

one finds [2]

$$< \Delta M^2 >= N < \Delta \mu_z^2 >= \frac{1}{4}g^2\mu_B^2 N \cosh^{-2}\left(\frac{1}{2}\frac{g\mu_B H}{k_B T}\right)$$

and then

$$\chi = \frac{1}{k_B T}< \Delta M^2 >= \beta < \Delta M^2 > .$$

---

[2] The single $\boldsymbol{\mu}$'s are uncorrelated i.e. $< \Delta \mu_n \Delta \mu_m >=< \Delta \mu_n >< \Delta \mu_m >$, for $n \neq m$.

Problem F.IV.9 Consider an ensemble of $N/2$ pairs of atoms at $S=1/2$ interacting through an Heisenberg-like coupling $K\mathbf{S}_1 \cdot \mathbf{S}_2$ with $K > 0$. By neglecting the interactions among different pairs, derive the magnetic susceptibility. Express the density matrix and the operator S_z on the basis of the singlet and triplet states. Finally derive the time-dependence of the statistical ensemble average $< S_1^z(0) \cdot S_1^z(t) >$, known as **auto-correlation function**.

Solution:

The eigenvalues are $E_s = (K/2)S(S + 1)$, with $S = 0$ and $S = 1$. The susceptibility is

$$\chi = \left(\frac{N}{2} \right) (p_0 \chi_0 + p_1 \chi_1),$$

where

$$p_S = \frac{(2S + 1)e^{-\frac{E_S}{k_B T}}}{e^{-\frac{E_0}{k_B T}} + 3e^{-\frac{E_1}{k_B T}}}$$

and

$$\chi_S = \frac{g^2 \mu_B^2 S(S + 1)}{3k_B T}.$$

Then

$$\chi = \frac{N g^2 \mu_B^2}{3k_B T} \frac{3e^{-\frac{K}{k_B T}}}{1 + 3e^{-\frac{K}{k_B T}}}.$$

On the basis given by the states

$$|1> = |+ + >, \quad |2> = |- - >, \quad |3> = \frac{1}{\sqrt{2}}(|+ - > +|- + >) \quad \text{and}$$

$$|4> = \frac{1}{\sqrt{2}}(|+ - > -|- + >)$$

omitting irrelevant constants, one has

$$< i|\mathcal{H}|j > = \begin{pmatrix} K & 0 & 0 & 0 \\ 0 & K & 0 & 0 \\ 0 & 0 & K & 0 \\ 0 & 0 & 0 & 0 \end{pmatrix}$$

Then the density matrix is

$$< i|\rho|j > = < i|e^{-\beta H}|j > = \begin{pmatrix} e^{-\beta K} & 0 & 0 & 0 \\ 0 & e^{-\beta K} & 0 & 0 \\ 0 & 0 & e^{-\beta K} & 0 \\ 0 & 0 & 0 & 1 \end{pmatrix}$$

By letting S_1^z act on the singlet and triplet states, one has

$$< i|S_1^z|j >= \frac{1}{2} \begin{pmatrix} 1 & 0 & 0 & 0 \\ 0 & -1 & 0 & 0 \\ 0 & 0 & 0 & 1 \\ 0 & 0 & 1 & 0 \end{pmatrix}$$

The autocorrelation-function is

$$g(t) =< \{S_1^z(t) \cdot S_1^z(0)\} >= Re[< S_1^z(t) \cdot S_1^z(0) >] \quad \text{where} \quad \{A, B\} = \frac{1}{2}(AB+BA),$$

$$< S_1^z(t) \cdot S_1^z(0) >= Tr\left[\frac{\rho}{Z} e^{-\frac{iHt}{\hbar}} S_1^z e^{\frac{iHt}{\hbar}} S_1^z\right]$$

By setting $\omega_e = \frac{K}{\hbar}$, (Heisenberg exchange frequency), one writes

$$< S_1^z(t) \cdot S_1^z(0) >= \frac{1}{4(1 + 3e^{-\beta K})} \times$$

$$Tr\{ \begin{pmatrix} e^{-\beta K} & 0 & 0 & 0 \\ 0 & e^{-\beta K} & 0 & 0 \\ 0 & 0 & e^{-\beta K} & 0 \\ 0 & 0 & 0 & 1 \end{pmatrix} \times$$

$$\times \begin{pmatrix} e^{-i\omega_e t} & 0 & 0 & 0 \\ 0 & e^{-i\omega_e t} & 0 & 0 \\ 0 & 0 & e^{-i\omega_e t} & 0 \\ 0 & 0 & 0 & 1 \end{pmatrix} \begin{pmatrix} 1 & 0 & 0 & 0 \\ 0 & -1 & 0 & 0 \\ 0 & 0 & 0 & 1 \\ 0 & 0 & 1 & 0 \end{pmatrix} \times$$

$$\begin{pmatrix} e^{i\omega_e t} & 0 & 0 & 0 \\ 0 & e^{i\omega_e t} & 0 & 0 \\ 0 & 0 & e^{i\omega_e t} & 0 \\ 0 & 0 & 0 & 1 \end{pmatrix} \begin{pmatrix} 1 & 0 & 0 & 0 \\ 0 & -1 & 0 & 0 \\ 0 & 0 & 0 & 1 \\ 0 & 0 & 1 & 0 \end{pmatrix} \},$$

and then

$$g(t) = Re[< S_1^z(t) \cdot S_1^z(0) >]$$

$$= \frac{1}{4(1 + 3e^{-\beta K})} [2e^{-\beta K} + e^{-\beta K} \cos(\omega_e t) + \cos(\omega_e t)].$$

For $k_B T \gg K$

$$g(t) = \frac{1}{8}[1 + \cos(\omega_e t)].$$

$1/\omega_e$ can be defined as the **correlation time**, in the infinite temperature limit.

Problem F.IV.10 By resorting to the **Bohr-Sommerfeld quantization** rule (Problem I.4.4) for the canonical momentum derive the **cyclotron frequency** and the energy levels for a free electron (without spin) moving in the xy plane, in the presence of a constant homogeneous magnetic field along the z axis.

Solution:
The canonical momentum is $\mathbf{p} = m\mathbf{v} - e\mathbf{A}/c$. From the quantization along the circular orbit

$$\oint \mathbf{p}.d\mathbf{q} = \oint (m\mathbf{v} - \frac{e}{c}\mathbf{A})d\mathbf{q} = mv2\pi R - \frac{e}{c}\pi R^2 H = \frac{\pi e R^2 H}{c}$$

(R radius of the orbit). The equilibrium condition along the trajectory implies $v = eHR/mc$ and then the quantization rule yields

$$\frac{\pi R_n^2 eH}{c} = nh, \ \ (\text{with } n = 1, 2, ...) \ .$$

The energy becomes

$$E_n = \frac{mv_n^2}{2} = \frac{e\hbar H}{mc}n \equiv \hbar\omega_c n \equiv 2\mu_B H$$

with

$$\omega_c = \frac{eH}{mc}$$

cyclotron frequency. For the quantum description, which includes $n = 0$ and the zero-point energy $\hbar\omega_c/2$, see Appendix XIII.1.

Problem F.IV.11 By referring to a Rydberg atom (§1.5) and considering that the diamagnetic correction to a given n-level rapidly increases on increasing n, discuss the limit of applicability of the perturbative approach, giving an estimate of the breakdown value of n in a magnetic field of 1 Tesla (see Eq. 4.36). Then discuss why the Rydberg atoms are highly polarizable and ionized by a relatively small electric field.

Solution:
From

$$\Delta E_n = \frac{e^2 H^2}{8mc^2}\frac{2}{3} < r_n^2 >$$

considering (see Table I.4.3)

$$< r^2 >_{nlm} = \frac{n^2}{2}(\frac{a_0}{Z})^2[5n^2 + 1 - 3l(l+1)] \simeq \frac{a_0^2 n^4}{2}$$

for large n and l and $Z = 1$, as for ideal total screen. Then if one assumes that the perturbation approach can be safely used up to a diamagnetic correction $\Delta E_n(H) \sim 0.2E_n^0$, one obtains

$$\frac{e^2 H^2}{12mc^2} \frac{a_0^2 n_{lim}^4}{2} \sim 0.2 \frac{e^2}{2a_0} \frac{1}{n_{lim}^2}$$

from which a limiting value of the quantum number n turns out around $n_{lim} \sim$ 70.

As regards the electric polarizability, by considering that in Eq. 4.8 the relevant matrix elements imply an increase of the numerator with n^2 while the difference in energy at the denominator varies as $1/n^3$ (remember the correspondence principle, Problem I.5.2) one can deduce that the electric polarizability must increase as n^7.

5

Nuclear moments and hyperfine interactions

Topics

Angular, magnetic and quadrupole moments of the nucleus
Magnetic electron-nucleus interaction
Quadrupolar electron-nucleus interaction
Hyperfine structure and quantum number F
Hydrogen atom re-examined: fine and hyperfine structure

5.1 Introductory generalities

Until now the nucleus has been often considered as a point charge with infinite mass, when compared to the electron mass. The **hyperfine structure** in high resolution optical spectra and a variety of experiments that we shall mention at a later stage, point out that the nuclear charge is actually distributed over a finite volume. Several phenomena related to such a charge distribution occur in the atom and can be described as due to **nuclear moments**. One can state the following:

i) most nuclei have an **angular momentum**, usually called **nuclear spin**. Accordingly one introduces a nuclear spin operator $\mathbf{I}\hbar$, with related quantum numbers I and M_I, of physical meaning analogous to the one of J and M_J for electrons.

Nuclei having **even** A and **odd** N have integer quantum spin number I (hereafter **spin**) while nuclei at **odd** A have semi-integer spin $I \leq 9/2$; nuclei with both A and N **even** have $I = 0$.

ii) associated with the angular momentum one has a **dipole magnetic moment**, formally described by the operator

$$\boldsymbol{\mu}_I = \gamma_I \mathbf{I}\hbar = g_n M_n \mathbf{I} \tag{5.1}$$

Nucleus	Z	N	I	μ/M_n	g_n
neutron	0	1	1/2	-1.913	-3.826
^{1}H	1	0	1/2	2.793	5.586
^{2}H	1	1	1	0.857	0.857
^{3}He	2	1	1/2	-2.12	-4.25
^{4}He	2	2	0	-	-
^{12}C	6	6	0	-	-
^{13}C	6	7	1/2	0.702	1.404
^{14}N	7	7	1	0.404	0.404
^{16}O	8	8	0	-	-
^{17}O	8	9	5/2	-1.893	-0.757
^{19}F	9	10	1/2	2.628	5.257
^{31}P	15	16	1/2	1.132	2.263
^{133}Cs	55	78	7/2	2.579	0.737

Table V.1. Properties of some nuclei

where g_n is the **nuclear Landé factor** and M_n the **nuclear magneton**, given by $M_n = \mu_B/1836.15 = e\hbar/2M_p c$, with M_p proton mass. γ_I is the **gyromagnetic ratio**. g_n (which depends on the intrinsic nuclear properties) in general is different from the values that have been seen to characterize the electron Landé factor. For instance, the proton has $I = 1/2$ and $\mu_I = 2.796 M_n$ and then $g_n = 5.59$. For deuteron one has $I = 1$ and $\mu_I = 0.86 M_n$. Since the angular momentum of the neutron is $I = 1/2$, from the comparison of the moments for proton and deuteron one can figure out a "vectorial" composition with the neutron and proton magnetic moments pointing along opposite directions.

At variance with most nuclei, for which g_n is **positive**, neutron as well as the nuclei ^{3}He, ^{15}N and ^{17}O have magnetic moment **opposite** to the angular momentum. Thus for them g_n is **negative**, similarly to electron. The pictorial composition indicated for deuteron does not account for a discrepancy of about 0.023 M_n, which is attributed to the fact that the ground state of the deuteron involves also the D excited state, with a little weight (about 4 percent). Properties of some nuclei are listed in Table V.1.

iii) nuclei with $I \geq 1$ are characterized by a charge distribution lacking spherical symmetry. Therefore, in analogy with classical concepts, they posses a **quadrupole electric moment**. For charge rotationally symmetric along the z axis the quadrupole moment is defined

$$Q = \frac{1}{Ze} \int [3z^2 - r^2]\rho(\mathbf{r})d\tau \tag{5.2}$$

For uniform charge density $\rho(\mathbf{r})$, one has $Q = (2/5)(b^2 - a^2)$, a and b being the axis of the ellipsoid (see Problem F.V.1). Since the average radius of the nucleus can be written $R_n = (a^2 b)^{1/3}$, by indicating with δR_n the departure from the sphere (i.e. $b = R_n + \delta R_n$) one has $Q = (6/5)R_n^2[\delta R_n/R_n]$.

iv) since proton and neutron ($I = 1/2$) are **fermions**, a nucleus with mass number A **odd** is a fermion while for A **even** the nuclei must obey to the **Bose-Einstein statistics**; in fact the exchange of two nuclei corresponds to the exchange of A pairs of fermions.

5.2 Magnetic hyperfine interaction - F states

The nuclear magnetic moment causes an interaction with the electrons of magnetic character, that can be thought to arise from the coupling between $\boldsymbol{\mu}_J$ and $\boldsymbol{\mu}_I$. In the framework of the vectorial model one can extend the usual assumptions (see §3.2) to yield a coupling Hamiltonian of the form

$$\mathcal{H}_{hyp}^{mag} = a_J \mathbf{I}.\mathbf{J} \tag{5.3}$$

In the quantum mechanical description the magnetic hyperfine interaction is obtained by considering the one-electron magnetic Hamiltonian (see §4.3)

$$\mathcal{H}_m = \frac{e}{2mc}\left(\mathbf{p}.\mathbf{A} + \mathbf{A}.\mathbf{p}\right) + 2\mu_B \mathbf{s}.\nabla \times \mathbf{A} \tag{5.4}$$

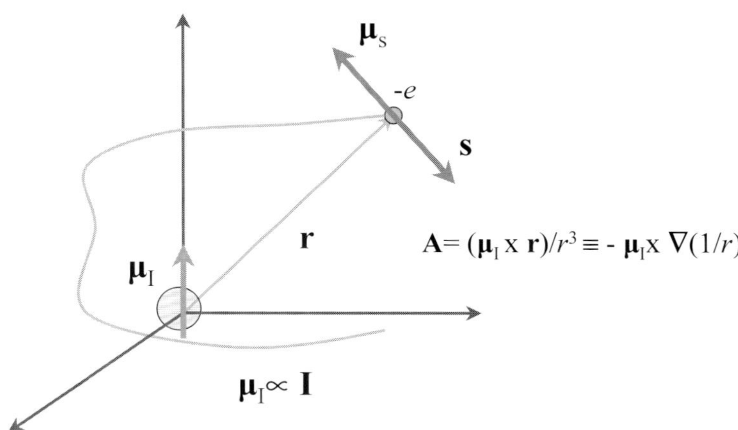

Fig. 5.1. Pictorial view of the nucleus-electron interaction of magnetic origin.

with a vector potential $\mathbf{A}$ due to the dipole moment $\boldsymbol{\mu}_I$ at the origin (see the sketchy description in Fig. 5.1).

By means of some vector algebra (see Problem F.V.14) and by singling out the terms having a singularity at the origin, as it could be expected on physical grounds, the magnetic hyperfine Hamiltonian can be written in the form

$$\mathcal{H}^{mag}_{hyp} = -\boldsymbol{\mu}_I.\mathbf{h}_{eff} \tag{5.5}$$

namely the one describing the magnetic moment $\boldsymbol{\mu}_I$ in an effective field given by

$$\mathbf{h}_{eff} = \mathbf{h}_1(\mathbf{l}) + \mathbf{h}_2(\mathbf{s}) + \mathbf{h}_3(sing.) \ . \tag{5.6}$$

The orbital term

$$\mathbf{h}_1 = -\frac{2\mu_B \mathbf{l}}{r^3} \tag{5.7}$$

is derived by considering the magnetic field at the nucleus due to the electronic current, in the a way similar to the deduction of the spin-orbit interaction (see §1.6), from $\mathbf{h}_1 = \mathbf{E} \times \mathbf{v}/c = -el\hbar/mcr^3$.

The field

$$\mathbf{h}_2 = \frac{2\mu_B}{r^3}\left(\mathbf{s} - 3\frac{(\mathbf{s}.\mathbf{r})\mathbf{r}}{r^2}\right) \tag{5.8}$$

is the classical field that a dipole at the position $\mathbf{r}$ induces at the origin:

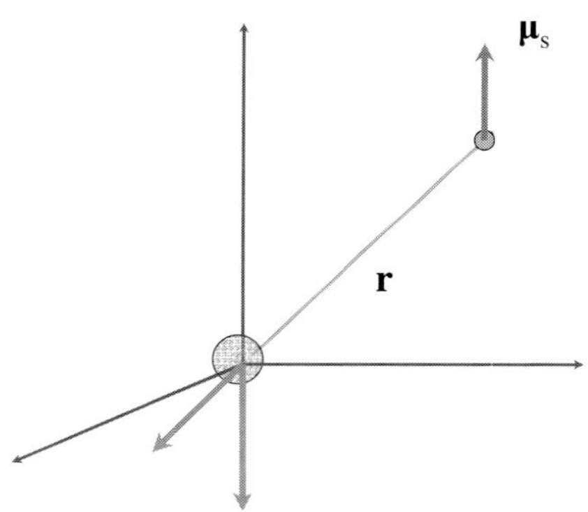

Finally

$$\mathbf{h}_3 = -\frac{2\mu_B 8\pi}{3}\mathbf{s}\delta(\mathbf{r}) \tag{5.9}$$

is a field that includes all the singularities at the origin related to the expectation values of operator of the form r^{-3} (see Table I.4.3), for s states. The **contact term** $\mathcal{H}_{cont} \propto (\mathbf{I.s})\delta(\mathbf{r})$ can be derived from a classical model where the nucleus is treated as a sphere uniformly magnetized (see Problem F.V.14).

It is remarked that an analogous contact term of the form $A\mathbf{s}_1.\mathbf{s}_2\delta(\mathbf{r}_{12})$, with $A = -(8\pi/3)(e\hbar/mc)^2$, is involved in the electron-electron **magnetic** interaction, as already recalled at Problem F.II.4.

The three fields in Eq. 5.6 are along different directions. However, by recalling the precessional motions and then the Wigner-Eckart theorem and the precession of $\mathbf{l}$ and $\mathbf{s}$ around $\mathbf{j}$, (see Eq. 4.25) one writes

$$a_j = -\gamma_I \hbar \frac{<\mathbf{h}_{eff}.\mathbf{j}>}{<j^2>} = -\gamma_I \hbar < l,s,j| \frac{\mathbf{h}_{eff}.(\mathbf{l}+\mathbf{s})}{j^2}|l,s,j> \qquad (5.10)$$

Since $\mathbf{r.l} = 0$ and $< |\mathbf{s.r}/r|^2 > = 1/4$, one obtains

$$a_j = \frac{2\mu_B \gamma_I \hbar}{j(j+1)} < l,s,j| \frac{l^2}{r^3} + \frac{8\pi}{3}\mathbf{s.j}\delta(\mathbf{r})|l,s,j> \quad . \qquad (5.11)$$

Then

$$a_j = \frac{16\pi}{3}\mu_B \gamma_I \hbar |\phi(0)|^2_{l=0} \qquad \text{for } s \text{ electrons} \qquad (5.12)$$

and

$$a_j = 2\mu_B \gamma_I \hbar \frac{l(l+1)}{j(j+1)} < r^{-3} >_{l\neq 0} \qquad \text{for } l \neq 0 \text{ , with } j = l \pm \frac{1}{2}. \qquad (5.13)$$

For $l = 0$ the angular average of $\mathbf{h}_2$ and $\mathbf{h}_1$ (Eqs. 5.7 and 5.8) yields zero: only the contact term related to $\mathbf{h}_3$ contributes to a_j once that the expectation values are evaluated (see Problem V.2.5).

From the Hydrogenic wavefunctions (§1.4) one evaluates

$$|\phi(0)|^2_{l=0} = Z^3/\pi a_0^3 n^3$$

and

$$< r^{-3} >_{l\neq 0} = Z^3/a_0^3 n^3 l(l+1/2)(l+1) \, , \text{ in Eq. 5.13.}$$

Therefore the effective field turns out of the order of $8 \times 10^4 (Z^3/n^3)$ Gauss for s electrons and of the order of $3 \times 10^4 (Z^3/n^5)$ Gauss for $l \neq 0$.

Values of the effective hyperfine field at the nucleus due to the optical electron, for the lowest energy states in alkali atoms, are reported in Table 5.1.

For two or more electrons outside the closed shells, in the **LS** coupling scheme one has to extend Eqs. 5.7-5.10 to total **L**, **S** and **J**, thus specifying a_J in Eq. 5.3.

The energy corrections related to the magnetic hyperfine interaction can be expressed by introducing the total angular momentum **F**

	$^2S_{1/2}$	$^2P_{1/2}$	$^2P_{3/2}$
Na	45	4.2	2.5
K	63	7.9	4.6
Rb	130	16	28.6
Cs	210	28	13

Table 5.1. Magnetic field (in Tesla) at the nucleus in alkali atoms, as experimentally obtained by direct magnetic dipole transitions between hyperfine levels (see Chapter 6) or by high-resolution irradiation in beams (see Fig. 5.3).

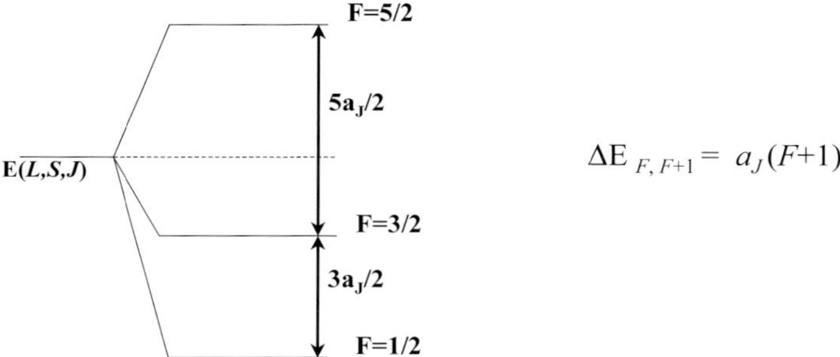

Fig. 5.2. Magnetic hyperfine structure for $J = 3/2$ and $I = 1$.

$$\mathbf{F} = \mathbf{I} + \mathbf{J} \tag{5.14}$$

with the related quantum numbers F (integer or half integer) and M_F taking the values from $-F$ to $+F$.

The structure of the **hyperfine energy levels** turns out

$$< L, S, J, F|a_J\mathbf{I}.\mathbf{J}|L, S, J, F > = \frac{a_J}{2}\left(F(F+1) - I(I+1) - J(J+1) \right) \tag{5.15}$$

The hyperfine structure for the electronic state $J = 3/2$ and nuclear spin $I = 1$ is illustrated in Fig. 5.2, showing the **interval rule** $\Delta_{F,F+1} = a_J(F+1)$. In Fig. 5.3 the hyperfine structure of the $D_{1,2}$ doublet in Na atom is reported.

It is reminded that the definition of the second as time unit and its metrological measure is obtained through the magnetic dipole $F = 4 \Leftrightarrow F = 3$

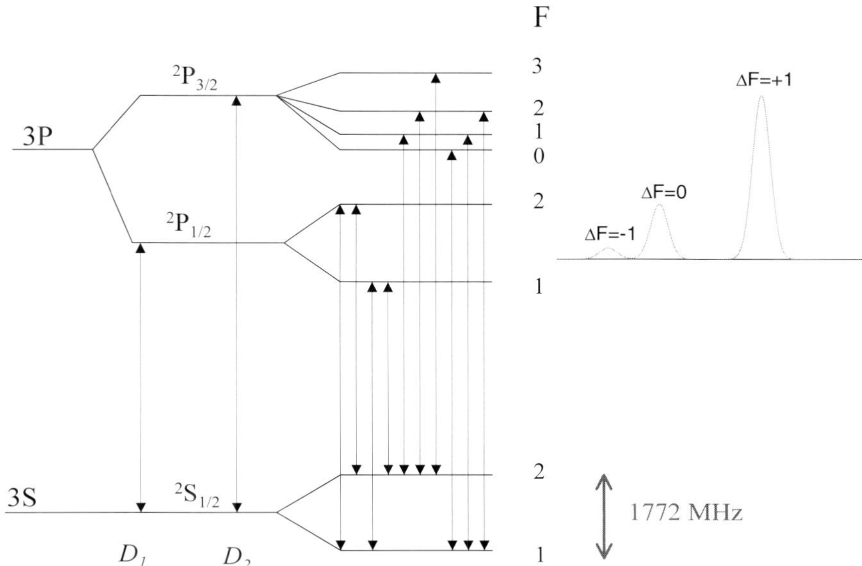

Fig. 5.3. Hyperfine magnetic structure of the low-energy states in Na atom, with the schematic illustration of the lines detected by means of resonance irradiation for the D_2 component in atomic beams (for details see Chapter 6) by using a narrow band variable-frequency dye laser (Problems F.V.2 and F.V.13).

transition, at 9172.63 MHz, in ^{133}Cs atom $(I = 7/2)$ in the ground state $^2S_{1/2}$.

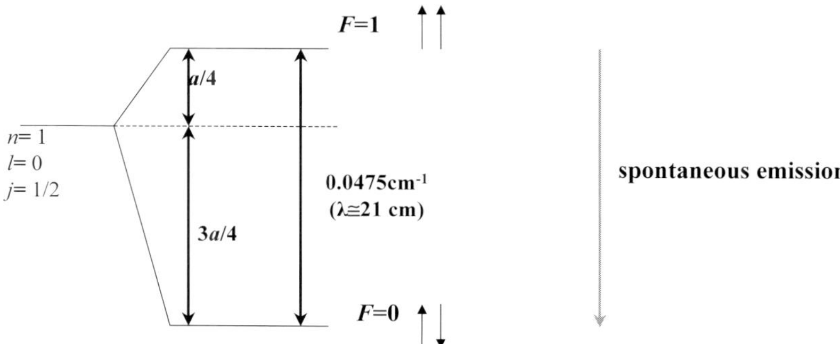

Fig. 5.4. Magnetic hyperfine structure of the ground state in Hydrogen and line at 21 cm resulting from the spontaneous emission from the $F=1$ state, the transition being driven by the magnetic dipole mechanism.

The hyperfine energy levels for the ground state of Hydrogen are sketched in Fig. 5.4, with the indication of the spontaneous emission line at 21 cm, largely used in the astrophysical studies of galaxies.

A complete description of the fine and hyperfine structure of the energy levels in Hydrogen, including the results by **Dirac** and from the **Lamb** electrodynamics, is given in Appendix V.1.

Finally we mention that the effect of an external magnetic field on the hyperfine states of the atom can be studied in a way strictly similar to what has been discussed at Chapter 4 in regards of the fine structure levels. **Zeeman** as well as **Paschen-Back** regimes are currently observed (see Problem F.V.3)

Problems V.2

Problem V.2.1 Evaluate the dipolar and the contact hyperfine splitting for the ground state of positronium. Estimate the effective magnetic field experimented by the electron.

Solution:
The hyperfine dipolar Hamiltonian is

$$\mathcal{H}_d = \frac{(2\mu_B)^2}{r^3} \left[\frac{3(\mathbf{r} \cdot \mathbf{s}_1)(\mathbf{r} \cdot \mathbf{s}_2)}{r^2} - \mathbf{s}_1 \cdot \mathbf{s}_2 \right],$$

while the Fermi contact term is

$$\mathcal{H}_F = \frac{8\pi}{3}(2\mu_B)^2|\psi(0)|^2 \mathbf{s}_1 \cdot \mathbf{s}_2$$

yielding

$$E_F(S) = \frac{8\pi}{3}(2\mu_B)^2|\psi(0)|^2 \frac{1}{2}S(S+1) + const.,$$

where $|\psi(0)|^2$ represents the probability of finding the electron and the positron in the same position, in the $1s$ state. Being zero the contribution from $\mathcal{H}_d$, the separation between the singlet and triplet levels is given by

$$E_F(S=0) - E_F(S=1) = \frac{8\pi}{3}\frac{(2\mu_B)^2}{\pi a_p^3} = \frac{4\mu_B^2}{3a_o^3} = 5 \cdot 10^{-4}\, eV,$$

with a_p Bohr radius for positronium. The magnetic field experimented by the electron is about 4×10^4 Gauss.

Problem V.2.2 Consider a pair of electrons and a pair of protons at the distance $e - e$ and $p - p$ of 2 Å. Evaluate the conditions maximizing and minimizing the dipole-dipole interaction, the energy corrections in both cases, and the magnetic field due to the second particle, for parallel orientation.

Solution:

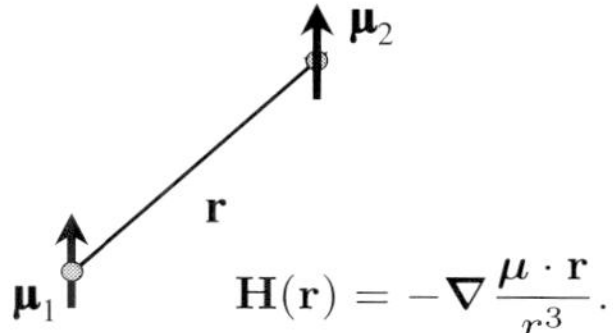

$$\mathbf{H}(\mathbf{r}) = -\nabla \frac{\boldsymbol{\mu} \cdot \mathbf{r}}{r^3}.$$

From the interaction energy

$$E_{int} = -\boldsymbol{\mu}_2 \cdot \mathbf{H}_1(\mathbf{r}) = \frac{\boldsymbol{\mu}_1 \cdot \boldsymbol{\mu}_2}{r^3} - \frac{3(\boldsymbol{\mu}_1 \cdot \mathbf{r})(\boldsymbol{\mu}_2 \cdot \mathbf{r})}{r^5}$$

$E_{int} = 0$ for $\boldsymbol{\mu}_2 \perp \mathbf{H}_1(\mathbf{r})$.

For parallel orientation of the $\boldsymbol{\mu}$'s $E_{int} = (\mu_1\mu_2/r^3)(1 - 3\cos^2\theta)$ and for $|\mathbf{r}|$ fixed the extreme values are

$$E' = -\frac{2\mu_1\mu_2}{r^3} \quad \text{for } \theta = 0, \pi$$

$$E'' = \frac{\mu_1\mu_2}{r^3} \quad \text{for } \theta = \frac{\pi}{2} \quad .$$

For two electrons at $|\mathbf{r}| = 2$ Å, for $\mu_s = 2\mu_B\sqrt{s(s+1)}$

$$E' = -6.45 \cdot 10^{-17} \text{erg} \ \text{ and } \ |\mathbf{H}'| = 4016 \text{ Oe}$$

$$E'' = +3.22 \cdot 10^{-17} \text{erg} \ \text{ and } \ |\mathbf{H}''| = 2008 \text{ Oe.}$$

For two protons

$$\mu_p = g_p\mu_N\sqrt{I(I+1)}, \ \ I = 1/2, \ \ \mu_N = \mu_B/1836, \ \ g_I = 5.6$$

and then

$$E' = -1.5 \cdot 10^{-22} \text{erg} \ \text{ and } \ |\mathbf{H}'| = 6 \text{ Oe}$$

$$E'' = 7.5 \cdot 10^{-23} \text{erg} \ \text{ and } \ |\mathbf{H}''| = 3 \text{ Oe.}$$

For the derivation of the eigenstates and eigenvalues see Problem F.V.8.

Problem V.2.3 Evaluate the magnetic field at the nuclear site in the Hydrogen atom for an electron in the states $1s$, $2s$ and $3s$. Estimate the energy difference between the states for parallel and antiparallel nuclear and electronic spins.

Solution:
From Eq. 5.15 with

$$a_j = \frac{g_I \mu_N}{\sqrt{j(j+1)}} h_J = \frac{8\pi}{3} g_e \mu_B \, g_I \, \mu_N \, |\phi(0)|^2$$

and $|\phi(0)|^2 = 1/\pi a_0^3 n^3$ the field can be written

$$h_j = \frac{8\pi}{3} g_e \mu_B \, |\phi(0)|^2 \, [j(j+1)]^{1/2} \, .$$

The energy separation a for $j \equiv s = 1/2$ (Fig. 5.4) and the field turn out

n	$|\phi(0)|^2(\mathrm{cm}^{-3})$	$a\ (\mathrm{cm}^{-1})$	$h_J\ (\mathrm{kGauss})$
1	$2.15 \cdot 10^{24}$	0.0474	289
2	$2.69 \cdot 10^{23}$	0.00593	36.1
3	$7.96 \cdot 10^{22}$	0.00176	10.7

Problem V.2.4 Consider a **muonic atom** (negative muon) and Hydrogen, both in the $2p$ state. Compare in the two atoms the following quantities:

i) expectation values of the distance, kinetic and potential energies;
ii) the spin-orbit constant and the separation between doublet due to $2p - 1s$ transition (see §1.6);
iii) the magnetic hyperfine constant and the line at 21 cm (note that the magnetic moment of the muon is about 10^{-2} Bohr magneton).

Solution:

i) $\langle r \rangle \propto a_0^*$ with $a_0^* = a_0/186$; $E_n^\mu = 186 E_n^H$,   with $E_n^H = -e^2/2a_0 n^2$.
By resorting to the **virial theorem**, $\langle V \rangle = 2 \cdot 186 E_n^H$ and $\langle T \rangle = -\frac{\langle V \rangle}{2}$.

ii) From

$$\mathcal{H}_{s.o.} = \frac{e^2 \hbar^2}{2m_\mu^2 c^2} \frac{1}{r^3} \mathbf{l} \cdot \mathbf{s}$$

by taking into account the scale factors for m_μ and for r, one finds $\xi_{2p}^\mu = 186\, \xi_{2p}^H$, with doublet separation $\frac{3}{2}\, \xi_{2p}^\mu$.
Alternatively, by considering the spin-orbit Hamiltonian in the form $\mu_l^e \, \mu_s^e \, \langle r^{-3} \rangle$, since $\mu_l^\mu \sim \mu_l^e/186$ and $\mu_s^\mu \sim 10^{-2}$, the order of magnitude of the correcting factor can be written $(186)^3/186 \cdot 100$.

iii) From $\mathcal{H}_{hyp} = -\boldsymbol{\mu}_n \cdot \mathbf{h}_{eff}$ and $|h_{eff}| \propto \mu_\mu |\psi(0)|^2$, since $\mu_\mu \sim 10^{-2}\mu_B$ and $|\phi(0)|^2 \sim (186)^3/a_0^3$, one has $a_{1s}^\mu \approx a_{1s}^H \cdot 6.5 \cdot 10^4$ and $\lambda_{21}^\mu = \lambda_{21}^H/6.5 \cdot 10^4$.

Problem V.2.5 Estimate the dipolar magnetic field that the electron in $2p_{1,0}$ states and spin eigenfunction α creates at the nucleus in the Hydrogen atom.

Solution:

From Eq. 5.8, since $\boldsymbol{\mu}_s = -2\mu_B\mathbf{s}$ and $-2\mu_B s_z = -\mu_B$, by taking into account that for symmetry reasons only the z-component is effective (the terms of the form $z.x$ and $z.y$ being averaged out) one has

$$h_z^{dip} = -\mu_B \left[-\frac{1}{r^3} + \frac{3z^2}{r^5} \right] = \frac{\mu_B}{r^3} \left[1 - 3cos^2\theta \right]$$

The expectation value of $3cos^2\theta/r^3$ on $R_{21}(r)Y_{11}(\theta, \varphi)$ reads

$$< \frac{1}{r^3} > \frac{9}{4} \int_{-1}^{1} d(cos\theta) sin^2\theta cos^2\theta = \frac{3}{2} < \frac{1}{r^3} >$$

Thus $< h_z^{dip} > = -(\mu_B/2) < 1/r^3 >$. By taking $< 1/r^3 >_{211}$ from Table I.4.3, $|h_z^{dip}| \simeq 1.5 \times 10^3$ Gauss, to be compared to Eq. 5.13.

For the electron in the $2p_0$ state one obtains $< (1 - 3cos^2\theta)/r^3 > = -(4/5) < 1/r^3 >$ (again considering as effective only the z component).

This condition is the one usually occurring in strong magnetic fields where only the z components of $\mathbf{s}$ and of $\mathbf{I}$ are of interest.

The vanishing of $< \mathbf{h}_2 >$ (Eq. 5.8) in s states arises from $\int (1 - 3cos^2\theta) sin\theta d\theta = 0$.

5.3 Electric quadrupole interaction

Since the first studies of the hyperfine structure by means of high resolution spectroscopy, it was found that in some cases the interval rule $\Delta_{F,F+1} = a_J(F+1)$ was not obeyed. The breakdown of the interval rule was ascribed to the presence of a further electron-nucleus interaction of **electrical character**, related to the electric quadrupole moment of the nucleus. As we shall see, this second hyperfine interaction is described by an Hamiltonian different from the form $a_J\mathbf{I}.\mathbf{J}$ which is at the basis of the interval rule.

To derive the electric quadrupole Hamiltonian one can start from the classical energy of a charge distribution in a site dependent electric potential:

$$E = \int \rho_n(\mathbf{r})V_P(\mathbf{r})d\tau_n \tag{5.16}$$

By expanding the potential V_P due to electrons around the center of charge, one writes

$$E = \int \rho_n(\mathbf{r})V_P(0)d\tau_n + \sum_\alpha (\frac{\partial V_p}{\partial x_\alpha})_0 \int \rho_n(\mathbf{r})x_\alpha d\tau_n + \frac{1}{2}\sum_{\alpha,\beta}(\frac{\partial^2 V_p}{\partial x_\alpha \partial x_\beta})_0 \int \rho_n(\mathbf{r})x_\alpha x_\beta d\tau \tag{5.17}$$

where one notices the **monopole interaction** (already taken into account as potential energy in the electron core Hamiltonian), a **dipole term** which is **zero** (the nuclei do not have electric dipole moment) and the **quadrupole term** of the form

$$E_Q = \frac{1}{2}\sum_{i,j} Q'_{i,j}V_{i,j}, \quad \text{with } V_{i,j} = \int \rho_{elec.}(\mathbf{r})\frac{3x_i x_j - \delta_{i,j}r^2}{r^5}d\tau_n . \tag{5.18}$$

In the quantum description

$$Q'_{i,j} = e \sum_{nucleons} (3x_i^n x_j^n - \delta_{i,j}r_n^2) \tag{5.19}$$

is the quadrupole moment operator, while $V_{i,j}$ is the electric field gradient operator, a sum of terms of the form $-e(3x_i x_j - \delta_{i,j}r^2)/r^5$.

Without formal derivation (for details see Problem F.V.15), we state that the correspondent Hamiltonian can be rewritten

$$\mathcal{H}^Q_{hyp} = b_J[3(\mathbf{I}.\mathbf{J})^2 + \frac{3}{2}\mathbf{I}.\mathbf{J} - I(I+1)J(J+1)] \tag{5.20}$$

where $b_J = eQV_{zz}/2I(2I-1)J(2J-1)$, with eQV_{zz} the **quadrupole coupling constant**. The z-component of the electric field gradient is

$$V_{zz} = <J,J| - e\sum_{elec.} \frac{(3z_e^2 - r_e^2)}{r_e^5}|J,J> \tag{5.21}$$

while

$$eQ = <I,I|eQ_z|I,I> = <I,I|e\sum_n (3z_n^2 - r_n^2)|I,I> \tag{5.22}$$

is the quantum equivalent of the classical quadrupole moment. Q is measured in cm^2 and a practical unit is 10^{-24} cm^2, called **barn**. Q positive means elongation of the nuclear charge along the spin direction while for negative Q the nuclear ellipsoid has its major axis perpendicular to $\mathbf{I}$. For $I = 0$ or $I = 1/2$, $Q = 0$.

It is remarked that in the Hamiltonian 5.20 the first term is the one implying the breakdown of the interval rule.

With the usual procedure to evaluate the coupling operators in terms of the correspondent squares of the angular momentum operators, one can derive

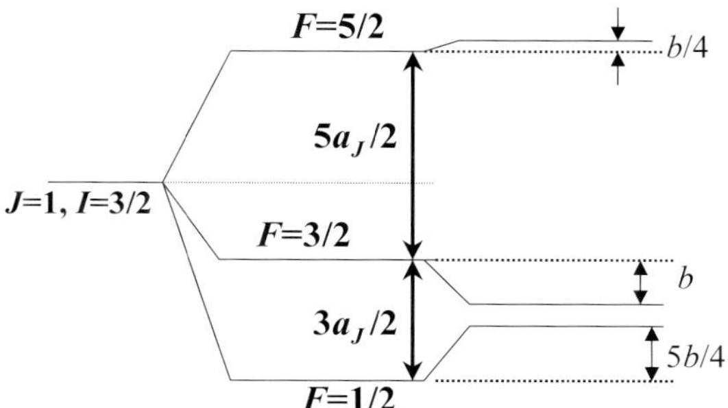

Fig. 5.5. Hyperfine magnetic and electric quadrupole energy levels for an atom with $I = 3/2$ and $J = 1$; a_J is the hyperfine constant while here $b = eQV_{zz}$ (see Eqs. 5.20- 5.22). Q has been assumed positive. The value of a_J/b is arbitrary.

the energy corrections associated with the Hamiltonian 5.20. In Fig. 5.5 the structure of the hyperfine (magnetic and electric) levels for $I = 3/2$ and $J = 1$ is shown.

The one-electron **electric field gradient** (Eq. 5.21) is written $< j, j|(3cos^2\theta - 1)/r^3|j, j >$.

For a wavefunction of the form $\varphi_{j,j} = R_{n,l}Y_{l,l}\chi^{spin}$, since

$$\int Y_{l,l}^*(3cos^2\theta - 1)Y_{l,l}sin\theta d\theta = -\frac{2l}{(2l + 3)}$$

for $\chi^{spin} \equiv \alpha$, one has

$$q \equiv -\frac{V_{zz}}{e} = \frac{2l}{2l + 3} < r^{-3} > . \tag{5.23}$$

In terms of j one can write

$$q_j = \frac{(2j - 1)}{(2j + 2)} < r^{-3} > \tag{5.24}$$

valid for $m_s = 1/2$ as well as for $m_s = -1/2$. For s states the spherical symmetry of the charge distribution implies $q = 0$.

For Hydrogenic wavefunctions the order of magnitude of the quadrupole coupling constant is

$$e^2qQ \sim 10^6Q\frac{Z^3}{n^6} \simeq 10^{-6}\frac{Z^3}{n^6}eV \tag{5.25}$$

for $Q \sim 10^{-24}$ cm^2.

In the condensed matter the operators V_{jk} can be substituted by the correspondent expectation values. The electric field gradient tensor has a principal axes frame of reference in which $V_{XY} = V_{XZ} = V_{YZ} = 0$, while $\sum_\alpha V_{\alpha\alpha} = 0$, with $|V_{ZZ}| > |V_{YY}| > |V_{XX}|$. $\eta = (V_{XX} - V_{YY})/V_{ZZ}$ is defined the **asymmetry parameter** (see Problems V.3.1 and V.3.2).

Problems V.3

Problem V.3.1 Find eigenvalues, eigenstates and transition probabilities for a nucleus at $I = 1$ in the presence of an electric field gradient at cylindrical symmetry (expectation values $V_{ZZ} = eq$, $V_{XX} = V_{YY} = -eq/2$).

Repeat for an electric field gradient lacking of the cylindrical symmetry ($V_{XX} \neq V_{YY}$).

Solution:

From Eq. 5.20 and by referring to the expectation values for the electric field gradient the Hamiltonian is

$$\mathcal{H}_Q = A \left\{ 3\hat{I}_z^2 - \hat{I}^2 + \frac{1}{2}\eta(\hat{I}_+^2 + \hat{I}_-^2) \right\}$$

where

$$A = \frac{e^2 qQ}{4I(2I - 1)} = \frac{1}{4}e^2 qQ, \quad eq = V_{ZZ}, \quad \eta = \frac{V_{XX} - V_{YY}}{V_{ZZ}}.$$

For cylindrical symmetry $\eta = 0$ and $\mathcal{H}_Q$ commutes with I_z and I^2. The eigenstates are

$$|1> = \begin{pmatrix} 1 \\ 0 \\ 0 \end{pmatrix}, |0> = \begin{pmatrix} 0 \\ 1 \\ 0 \end{pmatrix}, |-1> = \begin{pmatrix} 0 \\ 0 \\ 1 \end{pmatrix}.$$

In matrix form

$$\mathcal{H}_Q = A \begin{pmatrix} 1 & 0 & 0 \\ 0 & -2 & 0 \\ 0 & 0 & 1 \end{pmatrix}$$

and

$$\mathcal{H}_Q|\pm 1> = A|\pm 1> \quad \mathcal{H}_Q|0> = -2A|0>.$$

It can be noticed that **magnetic dipole transitions**, with $\Delta M_I = \pm 1$ and circular polarized radiation, are allowed (read §6.2).

For $\eta \neq 0$ the Hamiltonian in matrix form can be written [1]

$$\mathcal{H}_Q = A \begin{pmatrix} 1 & 0 & \eta \\ 0 & -2 & 0 \\ \eta & 0 & 1 \end{pmatrix}.$$

From

$$\mathrm{Det}(A^{-1}\mathcal{H}_Q - \epsilon I) = 0 \quad \Rightarrow \quad (2+\epsilon)(1+\epsilon^2 - 2\epsilon - \eta^2) = 0$$

the eigenvalues turn out

$$E = A\epsilon = -2A, \quad (1 \pm \eta)A$$

with corresponding eigenvectors

$$|-2A> = \begin{pmatrix} 0 \\ 1 \\ 0 \end{pmatrix} \qquad |(1 \pm \eta)A> = \frac{1}{\sqrt{2}} \begin{pmatrix} 1 \\ 0 \\ \pm 1 \end{pmatrix}.$$

The unitary transformation that diagonalizes $\mathcal{H}_Q$ is

$$\mathcal{H}'_Q = U\mathcal{H}_Q U^+$$

with

$$U = \frac{1}{\sqrt{2}} \begin{pmatrix} 1 & 0 & 1 \\ 0 & \sqrt{2} & 0 \\ 1 & 0 & -1 \end{pmatrix}.$$

---

[1] The matrices of the angular momentum operators for $I=1$ in a basis which diagonalizes I_z and I^2 are

$$I_x = \frac{1}{\sqrt{2}} \begin{pmatrix} 0 & 1 & 0 \\ 1 & 0 & 1 \\ 0 & 1 & 0 \end{pmatrix} \qquad I_y = \frac{1}{\sqrt{2}} \begin{pmatrix} 0 & -i & 0 \\ i & 0 & -i \\ 0 & i & 0 \end{pmatrix}$$

$$I_z = \begin{pmatrix} 1 & 0 & 0 \\ 0 & 0 & 0 \\ 0 & 0 & -1 \end{pmatrix} \qquad I^2 = 2 \begin{pmatrix} 1 & 0 & 0 \\ 0 & 1 & 0 \\ 0 & 0 & 1 \end{pmatrix}$$

$$I_z^2 = \begin{pmatrix} 1 & 0 & 0 \\ 0 & 0 & 0 \\ 0 & 0 & 1 \end{pmatrix} \qquad I_+ = \frac{1}{\sqrt{2}} \begin{pmatrix} 0 & 2 & 0 \\ 0 & 0 & 2 \\ 0 & 0 & 0 \end{pmatrix}$$

$$I_+^2 = \begin{pmatrix} 0 & 0 & 2 \\ 0 & 0 & 0 \\ 0 & 0 & 0 \end{pmatrix}$$

Then

$$\mathcal{H}'_Q = A \begin{pmatrix} 1+\eta & 0 & 0 \\ 0 & -2 & 0 \\ 0 & 0 & 1-\eta \end{pmatrix}.$$

The structure of the energy levels is

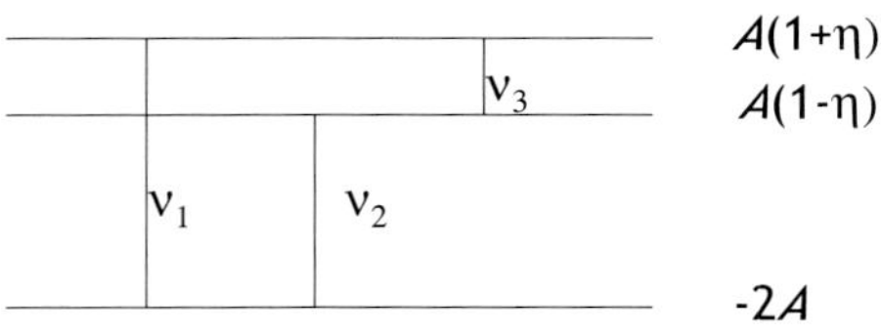

with transition frequencies

$$\nu_1 = \frac{3A}{h}\left(1+\frac{\eta}{3}\right), \quad \nu_2 = \frac{3A}{h}\left(1-\frac{\eta}{3}\right)$$

and

$$\nu_3 = 2A\eta/h.$$

From the interaction Hamiltonian with a radio frequency field $\mathbf{H}^{RF}$ (see Problem F.V.6 and for details Chapter 6)

$$\mathcal{H}_I = -\gamma\hbar\mathbf{H}^{RF}\cdot\mathbf{I}$$

and by taking into account that

$$\mathcal{H}'_I = U\mathcal{H}_I U^+$$

one finds

$$\mathcal{H}'_I = -\gamma\hbar \begin{pmatrix} 0 & H^{RF}_x & H^{RF}_z \\ H^{RF}_x & 0 & iH^{RF}_y \\ H^{RF}_z & -iH^{RF}_y & 0 \end{pmatrix}.$$

The transition amplitudes are

$$\langle A(1+\eta)|\mathcal{H}'_I|-2A\rangle = -\gamma\hbar H^{RF}_x$$

$$\langle -2A|\mathcal{H}'_I|A(1-\eta)\rangle = i\gamma\hbar H^{RF}_y$$

$$\langle A(1+\eta)|\mathcal{H}'_I|A(1-\eta)\rangle = -\gamma\hbar H_z^{RF}.$$

All the transitions are allowed, with intensity depending on the orientation of the radio frequency field with respect to the electric field gradient.

Problem V.3.2 Consider a ^{23}Na nucleus at a distance 1 Å from a fixed charge $-e$. Estimate the eigenvalues of the electric quadrupole interaction and the frequency of the radiation which induces transitions driven by the magnetic dipole mechanism (the electric quadrupole moment of ^{23}Na is $Q = +0.101 \cdot 10^{-24}$ cm^2).

Solution:

From the eigenvalues of the electric quadrupole Hamiltonian (see Problem V.3.1, for $\eta = 0$)

$$E_{I,M_I} = \frac{eQV_{ZZ}}{4I(2I-1)}(3M_I^2 - I(I+1)).$$

For $I = \frac{3}{2}$

$$E_{\pm 3/2} = \frac{1}{4}eQV_{ZZ} \qquad\qquad E_{\pm 1/2} = -\frac{1}{4}eQV_{ZZ}$$

The transition probabilities related to a perturbation Hamiltonian of the form $\mathcal{H}_P \propto \mathbf{H_x}^{RF} \cdot \mathbf{I} \propto I_{\pm}$ (see §6.2) involve the matrix elements

$$\left\langle \frac{3}{2},\frac{3}{2}\left|\hat{I}_+\right|\frac{3}{2},\frac{1}{2}\right\rangle = \sqrt{3}\left\langle \tfrac{3}{2},\tfrac{1}{2}\left|\hat{I}_+\right|\tfrac{3}{2},-\tfrac{1}{2}\right\rangle = 2$$

$$\left\langle \frac{3}{2},-\frac{1}{2}\left|\hat{I}_+\right|\frac{3}{2},-\frac{3}{2}\right\rangle = \sqrt{3}$$

Then $W_{\frac{3}{2},\frac{1}{2}} \propto 3$, $W_{\frac{1}{2},-\frac{1}{2}} \propto 4$, $W_{-\frac{1}{2},-\frac{3}{2}} \propto 3$. The transition frequencies turn out

$$\nu_{\frac{3}{2},\frac{1}{2}} = \frac{1}{h}\left(E_{\frac{3}{2}} - E_{\frac{1}{2}}\right) = \frac{1}{2}\frac{eQV_{zz}}{h}, \qquad \nu_{-\frac{3}{2},-\frac{1}{2}} = \frac{1}{h}\left(E_{-\frac{3}{2}} - E_{-\frac{1}{2}}\right) = \frac{1}{2}\frac{eQV_{zz}}{h}$$

From

$$V_{ZZ} = \frac{3z^2 - r^2}{r^5}e = \frac{2e}{r^3} \qquad \text{and} \qquad V_{XX} = V_{YY} = -\frac{e^2}{r^3}$$

(note that the Laplace equation holds) one estimates

$$\nu_{\frac{3}{2},\frac{1}{2}} = \nu_{-\frac{3}{2},-\frac{1}{2}} = \frac{e^2Q}{hr^3} \simeq 3.5\,MHz.$$

Appendix V.1 Fine and hyperfine structure in Hydrogen

Having introduced the various interaction terms (spin-orbit, relativistic corrections and hyperfine interaction) to be taken into account for one-electron states in atoms, it is instructive to reconsider the Hydrogen atom and to look at the detailed energy diagram (Fig. 5.6).

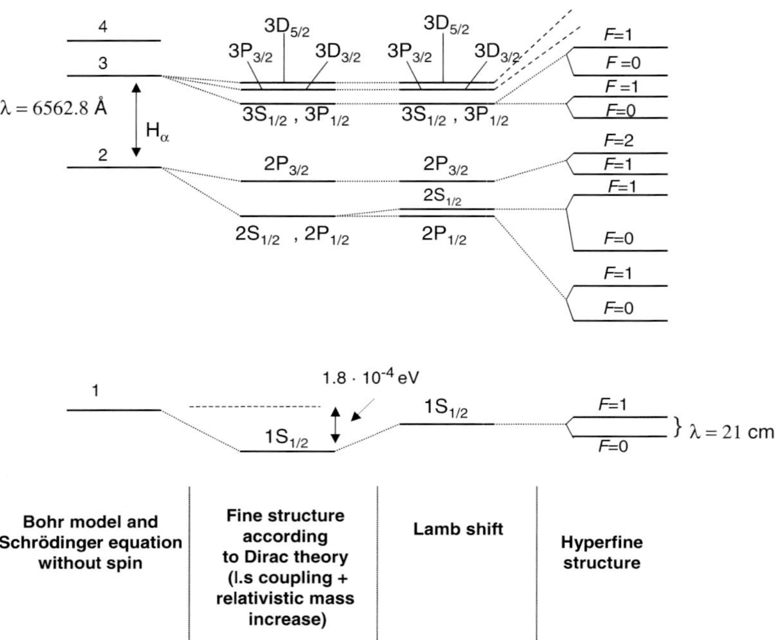

Fig. 5.6. Energy levels in Hydrogen including the effects contributing to its detailed structure. The scale is increased from left to right and some energy splittings are numerically reported to give an idea of the energy separations. The fine structure of the $n=2$ level is detailed in Figure 5.7.

The solution of the non-relativistic Schrodinger equation (§1.4) provided the eigenvalues $E_{n,l} = -R_H hc/n^2$. Then the spin-orbit Hamiltonian $\mathcal{H}_{so} = (e^2/2m^2c^2r^3)\mathbf{l.s}$ was introduced (§1.6). However we did not really discuss at that point the case of Hydrogen (where other relativistic effects are of comparable strength) dealing instead with heavier atoms (§2.2 and Chapter 3) where the most relevant contribution to the fine structure arises from $\mathcal{H}_{so}$. At §2.2 and Problem F.I.15 it was pointed out that a more refined relativistic description would imply a shift of the s-states (where $l = 0$ while at the same time a divergent behaviour for $r \to 0$ is related to the positional part of $\mathcal{H}_{so}$). Finally the hyperfine magnetic interaction was introduced (§5.2) where $\mathcal{H}_{hyp.} = a_j \mathbf{I.j}$, with $I = 1/2$, $j = l \pm 1/2$ and a_j given by Eqs. 5.12 and 5.13.

The simplest relativistic correction which could remove the accidental degeneracy in l was already deduced in the old quantum theory. As a consequence of the relativistic mass $m = m(v)$, for elliptical orbits in the Bohr model, **Sommerfeld** derived for the energy levels

$$E_{n,k} = -\frac{R_H hc}{n^2}\left[1 + \frac{\alpha^2}{n^2}\left(\frac{n}{k} - \frac{3}{4}\right) + ...\right]$$

where k is a second quantum number related to the quantization of the angular momentum $\int p_\theta d\theta = kh$ (θ polar angle)(see Problem I.4.4), while $\alpha = (e^2/\hbar c) \simeq 1/137$ is the **fine-structure constant**.

The **Dirac electrodynamical theory**, which includes spin-orbit interaction and classical relativistic effects (the relativistic kinetic energy being $c(p^2 + m^2 c^2)^{1/2} - mc^2 \simeq (p^2/2m) - (p^4/8m^3 c^2)$ (see Prob. F.I.15), provided the fine-structure eigenvalues

$$E_{n,j}^{fs} = -\frac{R_H hc\alpha^2}{n^3}\left[\frac{1}{j + 1/2} - \frac{3}{4n}\right] = E_n^0 \frac{\alpha^2}{n^2}\left[\frac{n}{j + 1/2} - \frac{3}{4}\right] \ ,$$

with the relevant findings that the quantum number j and not l is involved and the shift for the s-states is explicit. Accordingly, the ground state of Hydrogen atom is shifted by -1.8×10^{-4} eV and the $n = 2$ energy level is splitted in a doublet, the $p_{3/2}$ and $p_{1/2}$ states (this latter degenerate with $s_{1/2}$) being separated by an amount of 0.3652 cm^{-1}. The H$_\alpha$ line of the **Balmer series** (at 6562.8 Å) was then detected in the form of a doublet of two lines, since the Doppler broadening in optical spectroscopy prevented the observation of the detailed structure.

Giulotto and other spectroscopists, through painstaking measurements, noticed that the relativistic Dirac theory had to be modified and that a more refined description was required in order to account for the detailed structure of the H$_\alpha$ line. A few years later (1947) **Lamb**, by means of microwave spectroscopy (thus inducing magnetic dipole transitions between the levels) could directly observe the energy separation between terms at the same quantum number j. The energy difference between $2S$ and $2P$ states turned out 0.03528 cm^{-1} and the line had a fine structure of five lines, some of them broadened. Later on, by Doppler-free spectroscopy using dye lasers (**Hansch** et al., see Problem F.V. 13 for an example) the seven components of the H$_\alpha$ line consistent with the Lamb theory could be inferred. It was also realized that this result had to be generalized and the states with the same n and j quantum numbers, but different l, have different energy.

The **Lamb shift** (reported in detail in Fig. 5.7 for the $n = 2$ states) triggered the development of the **quantum electrodynamical theory**, which

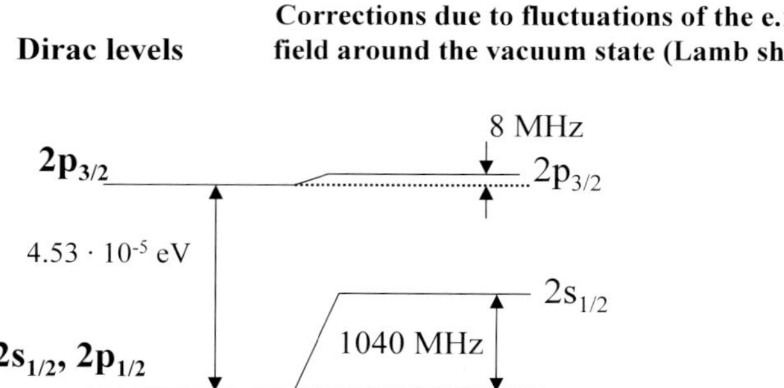

Fig. 5.7. Lamb shift for the $n=2$ levels in Hydrogen.

fully account for the fine structure of the levels on the basis of physical grounds that electrons are continuously emitting and adsorbing photons by transitions to virtual states. These states are poorly defined in energy due to their very short lifetimes. Qualitatively the Lamb shift can be considered the result of zero-point fluctuations of the set of harmonic oscillators describing the electromagnetic radiation field. These fluctuations induce analogous effects on the motion of the electron. Since the electric field in the atom is not uniform, the effective potential becomes different from the one probed by the electron in the average position.

The shift of the ground state due to the Lamb correction is about six times larger than the magnetic hyperfine splitting.

As regards the hyperfine splitting in the Hydrogen atom, at §5.2 it has been shown how the structure depicted in Figure 5.6 is originated.

Problems F.V

Problem F.V.1 The electric quadrupole moment of the deuteron is $Q = 2.8 \cdot 10^{-3}$ barn. By referring to an ellipsoid of uniform charge, evaluate the extent of departure of the nuclear charge distribution from the sphere. Assume for average nuclear radius $R_n \simeq 1.89 \cdot 10^{-13}$ cm.

Solution:

From

$$Q = \frac{1}{Ze}\rho \int_V (3z^2 - r^2)d\tau,$$

for the ellipsoid, defined by the equation $(x^2+y^2)/a^2+(z^2/b^2) = 1$, one obtains $Q = (2/5)(b^2 - a^2)$. If the average nuclear radius is taken to be $R_n^3 = a^2 b$ (the volume of the ellipsoid is $\frac{4}{3}\pi a^2 b$), with $R_n + \delta R_n = b$, then, for $\delta R_n \ll R_n$

$$a^2 = \frac{R_n^3}{R_n + \delta R_n} = \frac{R_n^2}{1 + \frac{\delta R_n}{R_n}} \approx R_n^2 \left(1 - \frac{\delta R_n}{R_n}\right)$$

and

$$b^2 - a^2 \approx R_n^2 \left[1 + 2\left(\frac{\delta R_n}{R_n}\right) + \left(\frac{\delta R_n}{R_n}\right)^2\right] - R_n^2 \left(1 - \frac{\delta R_n}{R_n}\right)$$

$$= R_n^2 \left[3\left(\frac{\delta R_n}{R_n}\right) + \left(\frac{\delta R_n}{R_n}\right)^2\right] \approx 3R_n^2 \left(\frac{\delta R_n}{R_n}\right).$$

Hence

$$Q = \frac{6}{5}R_n^2 \left(\frac{\delta R_n}{R_n}\right)$$

corresponding to $\left(\frac{\delta R_n}{R_n}\right) \approx 6.5 \cdot 10^{-2}$.

Problem F.V.2 The D_2 line of the Na doublet (see Fig. 5.3) displays an hyperfine structure in form of triplet, with separation between pairs of adjacent lines in the ratio not far from 1.5. Justify this experimental finding from the hyperfine structure of the energy levels and the selection rules (see Prob. F.V.13 for some detail on the experimental method).

Solution:

From the splittings in Fig. 5.3 and the selection rule $\Delta F = 0, \pm 1$ one can deduce that the hyperfine spectrum consists of three lines $\nu_{\{3,2,1\}\leftrightarrow 2}$ corresponding to the transitions $^2P_{\frac{3}{2}}(F = 3, 2, 1) \leftrightarrow {}^2S_{\frac{1}{2}}(F = 2)$ and of three lines

$\nu_{\{2,1,0\}\leftrightarrow 1}$ corresponding to the transitions $^2P_{\frac{3}{2}}(F = 2,1,0) \leftrightarrow \, ^2S_{\frac{1}{2}}(F = 1)$.
From the interval rule

$$\frac{\nu_{3,2} - \nu_{2,2}}{\nu_{2,2} - \nu_{1,2}} = \frac{3}{2}$$

$$\frac{\nu_{2,1} - \nu_{1,1}}{\nu_{1,1} - \nu_{0,1}} = 2.$$

The lines in Figure correspond to the transitions $^2P_{\frac{3}{2}}(F = 3,2,1) \leftrightarrow \, ^2S_{\frac{1}{2}}(F = 2)$.

Problem F.V.3 Plot the magnetic hyperfine levels for an atom in the state $^2S_{1/2}$ and nuclear spin $I = 1$. Then derive the corrections due to a magnetic field, in the **weak** and **strong field regimes** (with respect to the hyperfine energy). Classify the states in the two cases and draw a qualitative correlation between them.

Solution:

In the weak-field regime the hyperfine correction is

$$\Delta E = g_F \, \mu_B \, H_0 \, m_F$$

with

$$g_F = g_J \frac{F(F+1) + J(J+1) - I(I+1)}{2F(F+1)} - g_I \frac{\mu_N}{\mu_B} \frac{F(F+1) + I(I+1) - J(J+1)}{2F(F+1)}$$

where the second term can be neglected ($\mu_N \ll \mu_B$).

A relatively small field breaks up the **IJ** coupling. The hyperfine Zeeman effect then is replaced by the hyperfine Paschen-Back effect. The oscillating components in the x and y directions average to zero and the final result is that the nuclear angular momentum vector **I** is oriented along $\mathbf{H}_0$. The quantum number F is no longer defined while the quantum numbers m_I and m_J describe **I** and **J**. This splitting of the energy involves three terms, one being $g_J \, \mu_B \, H_0 \, m_J$, already considered in the Zeeman effect (§4.3.2), the other is $am_I m_J$ and the third one, $-\mu_N g_I m_I H_0$ is negligible.

See Figure:

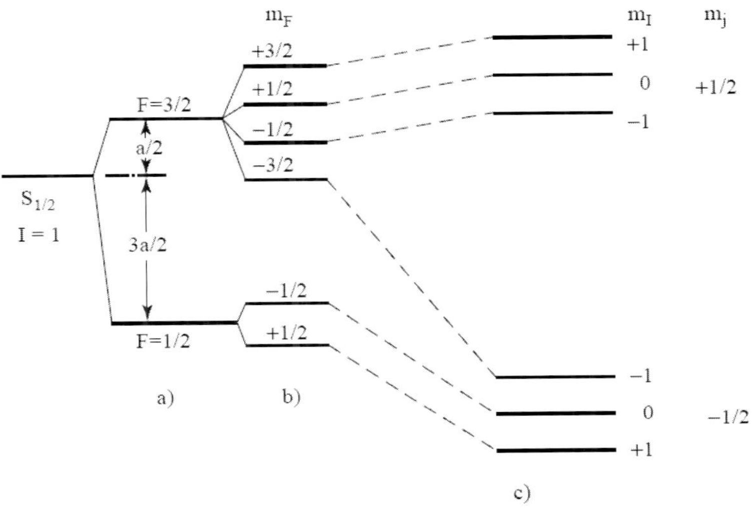

Fig. 5.8. Hyperfine structure of the $S_{1/2}$ state with $I = 1$: a) **in zero field**; b) **in weak field**, Zeeman regime; c) **in strong field**, Paschen-Back regime.

Problem F.V.4 In the Na atom the hyperfine interaction for the P state is much smaller than the one in the S ground state. In poor resolution the hyperfine structure is observed in the form of a doublet, with relative intensities 5 and 3. From this observation derive the nuclear spin (see also Prob. F.V.13).

Solution:
From

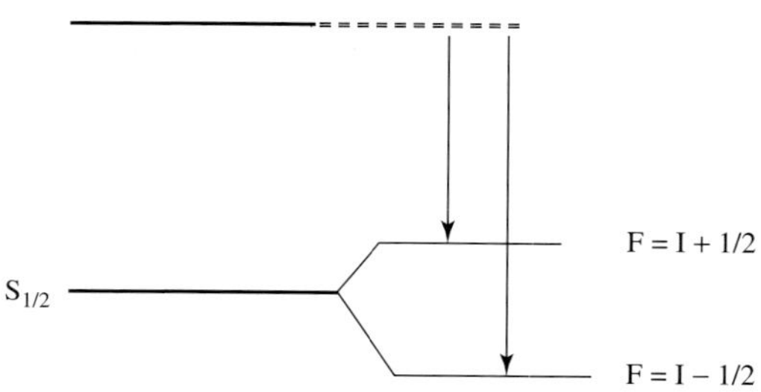

The intensity being $\propto (2F+1)$ and the ratio $(I+1)/I = 5/3$, then $I = 3/2$.

Problem F.V.5 For a solid ideally formed by a mole of non-interacting deuterons in an electric field gradient, derive the contributions to the entropy and to the specific heat, in the high temperature limit.

Solution:
The quadrupolar interaction $e^2qQ[3M^2 - I(I+1)]/4$ yields two energy levels, one doubly degenerate ($M = 0, \pm 1$).
By indicating with ϵ the separation between the levels, the partition function is

$$Z(\beta) = (1 + 2e^{-\beta\epsilon})^{N_A}, \ \text{with} \ \beta = 1/k_BT.$$

From the free energy

$$F(T) = -k_BT \ln Z = -RT \ln(1 + 2e^{-\epsilon/k_BT}) \ ,$$

the entropy turns out

$$S = -\frac{\partial F}{\partial T} = R\ln(1 + 2e^{-\epsilon/k_BT}) + \frac{2N_A\epsilon}{T}\frac{e^{-\epsilon/k_BT}}{1 + 2e^{-\epsilon/k_BT}}.$$

The internal energy is

$$U = 2N_A\epsilon\frac{1}{e^{\epsilon/k_BT} + 2}.$$

In the high temperature limit

$$U \simeq \frac{2}{3}N_A\epsilon\left(1 - \frac{\epsilon}{3k_BT}\right),$$

so that

$$C = \frac{\partial U}{\partial T} = \frac{2}{9} R \left(\frac{\epsilon}{k_B T} \right)^2 \propto T^{-2},$$

namely the high-temperature tail of a **Schottky anomaly** (a "bump" in the specific heat vs temperature), typical of two-levels systems.

Problem F.V.6 Consider the Hydrogen atom, in the ground state, in a magnetic field H_0 and write the magnetic Hamiltonian including the hyperfine interaction. First derive the eigenvalues and the spin eigenvectors in the limit $H_0 \to 0$ and estimate the frequencies of the transitions induced by an oscillating magnetic field perpendicular to the quantization axis.

Then derive the correction to the eigenvalues due to a weak magnetic field.

Finally consider the opposite limit of strong external magnetic field. Draw the energy levels with the appropriate quantum numbers, again indicating the possibility of inducing magnetic dipole transitions between the hyperfine levels (this is essentially the **EPR** experiment, see for details Chapter 6) and from the resulting lines show how the hyperfine constant can be extracted.

Figure out a schematic correlation diagram connecting the eigenvalues for variable external field.

Solution:
From the Hamiltonian

$$\mathcal{H}_s = 2\mu_B \mathbf{S} \cdot \mathbf{H}_0 - \gamma \hbar \mathbf{I} \cdot \mathbf{H}_0 + a\mathbf{I} \cdot \mathbf{S}$$

(γ nuclear gyromagnetic ratio, a hyperfine interaction constant, with $a = hc/\lambda$ and $\lambda = 21$ cm), for $H_0 \to 0$ the eigenstates are classified by S, I, F and M_F and for $I = S = 1/2$ two magnetic hyperfine states, $F = 0$ and $F = 1$, occur. Then $E_{\frac{1}{2},\frac{1}{2},1} = a/4$ and $E_{\frac{1}{2},\frac{1}{2},0} = -3/4a$.

The spin eigenvectors are the same of any two spins system, i.e.

$$\left. \begin{array}{l} |\alpha_e \alpha_p > \\ |\beta_e\ \beta_p > \\ \frac{1}{\sqrt{2}}|\alpha_e\beta_p + \alpha_p\beta_e > \end{array} \right\} \text{ defining the triplet } T_{1,1}\ T_{1,0}\ T_{1,-1}$$

and

$$\frac{1}{\sqrt{2}}|\alpha_e\beta_p - \alpha_p\beta_e > \text{ defining the singlet } S_{0,0}.$$

The magnetic perturbation implies an operator of the form $\mu_x = 2\mu_B S_x - \gamma\hbar I_x$ and the matrix elements involving the triplet and singlet states turn out

$$< 1,1|\mu_x|1,0 >= \frac{1}{2}\frac{1}{\sqrt{2}}(g\mu_B - \gamma\hbar)$$

$$< 1,1|\mu_x|0,0 >= \frac{1}{2}\frac{1}{\sqrt{2}}(-g\mu_B - \gamma\hbar)$$

$$< 1,0|\mu_x|1,-1 > = \frac{1}{2}\frac{1}{\sqrt{2}}(g\mu_B - \gamma\hbar)$$

$$< 0,0|\mu_x|1,-1 > = \frac{1}{2}\frac{1}{\sqrt{2}}(g\mu_B + \gamma\hbar)$$

$$< 1,0|\mu_x|0,0 > = < 1,1|\mu_x|1,-1 > = 0.$$

Therefore the allowed transitions are $S \to T_1$ and $S \to T_{-1}$ corresponding to the transition frequency $\nu = \frac{a}{h} = 1420$ MHz (and formally $T_{-1} \to T_0$, $T_0 \to T_{+1}$ at $\nu = 0$).

For weak field H_0, neglecting the interaction with the proton magnetic moment and considering that the perturbation acts on the basis where F^2, F_z, I^2 and S^2 are diagonal, the matrix for $\mathcal{H} = a\mathbf{I} \cdot \mathbf{S} + 2\mu_B \mathbf{S} \cdot \mathbf{H}_0$ is

$$\begin{pmatrix} a/4 + \mu_B H_0 & 0 & 0 & 0 \\ 0 & a/4 & 0 & -\mu_B H_0 \\ 0 & 0 & a/4 - \mu_B H_0 & 0 \\ 0 & -\mu_B H_0 & 0 & -3a/4 \end{pmatrix}$$

From the secular equation the eigenvalues are found by solving

$$\frac{a}{4} + \mu_B H_0 - E = 0 \qquad \frac{a}{4} - \mu_B H_0 - E = 0$$

$$(\frac{a}{4} - E)(-\frac{3a}{4} - E) - \mu_B^2 H_0^2 = 0$$

yielding $E_{1,2} = a/4 \pm \mu_B H_0$ and $E_{3,4} = -a/4 \pm (a/2)[1 + 4\mu_B^2 H_0^2/a^2]^{1/2}$, namely the **Breit-Rabi diagram** reported below:

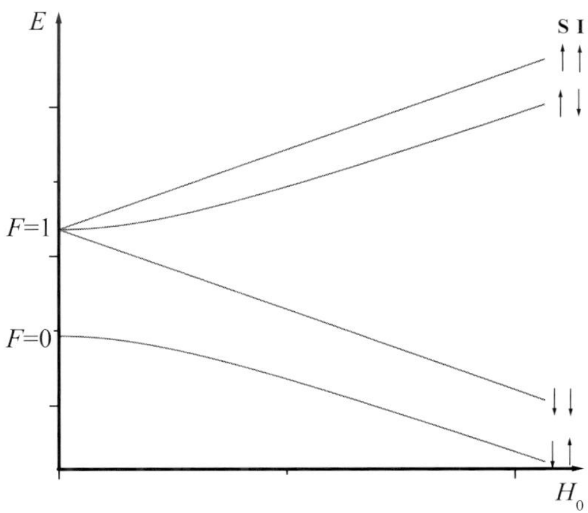

In the strong field regime the eigenvalues are the ones for S_z, $I_z\, S_z$ and I_z:

$$E = 2\mu_B \mathrm{H}_0 m_S + a m_S M_I - \gamma\hbar M_I \mathrm{H}_0$$

The first term is dominant and the diagram is

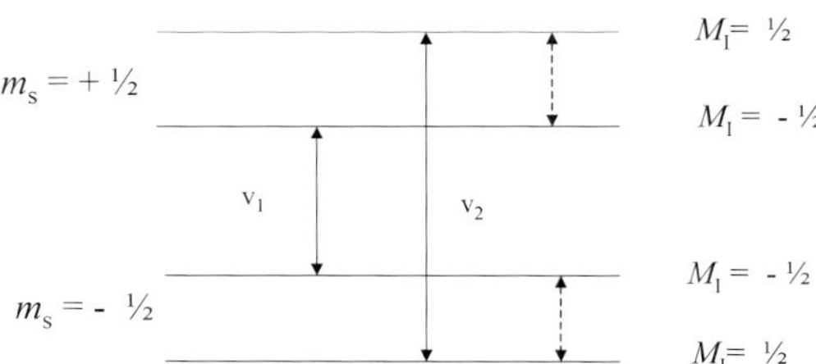

with the electronic transitions $\Delta m_S = \pm 1$ (and $\Delta M_I = 0$) at the frequencies

$$\nu_{1,2} = \frac{2\mu_B \mathrm{H}_0 \pm a/2}{h}.$$

The nuclear transitions $\Delta M_I = \pm 1$ (and $\Delta m_S = 0$) occur at $a/2h$.

Since the internal field due to the electron is usually much larger than H_0 the third term can be neglected (see the Figure below).

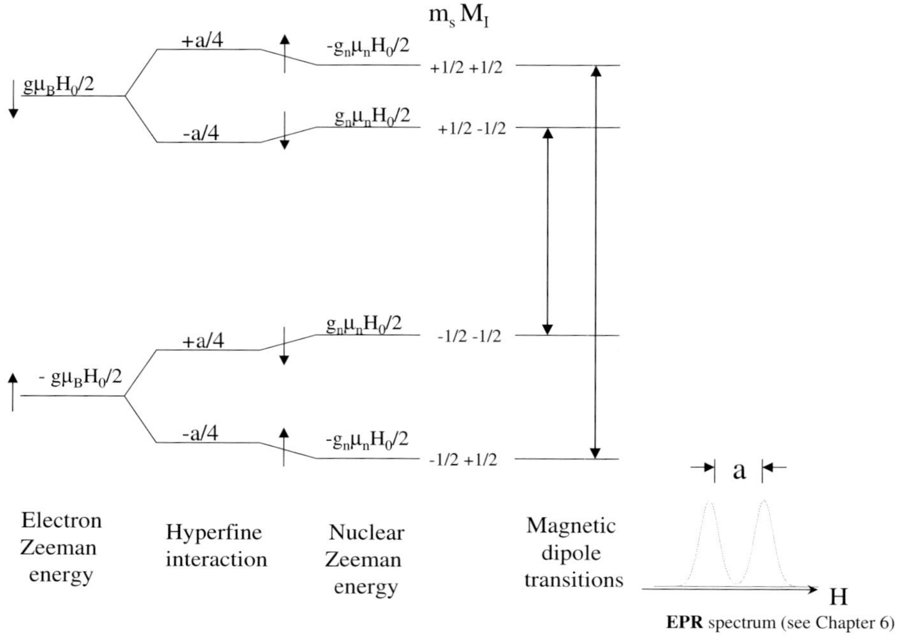

Problem F.V.7 The ^{209}Bi atom has an excited $^2D_{5/2}$ state, with 6 sublevels due to hyperfine interaction. The separations between the hyperfine levels are 0.23, 0.31, 0.39, 0.47 and 0.55 cm^{-1}. Evaluate the nuclear spin and the hyperfine constant.

Solution:

From $E(F, I, J) = (a/2)\left[F(F+1) - I(I+1) - J(J+1)\right]$ and
$E_{F+1} - E_F = a(F+1)$, one finds $a = 0.08$ cm^{-1} and $F_{\max} = 7$.
Therefore $F = 2, 3, 4, 5, 6, 7$ and since $J = 5/2$ the nuclear spin must be $I = \frac{9}{2}$.

Problem F.V.8 A proton and an anti-proton, at a given distance d, interact through the magnetic dipole-dipole interaction. Derive the total spin eigenstates and eigenvalues in term of the proton magnetic moment(it is reminded that the magnetic moment of the antiproton is the same of the proton, with **negative** gyromagnetic ratio).

Solution:

From

$$\mathcal{H} = \frac{\boldsymbol{\mu}_1 \cdot \boldsymbol{\mu}_2}{r^3} - 3\frac{(\boldsymbol{\mu}_1 \cdot \mathbf{r})(\boldsymbol{\mu}_2 \cdot \mathbf{r})}{r^5}.$$

with $\boldsymbol{\mu}_1 = 2\mu_p\mathbf{s}_1$ and $\boldsymbol{\mu}_2 = -2\mu_p\mathbf{s}_2$, by choosing the z axis along $\mathbf{r}$

$$\mathcal{H} = -4\,\frac{\mu_p^2}{d^3}\,\mathbf{s}_1 \cdot \mathbf{s}_2 + 12\,\frac{\mu_p^2}{d^3}\,s_1^z s_2^z.$$

Since $\mathbf{s}_1 \cdot \mathbf{s}_2 = S(S+1)/2 - 3/4$ and $s_1^z s_2^z = (1/2)M_S^2 - (1/2)\cdot(1/2)$, one finds

Eigenstates S M_S Energies

singlet   0   0   0

triplet   1 $\begin{array}{c} 1 \\ 0 \\ -1 \end{array}$ $\begin{array}{c} 2\mu_p^2/d^3 \\ -4\mu_p^2/d^3 \\ 2\mu_p^2/d^3 \end{array}$

i. e.

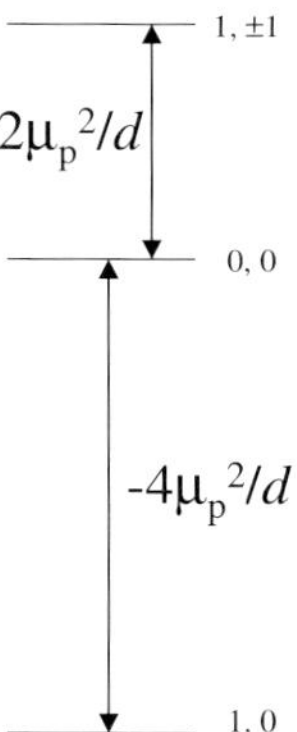

Problem F.V.9 Two electrons interact through the dipolar Hamiltonian. A strong magnetic field is applied along the z-direction, at an angle θ with the line connecting the two electrons. Find the eigenvalues and the corresponding eigenfunctions for the two spins system, in terms of the basis functions $\alpha_{1,2}$ and $\beta_{1,2}$.

Solution:

In the light of Eq. 5.8 for the dipolar field (see also Problem V.2.2) the total Hamiltonian is

$$\mathcal{H} = \mathcal{H}_0 + \mathcal{H}_d$$

$$= 2\mu_B H_0 (s_z^{(1)} + s_z^{(2)}) + \frac{4\mu_B^2}{r^3}\left\{ \mathbf{s}_1 \cdot \mathbf{s}_2 - \frac{3}{r^2}[(\mathbf{s}_1 \cdot \mathbf{r})(\mathbf{s}_2 \cdot \mathbf{r})] \right\}.$$

In order to evaluate the matrix elements it is convenient to write the perturbation Hamiltonian in the form (called **dipolar alphabet**)

$$\mathcal{H}_d = \frac{4\mu_B^2}{r^3}[A + B + C + D + E + F]$$

where

$$A = s_z^{(1)}s_z^{(2)}[1 - 3\cos^2\theta], \quad B = -\frac{1}{4}\left[s_+^{(1)}s_-^{(2)} + s_-^{(1)}s_+^{(2)}\right](1 - 3\cos^2\theta),$$

θ angle between $\mathbf{H}_0$ and $\mathbf{r}$. The terms C, D, E and F involve operators of the form $s_+^{(1)}s_z^{(2)}$, $s_-^{(1)}s_z^{(2)}$, $s_+^{(1)}s_+^{(2)}$, $s_-^{(1)}s_-^{(2)}$ and can be neglected. In fact these terms are off-diagonal and produce admixtures of the zero-order states to an amount of the order of $\left(\mu_B/r^3\right)/H_0$.

Thus the dipolar Hamiltonian is written in the form

$$\mathcal{H}_d = \underbrace{\frac{4\mu_B^2}{r^3}(1 - 3\cos^2\theta)}_{\mathcal{A}}\left[s_z^{(1)}s_z^{(2)} - \frac{1}{4}\left(s_+^{(1)}s_-^{(2)} + s_-^{(1)}s_+^{(2)}\right)\right],$$

most commonly used.

The complete set of the basis functions is $\alpha_1\alpha_2$, $\alpha_1\beta_2$, $\alpha_2\beta_1$ and $\beta_1\beta_2$ and the matrix elements are

$$< \alpha\alpha|\mathcal{H}_T|\alpha\alpha > = 2\,\mu_B\,H_0 + \frac{1}{4}\mathcal{A}$$

$$< \alpha\beta|\mathcal{H}_T|\alpha\beta > = < \beta\alpha|\mathcal{H}_T|\beta\alpha > = -\frac{\mathcal{A}}{4}$$

$$< \alpha\beta|\mathcal{H}_T|\beta\alpha > = < \beta\alpha|\mathcal{H}_T|\alpha\beta > = -\frac{\mathcal{A}}{4}$$

$$< \beta\beta|\mathcal{H}_T|\beta\beta > = -2\mu_B H_0 + \mathcal{A}/4$$

It is noted that while the term A is completely diagonal, the term B only connects $|m_s^{(1)}m_s^{(2)} >$ to states $< m_s^{(1)} + 1, m_s^{(2)} - 1|$ or $< m_s^{(1)} - 1, m_s^{(2)} + 1|$. B simultaneously flips one spin up and the other down.

The secular equation is

$$\begin{vmatrix} (+2\mu_B H_0 + \frac{\mathcal{A}}{4}) - E & 0 & 0 & 0 \\ 0 & -\frac{\mathcal{A}}{4} - E & -\frac{\mathcal{A}}{4} & 0 \\ 0 & -\frac{\mathcal{A}}{4} & -\frac{\mathcal{A}}{4} - E & 0 \\ 0 & 0 & 0 & (-2\mu_B H_0 + \frac{\mathcal{A}}{4}) - E \end{vmatrix} = 0.$$

and the eigenvalues turn out

$$E_1 = -2\mu_B \left(H_0 - \frac{\mu_B}{2r^3}(1 - 3\cos^2\theta) \right)$$

$$E_2 = 0$$

$$E_3 = -\frac{2\mu_B^2}{r^3}(1 - 3\cos^2\theta)$$

$$E_4 = +2\mu_B \left(H_0 - \frac{\mu_B}{2r^3}(1 - 3\cos^2\theta) \right).$$

The correspondent eigenfunctions being $\alpha_1\alpha_2$, $\beta_1\beta_2$ and $\frac{1}{\sqrt{2}}[\alpha_1\beta_2 \pm \alpha_2\beta_1]$, as expected.

Problem F.V.10 In the Ba atom the line due to the transition from the $6s\,6p\ J = 1$ to the $(6s)^2$ ground state in high resolution is evidenced as a triplet, with line intensities in the ratio 1, 2 and 3. Evaluate the nuclear spin.

Solution:

Since $\mathbf{F} = \mathbf{I} + \mathbf{J}$

$$\text{for } J = 0 \text{ one has } I = F \implies \text{ no splitting}$$

$$\text{for } J = 1 \implies \text{ splitting in } (2I + 1) \text{ or in } (2J + 1) \text{ terms.}$$

$$I = 0 \implies \text{ no splitting,}$$

$$\text{for } I = \frac{1}{2} \text{ and } J = 1\ \Delta F = 0, \pm 1 \implies \text{ two lines}$$

$$I = 1 \text{ or } I > 1 \implies \text{ three lines.}$$

Looking at the intensities, proportional to $e^{-E/k_B T}(2F + 1)$, where the energy E is about the same

$$\text{for } I = 1\ F = 0, 1, 2 \implies \text{ intensities: } 1, 3, 5$$

$$\text{for } I = \frac{3}{2}\ F = \frac{1}{2}, \frac{3}{2}, \frac{5}{2} \implies \text{ intensities: } 2, 4, 6.$$

Therefore $I = \frac{3}{2}$.

Problem F.V.11 In the assumption that in a metal the magnetic field on the electron due to the hyperfine interaction with $I = 1/2$ nuclei is $H_z = (a/N)\Sigma_n I_n^z$ (a constant and same population on the two states) prove that the odd moments of the distribution are zero and evaluate $< H_z^2 >$. Then evaluate $< H_z^4 >$ and show that for large N the distribution tends to be Gaussian, the width going to zero for $N \to \infty$.

Solution:

$\langle \hat{H}_z^{2n+1} \rangle = 0$ for symmetry. Since $(I_n^z)^2 = \frac{1}{4}$

$$\left\langle \left(\sum_n I_n^z \right)^2 \right\rangle = \sum_n \langle (I_n^z)^2 \rangle + \sum_{n \neq m} \langle I_n^z I_m^z \rangle = \frac{1}{4} N \ ,$$

and then $\langle H_z^2 \rangle = (\frac{a}{2N})^2 N$.

$$\langle H_z^4 \rangle = \left(\frac{a}{N} \right)^4 \sum_{i,j,k,l} \langle I_i^z I_j^z I_k^z I_l^z \rangle = \left(\frac{a}{N} \right)^4 3 \sum_{i,j} \langle (I_i^z)^2 (I_j^z)^2 \rangle - \left(\frac{a}{N} \right)^4 \sum_i \langle (I_i^z)^4 \rangle$$

$$= \left(\frac{a}{2N} \right)^4 (3N^2 - N).$$

In the thermodynamical limit one has $\langle H_z^4 \rangle \simeq 3 \left(\frac{a}{2N} \right)^4 N^2$: the first two even moments correspond to the Gaussian moments.

Problem F.V.12 Evaluate the transition probability from the state $M = -1/2$ to $M = +1/2$ by spontaneous emission, for a proton in a magnetic field of 7500 Oe.

Solution:

From the expression for A_{21} derived in App. I.3 and extending it to magnetic dipole transitions, one can write

$$A_{21} = \frac{4\omega_L^3}{3c^3 \hbar} | < 2|\boldsymbol{\mu}|1 > |^2$$

$$= \frac{4\omega_L^3}{3c^3 \hbar} \left\{ \left| < \frac{1}{2}|\mu_x| - \frac{1}{2} > \right|^2 + \left| < \frac{1}{2}|\mu_y| - \frac{1}{2} > \right|^2 \right\}$$

with $\boldsymbol{\mu} = \gamma \hbar \mathbf{I}$. From $I_{\pm} = I_x \pm i I_y$ one derives $A_{21} = (2/3)(\gamma^2 \hbar / c^3) \omega_L^3$ and for $\gamma = 42.576 \cdot 2\pi \cdot 10^2$ Hz/Gauss, $\omega_L = \gamma H_0 = 2\pi \cdot 31.9$ MHz, yielding

$$A_{21} \simeq 1.5 \times 10^{-25} \, s^{-1}.$$

Problem F.V.13 High-resolution laser spectroscopy allows one to evidence the hyperfine structure in the optical lines with almost total elimination of the **Doppler broadening**.

The Figure below

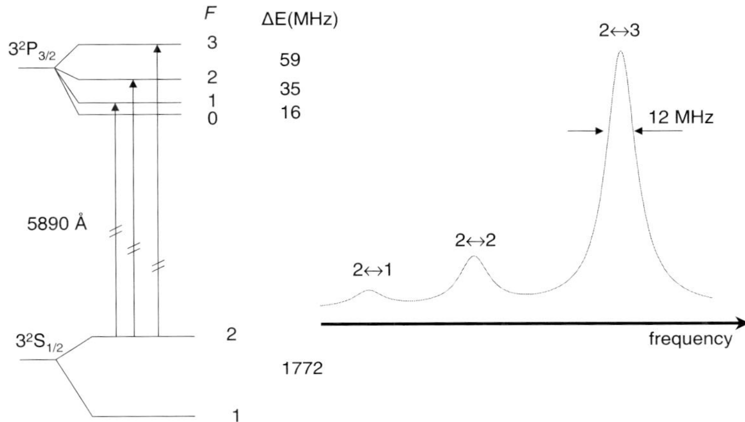

shows the hyperfine structure of the $^2S_{1/2} - ^2P_{3/2}$ D_2 line of Na at 5890 Å. (This spectrum is obtained by irradiating a collimated beam of sodium atoms at right angles by means of a narrow-band single-mode laser and detecting the fluorescent light after the excitation. This and other high resolution spectroscopic techniques are nicely described in the book by **Svanberg**, quoted in the Preface).

From the Figure, discuss how the magnetic and electric hyperfine constants could be derived and estimate the life-time of the $^2P_{3/2}$ state (in the assumption that is the only source of broadening).

Then compare the estimated value of the life time with the one known (from other experiments), $\tau = 1.6$ ns. In the assumption that the extra-broadening is entirely due to Doppler **second-order relativistic shift**, quadratic in (v/c) and independent from the direction of motion, estimate the temperature of the oven from which the thermal atomic beam is emerging, discussing the expected order of magnitude of the broadening (see Problem F.I.7).

Solution:
For the ground-state $^2S_{1/2}$, $I = 3/2$ and $J = S = 1/2$, the quadrupole interaction being zero, from Eq. 5.15 the separation between the $F = 2$ and $F = 1$ states directly yields the magnetic hyperfine constant $a = \Delta_{1,2}/(F+1) = 886$ MHz, corresponding to an effective magnetic field of about 45 Tesla (see Table V.2.1).

The sequence of the hyperfine levels for the $^2P_{3/2}$ state does not follow exactly the interval rule. In the light of Eq. 5.20 an estimate of the quadrupole coupling constant b can be derived (approximate, the correction being of the order of the intrinsic line-widths).

In the assumption that the broadening (12 MHz) is due only to the life-time one would have $\tau = 1/2\pi\Delta\nu \simeq 13.3 \times 10^{-9}$ s, a value close to the one ($\tau \simeq 16 \times 10^{-9}$s) pertaining to the $3^2P_{3/2}$ state ($\Delta\nu \simeq 10$ MHz).

The most probable velocity of the beam emerging from the oven is $v = \sqrt{3k_BT/M_{Na}}$, that for $T \simeq 500$ K corresponds to about 7×10^4 cm/s.

While the first-order Doppler broadening is in the range of a few GHz, scaling by a term of the order of v/c leads to an estimate of the **second-order Doppler broadening** in the kHz range. Thus the extra-broadening of a few MHz is likely to be due to the **residual first-order broadening** (for a collimator ratio of the beam around 100 being typically around some MHz).

Problem F.V.14 From the perturbation generated by nuclear magnetic moment on the electron, derive the effective magnetic field in the hyperfine Hamiltonian $\mathcal{H}_{hyp} = -\boldsymbol{\mu}_I.\mathbf{h}_{eff}$ (Eq. 5.6).

Solution:

From the vector potential (see Fig. 5.1 and Eq. 5.4) the magnetic Hamiltonian for the electron is

$$\mathcal{H}_{hyp} = 2\mu_B \frac{\mathbf{l}.\boldsymbol{\mu}_I}{r^3} + 2\mu_B \mathbf{s}.\boldsymbol{\nabla} \times [-\boldsymbol{\nabla} \times \frac{\boldsymbol{\mu}_I}{r}]$$

Since

$$\boldsymbol{\nabla} \times [-\boldsymbol{\nabla} \times \frac{\boldsymbol{\mu}_I}{r}] = -g_n M_n \{ \frac{\mathbf{I}}{r^3} - \frac{3(\mathbf{I}.\mathbf{r})\mathbf{r}}{r^5} \} + g_n M_n \mathbf{I} div(\frac{\mathbf{r}}{r^3}) \ ,$$

while $div(\mathbf{r}/r^3) = 4\pi\delta(\mathbf{r})$, one writes

$$\mathcal{H}_{hyp} = 2\mu_B g_n M_n \frac{\mathbf{I}.\mathbf{l}}{r^3} - 2\mu_B g_n M_n \{ \frac{\mathbf{s}.\mathbf{I}}{r^3} - \frac{3(\mathbf{I}.\mathbf{r})(\mathbf{s}.\mathbf{r})}{r^5} \} + 2\mu_B g_n M_n (\mathbf{s}.\mathbf{I})4\pi\delta(\mathbf{r}) \equiv$$

$$\equiv A + B + C,$$

To deal with the singularities at the origin involved in B and C, let us define with V_ε a little sphere of radius ε centered at $r = 0$. Then in the integral for the expectation values

$$I = \int_{V_\varepsilon} B\phi^*(\mathbf{r})\phi(\mathbf{r})d\tau \equiv \int_{V_\varepsilon} Bf(\mathbf{r})d\tau$$

one can expand $f(\mathbf{r})$ in Taylor series, within the volume V_ε

$$f(\mathbf{r}) = f(0) + \mathbf{r}.\boldsymbol{\nabla} f(\mathbf{r}) + \frac{1}{2}(\mathbf{r}.\boldsymbol{\nabla})(\mathbf{r}.\boldsymbol{\nabla})f(\mathbf{r})$$

In I there are two types of terms, one of the form

$$s_x I_x \frac{\partial^2}{\partial x^2}\left(\frac{1}{r}\right) \quad (a)$$

the other of the form

$$(s_x I_y + s_y I_x)\frac{\partial^2}{\partial x \partial y}\left(\frac{1}{r}\right) \quad (b)$$

In the expansion $(\mathbf{r}.\boldsymbol{\nabla} f)$ is odd while (a) terms are even, thus yielding zero in I. The product of (a) terms with the third term in $f(\mathbf{r})$ when even, contributes with a term quadratic in ε.

The terms of type (b) are odd in the two variables, while $\mathbf{r}.\boldsymbol{\nabla} f$ includes odd terms in a single variable. In the same way are odd (and do not give contribution) the terms $(b) f(0)$. Finally the terms (b) times the third term in the expression again contribute to I only to the second order in ε. Therefore, one can limit I only to

$$I = 2g_n M_n \mu_B f(0)\frac{1}{3}\int_{V_\varepsilon} (\mathbf{s}.\mathbf{I})\nabla^2\left(\frac{1}{r}\right)d\tau.$$

Since $\nabla^2(1/r) = -4\pi\delta(\mathbf{r})$ the magnetic hyperfine hamiltonian can be rewritten [2]

$$\mathcal{H}_{hyp} = 2\mu_B g_n M_n \frac{\mathbf{I}.\mathbf{l}}{r^3} - 2\mu_B g_n M_n\left[\frac{\mathbf{s}.\mathbf{I}}{r^3} - \frac{3(\mathbf{s}.\mathbf{r})(\mathbf{I}.\mathbf{r})}{r^5}\right]^* + \frac{16\pi}{3}\mu_B g_n M_n \mathbf{s}.\mathbf{I}\delta(\mathbf{r})$$

Thus the effective field $\mathbf{h}_{eff}$ in the form given in Eq. 5.6 is justified.

A model which allows one to derive similar results for the dipolar and the contact terms is to consider the nucleus as a small sphere with a uniform magnetization $\mathbf{M}$, namely a magnetic moment $\boldsymbol{\mu}_n = (4\pi R^3/3)\mathbf{M}$. For $r > R$ the magnetic field is the one of a point magnetic dipole. Inside the sphere $\mathbf{H}_{int} = (8\pi/3)\mathbf{M}$. By taking the limit $R \to 0$, keeping μ_n constant and then assuming that $\mathbf{M} \to \infty$, so that $\int_{r<R} H_{int} d\mathbf{r}_n = 8\pi\mu_n/3$, the complete expression of the field turns out

$$\mathbf{H} = -\frac{\boldsymbol{\mu}_n}{r^3} + 3\frac{(\boldsymbol{\mu}_n.\mathbf{r})\mathbf{r}}{r^5} + \frac{8\pi}{3}\mu_n\delta(\mathbf{r}).$$

Problem F.V.15 From the energy of the nuclear charge distribution in the electric potential due to the electron (Eq. 5.16) derive the hyperfine quadrupole Hamiltonian (Eq. 5.20).

---

[2] The star in the following equation means that in the expectation value a small sphere at the origin can be excluded in the integration and then ε set to zero. All singularities are included in the contact term.

Solution:

By starting from Eq. 5.18 a new tensor Q_{ij} so that $\sum_l Q_{ll} = 0$ is defined

$$Q_{ij} = 3Q'_{ij} - \delta_{ij} \sum_l Q'_{ll}$$

and in terms of Q'_{ij} one has

$$E_Q = \frac{1}{6} \sum_{ij} Q_{ij} V_{ij} + \frac{1}{6} \sum_l Q'_{ll} \sum_j V_{jj}$$

The second term can be neglected since $\sum_j V_{jj} \simeq 0$. Thus

$$\mathcal{H}^Q_{hyp} = \sum_{ij} \hat{Q}_{ij} \frac{\hat{V}_{ij}}{6}$$

where the operators are

$$\hat{Q}_{ij} = e \sum_n (3x_i^n x_j^n - \delta_{ij} r_n^2)$$

$$\hat{V}_{ij} = -e \sum_e \frac{(3x_i x_j - \delta_{ij} r_e^2)}{r_e^5}$$

This Hamiltonian can be simplified by expressing the five independent components of Q_{ij} in terms of one. Semiclassically this simplification originates from the precession of the nuclear charges around $\mathbf{I}$, yielding a charge distribution with cylindrical symmetry around the z direction of the nuclear spin.

Then

$Q_{ij} = 0$ for $i \neq j$ and being $\sum_l Q_{ll} = 0$, one has $Q_{11} = Q_{22} = -Q_{33}/2$ with $Q_{33} = \int \rho_n(\mathbf{r})(3z^2 - r^2)d\tau_n$.

In the quantum description the reduction of $\mathcal{H}^Q_{hyp}$ is obtained by considering that only the dependence from the orientation is relevant. Thus, for the matrix elements $< I, M'_I | \hat{Q}_{ij} | I, M_I >$ (other quantum numbers for the nuclear state being irrelevant), by using **Wigner-Eckart theorem** one writes

$$< I, M'_I | \hat{Q}_{ij} | I, M_I >= C < I, M'_I | \frac{3}{2}(I_i I_j + I_j I_i) - \delta_{ij} I^2 | I, M_I > .$$

By defining, in analogy to the classical description, the quadrupole moment Q in proton charge units as

$$Q =< II| \frac{\hat{Q}_{zz}}{e} |II >\equiv< II| \sum_n (3z_n^2 - r_n^2)|II >$$

the constant C is obtained:

$$C < II|3I_z^2 - I^2|II >= C[3I^2 - I(I+1)] = eQ$$

Therefore all the components Q_{ij} are expressed in terms of Q, which has the classical physical meaning (see Eq. 5.2 and Eq. 5.22). Then the quadrupole operator is

$$\hat{Q}_{ij} = \frac{eQ}{I(2I-1)}\{\frac{3}{2}(I_iI_j + I_jI_i) - \delta_{ij}I^2\}$$

Analogous procedure can be carried out for the electric field gradient operator:

$$\hat{V}_{ij} = \frac{eq_J}{J(2J-1)}\{\frac{3}{2}(J_iJ_j + J_jJ_i) - \delta_{ij}J^2\}$$

where

$$q_J =< JJ|\frac{\hat{V}_{zz}}{e}|JJ >=< JJ| - \frac{\sum_e(3z_e^2 - r_e^2)}{r_e^5}|JJ >$$

Finally, since

$$\sum_{ij} I_iI_jJ_iJ_j = (\sum_i I_iJ_i)^2 = (\mathbf{I.J})^2$$

$$\sum_{ij} I_iI_j\delta_{ij}J^2 = (\sum_i I_i)^2 J^2 = I^2J^2$$

$$\sum_{ij} I_iI_jJ_jJ_i = (\mathbf{I.J})^2 + (\mathbf{I.J})$$

the quadrupole hyperfine Hamiltonian is written

$$\mathcal{H}_{hyp}^Q = \frac{eq_JQ}{2I(2I-1)J(2J-1)}\{3(\mathbf{I.J})^2 + \frac{3}{2}(\mathbf{I.J}) - I^2J^2\} \ ,$$

as in Eq. 5.20 (see also Eq. 5.24).

Problem F.V.16 At §1.5 the isotope effect due to the **reduced mass** correction has been mentioned. Since two isotopes may differ in the nuclear radius R by an amount δR, once that a finite nuclear volume is taken into account a further shift of the atomic energy levels has to be expected. In the assumption of nuclear charge Ze uniformly distributed in a sphere of radius $R = r_F A^{1/3}$ (with Fermi radius $r_F = 1.2 \times 10^{-13}$ cm) estimate the **volume shift** in an hydrogenic atom and in a muonic atom. Finally discuss the effect that can be expected in muonic atoms with respect to ordinary atoms in regards of the hyperfine terms.

Solution:

The potential energy of the electron is $V(r) = -Ze^2/r$ for $r \geq R$, while (see Problem I.4.6)

$$V(r) = -3\frac{Ze^2}{2R}(1 - \frac{r^2}{3R^2}) \text{ for } r \leq R$$

The first-order correction, with respect to the nuclear point charge hydrogenic Hamiltonian, turns out

$$\Delta E = \frac{Ze^2}{2R} \int_0^R |R_{nl}(r)|^2 (-3 + \frac{r^2}{R^2} + \frac{2R}{r}) r^2 dr \simeq \frac{Ze^2}{10} R^2 |R_{nl}(0)|^2$$

The correction is negligible for non-s states, where $\mathcal{R}_{nl}(0) \simeq 0$, while for s states one has

$$\Delta E = \frac{2}{5} e^2 R^2 \frac{Z^4}{a_0^3 n^3}$$

In terms of the difference δR in the radii (to the first order) the shift turns out

$$\delta E \simeq \frac{4}{5} Z e^2 R^2 \frac{Z^4}{a_0^3 n^3} \frac{\delta R}{R}$$

In muonic atoms (see §1.5) because of the change in the reduced mass and in the Bohr radius a_0, the volume isotope effect is dramatically increased with respect to ordinary hydrogenic atoms.

As regards the hyperfine terms one has to consider the decrease in the Bohr radius and in the Bohr magneton ($\mu_B \propto 1/m$) (see Problem V.2.4). For the hyperfine quadrupole correction small effects have to be expected, since only states with $l \neq 0$ are involved.

Finally it is mentioned that an **isomeric shift**, analogous to the volume isotope shift, occurs when a radiative decay (e.g. from ^{57}Co to ^{57}Fe) changes the radius of the nucleus. The isomeric shift is experimentally detected in the **Mössbauer resonant absorption spectrum** (see §14.6).

6

Spin statistics, magnetic resonance, spin motion and echoes

Topics

Spin temperature and spin thermodynamics
Magnetic resonance and magnetic dipole transitions
NMR and EPR
Spin echo
Cooling at extremely low temperatures

This Chapter, dealing with nuclear and electronic angular momenta in magnetic fields, further develops topics already discussed in Chapters 4 and 5. The new arguments involve some aspects of **spin statistics** and of **magnetic resonance** (namely how to drive the angular and magnetic moments and to change their components along a magnetic field). The magnetic resonance experiment in most cases is equivalent to drive magnetic dipole transitions among Zeeman-like levels.

6.1 Spin statistics, spin-temperature and fluctuations

Let us refer to a number N (of the order of the Avogadro number) weakly interacting spins $S = 1/2$, each carrying magnetic moment $\boldsymbol{\mu} = -2\mu_B \mathbf{S}$, in static and homogeneous magnetic field $\mathbf{H}$ along the z-axis. At the thermal equilibrium the statistical distribution depicted in Fig. 6.1 occurs. The number of spins on the two energy levels (**statistical populations**) are

$$N_- = N \frac{e^{\frac{\mu_B H}{k_B T}}}{e^{\frac{\mu_B H}{k_B T}} + e^{\frac{-\mu_B H}{k_B T}}} \equiv \frac{N}{Z} e^{\frac{\mu_B H}{k_B T}} \tag{6.1}$$

and

$$N_+ = \frac{N}{Z} e^{\frac{-\mu_B H}{k_B T}} \ , \tag{6.2}$$

with Z the partition function (for reminds see §4.4, Problems F.I.1 and F.IV.8). The contribution to the thermodynamical energy is

$$U = N_-(-\mu_B H) + N_+(\mu_B H) = \mu_B H(2N_+ - N) \equiv [\frac{2N}{Z} e^{-\varepsilon/T} - N]\mu_B H \tag{6.3}$$

with

$$\varepsilon = \frac{\mu_B H}{k_B} \quad \text{the "magnetic temperature"}$$

(having assumed $U = 0$ in the absence of the magnetic field).

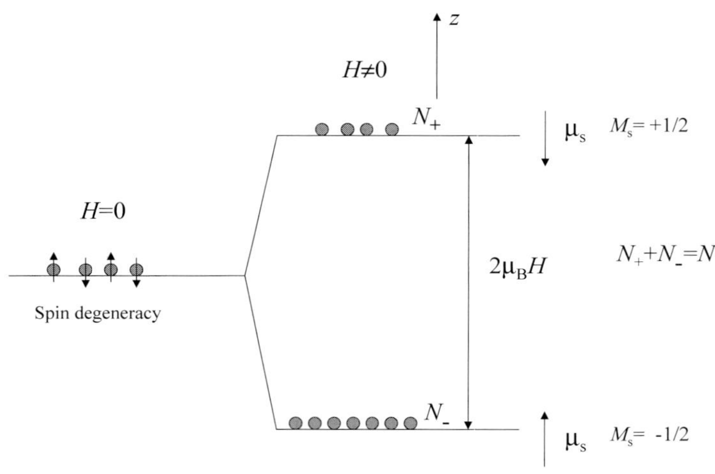

Fig. 6.1. Pictorial view of the statistical distribution of N spins $S = 1/2$ on the two "Zeeman levels" in a magnetic field, with $N_- > N_+$ at thermal equilibrium. In a field of 1 Tesla the separation energy $2\mu_B H$ is 1.16×10^{-4} eV, corresponding to the temperature $T = 2\mu_B H/k_B = 1.343$ K.
An equivalent description holds for protons, with $I = 1/2$, with the lowest energy level corresponding to quantum magnetic number $M_I = +1/2$, the gyromagnetic ratio being positive (§5.1). The energy separation between the two levels, for proton magnetic moments, is $2\mu_p H$, with $\mu_p = M_n g_n I$ and M_n the nuclear magneton, $g_n = 5.586$ the nuclear g-factor. In a field of 1 Tesla, for protons the separation turns out 1.76×10^{-7} eV (or 20.4×10^{-4} K).

The statistical populations N_+ or N_- are modified when the temperature (or the field) is changed and after some time a new equilibrium condition is attained. $N_\pm$ can be varied, while keeping the temperature of the thermal reservoir and the magnetic field constant, by proper irradiation at the transitional frequency $\nu = 2\mu_B H/h$, by resorting to the **magnetic dipole transition mechanism** (the methodology is known, in general, as **magnetic resonance**, described in some detail at §6.2).

When $N_\pm$ are modified, in principle the energy U can take any value in between $-N\mu_B H$ (corresponding to full occupation of the state at $M_S = -1/2$) and $+N\mu_B H$ (complete reversing of all the spins, with $N_+ = N$).

From thermodynamics, no volume variation being involved, the entropy of the spin system can be defined

$$S_{spin} \equiv S = \int \frac{1}{T}(\frac{\partial U}{\partial T})_V \, dT \tag{6.4}$$

and therefore, from Eq. 6.3,

$$S = 2\mu_B H \int \frac{1}{T}(\frac{\partial N_+}{\partial T})_V \, dT \tag{6.5}$$

When the statistical distribution on the levels is modified the entropy changes, in the way sketched in Fig. 6.2 in terms of the energy U.

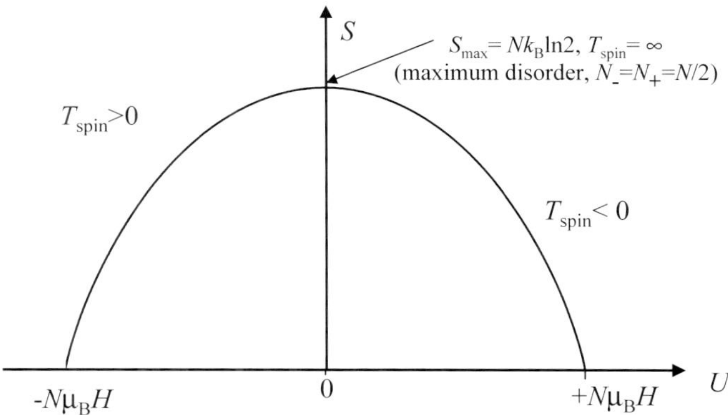

Fig. 6.2. Entropy S as a function of the energy U in a spin system. The statistical entropy is defined as the logarithm of the number of ways a given spin distribution can be attained (See §6.4). The zeroes at $U = \pm N\mu_B H$ correspond to all spins in a single state (see also Problems VI.1.2, VI.1.4 and F.VI.1).

Since the temperature can be expressed as

$$\frac{1}{T} = \frac{\partial S}{\partial U} \tag{6.6}$$

(in the partial differentiation keeping constant all the other thermodynamical variables), one can define a **spin temperature** T_{spin} in terms of N_+ and N_-. Thus a spin temperature is defined also for $U > 0$, eventhough there is not a correspondent thermal equilibrium temperature T of the reservoir. When, by means of magnetic resonance methods (or, for example, simply by suddenly reversing the magnetic field) the equilibrium distribution is altered, then $T_{spin} \neq T$. It should be remarked that this non-equilibrium situation can last for time intervals of experimental significance only when the probability of spontaneous emission (see Appendix I.2) is not so strong to cause fast restoring. This is indeed the case for states of magnetic moments in magnetic fields (see the estimate in Problem F.V.12). However, exchanges of energy with the thermal bath, related to the time-dependence of the Hamiltonians coupling the spin system to all other degrees of freedom (the **"lattice"**), usually occur. This is why a given non-equilibrium spin distribution rather fast attains the equilibrium condition, usually through an exponential process characterized by a time constant called **spin-lattice relaxation time** T_1 (see §6.2). The relaxation times T_1's, particularly at low temperatures, are often long enough to allow one to deal with non-equilibrium states.

Let us imagine to have prepared one spin system at $T_{spin} = -300$ K and to bring it in thermal contact with another one, strictly equivalent but at thermal equilibrium, namely at $T_{spin} = T = 300$ K. The two systems reach a common equilibrium by means of **spin-spin transitions** in which two spins exchange their relative orientations (this process involves a **spin-spin relaxation time** T_2 usually much shorter than T_1). The total energy is constant while the temperatures of both the two sub-systems evolve, as well as the entropy. The final spin temperatures are $+\infty$ and $-\infty$ and the entropy takes its maximum value. The internal equilibrium, with $T_{spin} = \pm\infty$, is attained in very short times (for $T_2 \ll T_1$). Then the spin-lattice relaxation process drives the system towards the thermodynamical equilibrium condition, where $T_{spin} = T$.

Now we return to the field induced magnetization

$$M = N < \mu_z >_H \ , \tag{6.7}$$

$< \mu_z >_H$ being the statistical average of the component of the magnetic moment along the field (see §4.4).

From

$$\frac{-g\mu_B \sum_{M_J} M_J e^{-g\mu_B M_J H/k_B T}}{Z} = k_B T \frac{\partial\left(ln(\sum_{M_J} e^{-g\mu_B M_J H/k_B T})\right)}{\partial H} \tag{6.8}$$

(See Eq. 4.29), the magnetization can be written

$$M = N k_B T \left(\frac{\partial ln Z}{\partial H}\right)_T \ . \tag{6.9}$$

For $J = S = 1/2$

$$M = \frac{N}{2} 2\mu_B tanh\left(\frac{\mu_B H}{k_B T}\right) \tag{6.10}$$

Let us now evaluate the mean square deviation of the magnetization from this average equilibrium value, i.e. its **fluctuations** $< (M- < M >)^2 >$ (now we have added the symbol $<>$ to M in Eqs. 6.9 or 6.10 to mean its average character). The magnetization has a Gaussian distribution around the average value $< M >$, zero for $H = 0$ (See Problem VI.1.1) and the one in Eq. 6.10 in the presence of the field.

For the fluctuations one has

$$< \Delta M^2 >=< (M- < M >)^2 >=$$
$$=< M^2 > -2 < M < M >> + < M >^2=< M^2 > - < M >^2 \tag{6.11}$$

The single $< \mu_z >$'s are **uncorrelated** and therefore $< \Delta M^2 >= N < \Delta \mu_z^2 >$ with $< \Delta \mu_z^2 >=< \mu_z^2 > -(< \mu_z >)^2$, yielding

$$< M^2 >= N < \mu_z^2 >_H = 4N\mu_B^2 \frac{\sum_{M_s} M_s^2 e^{-xM_s}}{Z}$$

with $x = (2\mu_B H/k_B T)$ and $M_S^2 = 1/4$.
Then $< M^2 >= N\mu_B^2$ and finally, from Eqs. 6.11 and 6.10

$$< \Delta M^2 >= N\mu_B^2 \left[1 - tanh^2 \left(\frac{\mu_B H}{k_B T}\right)\right] \tag{6.12}$$

Now we look for the relationship of the fluctuations to the **response function**, the magnetic susceptibility $\chi = \partial < M > /\partial H$. Again, form Eq. 6.10 one derives

$$\chi = N\mu_B \left[1 - tanh^2 \left(\frac{\mu_B H}{k_B T}\right)\right] \frac{\mu_B}{k_B T} \tag{6.13}$$

and therefore

$$< \Delta M^2 >= k_B T \chi \ . \tag{6.14}$$

This relationship is a particular case of the **fluctuation-dissipation theorem**, relating the spectrum of the fluctuations to the response functions (see Problem IV.8 for an equivalent derivation).

The considerations carried out in the present paragraph are a few illustrative examples of the topic that one could call **spin thermodynamics**. This field includes the method of **adiabatic demagnetization**, which allows one to reach the lowest temperatures (§6.4). A valuable introduction to statistical physics with paramagnets, leading step by step the reader to the concepts suited for extending the arguments recalled in the present paragraph, can be found in Chapters 4 and 5 of the book by **Amit** and **Verbin**, quoted in the Preface.

Problems VI.1

Problem VI.1.1 Express the probability distribution of the total "magnetization" along a given direction in a system of N independent spin $S = 1/2$, in zero magnetic field.

Solution:

Along the z-direction two values $\pm\mu_B$ are possible for the magnetic moment. The probability of a given sequence is $(1/2)^N$. A magnetization $M = n\mu_B$ implies $\frac{1}{2}(N + n)$ magnetic moments "up" and $\frac{1}{2}(N - n)$ magnetic moments "down" (see Fig. 6.1). The total number of independent sequences giving such a distribution is

$$W(n) = \frac{N!}{\left[\frac{1}{2}(N + n)\right]! \left[\frac{1}{2}(N - n)\right]!}.$$

The probability distribution for the magnetization is thus $W(M) = W(n)(1/2)^N$. From Stirling approximation and series expansion

$$ln\left(1 \pm \frac{n}{N}\right) \approx \pm\frac{n}{N} - \frac{n^2}{2N^2} \pm \ldots$$

one has

$$ln\, W(M) \approx -\frac{1}{2} ln\left(\frac{\pi N}{2}\right) - \frac{n^2}{2N}$$

so that

$$W(M) \approx \left(\frac{2}{\pi N}\right)^{1/2} exp\left[-\frac{n^2}{2N}\right]$$

namely a **Gaussian distribution** around the value $<M>=0$, at width about $(N)^{1/2}$. It is noted that the **fractional width** goes as $N^{-1/2}$, rapidly decreasing for large N.

Problem VI.1.2 Express the entropy of an ensemble of $S = 1/2$ spins in a magnetic field and discuss the spin temperature recalled in Fig. 6.2.

Solution:

The free energy is

$$F = -\frac{1}{\beta} ln[1 + exp(-2\beta\,\mu_B\,H)],$$

with $\beta \equiv 1/k_B T$, having set $U=0$ for the low-energy level. Then

$$S = -\left(\frac{\partial F}{\partial T}\right)_H = -k_B[(1 - u)\,ln(1 - u) + u\,ln\,u]$$

with $u = U/2\mu_B H$. The temperature as a function of the energy is obtained by inversion (see Eq. 6.6):

$$\frac{2\mu_B H}{k_B T} = ln(u^{-1} - 1) \quad \text{and then} \quad T = (2\mu_B H/k_B)/ln(u^{-1} - 1) \ ,$$

justifying the plot in Fig. 6.2. The maximum of S is for $u = 1/2$ (see Problem VI.1.4).

An equivalent derivation is obtained by considering the number of available states

$$W = \frac{N!}{(N_+)!(N_-)!}$$

Resorting to the Stirling approximation (see Prob. VI.1.1) one has

$$S = k_B ln W = k_B[NlnN - N_+ lnN_+ - N_- lnN_-]$$

and being $U = N_- \alpha$, with $\alpha \equiv 2\mu_B H$,

$$S = k_B \left\{ NlnN - \frac{U}{\alpha}ln\frac{U}{\alpha} - \left(N - \frac{U}{\alpha}\right) ln \left(N - \frac{U}{\alpha}\right)\right\}$$

From Eq. 6.6 the temperature turns out

$$T = \frac{\alpha}{k_B ln(\frac{\alpha N}{U} - 1)}$$

or

$$\frac{1}{T} = \frac{k_B}{\alpha}[lnN_+ - lnN_-]$$

Problem VI.1.3 Two identical spin systems at $S = 1/2$, prepared at spin temperatures $T_a = E \, / \, 2 \, k_B$ and $T_b = - \, E \, / \, k_B$ are brought into interaction. Find the energy and the spin temperature of the final state.
Solution:
By setting $E = 0$ for the low energy level, $U_x = U_a + U_b$ is written

$$U_x = 2NE\frac{exp(-E/k_B T_x)}{1 + exp(-E/k_B T_x)}.$$

Since

$$U_a = NE\frac{exp(-2)}{1 + exp(-2)}$$

and

$$U_b = NE\frac{exp(1)}{1 + exp(+1)},$$

one has

$$exp(E/k_BT_x) = \frac{e^2 + e^{-1} + 2e}{2 + e^2 + e^{-1}} \equiv z$$

and then

$$T_x = \frac{E}{k_B \, ln \, z} \approx 3.3 \, \frac{E}{k_B}.$$

Problem VI.1.4 Show that the entropy of a system can be written

$$S = -k_B \sum_n p_n \, ln \, p_n$$

where p_n is the probability that the system is found in the state at energy E_n, namely for a canonical ensemble

$$p_n = \frac{exp(-E_n/k_BT)}{Z},$$

with Z partition function.

Solution:

In fact

$$S = -\frac{k_B}{Z} \sum_n exp\left(-\frac{E_n}{k_BT}\right)\left[-\frac{E_n}{k_BT} - lnZ\right]$$

$$= k_B \frac{lnZ}{Z} \sum_n exp\left(-\frac{E_n}{k_BT}\right) + \frac{1}{T} \sum_n exp\left(-\frac{E_n}{k_BT}\right)\frac{E_n}{Z}$$

$$= k_B \, lnZ + \frac{1}{T} \sum_n exp\left(-\frac{E_n}{k_BT}\right)\frac{E_n}{Z}.$$

On the other hand, from

$$F = -k_B \, T \, lnZ$$

one can write

$$S = \frac{U - F}{T} = k_B \frac{\partial(T \, lnZ)}{\partial T}$$

and

$$U = k_B \, T^2 \frac{\partial \, lnZ}{\partial T}.$$

Then

$$S = -\left[\frac{\partial F}{\partial T}\right]_{v,H} = k_B \, lnZ + k_B \, T \frac{\partial \, lnZ}{\partial T}.$$

Since

$$\frac{\partial \, lnZ}{\partial T} = \frac{1}{k_B \, T^2} \sum_n exp\left(-\frac{E_n}{k_BT}\right)\frac{E_n}{Z}$$

one has

$$S = k_B \, lnZ + \frac{1}{T} \sum_n exp\left(-\frac{E_n}{k_B T}\right) \frac{E_n}{Z} \ .$$

Problem VI.1.5 A model widely used in statistics and in magnetism is the **Ising** model, for which an Hamiltonian of the form $\mathcal{H} = -K \sum_{i,j} s_i \, s_j$ is assumed, with the spin variables s_i taking the values $+1$ and -1. K is the **exchange integral** (see § 2.2.2). Derive the partition function Z, the free energy F, the thermodynamical energy U and the specific heat C_V, for a system of N spins.

Solution:

By indicating with $N_{p,a}$ the number of parallel (p) and antiparallel (a) spins with $N_p + N_a = N - 1$ the number of interacting pairs, the energy of a given spin configuration is $E = -K(N_p - N_a) = -K(2N_p + 1 - N)$.

The number of permutations of the $(N - 1)$ pairs is $(N - 1)!$, of which $(N - 1)!/N_a! \, N_p!$ are distinguishable. Therefore the sum over the states reads

$$Z = 2 \sum_{N_p=0}^{N-1} \left[\frac{(N-1)!}{N_a! \, N_p!}\right] exp\left[+\frac{K(2N_p + 1 - N)}{k_B T}\right]$$

$$= 2\, exp\left[+\frac{K(1-N)}{k_B T}\right] \sum_{N_p} \left[\frac{(N-1)!}{(N-1-N_p)! N_p!}\right] exp\left[+\frac{K(2N_p)}{k_B T}\right]$$

(the factor 2 accounts for the configurations arising under the reversing of all the spins without changing N_p or N_a). The sum is the expansion of $\{1+\exp [(2 \, K/k_B T)]\}^{N-1}$ and therefore

$$Z = 2^N \left[\cosh \frac{K}{k_B T}\right]^{N-1} \ .$$

Then

$$F = -k_B \, T \, lnZ = -k_B \, T \left[N \, ln2 + (N-1)\, ln(\cosh \frac{K}{k_B T})\right] ,$$

$$U = -\frac{\partial(lnZ)}{\partial \beta} = -(N-1)K \tanh(\beta K) \ , \qquad \text{with} \qquad \beta = \frac{1}{k_B T}$$

and

$$C_V = (\frac{\partial U}{\partial T})_N = (N-1)\left(\frac{K^2}{k_B T^2}\right)\left[\frac{1}{(\cosh \beta K)^2}\right] .$$

6.2 The principle of magnetic resonance and the spin motion

Transitions involving hyperfine states or nuclear and/or electronic Zeeman-states in magnetic fields are carried out by resorting to the magnetic dipole mechanism. These transitions are usually performed by exploiting the phenomenon elsewhere called **magnetic resonance**, which allows one to drive electronic or nuclear magnetic moments. This type of experiments are at the core of modern **microwave** and **radiofrequency spectroscopies**.

The first experiment of magnetic resonance, by **Rabi**, involved molecular beams (see Fig. 6.3).

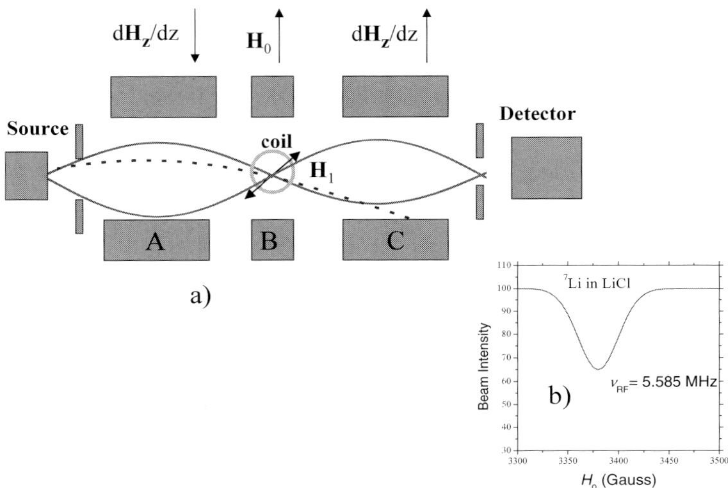

Fig. 6.3. a) Sketch of the experimental setup for magnetic resonance in beams (**ABMR**). The magnetic fields A and C have gradients along opposite directions. In region B the magnetic field $\mathbf{H}_0 \parallel z$ is homogeneous. The radiofrequency (or the microwave) field H_1 in region B is perpendicular to $\mathbf{H}_0$.
In part b) of the Figure the typical signal of magnetic resonance, detected as a minimum in the arrival of the atoms when in region C the refocusing of the deviations is inhibited (dotted line) by the resonance driven by $\mathbf{H}_1$ in region B (see text).

The vectorial description, with classical equation of motion (Chapter 3) is the following (see Fig. 6.4). The motion of the angular momentum $\mathbf{L}$ in $\mathbf{H}_0$ is described by

$$\frac{d\mathbf{L}}{dt} = \boldsymbol{\mu}_L \times \mathbf{H}_0 \tag{6.15}$$

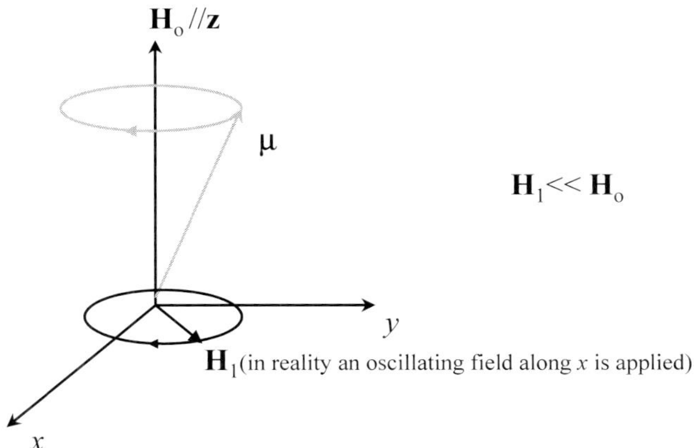

Fig. 6.4. Precessional motion of the magnetic moment $\boldsymbol{\mu}$ at the angular frequency $\omega_L = \gamma\, H_0$ and rotation of the radio frequency field H_1 at ω_{RF}. For $\omega_L = \omega_{RF}$ the magnetic resonance occurs. The gyromagnetic ratio γ is $\mu_I/I\hbar$ for nuclear moment (see §5.1) or $\gamma = \mu_J/J\hbar$ for electron magnetic moment.

implying the precession at the Larmor frequency ω_L (see §3.2 and Problem III.2.4). In a frame of reference rotating at angular frequency ω, Eq. 6.15 becomes[1]

$$\frac{d\boldsymbol{\mu}_L}{dt} = -\gamma\left(\mathbf{H}_0 + \frac{\omega}{\gamma}\right) \times \boldsymbol{\mu}_L. \tag{6.16}$$

Thus in the presence of the radiofrequency (or microwave) irradiation the effective field is

$$\mathbf{H}_{eff} = \left(H_0 + \frac{\omega_{RF}}{\gamma}\right)\hat{k} + H_1\hat{i} \tag{6.17}$$

When $\omega_{RF} = -\gamma H_0$ (the sign minus refers to clockwise precession), in the rotating frame of reference only $\mathbf{H}_1$ is active and the magnetic moment precesses around it, thus changing its component with respect to $\mathbf{H}_0$. As a consequence of the change in the z-component of the magnetic moment in the

---

[1] It is reminded that

$$\left(\frac{d\boldsymbol{\mu}}{dt}\right)_{lab.frame} = \left(\frac{\partial\boldsymbol{\mu}}{\partial t}\right)_{relative\ to\ rot.\ frame} + \left(\frac{d\boldsymbol{\mu}}{dt}\right)_{rot.frame}$$

the latter being $\boldsymbol{\omega} \times \boldsymbol{\mu}$.

region B of the Rabi experimental set up (Fig. 6.3) the compensation of the deviations due to $\mathbf{F} = \pm\mu_z(d\mathbf{H}/dz)$ in the regions A and C does no longer occur. Then a minimum in the number of atoms (or molecules) reaching the detector is observed.

The quantum mechanical description of the magnetic resonance corresponds to the situation depicted in Fig. 6.5 for nuclear spin $I = 1/2$, already discussed in other circumstances.

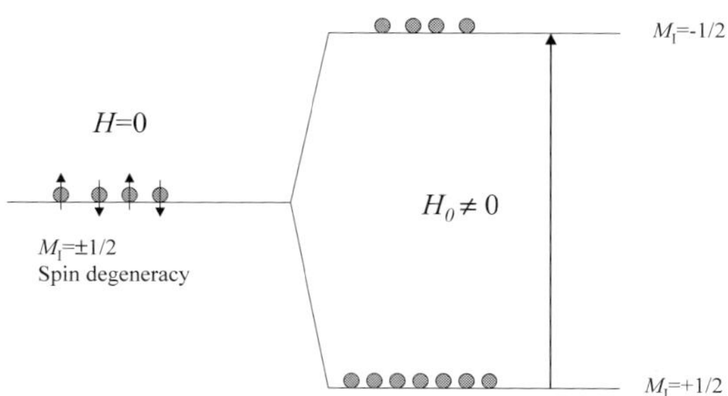

Fig. 6.5. Quantum magnetic levels for magnetic moment $\boldsymbol{\mu}_I = gM_n\mathbf{I} = \gamma\mathbf{I}\hbar$, for $I = 1/2$ in a magnetic field. The resonance corresponds to transitions from $M_I = +1/2$ to $M_I = -1/2$, driven by the magnetic dipole mechanism.

The eigenvalues are $\pm M_I g_n M_n H_0$ and magnetic dipole transitions, with selection rule $\Delta M_I = \pm 1$, are possible when the condition $h\nu_{RF} = g_n M_n H_0 \equiv \hbar\omega_L$ is verified, the perturbation Hamiltonian being $\mathcal{H}_P = -\boldsymbol{\mu}_I.\mathbf{H}_1 = g_n M_n H_1 I_x cos(\omega_{RF} t)$. In fact, by extending the description in Appendix I.3, the transition probability has to be written

$$W_{RF} \propto |<I, M_I'|I_x|I, M_I>|^2 \propto |<I, M_I'|I_+ + I_-|I, M_I>|^2 \ . \qquad (6.18)$$

According to the properties of the $I_\pm$ operators and to the orthogonality of states at different M_I, Eq. 6.18 leads to the selection rule $\Delta M_I = \pm 1$. The circular polarization required for $\Delta M_I = \pm 1$ transitions is the counterpart of the rotating field $\mathbf{H}_1$ perpendicular to the quantization axes.

A description of quantum character is possible by considering the time evolution of the expectation values for the spin components (Problem VI.2.1).

The description in terms of spin motion is particularly suited for understanding the modern **pulse resonance techniques**, which allow one to drive the magnetic moments along a given direction by controlling the length of the radiofrequency irradiation. Examples are shown in Fig. 6.6.

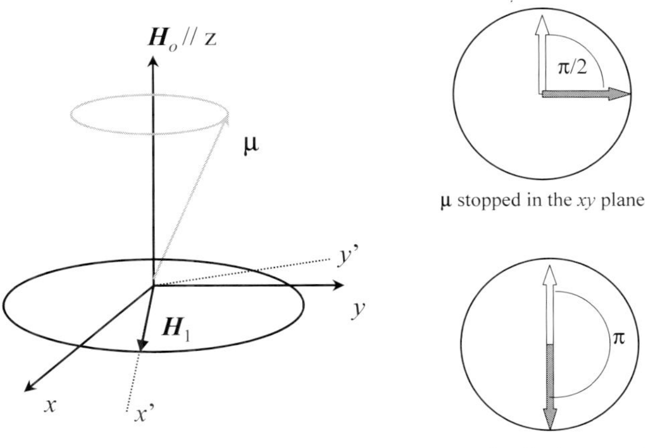

Fig. 6.6. Illustrative examples of the spin motions induced in pulse magnetic resonance, by stopping the irradiation after a given time. For the so-called $\pi/2$ **pulse** the time of irradiation turns out (see text for the precession around H_1) $\tau = (\pi/2)/\omega_1 = (\pi/2)/\gamma H_1 = (\pi/4)\hbar/H_1\mu_I$ for nuclear spin $I = 1/2$ and $\tau = (\pi/2)\hbar/H_1\mu_B$ for electron at $S = 1/2$. The π **pulse** requires an irradiation time 2τ and it corresponds to the complete reversing of the spins in the magnetic field H_0 (x', y', z' is the rotating frame).

Finally we mention that resonance experiments (**NMR** for nuclear, **EPR** for electron) nowadays are generally carried out in condensed matter, with a number of interesting applications.

In condensed matter the interactions with the other degrees of freedom (the **"lattice"**) or among spins themselves, play a relevant role. Phenomenologically the interactions are taken into account by the **Bloch equations**, that for the expectation values of the spin components, averaged over the statistical ensemble, in the rotating frame can be written

$$\frac{d < \overline{I_x} >}{dt} = -\frac{< \overline{I_x} >}{T_2} \tag{6.19a}$$

$$\frac{d < \overline{I_y} >}{dt} = -\frac{< \overline{I_y} >}{T_2} . \tag{6.19b}$$

These equations account for the decay of the transverse components of $< \mathbf{\overline{I}} >$, that at long time must vanish. For the longitudinal component the Bloch equation is

$$\frac{d<\overline{I_z}>}{dt} = \frac{(I_z^0 - <\overline{I_z}>)}{T_1} \tag{6.20}$$

(where I_z^0 is the expectation value of the z-component at the thermal equilibrium). This equation describes the relaxation process towards equilibrium, after a given alteration of the statistical populations (see §4.4 for a qualitative definition of the relaxation time T_1).

In order to have a complete description of the spin motions Eqs. 6.19 and 6.20 must be coupled to the equation

$$\frac{d<\overline{\mathbf{I}}>}{dt} = -\frac{g_n M_n}{\hbar} <\overline{\mathbf{I}}> \times \mathbf{H}_{eff} \tag{6.21}$$

where the effective field is defined in Eq. 6.17. Then one has a system of equations (6.19-6.21) for the expectation values of the spin components (often written in terms of the **nuclear magnetization** $\mathbf{M}_{nuclear} \propto \sum_i <\overline{\mathbf{I}_i}>$). These equations can be solved under certain approximations, to yield the time evolution of $<\overline{\mathbf{I}}>$ or of $\mathbf{M}_{nuclear}$.

The quantum description of the time evolution of the spin operators in magnetic resonance experiments, in the presence of the relaxation processes imbedded in Eqs. 6.19 and 6.20, is usually based on a variant of the time-dependent perturbation theory, the **density matrix method**. The textbook by **Slichter**, reported in the Preface, masterly deals with this matter to the due extent. We shall limit ourselves, in the next paragraph, to describe a very important phenomenon, the **spin echo**, that in simple circumstances can satisfactorily be treated on the basis of the semiclassical motions of the spin operators and of the Bloch equations.

Problems VI.2

Problem VI.2.1 Consider a single spin **s** in a constant and homogeneous magnetic field along the z-direction. From the time-dependent Schrödinger equation derive the expectation values of the spin components and show that similarly to the vectorial description, the precessional motion occurs.
Then consider a small oscillating magnetic field along the x-direction and prove that at the resonance reversing of **s** with respect to the static field occurs. Discuss the cases of pulse application of the oscillating field for time intervals so that the rotation of **s** is by angles $\pi/2$ and π. Qualitatively figure out what happens if spin-spin and spin-lattice interactions are taken into account.

Solution:
From

$$\mu_B \mathbf{H} \begin{pmatrix} 1 & 0 \\ 0 & -1 \end{pmatrix} |\phi(t)> = i\hbar \frac{d|\phi(t)>}{dt}$$

with

$$|\phi(t)> = \alpha(t)| \uparrow> + \beta(t)| \downarrow> \ ,$$

one derives

$$\alpha(t) = a\, exp\,(-i\omega_L t) \qquad\qquad \beta(t) = b\, exp\,(i\omega_L t)$$

with ω_L Larmor frequency. The expectation values are

$$< \phi(t)|s_z|\phi(t) > = \frac{\hbar}{2}[|\alpha|^2 - |\beta|^2], \quad \text{time-independent,}$$

while

$$< \phi(t)|s_x|\phi(t) > = (a\,b\,\hbar)\cos(\omega_L t)$$
$$< \phi(t)|s_y|\phi(t) > = (a\,b\,\hbar)\sin(\omega_L t) \ ,$$

indicating the precession depicted in Figs. 6.4 and 6.6.

In the presence of H_1 rotating in the (xy) plane

$$H_1(t) = H_1 exp[\pm i\omega t] \ ,$$

from the Schrödinger equation one derives

$$\frac{\hbar}{2}\omega_L \alpha + \mu_B H_1 exp[-i\omega t]\beta = i\hbar \frac{d\alpha}{dt}$$

$$\mu_B H_1 exp[+i\omega t]\alpha - \frac{\hbar}{2}\omega_L \beta = i\hbar \frac{d\beta}{dt}.$$

By writing the coefficients α and β in the form

$$\alpha = \Gamma(t)exp\left[-\frac{i\omega_L t}{2}\right] \qquad \beta = \Delta(t)exp\left[+\frac{i\omega_L t}{2}\right]$$

those equations are rewritten

$$\mu_B H_1 exp[-i(\omega - \omega_L)t]\Delta = i\hbar \frac{d\,\Gamma}{dt}$$

$$\mu_B H_1 exp[+i(\omega - \omega_L)t]\Gamma = i\hbar \frac{d\Delta}{dt}.$$

At the resonance

$$\mu_B H_1 \Delta = i\hbar \frac{d\,\Gamma}{dt} \quad \text{and} \quad \mu_B H_1 \Gamma = i\hbar \frac{d\Delta}{dt}.$$

From the derivative of the first, substituted in the second, one finds

$$\Gamma = \sin(\Omega t + \psi) \quad \text{and} \quad \Delta = i\cos(\Omega t + \psi) \quad \text{where} \quad \Omega = \frac{\mu_B H_1}{\hbar}.$$

By setting $\psi = 0$, by repeating the derivation of the expectation values one has

$$< \phi(t)|s_z|\phi(t) > = -\frac{\hbar}{2}\cos(2\Omega t)$$

$$< \phi(t)|s_x|\phi(t) > = -\frac{\hbar}{2}\sin(2\Omega t)\sin(\omega_L t)$$

$$< \phi(t)|s_y|\phi(t) > = \frac{\hbar}{2}\sin(2\Omega t)\cos(\omega_L t).$$

These equations can be interpreted in terms of the motion of **s** as the superposition of the **precession around** z at the Larmor frequency and the **rotation around** H_1 at the angular frequency $2\mu_B H_1/\hbar$.

In the rotating frame (see Fig. 6.6) where $\mathbf{H}_1$ is fixed, one has the rotation of **s** by a given angle, depending on the duration of the irradiation. Thus, in principle, one can prepare the magnetization $\mathbf{M} \propto \sum_i < \overline{\mathbf{s}}_i >$ along any direction, as schematically illustrated below:

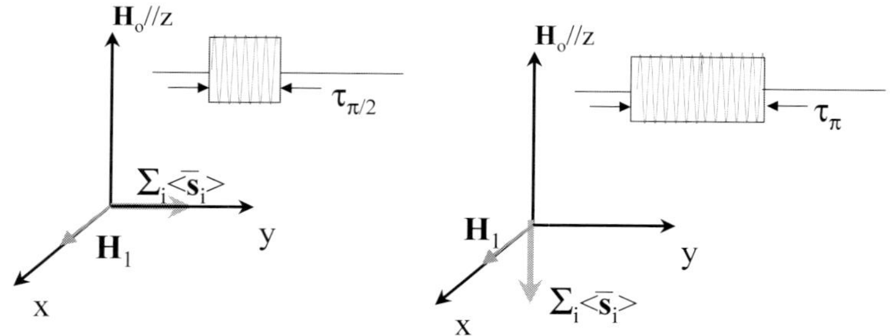

It is noted that the $(\pi/2)$ pulse corresponds to equalize of the statistical populations in the two Zeeman levels and magnetization in the xy plane. The π pulse corresponds to the inversion of $\mathbf{M}$ and therefore to **negative spin temperature** (see §6.1).

The spin-spin interaction implies the decay to zero in a time of the order of T_2 of the transverse components of $\mathbf{M}$. The spin-lattice interaction, with transfer of energy to the reservoir, drives the relaxation process towards the thermal equilibrium distribution, with $\mathbf{M}$ along $\mathbf{H_0}$, attained in a time of the order of T_1 (see Problem F.VI.5).

6.3 Spin and photon echoes

Let us imagine that a system of electronic or nuclear spins has been brought in the xy plane (perpendicular to the z-axis along the field $\mathbf{H}_0$) by a $\pi/2$ pulse, by means of the experimental procedure described at §6.2 and Problem VI.2.1. Once in the plane, the transverse components have to decay towards zero according to Eqs. 6.19, in a time of the order of T_2, yielding in a proper receiver a signal called **free induction decay (FID)**.

Now let us suppose that in times much shorter than T_2 another mechanism, different from the spin-spin interaction, causes a **distribution of precessional frequencies**. This mechanism could be due for instance to magnetic field inhomogeneities, to spatially varying diamagnetic or paramagnetic corrections to the external field H_0 or to a field gradient created by external coils. Because of the spread in the precessional frequencies, in a time usually called T_2^* and much shorter than T_2, the transverse components of the total magnetization $\mathbf{M}_{x,y} \propto \sum_i < \bar{\mathbf{I}}_{x,y}(i) >$ decay to value close to zero. After a time t_1 larger than T_2^* but shorter than T_2, a second pulse, of duration π, is applied (see Fig. 6.7). Since all the spins are flipped by 180° around the x'-axis, the ones precessing faster now are forced to return in phase with the ones precessing slower.

Thus after a further time interval t_1, refocussing of all the spins along a common direction occurs, yielding the "original" strength of the signal (only the reduction due to the intrinsic T_2-driven process is now acting, but $2t_1 \ll T_2$). This is called the **echo signal**. By repeating the π-pulses the envelope of the echoes tracks the real, **irreversible decay** of the $M_{x,y}$ components, as depicted in part b) of Fig. 6.7.

The relevance of pulse magnetic resonance experiments in the development of modern spectroscopies can hardly be over estimated. Besides the enlightenment of fundamental aspects of the quantum machinery, the echo experiments, first devised by **Hahn**, have been instrumental in a number of applications in solid state physics, in chemistry and in medicine (**NMR imaging**).

Furthermore the pulse magnetic resonance methodology has been transferred in the field of the **optical spectroscopy**, by using lasers. In this case special techniques are required, because in the optical range the "dipoles" go very fast out of phase (the equivalent of T_2 is very short).

In this respect we only mention that the **pseudo-spin formalism** can be applied to any system where approximately only two energy levels, corresponding to the spin-up and spin-down states, can be considered relevant. For a pair of states in atoms, to a certain extent **coherent electric radiation** can be used to induce the analogous of the inversion of the magnetization described at §6.2 and Problem VI.2.1, in terms of the populations on the lower and on the upper atomic or molecular levels. The "oscillating " electric dipole moment $\mathbf{R}_{21}$ (see Appendix I.3) plays the role analogous to the magnetic moment in the magnetic resonance phenomenon. After the "saturation of the

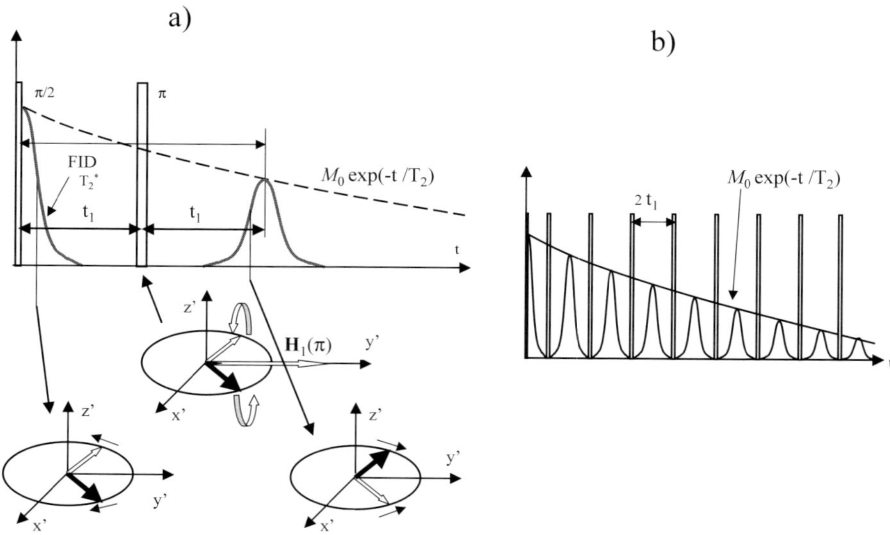

Fig. 6.7. Schematic representation of the spin motions generating the echo signal upon application of a sequence of $\pi/2$ and π pulses. Part a) shows the FID signal following the $\pi/2$ pulse and how the echo signal is obtained at the time $2t_1$ owing to the **reversible decay** of the magnetization in a time shorter than T_2. The rotation of the spins in the (x, y) plane, as seen in the rotating frame of reference, evidences how refocussing generates the echo.

It should be remarked that with pulse techniques, by switching the phase of the RF field it is possible to apply the second pulse (at time t_1) along a direction different from the one of the first pulse at $t = 0$ (e.g. from x' to y' in the rotating frame, see Fig. 6.6).

Part b) shows the effect of a sequence of π pulses (after the initial $\pi/2$), with a train of echoes, the envelope yielding the intrinsic **irreversible decay** of the transverse magnetization due to the T_2-controlled mechanism.

line" corresponding to the equalization of the two levels (to a $\pi/2$-pulse), a second pulse π at a time t_1 later, can force the diverging phases of the oscillating electric dipoles to come back in phase: a "light pulse", the **photon echo**, is observed at the time $2t_1$.

The analogies of two-levels atomic systems in interaction with coherent radiation with the spin motions in magnetic resonance experiments, are nicely described in the textbook by **Haken** and **Wolf**, quoted in the Preface.

6.4 Ordering and disordering in spin systems: cooling by adiabatic demagnetization

As already shown (see Problem VI.1.4), the entropy of an ensemble of magnetic moments in a magnetic field is related to the partition function Z:

$$S = -(\frac{\partial F}{\partial T})_H = [\frac{\partial(Nk_BTlnZ)}{\partial T}]_H \qquad (6.22)$$

F being the **Helmholtz free energy**.

From the statistical definition the entropy involves the number of ways W in which the magnetic moments can be arranged: $S = k_B lnW$. For angular momenta $\mathbf{J}$, in the high temperature limit the M_J states are equally populated and $W = (2J + 1)^N$. The statistical entropy is

$$S = Nk_B ln(2J + 1) \qquad (6.23)$$

For $T \to 0$, in finite magnetic field, there is only one way to arrange the magnetic moments (see §6.1) and then the spin entropy tends to zero. In general, since the probability $p(M_J)$ that J_z takes the value M_J is given by

$$p(M_J) = \frac{e^{-M_J g\mu_B H/k_B T}}{Z} \quad , \qquad (6.24)$$

the statistical entropy has to be written (see Problems VI.1.4 and F.VI.1)

$$S = -Nk_B \sum_{M_J} p(M_J)ln(p(M_J)) \qquad (6.25)$$

By referring for simplicity to non interacting magnetic ions with $J = S = 1/2$, at finite temperature the entropy is

$$S(T) = Nk_B \left((ln2)cosh(\frac{\epsilon}{T}) - \frac{\epsilon}{T}tanh(\frac{\epsilon}{T}) \right), \qquad (6.26)$$

where $\epsilon = \mu_B H/k_B$ is the **magnetic temperature** and $S(T \to \infty) = Nk_B ln2$ (Eq. 6.23).

The temperature dependence of the entropy is plotted below:

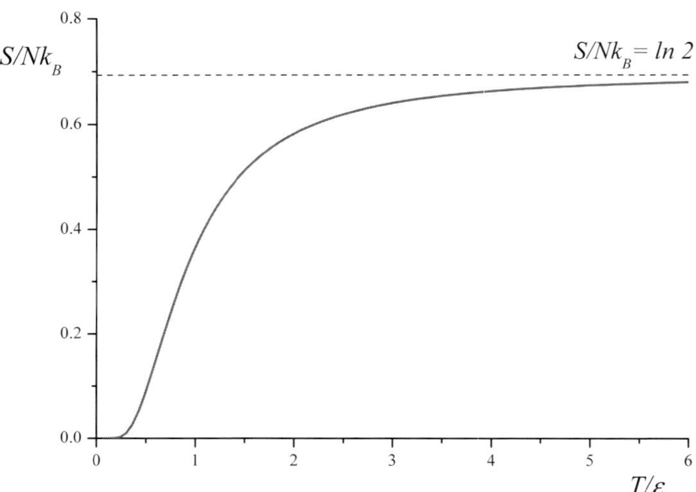

Now we describe the basic principle of the process called **adiabatic de-magnetization**, used in order to achieve extremely low temperatures.

A crystal with magnetic ions, almost non-interacting (usually a paramagnetic salt) is in thermal contact by means of an exchange gas (typically low-pressure Helium) with a reservoir, generally a bath of liquid Helium at $T = 4.2$ K. (This temperature can be further reduced, down to about 1.6 K, by pumping over the liquid so that the pressure is decreased).

In zero external field the spin entropy is practically given by $Nk_B ln2$. Only at very low temperature the residual internal field (for instance the one due to dipolar interaction or to the nuclear dipole moments) would anyway induce a certain ordering. The schematic form of the temperature dependence of the magnetic entropy is the one given by curve 1 in Fig. 6.8. Then the external field is applied, in isothermal condition at $T = T_{init}$, up to a certain value H_m. In a time of the order of the spin-lattice relaxation time T_1, spin alignment is achieved, the magnetic temperature is increased and $(T/\epsilon) \ll 1$. Therefore the magnetic entropy is decreased down to S_{init} (curve 2 in the Figure), at the same temperature of the thermal bath and of the crystal. The value S_{init} in Figure corresponds to Eq. 6.26 for $T \ll \epsilon$ and implies a large difference in the populations N_+ and N_- (see Fig. 6.1, where now the point at the energy $E = -N\mu_B H$ is approached). The external bath (the liquid Helium) absorbs the heat generated in the process, while the magnetic energy is decreased. The "internal" reservoir of the sample (namely all the other degrees of freedom besides the spins, already defined **"lattice"**) has its own entropy $S_{lattice}$ related to the vibrational excitations of the ions (in number N', ten or hundred times the number N of the magnetic ions).

Since in general the entropy is

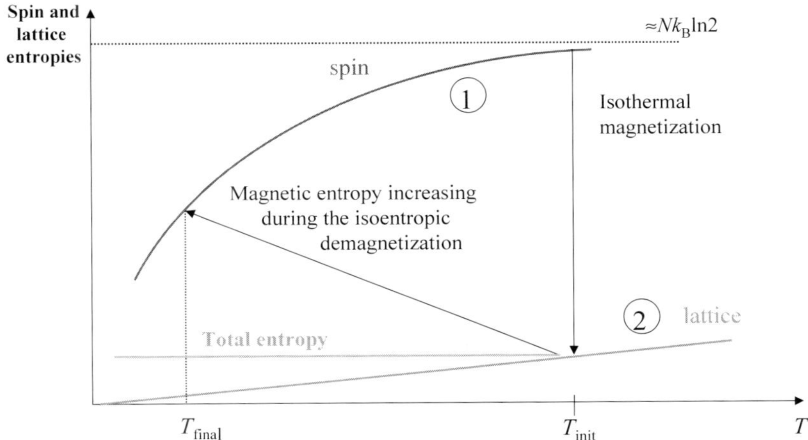

Fig. 6.8. Schematic temperature dependences of the **magnetic** and **lattice entropies** and of the decrease of the lattice temperature as a consequence of the demagnetization process. The order of magnitude of the lattice entropy is $S_{lattice} \sim 10^{-6}N'k_BT^3$ (with N' say 10 or $100N$, N being the number of magnetic ions). The initial lattice entropy, at $T = T_{init}$ has to be smaller than the spin entropy.

$$S \propto \int \frac{\delta Q}{T} = \int \frac{C_V}{T}dT \ ,$$

by considering that at low temperature the specific heat C_V of the lattice goes as T^3 (see the Debye contribution from acoustical vibrational modes at §14.5) one approximately has

$$S_{lattice} \sim 10^{-6}N'k_BT^3$$

(curve 2 in Fig. 6.8).

Now the exchange gas is pumped out and the sample remains in poor thermal contact with the external bath. The magnetic field is slowly decreased towards zero and the **demagnetization** proceed ideally in **isoentropic** condition. The total entropy stays constant while the magnetic entropy, step after step, each in time of the order of T_1, has to return to curve 1.

Therefore $S_{lattice}$ has to **decrease** of the same amount of the **increase** of the magnetic entropy S. Then the temperature of the "internal" thermal bath has to decrease to $T_{final} \ll T_{init}$.

The amount of cooling depends from the initial external field, from the lattice specific heat and particularly from the internal residual field H_{res} that limits the value of the magnetic entropy at low temperature. In fact, it prevents the total randomization of the magnetic moments. As an order of magnitude one has $T_{final} = T_{init}(H_{res}/H_{init})$.

The adiabatic demagnetization corresponds to the exchange of entropy between the spin system and the lattice excitations. In the picture of the spin temperature (§6.1) one has an increase of the spin temperature at the expenses of the lattice temperature. The final temperature usually is in the range of milliKelvin, when the electronic magnetic moments are involved in the process. Nuclear magnetic moments are smaller than the electronic ones by a factor $10^{-3} - 10^{-4}$ and then sizeable ordering of the nuclear spins can require temperature as low as 10^{-6} K or very strong fields. In principle, by using the nuclear spins the adiabatic demagnetization could allow one to reach extremely low temperatures. However, one has to take into account that the relaxation times T_1 become very long at low temperatures (while the spin-spin relaxation time T_2 remains of the order of milliseconds). The experimental conditions are such that **negative spin temperature** can easily be attained, for instance by reversing the magnetic field.

From these qualitative considerations it can be guessed that a series of experiments of thermodynamical character based on **spin ordering** and **spin disordering** can be carried out, involving non equilibrium states when the characteristic times of the experimental steps are shorter than T_1 or T_2.

We shall limit ourself to mention that by means of adiabatic demagnetization temperature as low as 2.8×10^{-10} K have been obtained. The **nuclear** moments of Copper have been found to order antiferromagnetically at 5.8×10^{-8} K , while in Silver they order antiferromagnetically at $T_N = 5.6 \times 10^{-10}$ K and ferromagnetically at $T_c = -1.9 \times 10^{-9}$ K.

Problems F.VI

Problem F.VI.1 A magnetic field H of 10 Tesla is applied to a solid of 1 cm^3 containing $N = 10^{20}$, $S = 1/2$ magnetic ions. Derive the magnetic contribution to the specific heat C_V and to the entropy S. Then estimate the order of magnitude of C_V and S at $T = 1$ K and $T = 300$ K.

Solution:

The thermodynamical quantities can be derived from the partition function Z. From the single particle statistical average the energy is

$$< E >= \sum_i p_i \, E_i$$

and from Maxwell-Boltzmann distribution function the probability of occupation of the i-th state is

$$p_i = \frac{exp(-E_i/k_BT)}{\sum_i exp(-E_i/k_BT)} \equiv \frac{exp(-E_i\beta)}{Z} .$$

$\sum_i p_i = 1$ and the partition function normalizes the probability p_i.

The total contribution from the magnetic ions to the thermodynamical energy U (per unit volume) is

$$U = N < E >$$

and

$$< E >= \frac{\sum_i E_i\, exp(-E_i/k_BT)}{Z} = -\frac{1}{Z}\frac{\partial Z}{\partial \beta} = -\frac{\partial lnZ}{\partial \beta}$$

thus yielding

$$U = -N\frac{\partial lnZ}{\partial \beta} .$$

For $\boldsymbol{\mu} = -g\,\mu_B\,\mathbf{S}$ and $g = 2$ (see §4.4) one has

$$Z = exp\left(\frac{\mu_B H}{k_B T}\right) + exp\left(-\frac{\mu_B H}{k_B T}\right) \equiv 2\cosh x$$

with

$$x = \frac{\mu_B H}{k_B T} \equiv \beta\mu_B H.$$

For independent particles $Z_{total} = Z^N$. Then $U = -N\mu_B H \tanh x$ and

$$C_V = \left(\frac{\partial U}{\partial T}\right)_H = \left(\frac{\partial \beta}{\partial T}\right)\left(\frac{\partial E}{\partial \beta}\right)_H = -k_B\,\beta^2\left(\frac{\partial U}{\partial \beta}\right)_H ,$$

i.e.

$$C_V = Nk_B \, x^2 \mathrm{sech}^2 x \equiv \frac{N\,k_B\,x^2}{\cosh^2 x}$$

plotted below.

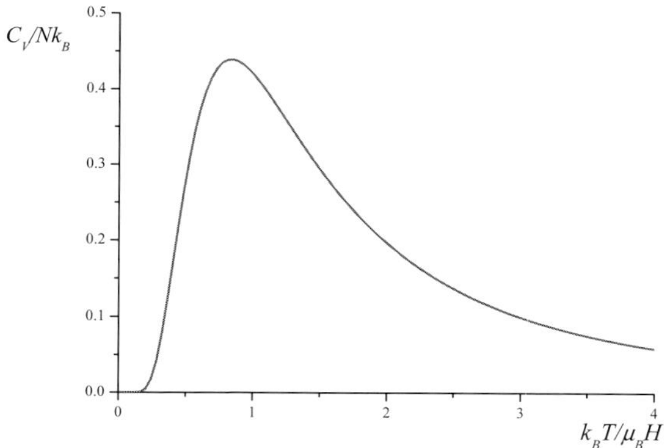

For the entropy, since (see Problem VI.1.4) $S = -k_B \sum_i p_i \, ln \, p_i$

$$S = -k_B \sum_i \left[\frac{exp(-\beta E_i)}{Z} \right] (-\beta E_i - lnZ) = \frac{<E>}{T} + k_B \, lnZ \ ,$$

so that

$$S = N\,k_B[ln\,(2\cosh x) - x\tanh x] \ ,$$

as it could also be obtained from $S = -(\partial F/\partial T)_H$ with $F = -Nk_BTlnZ$ (see the plot at §6.4).

Numerically, for $H = 10$ Tesla , $T = 1$ K corresponds to $x \gg 1$ and $T = 300$K to $x \ll 1$, so that

$$T = 1\,\mathrm{K} \ \ C_V \approx 0 \ , \ S \approx 0$$
$$T = 300\,\mathrm{K} \ \ C_V \approx 0 \ , \ S \approx k_B N ln2$$

Problem F.VI.2 A spin system $(S = 1/2)$ in a magnetic field of 10 Tesla, is prepared at a temperature close to 0 K and then put in contact with an identical spin system prepared in the condition of equipopulation of the two spin states. Find the spin temperature reached by the system after spin-spin exchanges, assuming that meantime no exchange of energy with the lattice occurs. Discuss the behavior of the entropy.

Solution:
The thermodynamical energies are

$$U_1 = 0 \qquad\qquad U_2 = \frac{N}{2}\,\mathcal{E}, \qquad \text{with } \mathcal{E} \text{ energy separation between the two spin states.}$$

From the final energy

$$U_{final} = U_1 + U_2 = \frac{N}{2}\,\mathcal{E}$$

the spin temperature is obtained by writing

$$U_{final} = \frac{2N\mathcal{E}}{Z}\, exp\left(-\frac{\mathcal{E}}{k_B\,T_{spin}}\right)\;.$$

Thus

$$1 = \frac{4}{exp(\mathcal{E}/k_B\,T_{spin}) + 1}$$

and $T_{spin} \simeq \mathcal{E}/(k_B \cdot 1.1)$. For $\mathcal{E} = 2\mu_B H$ one has $T_{spin} = 12.2$ K.
For the entropy see Problem F.VI.1 and look at Figure 6.1, by taking into account that no energy exchange with the reservoir is assumed to occur.

It is noted that the increase of the entropy can be related to the irreversibility of the process.

Problem F.VI.3 Prove that the mean square deviation of the energy of a system from its mean value (due to exchange of energy with the reservoir) is given by $k_B\,T^2\,C_V$, C_V being the heat capacity.

Solution:
The mean square deviation is

$$<(E - <E>)^2> = <E^2 - 2E<E> + <E>^2> = <E^2> - <E>^2$$

where

$$<E> = \sum_i \frac{E_i\,exp(-E_i/k_B T)}{Z} = -\frac{1}{Z}\frac{\partial Z}{\partial \beta}$$

while

$$<E^2> = \sum_i \frac{E_i^2\,exp(-E_i/k_B T)}{Z} = \frac{1}{Z}\frac{\partial^2 Z}{\partial \beta^2}.$$

Therefore

$$\langle E^2\rangle - \langle E\rangle^2 = \frac{\partial}{\partial\beta}\left[\frac{1}{Z}\frac{\partial Z}{\partial\beta}\right] = -\frac{\partial\langle E\rangle}{\partial\beta}.$$

Since

$$\frac{\partial}{\partial\beta} = -k_B T^2 \frac{\partial}{\partial T} \quad \text{and} \quad \frac{\partial\langle E\rangle}{\partial T} = C_V$$

one has

$$\langle (E - \langle E\rangle)^2\rangle = k_B T^2 C_V \ ,$$

another example of fluctuation-dissipation relationships (see Eq. 6.14).

The **fractional deviation** of the energy $\left[(\langle E^2\rangle - \langle E\rangle^2)/\langle E\rangle^2\right]^{\frac{1}{2}}$ at high temperatures, where $\langle E\rangle \approx Nk_B T$ and $C_V \approx Nk_B$ is of the order of $N^{-1/2}$, a very small number for N of the order of the Avogadro number (see Problem VI.1.1).

Problem F.VI.4 Compare the magnetic susceptibility of non-interacting magnetic moments $S = 1/2$ with the classical $S = \infty$ limit (where any orientation with respect to the magnetic field is possible).
Solution:
For $S = \infty$

$$\langle\mu\cos\theta\rangle = \mu\int\cos\theta\,exp\left(\frac{\mu H\cos\theta}{k_B T}\right)\sin\theta\,d\theta \Big/ \int exp\left(\frac{\mu H\cos\theta}{k_B T}\right)\sin\theta\,d\theta$$

$$= \mu\left[\coth\frac{\mu H}{k_B T} - \left(\frac{\mu H}{k_B T}\right)^{-1}\right]$$

$$\approx \frac{\mu^2 H}{3k_B T}$$

yielding the Langevin-like susceptibility.
For $S = 1/2$ (see §4.4)

$$\langle\mu\rangle = \mu\frac{\left[exp\left(\frac{\mu H}{k_B T}\right) - exp\left(-\frac{\mu H}{k_B T}\right)\right]}{\left[exp\left(\frac{\mu H}{k_B T}\right) + exp\left(-\frac{\mu H}{k_B T}\right)\right]} \approx \frac{\mu^2 H}{k_B T}$$

Problem F.VI.5 By taking inspiration from Fig. 6.7, devise an experimental procedure suitable to measure the spin-lattice relaxation time T_1.
 Solution:

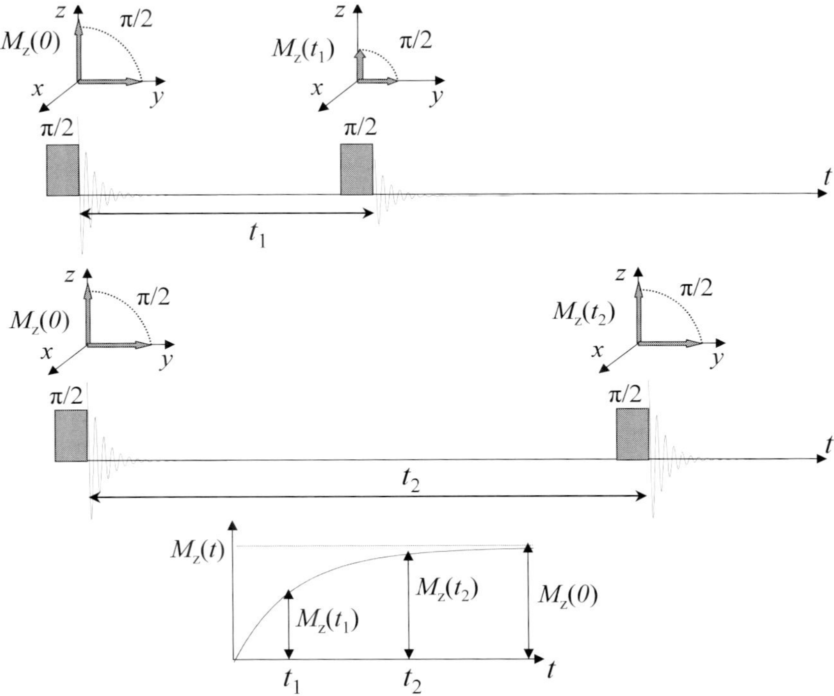

At $t{=}0$ the $(\pi/2)$ pulse brings the magnetization along y, saturating the populations of the two levels and yielding the **FID** signal. After a time t_1 a second $(\pi/2)$ pulse measures the magnetization $M_z(t_1)$ during the recovery towards the equilibrium value. By applying pairs of pulses with different t_1's (e.g. t_2) the recovery plot towards the equilibrium is constructed and T_1 is extracted.

Problem F.VI.6 For an ensemble of particles with a ground state at spin $S = 0$ and the excited state at energy Δ and spin $S = 1$, derive the paramagnetic susceptibility.

Then, by resorting to the fluctuation-dissipation theorem (see Eq. 6.14 and Problem F.IV.8) show that the same result is obtained.

Solution:

The energy levels are sketched below:

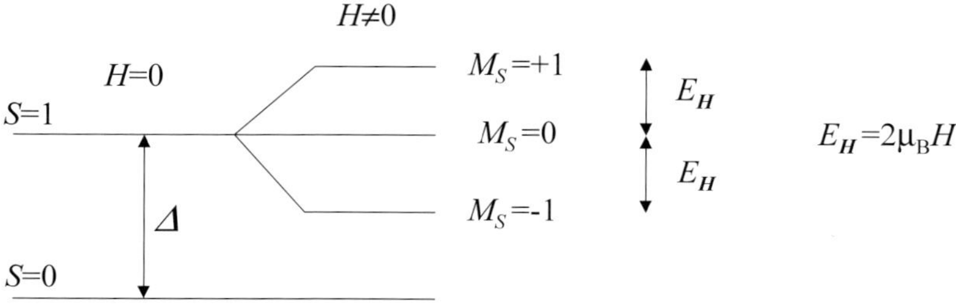

The direct expression for the single particle susceptibility is

$$\chi = \chi_0 p_0 + \chi_1 p_1$$

where $\chi_0 = 0$, $\chi_1 = \mu_B^2 g^2 S(S+1)/3k_B T = 8\mu_B^2/3k_B T$ and

$$p_{0,1} = \frac{(2S+1)e^{-\beta E_{0,1}}}{Z} \quad ,$$

with Z partition function. For $E_0 = 0$ and $E_1 = \Delta$ one has

$$\chi = \frac{8\mu_B^2}{k_B T} \frac{e^{-\beta\Delta}}{(1 + 3e^{-\beta\Delta})}$$

It is noted that the above equation is obtained in the limit of **evanescent field**, condition that will be retained also in the subsequent derivation. The magnetization is

$$M = N_{-1}\mu_z + N_0.0 - N_{+1}\mu_z$$

with $\mu_z = 2\mu_B$. Then

$$M = \frac{N2\mu_B}{Z}\{e^{-\beta(\Delta - E_H)} - e^{-\beta(\Delta + E_H)}\}$$

with $Z = 1 + e^{-\beta(\Delta - E_H)} + e^{-\beta\Delta} + e^{-\beta(\Delta + E_H)}$. Therefore,

$$M = N2\mu_B \frac{e^{-\beta\Delta}[e^{\beta E_H} - e^{-\beta E_H}]}{1 + e^{-\beta\Delta}[e^{\beta E_H} + 1 + e^{-\beta E_H}]}$$

and for $\beta E_H \ll 1$

$$M = 2\mu_B N e^{-\beta\Delta} \frac{2\beta E_H}{1 + 3e^{-\beta\Delta}} = \frac{8\mu_B^2 N}{k_B T} \frac{e^{-\beta\Delta}}{1 + 3e^{-\beta\Delta}} H,$$

yielding the susceptibility obtained from the direct expression.

From the fluctuation-dissipation relationship (see Eqs. 6.11 - 6.14), being the fluctuations uncorrelated
$< \Delta M^2 >= N < \Delta\mu_z^2 >$ with $< \Delta\mu_z^2 >=< \mu_z^2 > - < \mu_z >^2$, and

$$< \mu_z^2 >= 4\mu_B^2 \frac{\sum_{M_S,S} M_S^2 e^{-\beta E(M_s,S)}}{Z} = 4\mu_B^2 \{ \frac{e^{-\beta(\Delta - E_H)} + e^{-\beta(\Delta + E_H)}}{Z} \} \ .$$

From M

$$< \mu_z^2 >= 4\mu_B^2 \frac{[e^{-\beta(\Delta - E_H)} + e^{-\beta(\Delta + E_H)}]^2}{Z^2}$$

Thus

$$< \Delta M^2 >= 4N\mu_B^2 \frac{e^{-\beta\Delta}}{Z} \{ e^{\beta E_H} + e^{-\beta E_H} - \frac{e^{-\beta\Delta}}{Z}(e^{\beta E_H} - e^{-\beta E_H})^2 \}$$

and again for $\beta E_H \ll 1$

$$< \Delta M^2 >= 4N\mu_B^2 \frac{e^{-\beta\Delta}}{Z} \{ 2 - \frac{e^{-2\beta\Delta}}{Z}(\beta E_H)^2 \} \simeq 8N\mu_B^2 \frac{e^{-\beta\Delta}}{1 + 3e^{-\beta\Delta}} = k_B T \chi.$$

Problem F.VI.7 Consider an ideal paramagnet, with $S = 1/2$ magnetic moments. Derive the expression for the relaxation time T_1 in terms of the transition probability W (due to the time-dependent spin-lattice interaction) driving the recovery of the magnetization to the equilibrium, after a perturbation leading to a spin temperature T_s, different from the temperature $T = 300$ K of the thermal reservoir. Find the time-evolution of the spin temperature starting from the initial condition $T_s = \infty$.
   Solution:
The instantaneous statistical populations are

$$N_- = \frac{N}{Z} e^{\mu_B H \beta_s} \simeq \frac{N}{Z}(1 + \beta_s E_-)$$

with $\beta_s = 1/k_B T_s$

$$N_+ = \frac{N}{Z} e^{-\mu_B H \beta_s} \simeq \frac{N}{Z}(1 - \beta_s E_+) \ ,$$

while at the thermal equilibrium

$$N_{\mp}^{eq} \simeq \frac{N}{Z}(1 \pm \beta E_{\mp}) \simeq \frac{N}{2}(1 \pm \beta \Delta E)$$

with $\beta = 1/k_B T$ and $\Delta E = 2\mu_B H$.

From the equilibrium condition $N_- W_{-+} = N_+ W_{+-}$ one deduces

$$W_{+-} = W_{-+}\frac{N_-}{N_+} \simeq W_{-+}\frac{1 + \beta E_-}{1 - \beta E_+}$$

and $W_{+-} \simeq W(1 + \beta \Delta E)$, with $W_{-+} \equiv W$.

Since

$$\frac{dN_-}{dt} = -N_- W + N_+ W(1+\beta\Delta E) = -N_- W + (N-N_-)W(1+\beta\Delta E) = -2N_- W + 2N$$

then $N_-(t) = ce^{-2Wt} + N_-^{eq}$ and from the initial condition

$$N_-(t) = (N_-^{init} - N_-^{eq})e^{-2Wt} + N_-^{eq}$$

Evidently $dN_+/dt = -dN_-/dt$.

From the magnetization $M_z(t) \propto (N_- - N_+)$ one has $dM_z/dt \propto 2(dN_-/dt)$ and

$$M_z(t) = (M_z^{init} - M_z^{eq})e^{-2Wt} + M_z^{eq}$$

implying $1/T_1 = 2W$.

M_z is also inversely proportional to T_s and then one approximately writes

$$\beta_s(t) = (\beta_s^{init} - \beta)e^{-2Wt} + \beta$$

and for $\beta_s^{init} = 0$, $\beta_s(t) = \beta(1 - e^{-2Wt})$

$$T_s(t) = \frac{T}{1 - e^{-2Wt}} \qquad\qquad (a)$$

For exact derivation, over all the temperature range, from Problem VI.1.2 $T_s = (2\mu_B H/k_B)/ln(u^{-1} - 1)$ with $u = N_+/N$. Then the exact expression of the spin temperature is

$$T_s = \frac{2\mu_B H}{k_B}\left[ln\left(\frac{(N/2 - N_-^{eq})e^{-2Wt} + N_-^{eq}}{N - (N/2 - N_-^{eq})e^{-2Wt} - N_-^{eq}}\right)\right]^{-1} \qquad\qquad (b)$$

See plots below.

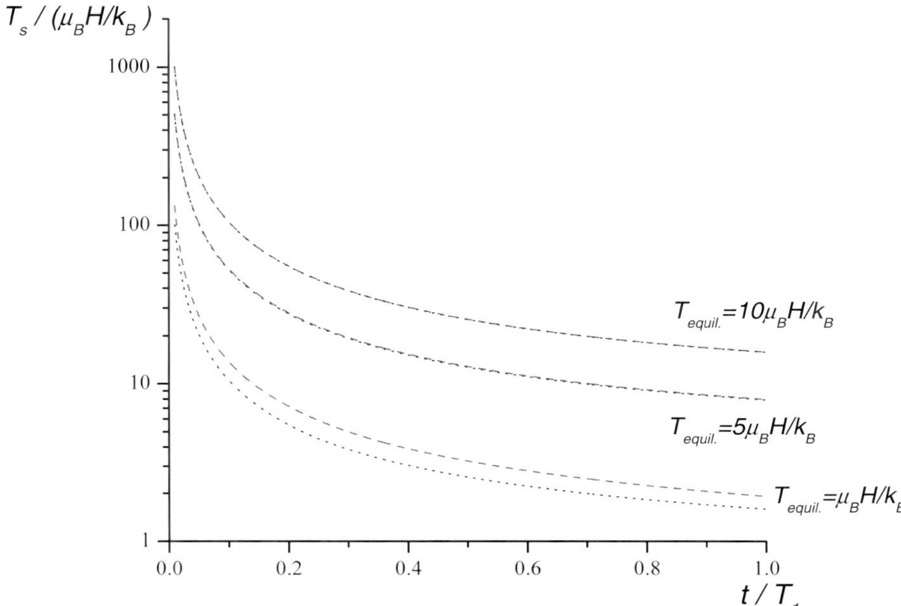

Plot of Eq. a) (dotted line) and Eq. b) (dashed line) showing the equivalence of the two procedures for $T > \mu_B H/k_B$. (In plotting Eq. b) keep at least three significant digits in the expansion.)

Problem F.VI.8 An hypothetical crystal has a mole of Na atoms, each at distance $d = 1$ Å from a point charge ion of charge $-e$ (and no magnetic moment). By taking into account the quadrupole interaction (§5.2) derive the energy, the entropy and the specific heat of the crystal around room temperature (Na nuclear spin $I = 3/2$ and nuclear quadrupole moment $Q = 0.14 \times 10^{-24}$ cm^2).

Solution:
The eigenvalues being $E_{\pm 1/2} = 0$ and $E_{\pm 3/2} = eQV_{zz}/2 = E$ (see Prob. V.3.2), the partition function is written

$$Z = [\sum_{M_I} e^{-\beta E_{M_I}}]^{N_A} = [2(1 + e^{-\beta E})]^{N_A} .$$

Then the free energy is

$$F = -k_B T \ln Z = -N_A k_B T \ln(1 + e^{-\beta E}) - N_A k_B T \ln 2$$

and

$$U = -\frac{\partial}{\partial\beta}lnZ = \frac{N_A E e^{-\beta E}}{(1 + e^{-\beta E})}$$

and

$$S = -\frac{\partial F}{\partial T} = N_A k_B ln(1 + e^{-\beta E}) + \frac{N_A E}{T}\frac{e^{-\beta E}}{(1 + e^{-\beta E})} + k_B N_A ln2$$

Since $E \sim 10^{-8}$ eV $\ll k_B T$, U and S can be written

$$U = \frac{N_A E}{2}\left(1 - \frac{1}{2}\frac{E}{k_B T}\right)$$

and

$$S \simeq R\left(2ln2 - \frac{1}{8}\frac{E^2}{k_B^2 T^2}\right)$$

(see Prob. F.VI.1 for the analogous case). From $C_V = \partial U/\partial T$, in the high temperature limit $C_V \simeq (1/4)R(E/k_B T)^2$, the high-temperature tail of the Schottky anomaly already recalled at Problem F.V.5.

7

Molecules: general aspects

Topics

Separation of electronic and nuclear motions
Symmetry properties in diatomic molecules
Labels for electronic states
One electron in axially symmetrical potential

In this Chapter we shall discuss the general aspects of the first state of "bonded matter", the aggregation of a few atoms to form a molecule. The related issues are also relevant for biology, medicine, astronomy etc. The knowledge of the quantum properties of the electronic states in molecules is the basis in order to create new materials, as the ones belonging to the "artificial matter", often obtained by means of subtle manipulations of atoms by means of special techniques.

We shall understand why the molecules are formed, why the H_2 molecule exists while two He atoms do not form a stable system, why the law of definite proportions holds or why there are multiple valences, what controls the geometry of the molecules. These topics have to follow as natural extension of the properties of atoms. Along this path new phenomena, typical of the realm of the molecular physics, will be emphasized.

In principle, the Schrödinger equation for nuclei and electrons contains all the information we wish to achieve. In practice, even the most simple molecule, the Hydrogen molecule-ion H_2^+, cannot be exactly described in the framework of such an approach: the Schrödinger equation is solved only when the nuclei are considered fixed. Therefore, in most cases we will have to deal with simplifying assumptions or approximations, which usually are not of mathematical character but rather based on the physical intuition and that must be supported by experimental findings.

The first basic assumption we will have to take into account is the **Born-Oppenheimer** approximation, essentially relying on the large ratio of the nuclear and electronic masses. It allows one to deal with a kind of separation between the motions of the electrons and of the nuclei. Another approximation that often will be used involves tentative wavefunctions for the electronic states as linear combination of a set of basis functions, that can help in finding appropriate solutions. For instance, a set of wavefunctions **centered at the atomic sites** will allow one to arrive at the **secular equation** for the approximate eigenvalues.

Finally in this Chapter we have to find how to label the electronic states in terms of **good quantum numbers**. This will be done in a way similar to the one in atoms, by relying on the **symmetry properties** of the potential energy (for example, the cylindrical symmetry) and by referring to the limit atomic-like situations of united-atoms or of separated-atoms.

7.1 Born-Oppenheimer separation and the adiabatic approximation

For a system of nuclei and electrons the Hamiltonian is written (see Fig. 7.1)

$$\mathcal{H} = -\frac{\hbar^2}{2} \sum_{\alpha} \frac{\nabla_{\alpha}^2}{M_{\alpha}} - \frac{\hbar^2}{2m} \sum_{i} \nabla_i^2 + \sum_{i<j} \frac{e^2}{r_{ij}} + \sum_{\alpha<\beta} \frac{Z_{\alpha} Z_{\beta} e^2}{R_{\alpha\beta}} - \sum_{\alpha,i} \frac{Z_{\alpha} e^2}{r_{i\alpha}} \equiv$$

$$\equiv T_n + T_e + V_{ee} + V_{nn} + V_{ne} \tag{7.1}$$

The corresponding wave function $\phi(\mathbf{R}, \mathbf{r})$ involves both the group $\mathbf{R}$ of the nuclear coordinates and the group $\mathbf{r}$ for the electrons. In the Hamiltonian the spin-orbit interactions and the hyperfine interactions have not been included, since at a first stage they can be safely neglected.

In order to solve the Schrödinger equation for $\phi(\mathbf{R}, \mathbf{r})$ one observes the large difference in nuclear and electronic masses (and the related differences in the electronic and roto-vibrational energies, as it will appear in subsequent Chapters). This difference suggests that in time intervals much shorter than the ones required for the nuclei to sizeably change their positions, the electrons have been able to take the quantum configuration pertaining to ideally fixed coordinates $\mathbf{R}$. Then one can attempt an eigenfunction of the form

$$\phi(\mathbf{R}, \mathbf{r}) = \phi_n(\mathbf{R})\phi_e(\mathbf{R}, \mathbf{r}), \tag{7.2}$$

where $\phi_n(\mathbf{R})$ pertains to the nuclei, while the electronic wavefunction ϕ_e involves only **parametrically** the nuclear coordinates, these latter ideally **frozen** in the configuration specified by $\mathbf{R}$. When such a function is included in the Schrödinger equation for the Hamiltonian given by Eq. 7.1 and the following equivalences are taken into account

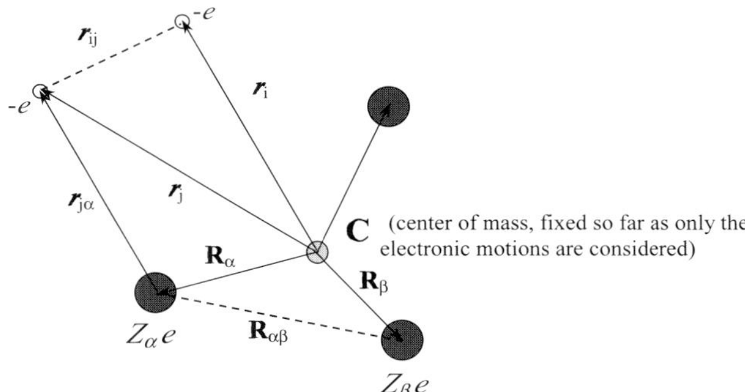

Fig. 7.1. Nuclear and electronic coordinates used in Eq. 7.1.

$$T_e \phi_n \phi_e = \phi_n T_e \phi_e \ ,$$

$$T_n \phi_n \phi_e \equiv -\hbar^2 \sum_\alpha \frac{\nabla_\alpha^2}{2M_\alpha} \phi_n \phi_e = -\hbar^2 \sum_\alpha \frac{1}{2M_\alpha} \boldsymbol{\nabla}_\alpha \cdot \{\phi_e \boldsymbol{\nabla}_\alpha \phi_n + \phi_n \boldsymbol{\nabla}_\alpha \phi_e \ \} =$$

$$= \phi_e T_n \phi_n + \phi_n T_n \phi_e - 2\hbar^2 \sum_\alpha \frac{1}{2M_\alpha} \boldsymbol{\nabla}_\alpha \phi_e \cdot \boldsymbol{\nabla}_\alpha \phi_n \ ,$$

then one has

$$\left[-\sum_\alpha \frac{\hbar^2}{2M_\alpha} (2\boldsymbol{\nabla}_\alpha \phi_e \cdot \boldsymbol{\nabla}_\alpha \phi_n) - \sum_\alpha \frac{\hbar^2}{2M_\alpha} \phi_n \nabla_\alpha^2 \phi_e \right] + \tag{7.3}$$

$$+\phi_e T_n \phi_n + \phi_n T_e \phi_e + (V_{nn} + V_{ne} + V_{ee})\phi_e \phi_n = E\phi_e \phi_n.$$

Let us assume that the terms included in the square brackets can be neglected (the conditions for such an approximation, essentially corresponding to the so-called **adiabatic approximation**, shall be discussed subsequently). For the electronic wavefunction one can write

$$T_e \phi_e + (V_{ne} + V_{ee})\phi_e = E_e^{(g)} \phi_e, \tag{7.4}$$

where $E_e^{(g)}(\mathbf{R})$ is the eigenvalue for the electrons in a frozen nuclear configuration. Then, from Eq. 7.3, by neglecting the terms in square-brackets, after dividing by ϕ_e one obtains the equation for the nuclear motions:

$$\left[T_n + V_{eff}(\mathbf{R}) \right] \phi_n = E\phi_n \tag{7.5}$$

with $V_{eff}(\mathbf{R}) = V_{nn}(\mathbf{R}) + E_e^{(g)}(\mathbf{R})$. In Eq. 7.5 the effective Hamiltonian includes the eigenvalue for the electrons $E_e^{(g)}$, for given $\mathbf{R}$'s, as **effective potential energy**.

Thus, by assigning to the nuclear and electronic states the appropriate set of **quantum numbers** ν and g, under the approximations discussed above the wavefunction solution for the Hamiltonian 7.1 is

$$\phi^{(g,\nu)}(\mathbf{r}, \mathbf{R}) = \phi_e^{(g)}(\mathbf{r}, \mathbf{R})\phi_n^{(\nu)}(\mathbf{R}), \tag{7.6}$$

with ϕ_e and ϕ_n eigenfunctions from Eqs. 7.4 and 7.5 respectively.

The electronic eigenvalue $E_e^{(g)}$, entering the effective potential energy in Eq. 7.5, is not a number as in atoms but **parametrically** depends from the nuclear coordinates. The total energy of the molecule can be written

$$E^{(g,\nu)} = E_e^{(g)}(\mathbf{R}_m) + V_{nn}(\mathbf{R}_m) + E_n^{(\nu)} , \tag{7.7}$$

where $\mathbf{R}_m$ means the nuclear configuration corresponding to the minimum for $V_{eff}(\mathbf{R})$ (see Eq. 7.5).

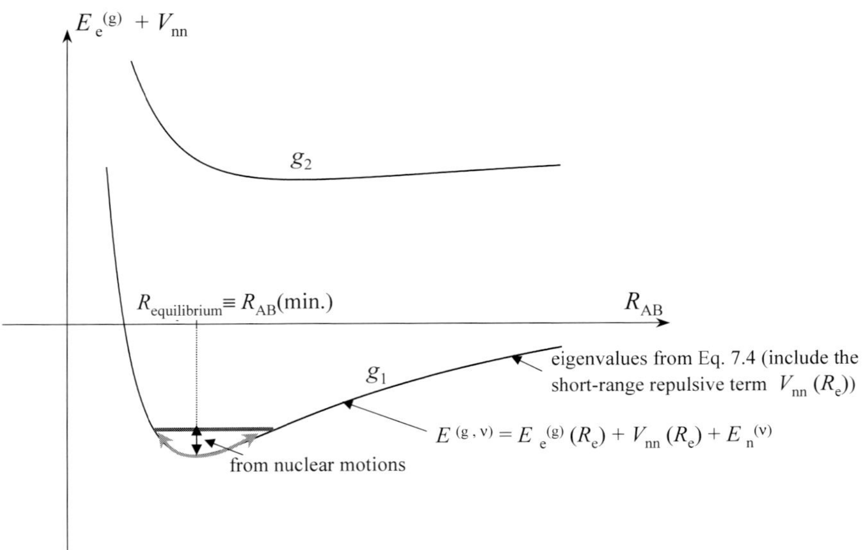

Fig. 7.2. Schematic view of the separation of the electronic and vibrational energies in a diatomic molecule and of the role of $E_e^{(g)}$ as effective potential energy for the nuclear motion, within the adiabatic approximation. The vibrational motion occurs in an effective potential energy, while the electrons follow **adiabatically** this motion.

The physical contents of such a framework are more easily grasped by referring to a diatomic molecule, where in practice the only parameter required to fix the nuclear configuration is the distance R_{AB} between the two nuclei, since as a first approximation the electronic states can be considered unaffected by the rotation of the molecule. A schematic view of the energy of the molecule for the electronic states as a function of R_{AB} and of the effect of the eigenvalue for the vibrational motion of the nuclei (corresponding to a variation of R_{AB}) is given in Fig. 7.2. Complete understanding of this illustration will be achieved after reading §8.1 and §10.3.1.

Let us briefly comment on the possibility to neglect the terms in square brackets in Eq. 7.3, corresponding to the validity of the adiabatic approximation. The order of magnitude of the contribution of those terms to the energy can be estimated by looking at the expectation values

$$< \phi_e \phi_n | \left[\cdots \right] | \phi_e \phi_n >,$$

Therefore a first term is

$$-\frac{\hbar^2}{M_\alpha} \int \phi_e^* \phi_n^* \boldsymbol{\nabla}_n \phi_e \cdot \boldsymbol{\nabla}_n \phi_n d\tau_n d\tau_e$$

that for $\phi_e(\mathbf{r}, \mathbf{R})$ in real form, becomes proportional to

$$\int \phi_e^* \boldsymbol{\nabla}_n \phi_e d\tau_e \propto \boldsymbol{\nabla}_n \int \phi_e^* \phi_e d\tau_e$$

which is zero for a given electronic state g_1. The second term is

$$-\frac{\hbar^2}{M_\alpha} \int \phi_n^* \phi_n d\tau_n \int \phi_e^* \nabla_n^2 \phi_e d\tau_e,$$

and by taking into account that the electronic wavefunction depends on $(\mathbf{r} - \mathbf{R})$, one can write

$$-\frac{\hbar^2}{M_\alpha} \int \phi_e^* \nabla_n^2 \phi_e d\tau_e \simeq \frac{m}{M_\alpha} < |T_e| >,$$

which is of the order of the contribution to the energy from the electronic kinetic term scaled by the factor m/M_α and thus negligible.

Finally one would have to consider the non-diagonal terms involving the operator $\boldsymbol{\nabla}_n$, of the form

$$\int \phi_e^{(g_2)*} \boldsymbol{\nabla}_n \phi_e^{(g_1)} d\tau_e. \tag{7.8}$$

These terms can be different from zero and in principle they drive transitions between electronic states associated with the nuclear motions, in other words to the **non-adiabatic** contributions. For large separation between the electronic states compared to the energy of the thermal motions, the transition probability is expected to be small.

One should remark that relevant effects in molecules (and in solids) actually originate from the non-adiabatic terms. We just mention **pre-dissociation** (spontaneous separation of the atoms), some **removal of degeneracy** in electronic states, the **Jahn-Teller** effect and, in solids, **resistivity** and **superconductivity**, related to the interaction of the electrons with the vibrations of the ions around their equilibrium positions (see Chapters 13 and 14).

7.2 Classification of the electronic states

7.2.1 Generalities

As in atoms, also in the molecules first one has to find how to label the electronic states in terms of constants of motions, namely derive the **good quantum numbers**. In atoms $\mathbf{l}^2$ and $\mathbf{l}_z$ commute with the central field Hamiltonian and then n,l, and m have been used to classify the one-electron states. Also for molecules the symmetry arguments play a relevant role: a rigorous classification is possible only for diatomic (or at least linear molecules) so that one axis of rotational symmetry is present.

Let us refer to Fig. 7.3. When the z axis is aligned along the molecular axis the potential energy V is a function of the cylindrical coordinates z and ρ, while it does not involve the angle φ. Then the l_z operator $-i\hbar\frac{\partial}{\partial\varphi}$ commutes with the Hamiltonian:

$$\left[l_z, V\right] \propto (x\frac{\partial}{\partial y} - y\frac{\partial}{\partial x})V\phi - V(x\frac{\partial}{\partial y} - y\frac{\partial}{\partial x})\phi = (\mathbf{r} \times \boldsymbol{\nabla}V)_z,$$

which is zero when the z axis is along the molecular axis.
For **homonuclear** molecules $(A = B)$ in terms of the positional vector $\mathbf{r}$ one has

$$|\phi(\mathbf{r})|^2 = |\phi(-\mathbf{r})|^2 \quad i.e. \quad \phi(\mathbf{r}) = -\phi(-\mathbf{r}) \quad \text{or} \quad \phi(\mathbf{r}) = +\phi(-\mathbf{r})$$

and one can classify the states with a letter **g** (from **gerade**) or **u** (from **ungerade**) according to the **even** or **odd** parity under the inversion of $\mathbf{r}$ with respect to the center of the molecule (see Fig. 7.3). One should also remark that the reflection with respect to the yz plane, bringing x in $-x$, changes the sign of the z-component of the angular momentum while the Hamiltonian is invariant. It follows that the energy must depend on the square of the l_z-eigenvalue while this operator has to convert the eigenfunction in the one having eigenvalue of opposite sign. The electronic states with l_z-eigenvalue different from zero must be **double degenerate**, each of the two states corresponding to different direction of the projection of the orbital angular momentum along the z-axis. On the other hand, for $\mathbf{l}_z$-eigenvalue equal to zero a further $-$ or $+$ sign has to be used to describe the behavior of the wavefunction upon reflection with respect to the planes containing the molecular axis.

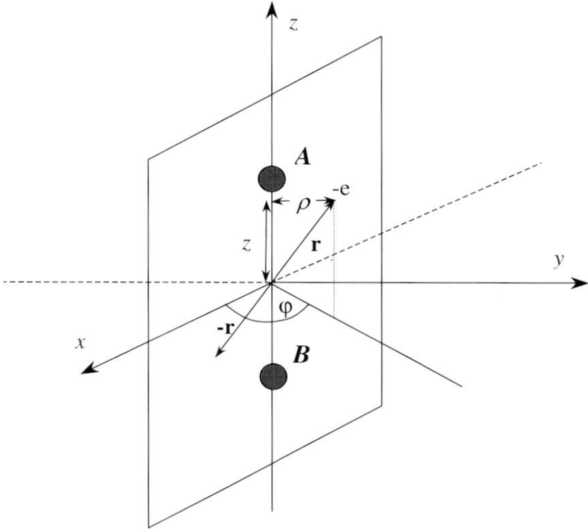

Fig. 7.3. Schematic view for the discussion of the symmetry arguments involved in the classification of one-electron states in a diatomic molecule. A and B are nuclei dressed by the electrons uninvolved in the bonding mechanism. When $A = B$ the molecule is homonuclear and it acquires the **inversion symmetry** with respect to the center. Then $\phi_e(\mathbf{r}) = \pm\phi_e(-\mathbf{r})$ and the classification **gerade** or **ungerade**, according to the sign of the wavefunction upon inversion (parity), becomes possible.

Finally, in these introductory remarks it is noted that the z-component of the total angular momentum, implying an algebric sum $L_z = \sum_i l_z^i$, is also a constant of motion, with associated a good quantum number M_L (see §7.2.3).

7.2.2 Schrödinger equation in cylindrical symmetry

By referring again to Fig. 7.3 and in the framework of the Born-Oppenheimer separation, the Schrödinger equation for the one-electron wavefunction is

$$\frac{-\hbar^2}{2m}\nabla^2_{z,\rho,\varphi}\phi + V\phi = E(R_{AB})\phi \ , \tag{7.9}$$

where $V = V(z,\rho)$. One should remark that if A and B are protons, namely we are dealing with the Hydrogen molecule ion, then

$$V = -\frac{e^2}{r_A} - \frac{e^2}{r_B}. \tag{7.10}$$

By using ellipsoidal coordinates with the nuclei at the foci of the ellipse, then Eq. 7.9, with V as in Eq. 7.10, is exactly solvable in a way similar to

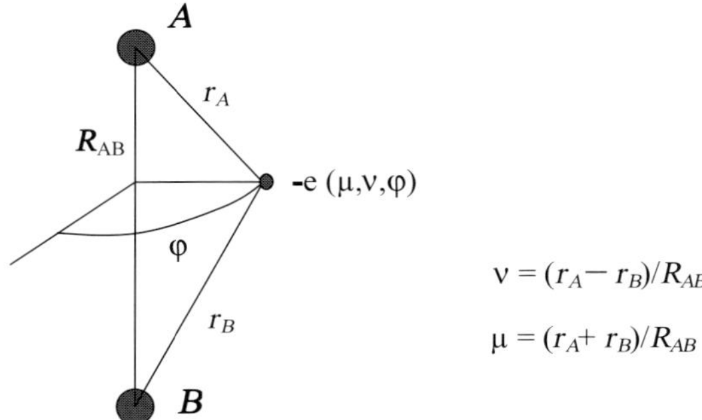

Hydrogen atom, with separation of the variables.

This solution would not be of much help, since when diatomic molecules with the nuclei dressed by the atomic (**core**) electrons have to be considered, the potential is no longer of the form in Eq. 7.10 and therefore relevant modifications can be expected. A similar modification in the atom is the removal of the accidental degeneracy upon abandoning the Coulomb potential. Thus we prefer to disregard the formal solution of Eq. 7.9 for strictly Coulomb-like potential and first give the general properties of electronic states just by referring to the cylindrical symmetry of V (again in a way analogous to atoms, where only the spherical symmetry of the Hamiltonian in the central field approximation was taken into account). Subsequently approximate methods will allow us to derive specific forms of the wavefunctions of more general use, rather than the exact expressions pertaining to the Hydrogen molecule ion.

The kinetic energy operator in cylindrical coordinates reads

$$\nabla^2_{z,\rho,\varphi} = \frac{\partial^2}{\partial\rho^2} + \frac{1}{\rho}\frac{\partial}{\partial\rho} + \frac{\partial^2}{\partial z^2} + \frac{1}{\rho^2}\frac{\partial^2}{\partial\varphi^2} \tag{7.11}$$

and by factorizing ϕ in the form $\phi = \chi(z,\rho)\Phi(\varphi)$ Eq. 7.9 is rewritten

$$\frac{2m\rho^2}{\hbar^2}\left[E(R_{AB}) - V(z,\rho)\right] + \frac{\rho^2}{\chi}\left[\frac{\partial^2\chi}{\partial z^2} + \frac{\partial^2\chi}{\partial\rho^2}\right] + \frac{\rho}{\chi}\frac{\partial\chi}{\partial\rho} = -\frac{1}{\Phi}\frac{\partial^2\Phi}{\partial\varphi^2}, \tag{7.12}$$

where at the first member one has only operators and functions of z and ρ while at the second member only of φ. As a consequence, Eq. 7.12 leads to solutions of the form $\phi = \chi\Phi$, where χ and Φ originate from the separate equations in which both members are equal to a constant independent on z, ρ and φ. We label that constant λ^2 and then

$$\frac{\partial^2 \Phi}{\partial \varphi^2} = -\lambda^2 \Phi$$

so that

$$\Phi = A e^{i\lambda\varphi} + B e^{-i\lambda\varphi}. \tag{7.13}$$

The boundary condition for Φ is

$$\Phi(\varphi + 2\pi n) = \Phi(\varphi)$$

and $exp(i\lambda n 2\pi) = 1$, thus yielding λ **integer**.

The meaning of the number λ can be directly grasped by looking for the eigenvalue of the z component of the angular momentum of the electron:

$$l_z \phi = a\phi \quad i.e. \quad -i\hbar \frac{\partial}{\partial \varphi} \chi \Phi = \chi(-i\hbar \frac{\partial e^{\pm i\lambda\varphi}}{\partial \varphi}) = \pm \lambda \hbar \chi \Phi = \pm \lambda \hbar \phi \; ,$$

namely λ measures in $\hbar$ unit the component of $\mathbf{l}$ along the molecular axis, a constant of motion, as it was anticipated.

From Eq. 7.12 it is realized that the eigenvalue $E(R_{AB})$ depends on λ^2. Therefore we understand that from a given atomic-like state of angular momentum $\mathbf{l}$, the presence of the second atom at the distance R_{AB} generates $(l+1)$ states of different energy. These states correspond to $l_z = 0, \pm 1, \pm 2...$ and are in general double degenerate, in agreement with the fact that the energy cannot depend on the sign of l_z, as we have previously observed. These one-electron states are labelled by the letters $\sigma, \pi, \delta.....$ in correspondence to $0, 1, 2$, etc... similarly to the atomic states $s, p, d...$

7.2.3 Separated-atoms and united-atoms schemes and correlation diagram

Other good quantum numbers for the electronic states to be associated with (z, ρ) in Eq. 7.12 can be introduced only when $\chi(z, \rho)$ can be factorized in two functions, involving separately z and ρ. This happens when one refers to the limit situations of united atoms (i.e. $R_{AB} \to 0$) or of separated atoms (i.e. $R_{AB} \to \infty$). In the **united-atoms** classification scheme (for example the Hydrogen molecule ion H_2^+ tends to become the He^+ atom) the two further quantum numbers are n and l, while λ tends to become m. Then the sequence of the states is

$$1s\sigma, 2s\sigma, 2p\sigma, 2p\pi...$$

and the parity **g** or **u** is fixed by the value of l, namely l even **g**, l odd **u**.

For $R_{AB} \to \infty$ the atoms are far away (H_2^+ becomes H with a proton at large distance) and for heteronuclear molecule one has

$$\sigma 1s_A, \sigma 1s_B, \sigma 2s_A, \sigma 2s_B, \sigma 2p_A...$$

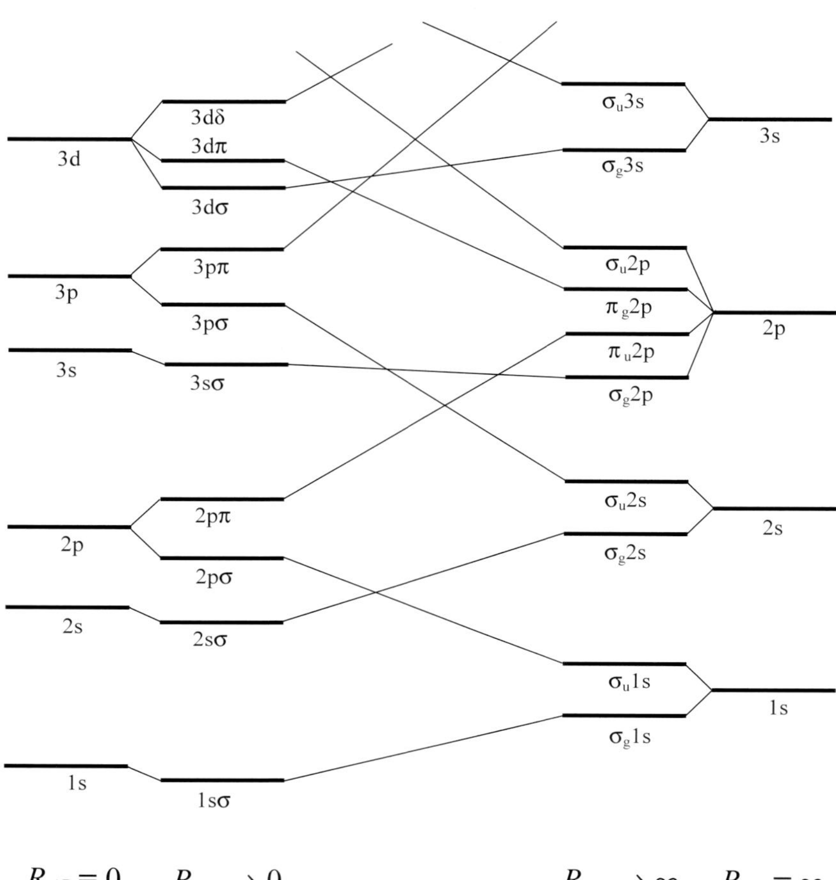

Fig. 7.4. Classification schemes for diatomic homonuclear molecules and correlation lines yielding a sketchy behavior of the eigenvalues $E(R_{AB})$ as a function of the interatomic distance.

For $A = B$ (homonuclear molecule) $(nl)_A = (nl)_B$, the splitting of the level due to the perturbing effect of the other nucleus (e.g. H^+ in the Hydrogen molecule ion) removes the degeneracy and the character **g** or **u** can be assigned.

The two classification schemes are obviously correlated. For the lowest energy levels the correlation can be established by direct inspection, by taking into account that λ and the **g** or **u** character do not depend on the distance R_{AB} (see Fig. 7.4). A pictorial view of the correlation diagram in terms of transformation of the orbitals upon changing the distance R_{AB} is given in

Fig. 7.5, having assumed the one-electron wavefunction in the form of linear combination of $1s$-atomic like wave functions and $2p_x$-wavefunctions, centered at the two sites A and B (for the proper description see §8.1).

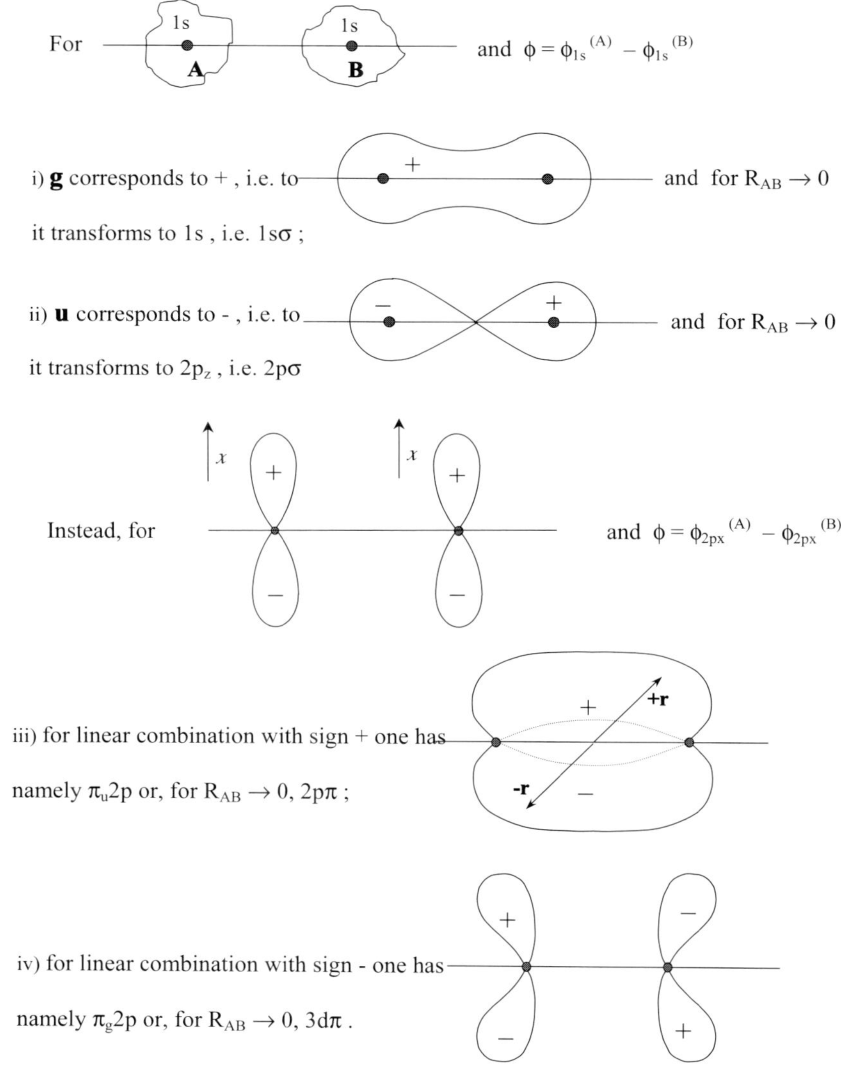

Fig. 7.5. Schematic view of the correlation diagram by referring to the transformation with the interatomic distance R_{AB} of the shape of the molecular orbitals generated by linear combination of atomic $1s$ (cases i) and ii)) and $2p_x$ orbitals (cases iii) and iv)) centered at the A and B sites (see §8.1 for details).

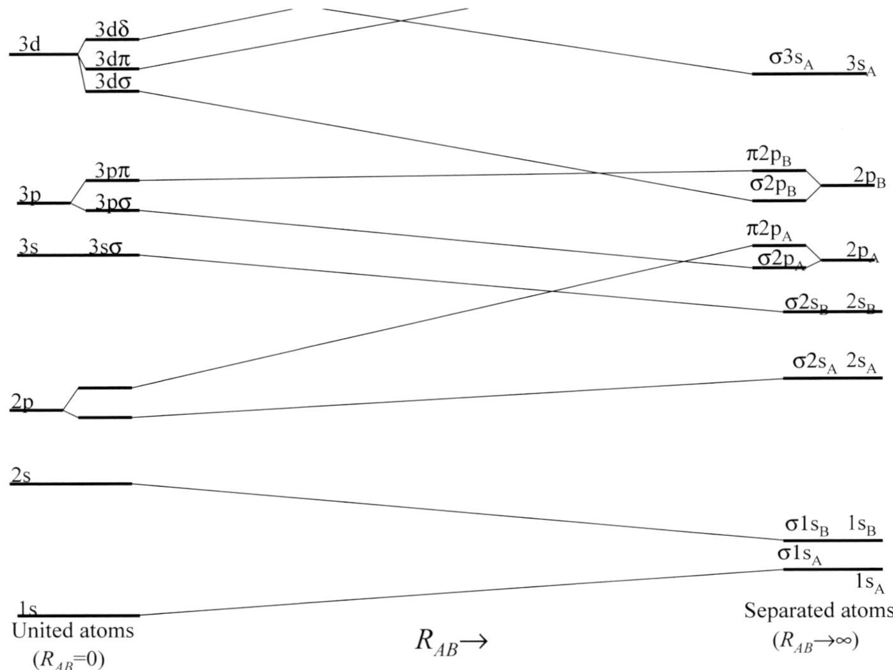

Fig. 7.6. Correlation diagram for heteronuclear diatomic molecules. For $R_{AB} \to \infty$ the assumption of effective nuclear charge $Z_A > Z_B$ has been made.

The correlation diagram for heteronuclear diatomic molecules is shown in Fig. 7.6.

It should be observed that there is a rule that helps in establishing the correlation diagram, the so-called **non-intersection** or **non-crossing** rule (**Von Neumann-Wigner rule**). This rule states that two curves $E_1(R_{AB})$ and $E_2(R_{AB})$ cannot cross if the correspondent wavefunctions ϕ_1 and ϕ_2 belong to the same symmetry species. In other words they can cross if they have different values either of λ or of the parity (**g** and **u**) or different multiplicities.

Finally we mention that the electronic states in a multielectron molecule can be classified in a way similar to the one used in the **LS** scheme for the atom (Chapter 3). From the algebraic sum $S_z = \sum_i s_z^{(i)}$ we construct M_S, while to $L_z = \sum_i l_z^{(i)}$ M_L is associated. The symbols $\Sigma, \Pi, \Delta...$ (generic Λ) are used for $M_L = 0, 1, 2$ etc. Then the state is labelled as

$$^{2S+1}\Lambda_{g,u} \ ,$$

g and **u** for homonuclear molecule.

For the state Σ, namely the one with zero component of the total angular momentum along the molecular axis, in view of the consideration on the property upon reflection with respect to a plane containing the axis, one adds the symbol $+$ or $-$ as right apex. Illustrative examples shall be given in dealing with particular diatomic molecules (§8.2).

Problems F.VII

Problem F.VII.1 From order of magnitude estimates of the frequencies to be associated with the motions of the electrons of mass m and of the nuclei of mass M in a molecule of "size" d, derive the correspondent velocities by resorting to the Heisenberg principle. By using analogous arguments derive the amplitude of the vibrational motion.

Solution:
From Heisenberg principle $p \sim \hbar/d$. The electronic frequencies can be defined

$$\nu_{elect} \simeq \frac{E_{elect}}{h} \sim \frac{1}{h}\frac{p^2}{2m} \sim \frac{1}{h}\frac{\hbar^2}{2d^2 m} = \frac{\hbar}{4\pi m d^2}$$

For the vibrational motion, by assuming for the elastic constant K $K d^2 \sim E_{elect}$ (a crude approximation, see §10.3 and Problem VIII.1.3)

$$E_{vib} \sim h\nu_{vib} \sim \left(\frac{m}{M}\right)^{\frac{1}{2}} E_{elect}$$

and

$$\nu_{vib} \sim \frac{\hbar}{d^2 \sqrt{mM}}\frac{1}{2\pi}.$$

Approximate expressions for the correspondent velocities are

$$v_{elect} \sim \sqrt{\frac{E_{elect}}{m}} \sim \frac{\hbar}{md} \ ,$$

$$v_{vib} \sim \sqrt{\frac{E_{vib}}{M}} \sim \left[\frac{h\,\hbar}{\sqrt{mM^3}d^2}\right]^{\frac{1}{2}} \sim \frac{\hbar}{m^{\frac{1}{4}} M^{\frac{3}{4}} d}$$

yielding

$$\frac{v_{vib}}{v_{elect}} \sim \left(\frac{m}{M}\right)^{\frac{3}{4}} \ll 1 \ .$$

For the rotational motion (see §10.1)

$$E_{rot} \simeq \frac{P^2}{2I} \sim \frac{\hbar^2}{Md^2} \sim \frac{m}{M}E_{elect}$$

(P angular momentum and I moment of inertia, see §10.1) and then

$$v_{rot} \sim \left(\frac{E_{rot}}{M}\right)^{\frac{1}{2}} \sim \left[\frac{\left(\frac{m}{M}\right) m\, v^2_{elect}}{M}\right]^{\frac{1}{2}} \sim v_{elect} \cdot \left(\frac{m}{M}\right) .$$

Estimates for the amplitude a of the vibrational motion can be obtained from

$$h\nu_{vib} \sim a^2\, K \sim a^2\, M(2\pi)^2 \nu^2_{vib}$$

so that

$$\frac{a^2}{d^2} = \frac{h\nu_{vib}}{E_{elect}} \sim \left(\frac{m}{M}\right)^{\frac{1}{2}} \qquad \text{and} \qquad a \simeq d\left(\frac{m}{M}\right)^{\frac{1}{4}} .$$

8

Electronic states in diatomic molecules

Topics

H_2^+ as prototype of the molecular orbital approach (MO)
H_2 as prototype of the valence bond approach (VB)
How MO and VB become equivalent
The quantum nature of the bonding mechanism
Some multi-electron molecules (N_2, O_2)
The electric dipole moment

In this Chapter we specialize the concepts given in Chapter 7 for the electronic states by introducing specific forms for the wavefunctions in diatomic molecules. Two main lines of description can be envisaged. In the approach known as **molecular orbital (MO)** the molecule is built up in a way similar to the *aufbau* method in atoms, namely by ideally adding electrons to one electron states. The prototype for this description is the Hydrogen molecule ion H_2^+. In the **valence bond (VB)** approach, instead, the molecule results from the interaction of atoms dressed by their electrons. The prototype in this case is the Hydrogen molecule H_2.

8.1 H_2^+ as prototype of MO approach

8.1.1 Eigenvalues and energy curves

In the Hydrogen molecule ion the Schrödinger equation for the electronic wavefunction $\phi(r, \theta, \varphi)$, or equivalently $\phi(z, \rho, \varphi)$ (see Fig. 8.1) is written

$$\mathcal{H}\phi = \{\frac{-\hbar^2}{2m}\nabla^2 - \frac{e^2}{r_A} - \frac{e^2}{r_B} + \frac{e^2}{R_{AB}}\}\phi = E(R_{AB})\phi. \tag{8.1}$$

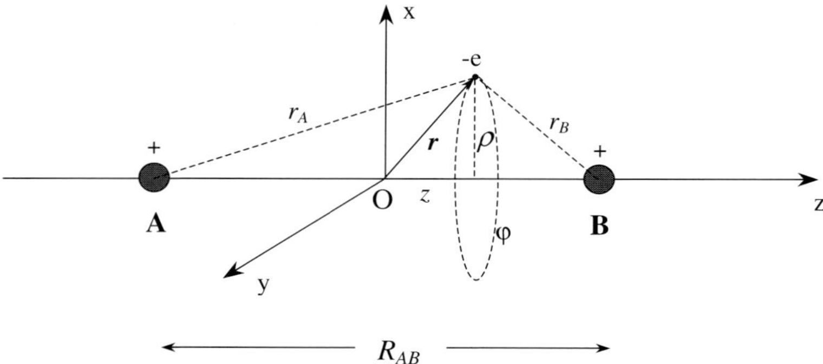

Fig. 8.1. Schematic view of the Hydrogen molecule ion H_2^+ and definition of the coordinates used in the MO description of the electronic states.

As already mentioned the exact solution of this Equation can be carried out in elliptic coordinates. Having in mind to describe H_2^+ as prototype for more general cases we shall not take that procedure.

It should be remarked that in the Hamiltonian in Eq. 8.1 the proton-proton repulsion e^2/R_{AB} (V_{nn} in Eq. 7.1; see Fig. 7.2) has been included, so that the total energy of the molecule, for a given inter-proton distance R_{AB}, will be found.

By taking into account that for $R_{AB} \rightarrow \infty$ the molecular orbital must transform into the atomic wavefunction ϕ_{1s} centered at the site A or at the site B, one can tentatively write

$$\phi = c_1 \phi_{1s}^{(A)} + c_2 \phi_{1s}^{(B)} . \tag{8.2}$$

This is a particular form of the **molecular orbital**, written as in the so-called **MO-LCAO method**, namely with the wavefunction as linear combination of atomic orbitals[1].

From the variational procedure, by deriving with respect to c_i the energy function

$$E(c_1, c_2) = \frac{\int \phi^* \mathcal{H} \phi \, d\tau}{\int \phi^* \phi \, d\tau} \tag{8.3}$$

with the tentative wavefunction given by Eq. 8.2, the usual equations

---

[1] A similar method is used also in more complex molecules, by writing $\phi = \sum_i c_i \phi_i$ and constructing the energy function $E = E(c_i)$ on the basis of the complete electronic Hamiltonian $\mathcal{H} = \sum_i (-\hbar^2/2m)\nabla_i^2 - e^2 \sum_{\alpha,i} Z_\alpha/R_{i\alpha} + e^2 \sum_{i,j}' 1/r_{ij}$, by iterative procedure evaluating the self-consistent coefficients c_i. This is the MO-LCAO-SCF method (see §9.1).

$$c_1(H_{AA} - E) + c_2(H_{AB} - ES_{AB}) = 0 \qquad (8.4)$$

$$c_1(H_{AB} - ES_{AB}) + c_2(H_{BB} - E) = 0$$

are obtained. Here

$$H_{AA} = H_{BB} = \int \phi_{1s}^{(A)*} \mathcal{H} \phi_{1s}^{(A)} d\tau$$

represents the energy of the H^+H or of the HH^+ configuration.

$$H_{AB} = H_{BA} = \int \phi_{1s}^{(B)*} \mathcal{H} \phi_{1s}^{(A)} d\tau = \int \phi_{1s}^{(A)*} \mathcal{H} \phi_{1s}^{(B)} d\tau \qquad (8.5)$$

called **resonance integral**, will be discussed at a later stage.

$$S_{AB} = \int \phi_{1s}^{(A)*} \phi_{1s}^{(B)} d\tau$$

is the **overlap integral**, a measure of the region where $\phi_{1s}^{(A)}$ and $\phi_{1s}^{(B)}$ are simultaneously different from zero:

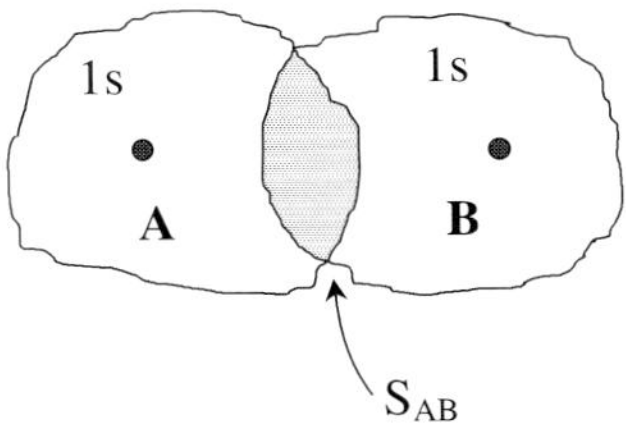

From Eqs. 8.4 the secular equations yields

$$E_\pm = \frac{H_{AA} \pm H_{AB}}{1 \pm S_{AB}} , \qquad (8.6)$$

with $c_1 = c_2$ for the sign $+$ and $c_1 = -c_2$ for the sign $-$. Thus

$$\phi_+ = \frac{1}{\sqrt{2(1 + S_{AB})}} \{\phi_{1s}^{(A)} + \phi_{1s}^{(B)}\} \qquad (8.7)$$

$$\phi_- = \frac{1}{\sqrt{2(1 - S_{AB})}} \{\phi_{1s}^{(A)} - \phi_{1s}^{(B)}\}$$

In order to discuss the dependence of the approximate eigenvalues $E_\pm$ on the interatomic distance R_{AB} one has to express H_{AA}, H_{AB} and S_{AB}. One writes

$$\overbrace{}^{\varepsilon_{AA}}$$

$$H_{AA} = \int \phi_{1s}^{(A)*} \{\mathcal{H}_{hydr.}\} \phi_{1s}^{(A)} d\tau + \int \phi_{1s}^{(A)*} \frac{e^2}{R_{AB}} \phi_{1s}^{(A)} d\tau - \overbrace{\int \phi_{1s}^{(A)*} \frac{e^2}{r_B} \phi_{1s}^{(A)} d\tau}^{\varepsilon_{AA}}$$

$$(8.8)$$

The first term is $-R_H hc$ (with R_H Rydberg constant), the second is e^2/R_{AB}. The third term, ε_{AA}, represents the somewhat classical interaction energy of an electron centered at A with the proton at B:

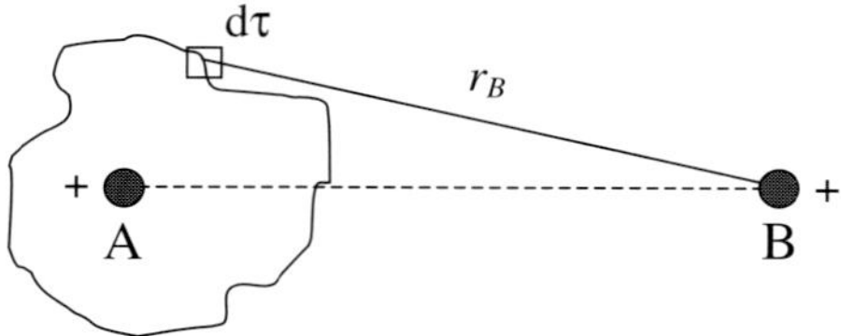

ε_{AA} can be evaluated by introducing confocal elliptic coordinates (see §7.2.2).

Then

$$\frac{1}{\pi a_0^3} \int_0^{2\pi} \int_1^{\infty} \int_{-1}^{+1} \frac{R_{AB}^3(\mu^2 - \nu^2)\, e^{-(\mu+\nu)R_{AB}/2a_0}}{4\, R_{AB}\, (\mu - \nu)}\, d\phi\, d\mu\, d\nu =$$

$$= \frac{1}{R_{AB}} \left[1 - \left(1 + \frac{R_{AB}}{a_o}\right) e^{\frac{-2R_{AB}}{a_o}} \right],$$

plotted below as a function of the internuclear distance in a_0 units:

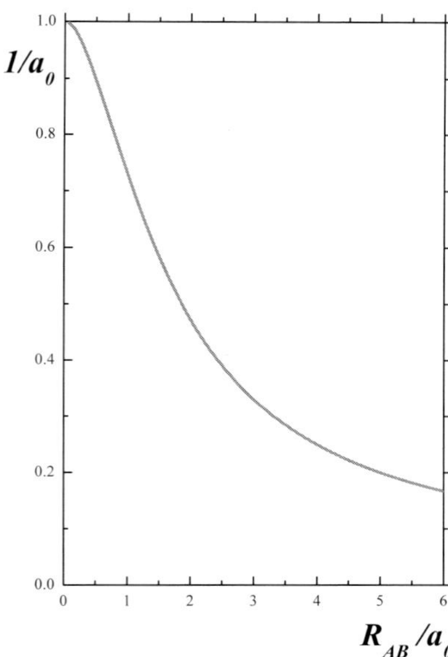

Therefore

$$\epsilon_{AA} = -\frac{e^2}{R_{AB}} \cdot \left[1 - e^{\frac{-2R_{AB}}{a_o}} \left(1 + \frac{R_{AB}}{a_o}\right) \right]$$

In analogous way the overlap integral S_{AB} and the resonance integral H_{AB} are evaluated.

$$S_{AB} = \frac{R_{AB}^3}{8\,\pi\,a_0^3} \int_0^{2\pi} \int_1^\infty \int_{-1}^{+1} (\mu^2 - \nu^2)\, e^{-\mu\,R_{AB}/a_0}\, d\mu\, d\nu\, d\varphi =$$

$$= \left[1 + \frac{R_{AB}}{a_o} + \frac{1}{3} \left(\frac{R_{AB}}{a_0}\right)^2 \right] e^{\frac{-R_{AB}}{a_o}} \tag{8.9}$$

is plotted below

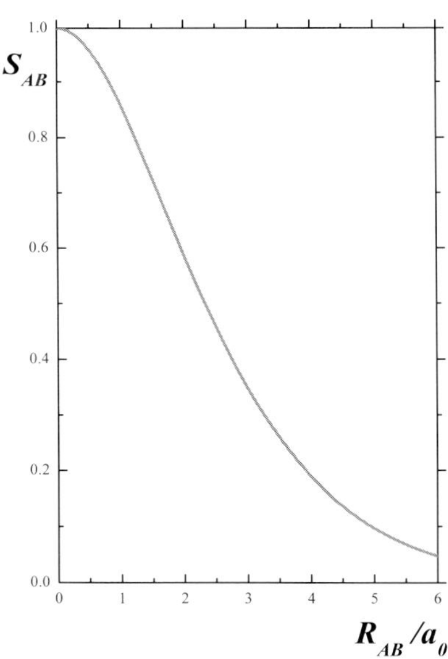

while

$$H_{AB} = \int \phi_{1s}^{(B)*}\, \{\mathcal{H}_{hydr.}\}\, \phi_{1s}^{(A)} d\tau + \frac{e^2}{R_{AB}} S_{AB} - \int \phi_{1s}^{(B)*}\, \frac{e^2}{r_B}\, \phi_{1s}^{(A)} d\tau =$$

$$= S_{AB} \left(\frac{e^2}{R_{AB}} - R_H hc \right) + \varepsilon_{AB},$$

with

$$\varepsilon_{AB} = -\int \phi_{1s}^{(B)*}\,\frac{e^2}{r_B}\,\phi_{1s}^{(A)}\,d\tau = -\frac{e^2}{a_o}\,e^{\frac{-R_{AB}}{a_o}}\left(1+\frac{R_{AB}}{a_o}\right), \qquad (8.10)$$

is approximately proportional to S_{AB}.

From Eq. 8.6 and the expressions for H_{AA}, S_{AB} and H_{AB}, the energy curves $E_{\pm}(R_{AB})$ are obtained. In Fig. 8.2 $E_{+}(R_{AB})$ is compared to the exact eigenvalue for the ground-state that could be obtained from the solution of Eq. 8.1 through elliptic coordinates.

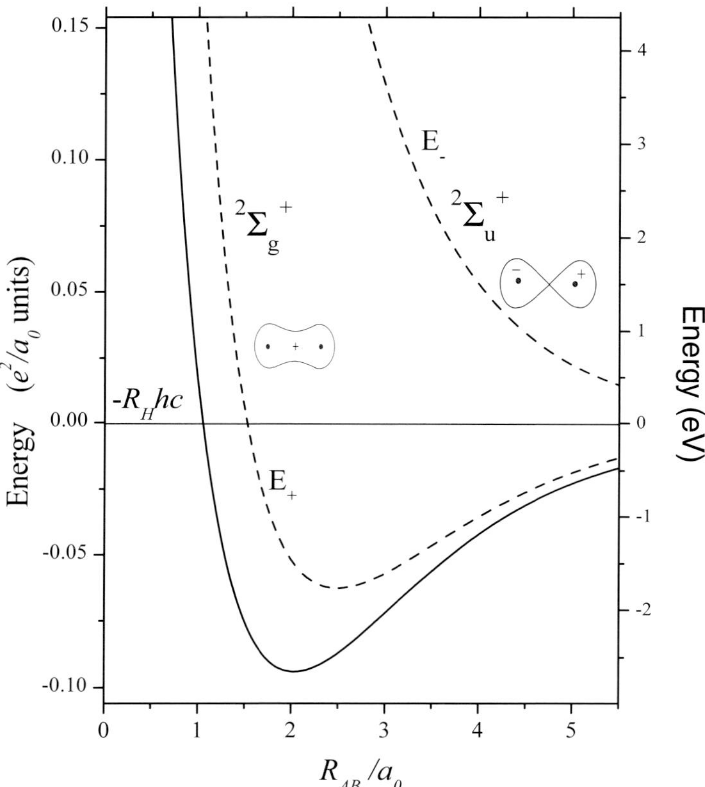

Fig. 8.2. Energy curve for the ground and first excited state of Hydrogen molecule ion as a function of the inter-proton distance R_{AB}, according to MO-LCAO orbital (dotted lines), with the classification of the electronic states in the separated-atoms scheme (see §7.3.3) and sketchy forms of the correspondent molecular orbitals. The bonding character of the $\sigma_g 1s$ state grants a minimum of the energy (in qualitative agreement with the exact calculation, solid line) while the $\sigma_u 1s$ orbital, for which $E_- > -R_H hc \equiv E(R_{AB} \to \infty)$, is **anti-bonding**. The exact result for E_- (not reported in figure) is well above the approximate energy E_- (dotted line).

The minimum in E_+ indicates that when the electron occupies the lowest energy state (the $\sigma_g 1s$ according to §7.3.3) bonding does occur.

Starting from atomic orbitals pertaining to excited states, e.g. the $2p_x$ Hydrogen states, one can obtain the molecular orbitals for the excited states, as sketched below (see also Fig. 7.3.3):

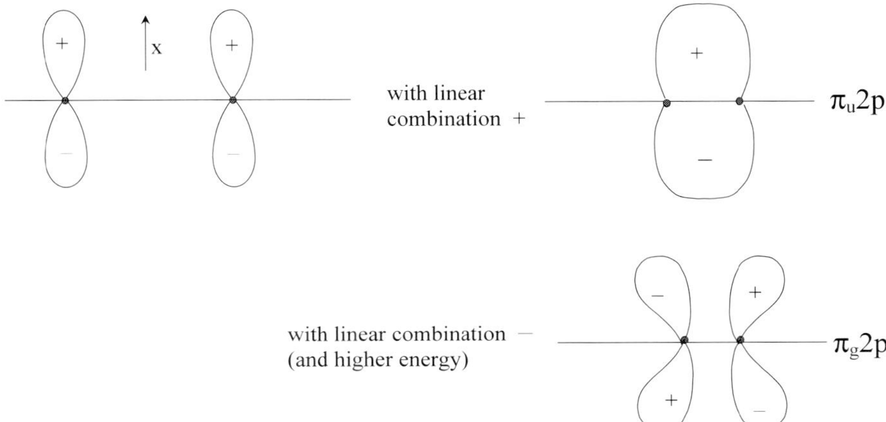

A better evaluation of eigenvalues and eigenfunctions (although still approximate) could be obtained by using more refined atomic orbitals. For instance, in order to take into account the polarization of the atomic orbitals due to the proton charge nearby, one could assume a wavefunction $\phi^{(A)}$ of the form

$$\phi^{(A)} = \phi_{1s}^{(A)} + a\,z\,e^{-Z_e\,r_A/a_0}, \tag{8.11}$$

with Z_e an effective charge. Along these lines of procedure one could derive values of the bonding energy and of the equilibrium interatomic distance R_{AB}^{eq} close to the experimental ones, which are

$$E(R_{AB}^{eq}) = -2.79\,eV, \quad R_{AB}^{eq} = 1.06\,\text{Å}. \tag{8.12}$$

Rather than pursuing a quantitative numerical agreement with the experimental data, now we shall move to the discussion of the physical aspects of the bonding mechanism.

Problems VIII.1

Problem VIII.1.1 Consider a μ-molecule formed by two protons and a muon. In the assumption that the muon behaves as the electron in the H_2^+ molecule, by means of scaling arguments evaluate the order of magnitude of the internuclear equilibrium distance, of the bonding energy and of the zero-point energy in the μ-molecule. (The zero-point energy $h\nu/2$ in H_2^+ is $0.14\,eV$ and it is reminded that $\nu = 1/2\pi\sqrt{k/M}$, with k the elastic constant and M the reduced mass).

Solution:

R_{AB}^{eq} is controlled by the analogous of the Bohr radius a_0, which is reduced with the mass by factor m_μ/m_e. Then

$$R_{AB}^{eq} \approx \frac{1}{200} R_{AB}^{eq}(H_2^+) \simeq 5 \times 10^{-3}\text{Å} .$$

and

$$E \approx 200\, E^{(H_2^+)} \simeq 500\text{eV} .$$

The force constants can approximately be written $k \approx \frac{e^2}{R^3}$ and then $k(\mu) = k(H_2^+)/8 \times 10^6$.

The vibrational energies scale with $\sqrt{k}$, so that

$$E_{v=0}^{vib} \approx 3 \cdot 10^3\, E_{v=0}^{vib}(H_2^+) \simeq 400\text{eV} .$$

Problem VIII.1.2 Derive the behavior of the probability density for the electron in H_2^+ at the middle of the molecular axis as a function of the inter-proton distance, for the ground MO-LCAO state, for $S_{AB} \ll 1$.

Solution:

From

$$\rho = |\phi_+|^2 \propto 2e^{-\frac{r_A+r_B}{a_0}} + e^{-\frac{2r_A}{a_0}} + e^{-\frac{2r_B}{a_0}} ,$$

for $r_A = r_B = R_{AB}/2$, $\rho \propto 4exp[-R_{AB}/a_0]$.

Problem VIII.1.3 In the harmonic approximation the vibrational frequency of a diatomic molecule is given by

$$\frac{1}{2\pi}\sqrt{\frac{(d^2E/dR^2)_{R_e}}{\mu}} ,$$

where μ is the reduced mass and R the interatomic distance (for detail see §10.3). Derive the vibrational frequency for H_2^+ in the ground-state.

Solution:

From $E(R) = (H_{AA} + H_{AB})/(1 + S_{AB})$

$$\frac{\partial^2 E}{\partial R^2} = \frac{[\frac{\partial^2 (H_{AA}+H_{AB})}{\partial R^2}(1 + S_{AB}) - \frac{\partial^2 S_{AB}}{\partial R^2}(H_{AA} + H_{AB})](1 + S_{AB})}{(1 + S_{AB})^3} -$$

$$- \frac{2[\frac{\partial (H_{AA}+H_{AB})}{\partial R}(1 + S_{AB})\frac{\partial S_{AB}}{\partial R} - (\frac{\partial S_{AB}}{\partial R})^2(H_{AA} + H_{AB})]}{(1 + S_{AB})^3}$$

From Eqs. 8.8-8.10, for $x = R/a_0$ and $k_1 = e^2/a_0$, one writes

$$S_{AB}(x) = e^{-x}(1 + x + \frac{x^2}{3}),$$

$$H_{AA}(x) = k_1 e^{-2x}(1 + \frac{1}{x}) - \frac{k_1}{2},$$

$$H_{AB}(x) = k_1 e^{-x}(\frac{1}{x} - \frac{1}{2} - \frac{7x}{6} - \frac{x^2}{6}).$$

Since $\partial E/\partial R = (\partial E/\partial x)(1/a_0)$ and $\partial^2 E/\partial R^2 = (\partial^2 E/\partial x^2)(1/a_0^2)$, one can conveniently express the second derivative of $E(R)$ in terms of x.

The curves for $E(R)$ and for $(d^2 E/dR^2)$ are reported below (dashed line $(d^2 E/dR^2)$, in $e^2/a^3{}_0$ unit, dotted line $E(R)$ referred to $-R_H hc$).

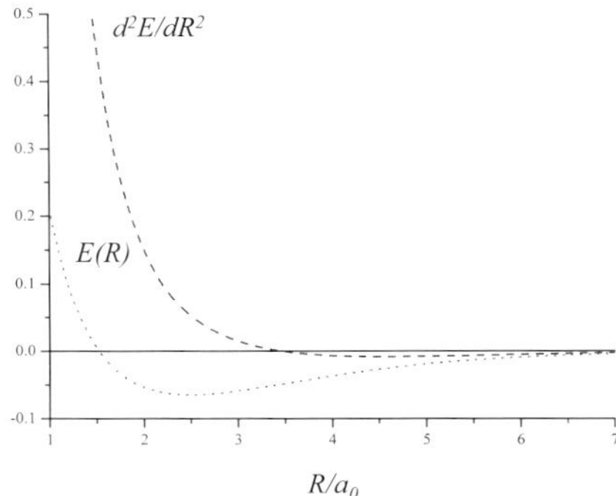

At $R_{eq} = 2.49a_0$ one finds $\partial^2 E/\partial R^2 = 0.054e^2/a_0^3 = 0.839 \times 10^5$ dyne/cm, yielding a vibrational frequency $\nu = 5.04 \times 10^{13}$ Hz (return to Problem F.VII.1).

8.1.2 Bonding mechanism and the exchange of the electron

How the bonded state of the Hydrogen molecule ion is generated? Why the bonding orbital is the $\sigma_g 1s$ while $\sigma_u 1s$ is antibonding? Which is the substantial role of the resonance integral H_{AB}?

A first way to answer to these questions is to look at the electronic charge distribution, controlled by $\rho_\pm = |\phi_\pm|^2$, where ϕ can be taken as in Eqs. 8.7. The intersection of ρ with a plane containing the molecular axis is sketched in Fig. 8.3.

In order to minimize the Coulomb energy one has to place the electron in the middle of the molecule. Thus one understands why only the $\sigma_g 1s$ state has a minimum in the energy E vs. R_{AB}.

It may be remarked that this consideration of forces between nuclei according to "classical" Coulomb-like estimate of the energies is not in contrast with the quantum character of the system. In fact, as stated by the **Hellmann-Feynman theorem** the forces can actually be evaluated "classically" provided that the charge distribution is made according to the result of the quantum description.

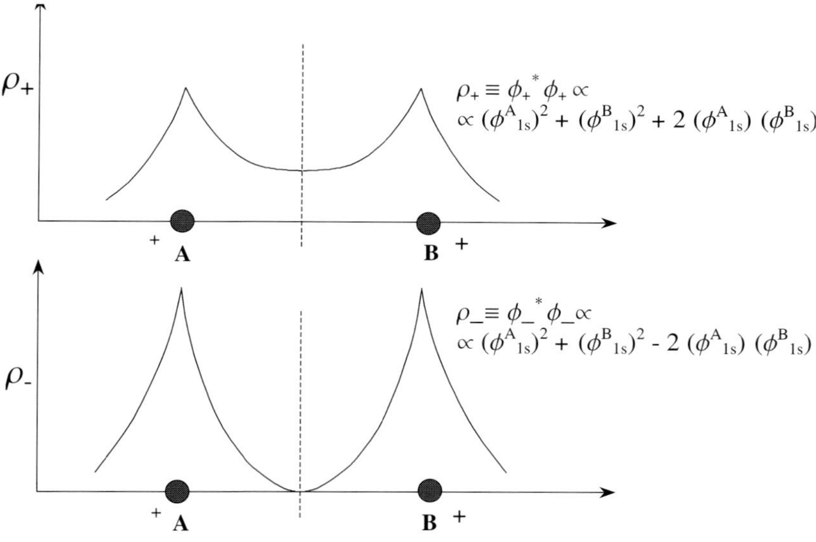

Fig. 8.3. Sketches of the charge distribution according to the bonding and antibonding molecular orbitals in H_2^+. For ϕ_- there is no electronic charge in the plane perpendicular to the molecular axis at the center of the molecule. On the other hand, in order to avoid repulsion between the protons, the negative charge must be placed right in the middle of the molecule, as a classical estimate of the energy indicates (see Problem VIII.2.1).

Now we are going to discuss the role of the resonance integral (Eq. 8.10) that is the source of the minimum in the energy at a given inter-proton distance. A suggestive interpretation of the role of H_{AB} can be given in terms of the exchange of the electron between the two equivalent $1s$ states centered at the proton A and at the proton B.

According to the model developed in Appendix I.2, by considering the basis states $|1\rangle$ and $|2\rangle$, as sketched below

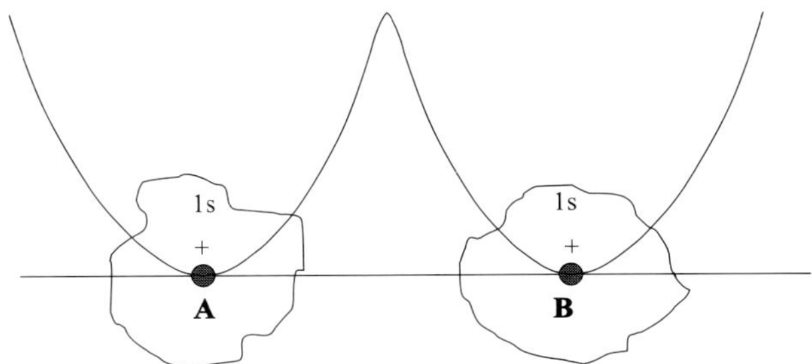

and by writing the generic state in term of linear combination $|\psi\rangle = c_1|1\rangle + c_2|2\rangle$, the coefficients obey the equations

$$i\hbar\dot{c}_1 = H_{11}c_1 + H_{12}c_2 \tag{8.13}$$
$$i\hbar\dot{c}_2 = H_{21}c_1 + H_{22}c_2$$

with $H_{11} = H_{22} = E_o$. H_{12} is the probability amplitude that the electron moves from state $|1\rangle$ to state $|2\rangle$.

By labeling A the value (negative) of H_{12}, from Eqs. 8.13 by taking sum and difference, one has

$$c_1(t) = \frac{a}{2}\,e^{-i\left(\frac{E_o-A}{\hbar}\right)t} + \frac{b}{2}\,e^{-i\left(\frac{E_o+A}{\hbar}\right)t} \tag{8.14}$$
$$c_2(t) = \frac{a}{2}\,e^{-i\left(\frac{E_o-A}{\hbar}\right)t} - \frac{b}{2}\,e^{-i\left(\frac{E_o+A}{\hbar}\right)t}.$$

It is noted that for the choice of the integration constant $a = 0$ or $b = 0$, stationary states $|\pm\rangle$ are obtained, correspondent to $\sigma_g 1s$ and to $\sigma_u 1s$, i.e.

$$|+\rangle = \frac{1}{\sqrt{2}}\Big[|1\rangle + |2\rangle\Big], \quad |-\rangle = \frac{1}{\sqrt{2}}\Big[|1\rangle - |2\rangle\Big]$$

with energies $E = E_o \pm A$.

The constants a and b in Eqs. 8.14 can be written in terms of the initial conditions for $c_1(t)$ and $c_2(t)$. By setting $c_1(0) = 1$ and $c_2(0) = 0$, one has

$$c_1(t) = e^{-i\frac{E_o}{\hbar}t}\cos\left(A\,t/\hbar\right)$$

$$c_2(t) = ie^{-i\frac{E_o}{\hbar}t}\sin\left(A\,t/\hbar\right)$$

with the behavior of the correspondent probabilities of presence $P_{1,2} = |c_{1,2}|^2$ shown in Fig. 8.4.

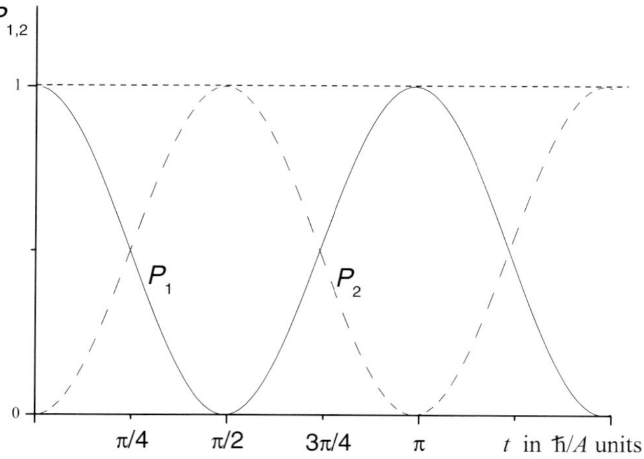

Fig. 8.4. Time dependence of the probability of presence of the electron on the sites A and B according to the description of two-levels states for H_2^+.

Thus one can idealize the formation of the molecule as due to the exchange of the electron from left to right and back, with the related decrease of the energy.

This description has some correspondence in classical systems, such as two weakly-coupled mechanical oscillators or LC circuits, with their two normal modes and the correspondent exchange of energy. Scattering experiments of protons on Hydrogen atoms confirm that the exchange process of the electron is real. When a proton is in the neighborhood (distance of the order of a_0) of an Hydrogen atom for a time of the order of $\hbar/2A$, with $A = (E_+ - E_-)$ (or multiple), an Hydrogen atom comes out after the scattering process.

8.2 Homonuclear molecules in the MO scenario

From the MO description of the states in H_2^+ it is now possible to analyze multi-electron **homonuclear** diatomic molecules. In a way analogous to the *auf bau* method in atoms, to build up the molecule in a first approximation one has to accommodate the electrons on the one-electron states derived for the prototype. This procedure is particularly simple if *a priori* one does not take into account the inter-electron interactions (e^2/r_{ij}), thus ideally assuming independent electrons. Then the energy is evaluated on the basis of the complete Hamiltonian, for $\Phi_{total} = \prod_i \phi_{MO}(r_i)$, by considering the dynamical equivalence of the electrons when different states are hypothesized. At §8.4 we shall discuss the hydrogen molecule to some extent, by taking into account the spin states and the antisymmetry requirement. For the moment, let us proceed to a qualitative description of some homonuclear diatomic molecules by referring most to the ground states.

For H_2 the ground state has the electronic configuration $(\sigma_g 1s)^2$, it is labelled $^1\Sigma_g^+$ (see §7.2.3) and the MO wavefunction is

$$\phi_{(\sigma_g 1s)^2}(\mathbf{r}_1, \mathbf{r}_2) = \sigma_g 1s(\mathbf{r}_1)\, \sigma_g 1s(\mathbf{r}_2), \tag{8.15}$$

that in the LCAO approximation is written (see Eq. 8.7)

$$\phi_{(\sigma_g 1s)^2}(\mathbf{r}_1, \mathbf{r}_2) = \frac{1}{2(1 + S_{AB})}\left[\phi_{1s}^{(A)}(\mathbf{r}_1) + \phi_{1s}^{(B)}(\mathbf{r}_1)\right] \cdot \left[\phi_{1s}^{(A)}(\mathbf{r}_2) + \phi_{1s}^{(B)}(\mathbf{r}_2)\right]$$
$$\tag{8.16}$$

The energy $E(R_{AB})$, evaluated by including in the Hamiltonian the term (e^2/r_{12}) by means of calculations strictly similar to the ones detailed for H_2^+ at §8.1, is sketched below:

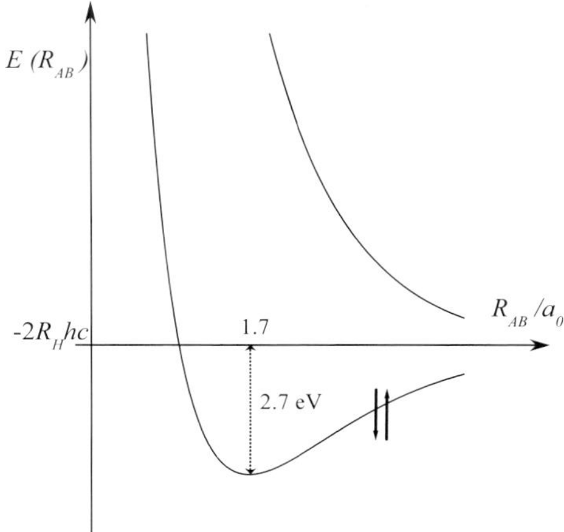

In He_2^+ the ground state has the electronic configuration $(\sigma_g 1s)^2(\sigma_u 1s)$ and the notation is $^2\Sigma_u$. The third electron has to be of **u** character, because of the Pauli principle.

The He_2 molecule cannot exist in state a stable state[2]. In fact, the electronic configuration should be $(\sigma_g 1s)^2(\sigma_u 1s)^2$, with the pictorial representation sketched below:

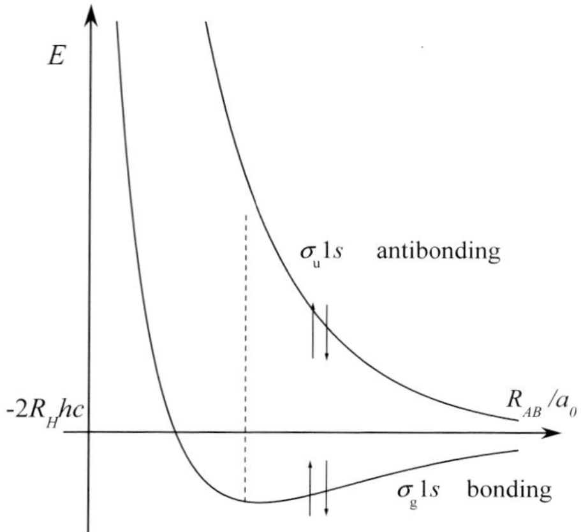

Since for $R_{AB} \simeq R_{AB}^{eq.}$ one has $E_- > |E_+|$ (see Eq. 8.6) the two antibonding electrons force the nuclei apart in spite of the bonding role of the electrons placed in the ground energy state.

Now we are going to discuss a pair of molecules exhibiting some aspects not yet encountered until now. In the N_2 molecule we have an example of **"strong bond"** due to σ MO orbital at large overlap integral and of **"weak bond"** due to π MO orbitals involving p atomic states, with little overlap. In fact one can depict the formation of the molecule as below

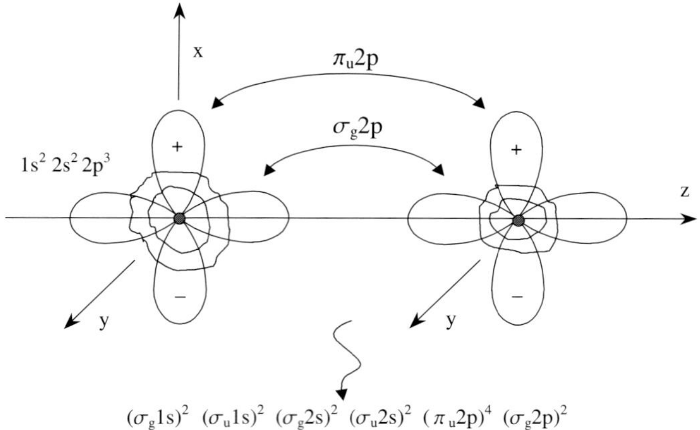

---

[2] Van der Waals interactions (described at §13.2.2), leading to very weak bonds at large distances, are not considered here.

where it is noted that the linear combination with the sign $+$ again implies electronic charge in the central plane (and therefore is a **bonding orbital**) although now the inversion symmetry is **u**. The $\sigma_g 2p$ orbital, ideally generated from the combination of $2p_z$ atomic orbitals, implies strong overlap. Since H_{AB} is somewhat proportional to S_{AB} (see Eqs. 8.9 and 8.10) one has a deep minimum in the energy and then a strong contribution to the bonding mechanism. On the contrary, from the combination of $2p_{x,y}$ atomic orbitals to generate the π MO's the overlap region is small and then one can expect a weak contribution to bond. The electronic state of the N_2 molecule is labelled $^1\Sigma_g^+$ and the molecular orbitals are fully occupied. Thus the molecule is somewhat equivalent to atoms at closed shells, explaining its stability and scarcely reactive character.

Another instructive case of homonuclear diatomic molecule is O_2. Here there are two further electrons to add to the configuration of N_2. These electrons must be set on the $\pi_g 2p$ orbital, in view of the Pauli principle. The $\pi_g 2p$ orbital is not fully occupied and one has to deal with **LS** coupling procedure, similar to the one discussed for atoms for non-closed shells. In principle there are the possibilities sketched below:

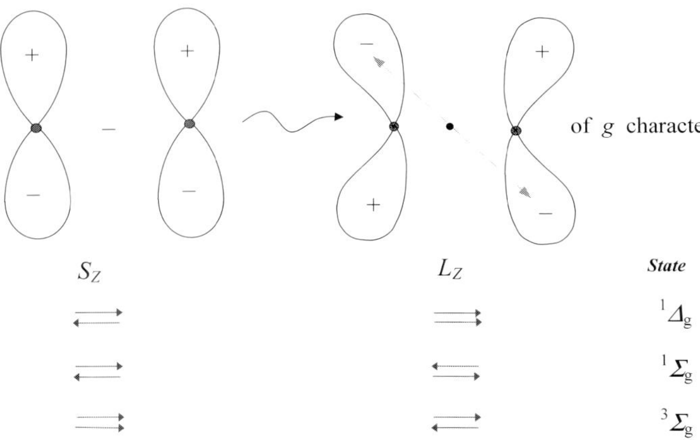

According to Hund rules, that hold also in molecules, the ground state is $^3\Sigma_g^-$ corresponding to the the maximization of the total spin. The g and - characters can be understood by inspection: in Fig. 8.5 it is shown how the property under the reflection in a plane containing the molecular axis results from the symmetry of the π orbitals.

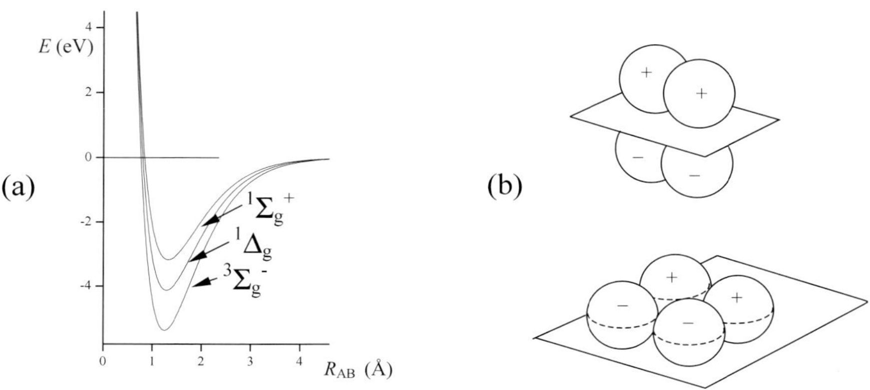

Fig. 8.5. Energy curves for the low energy states in the O_2 molecule (a) and sketchy illustration of the $(+\,-)$ symmetries for π^+ and π^- orbitals (b). The Σ state requires a label to characterize the behavior under reflection with respect to a plane containing the molecular axis. Since the two electrons occupy different π orbitals, one of them is $+$ and the other $-$, implying the overall $-$ character of the configuration.

The molecule is evidently **paramagnetic** and because of the partially empty external orbital has a certain reactivity, at variance with N_2. In fact the O_3 molecule (ozone) is known to exist.

Problems VIII.2

Problem VIII.2.1 Evaluate the amount of electronic charge that should be placed at the center of the molecular axis in H_2 in order to justify the dissociation energy ($\simeq 4.5\,\text{eV}$) at the interatomic equilibrium distance $R_{AB}^{eq} = 0.74$ Å.

Solution:

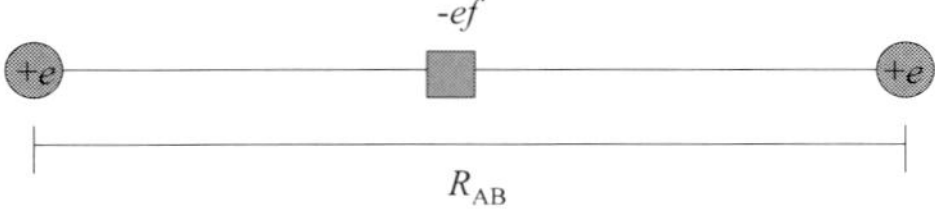

$$\frac{e^2}{R_{AB}} - ef\left[\frac{e}{R_{AB}/2} + \frac{e}{R_{AB}/2}\right] = -4.5\,\text{eV}\,.$$

From

$$\frac{e^2}{R_{AB}^{eq}} = 19.5\,\text{eV}\ ,$$

$$f = \frac{-4.5\,eV - 19.5\,eV}{-77.8\,\text{eV}} = 0.3\,.$$

Problem VIII.2.2 Write the spectroscopic terms for the ground state of the molecules Li$_2$, B$_2$, C$_2$, Br$_2$, N$_2^-$, N$_2^+$, F$_2^+$ and Ne$_2^+$.

Solution:

For Li$_2$, $(\sigma_g 1s)^2(\sigma_u 1s)^2(\sigma_g 2s)^2$, $^1\Sigma_g^+$.

For B$_2$, $(\sigma_g 1s)^2(\sigma_u 1s)^2(\sigma_g 2s)^2(\sigma_u 2s)^2(\pi_u 2p)^2$, $^3\Sigma_g^-$.

For C$_2$ the proper sequence of the energy levels has to be taken into account (one electron could be promoted from π_u to the σ_g state, see Fig. 7.4). However the electronic configuration $(\sigma_g 2s)^2(\sigma_u 2s)^2(\pi_u 2p)^4$ seems to be favored and the ground state term is $^1\Sigma_g^+$.

For

$$\text{Br}_2 \quad (\text{atoms in } {}^2P \text{ state}) \qquad\qquad {}^1\Sigma_g^+$$

$$\text{excited states} \qquad\qquad {}^1\Sigma_u^-,\ {}^1\Pi_g,\ {}^1\Pi_u,\ {}^1\Delta_g$$

$$\text{N}_2^+ \qquad\qquad {}^2\Sigma_g^+$$

$$\text{N}_2^- \qquad\qquad {}^2\Pi_g$$

$$\text{F}_2^+ \qquad\qquad {}^2\Pi_g$$

$$\text{Ne}_2^+ \qquad\qquad {}^2\Sigma_u^+$$

8.3 H$_2$ as prototype of the VB approach

In the framework of the **valence bond (VB)** method, where the molecule results from the interaction of atoms dressed by their electrons, the prototype is the Hydrogen molecule.

The Hamiltonian is written (see Fig. 8.6)

$$\mathcal{H} = \left[-\frac{\hbar^2}{2m}\nabla_1^2 - \frac{e^2}{r_{A1}}\right] + \left[-\frac{\hbar^2}{2m}\nabla_2^2 - \frac{e^2}{r_{B2}}\right] + \left[-\frac{e^2}{r_{A2}} - \frac{e^2}{r_{B1}}\right] + \left[\frac{e^2}{r_{AB}} + \frac{e^2}{r_{12}}\right] \equiv$$

$$\equiv [a] + [b] + [c] + [d] \tag{8.17}$$

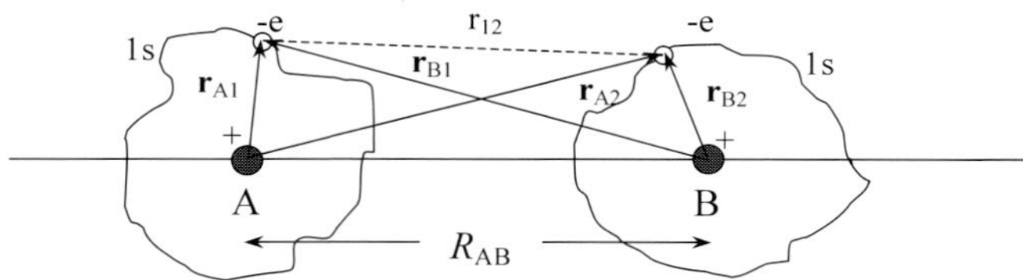

Fig. 8.6. Definition of the coordinates involved in the Hamiltonian for the H$_2$ molecule.

A tentative wavefunction could be $\phi(\mathbf{r}_1, \mathbf{r}_2) = \phi^A_{1s}(\mathbf{r}_1)\,\phi^B_{1s}(\mathbf{r}_2)$, corresponding to the situation in which the two electrons keep their atomic character and only Coulomb-like interactions with classical analogies are supposed to occur. However, this wave-function does not lead to the formation of the real bonded state. In that case, in fact, for the [a] and [b] terms in the Hamiltonian one obtains $-2R_H hc$ and for the interaction terms [c] and [d] one has

$$J = \frac{e^2}{R_{AB}} + \int |\phi^A_{1s}(\mathbf{r}_1)|^2 \, \frac{e^2}{r_{12}} \, |\phi^B_{1s}(\mathbf{r}_2)|^2 \, d\tau_1 \, d\tau_2 - 2 \int \frac{e^2}{r_{A2}} \, |\phi^B_{1s}(\mathbf{r}_2)|^2 \, d\tau_2 \quad .$$

$$(8.18)$$

The latter term in Eq. 8.18 is twice the attractive interaction between electron A-proton B, as sketched below

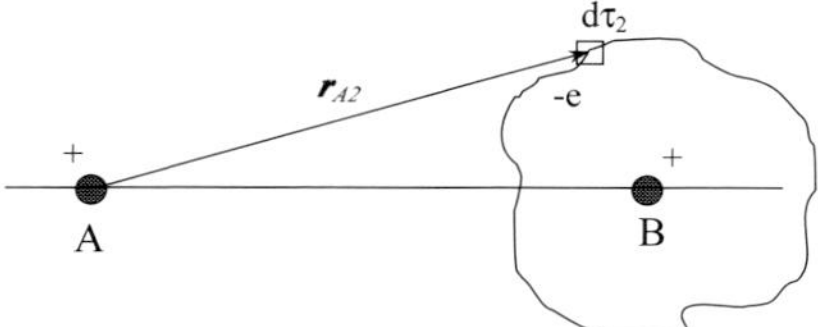

All the above terms in Eq. 8.18 correspond to classical electrostatic interactions and therefore J is usually called **Coulomb integral**. From the evaluation of J through elliptic coordinates, as described for ε_{AA} at §8.1.1, one could figure out that the energy curve $E(R_{AB})$ displays only a slight minimum, around 0.25 eV, in large disagreement with the experimental findings (see Figure 8.7). On the other hand, by recalling the description of the two electrons in Helium atom (§ 2.2) the inadequacy of the wavefunction $\phi^A_{1s}(\mathbf{r}_1)\phi^B_{1s}(\mathbf{r}_2)$ can be expected, since the indistinguishability of the electrons, once that the atoms are close enough to form a molecule, is not taken into account.

Then one rather writes

$$\phi_{VB}(\mathbf{r}_1, \mathbf{r}_2) = c_1 \, \phi^A_{1s}(\mathbf{r}_1)\, \phi^B_{1s}(\mathbf{r}_2) + c_2 \, \phi^A_{1s}(\mathbf{r}_2)\, \phi^B_{1s}(\mathbf{r}_1) \equiv c_1 \, |1\rangle + c_2 \, |2\rangle \quad (8.19)$$

By deriving the energy function with the usual variational procedure (see Eqs. 8.3-8.6) one obtains $c_1 = \pm c_2$ and

$$E_{\pm} = \frac{H_{11} \pm H_{12}}{1 \pm S_{12}} \quad , \qquad (8.20)$$

where $H_{11} \equiv \langle 1|H|1 \rangle = \langle 2|H|2 \rangle$, $H_{12} \equiv \langle 2|H|1 \rangle$, $S_{12} = S_{AB}^2$ and

$$\phi_{\pm} = \frac{1}{\sqrt{2\left(1 \pm S_{12}\right)}}\left[|1\rangle \pm |2\rangle\right]. \qquad (8.21)$$

The eigenvalues turn out

$$E_{\pm}(R_{AB}) = -2R_H hc + \frac{J}{1 \pm S_{AB}^2} \pm \frac{K}{1 \pm S_{AB}^2} \qquad (8.22)$$

where J is given by Eq. 8.18, while

$$K = \int \phi_{1s}^{A*}(\mathbf{r}_1)\,\phi_{1s}^{B*}(\mathbf{r}_2)\left[-\frac{e^2}{r_{A2}} - \frac{e^2}{r_{B1}} + \frac{e^2}{R_{AB}} + \frac{e^2}{r_{12}}\right]\phi_{1s}^{A}(\mathbf{r}_2)\,\phi_{1s}^{B}(\mathbf{r}_1)\,d\tau_1\,d\tau_2$$

$$(8.23)$$

is the **extended exchange integral**, with no classical analogy and related to the quantum character of the wavefunction. K can be rewritten

$$K = \frac{e^2}{R_{AB}}\,S_{AB}^2 - 2\,S_{AB}\,\epsilon_{AB} + \int \phi_{1s}^{A*}(\mathbf{r}_1)\,\phi_{1s}^{B*}(\mathbf{r}_2)\,\frac{e^2}{r_{12}}\,\phi_{1s}^{A}(\mathbf{r}_2)\,\phi_{1s}^{B}(\mathbf{r}_1)\,d\tau_1\,d\tau_2$$

$$(8.24)$$

where again one finds the resonance integral ε_{AB} (Eq. 8.10) and a **reduced exchange integral**

$$K_{red} = \int \phi_{1s}^{A*}(\mathbf{r}_1)\,\phi_{1s}^{B*}(\mathbf{r}_2)\,\frac{e^2}{r_{12}}\,\phi_{1s}^{A}(\mathbf{r}_2)\,\phi_{1s}^{B}(\mathbf{r}_1)\,d\tau_1\,d\tau_2 \qquad (8.25)$$

analogous to the one in Helium atom and **positive**.

From the evaluation of J and K the energy curves can be obtained, as depicted in Fig. 8.7.

It should be remarked that most of the bond strength is due to the exchange integral K.

As for any two electron systems (see §2.2) the spin wave functions are

$$\chi_{symm}^{S=1} \quad i.e. \quad \alpha(1)\alpha(2), \quad \beta(1)\beta(2) \quad \text{and} \quad \frac{1}{\sqrt{2}}\left[\alpha(1)\beta(2) + \alpha(2)\beta(1)\right]$$

$$\chi_{ant}^{S=0} \quad i.e. \quad \frac{1}{\sqrt{2}}\left[\alpha(1)\beta(2) - \alpha(2)\beta(1)\right]$$

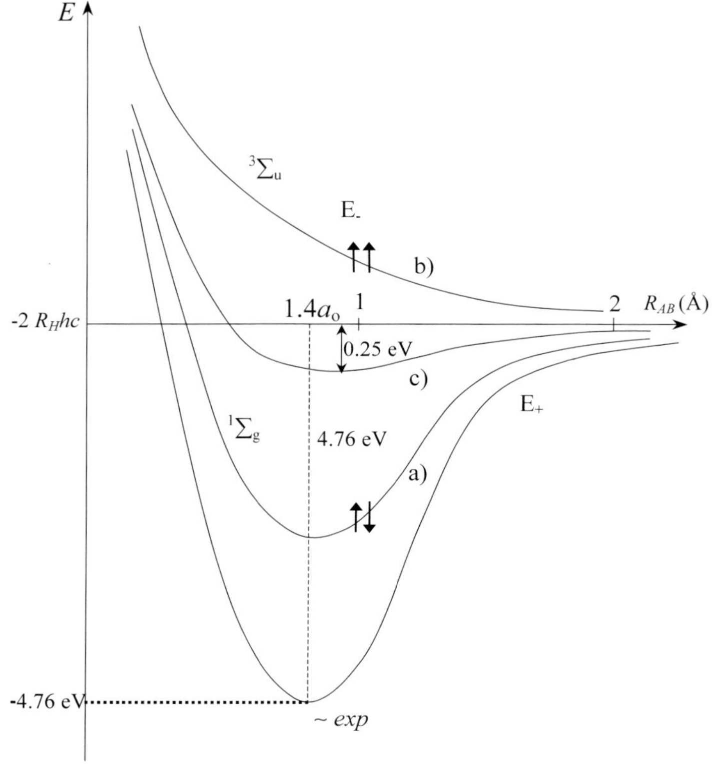

Fig. 8.7. Sketch of the energy curves of the Hydrogen molecule in the VB scheme as a function of the interatomic distance R_{AB}. The real curve (reconstructed by a variety of experiments) is indicated as *exp*, while curve c) illustrates the behavior expected from the Coulomb integral only (Eq. 8.18 in the text). Curves a) and b) illustrate the approximate eigenvalues $E_\pm$ in Eq. 8.22.

and the antisymmetry requirement implies that χ_{ant} is associated with ϕ_+, corresponding to the ground state ${}^1\Sigma_g$, while for the eigenvalue E_- one has to associate χ_{symm} with ϕ_-, to yield the state ${}^3\Sigma_u$.

At this point one may remark that the **VB** ground state for H_2 (see Eq. 8.21) is proportional to the MO state

$$\phi_{MO}(1,2) \propto \phi_A(1)\phi_B(2) + \phi_A(2)\phi_B(1) + \phi_A(1)\phi_A(2) + \phi_B(1)\phi_B(2)$$

(1,2 for $\mathbf{r}_1$ and $\mathbf{r}_2$) The "ionic" configurations $\phi_A(1)\phi_A(2)$ and $\phi_B(2)\phi_B(1)$ (Eqs. 8.7 and 8.16) are present in the MO orbital with the same coefficients, in order to account for the symmetry and to prevent electric charge transfer that would lead to a molecular dipole moment. A more detailed comparison of the electronic states for the Hydrogen molecule within the **MO** and **VB** approaches is discussed in the next Section.

Problems VIII.3

Problem VIII.3.1 Reformulate the description of the H_2 molecule in the VB approach in the assumption that the two Hydrogen atoms in their ground state are at a distance R so that exchange effects can be neglected. Prove that for large distance the interaction energy takes the dipole-dipole form and that by using the second order perturbation theory an attractive term going as R^{-6} is generated (see §13.2.2 for an equivalent formulation). Then remark that for degenerate $n = 2$ states the interaction energy would be of the form R^{-3}.

Solution:

From Fig. 8.6 and Eq. 8.17 the interaction is written

$$V = \frac{e^2}{R} - \frac{e^2}{r_{A2}} - \frac{e^2}{r_{B1}} + \frac{e^2}{r_{12}}.$$

Expansions in spherical harmonics (see Prob. II.2.1) yield

$$\frac{1}{r_{B1}} = \frac{1}{|R\boldsymbol{\rho} - \mathbf{r}_{1A}|} = \sum_{\lambda=0} \frac{r_{A1}^{\lambda}}{R^{\lambda+1}} P_\lambda(\cos\theta) = \frac{1}{R} + \frac{(\mathbf{r}_{A1}\cdot\boldsymbol{\rho})}{R^2} + \frac{3(\mathbf{r}_{A1}\cdot\boldsymbol{\rho})^2 - r_{A1}^2}{2R^3} + \cdots,$$

$$\frac{1}{r_{12}} = \frac{1}{|R\boldsymbol{\rho} + \mathbf{r}_{B2} - \mathbf{r}_{A1}|} = \frac{1}{R} + \frac{(\mathbf{r}_{A1} - \mathbf{r}_{B2})\cdot\boldsymbol{\rho}}{R^2} + \frac{3[(\mathbf{r}_{A1} - \mathbf{r}_{B2})\cdot\boldsymbol{\rho}]^2 - (\mathbf{r}_{A1} - \mathbf{r}_{B2})^2}{2R^3} + \cdots,$$

$$\frac{1}{r_{A2}} = \frac{1}{R} + \frac{(\mathbf{r}_{B2}\cdot\boldsymbol{\rho})}{R^2} + \frac{3(\mathbf{r}_{B2}\cdot\boldsymbol{\rho})^2 - r_{B2}^2}{2R^3} + \cdots$$

($\boldsymbol{\rho}$ unit vector along the interatomic axis).

Thus the dipole-dipole term (see §13.2.2) is obtained

$$V = -\frac{2z_1 z_2 - x_1 x_2 - y_1 y_2}{R^3} e^2,$$

the z-axis being taken along $\boldsymbol{\rho}$. By resorting to the second-order perturbation theory and taking into account the selection rules (App.I.3 and §3.5), the interaction energy turns out

$$E^{(2)} = \frac{e^4}{R^6} \sum_{mn} \frac{4z_{0m}^2 z_{0n}^2 + x_{0m}^2 x_{0n}^2 + y_{0m}^2 y_{0n}^2}{2E_0 - E_m - E_n}$$

where x_{om}, x_{on} etc. are the matrix elements connecting the ground state (energy E_0) to the excited states (energies E_m, E_n). $E^{(2)}$ being negative, the two atoms attract each other (**London** interaction, see §13.2.2 for details).

For the states at $n = 2$ the perturbation theory for degenerate states has to be used. From the secular equation a first order energy correction is found (see the similar case for Stark effect at §4.2). Thus the interaction energy must go as R^{-3}.

8.4 Comparison of MO and VB scenarios in H$_2$: equivalence from configuration interaction

Going back to the MO description for the H$_2$ molecule, by considering the possible occurrence of the first excited σ_u one-electron state and by taking into account the indistinguishability, four possible wavefunctions are:

$$\Phi_I(g,g) \equiv \phi_g(1)\phi_g(2) \qquad \mathbf{gg} \qquad a)$$
$$\Phi_{II}(u,u) \equiv \phi_u(1)\phi_u(2) \qquad \mathbf{uu} \qquad b)$$
$$\Phi_{III}(g,u) \equiv \frac{1}{\sqrt{2}}\left[\phi_g(1)\phi_u(2) + \phi_g(2)\phi_u(1)\right] \mathbf{ug}^{\mathbf{symm}} \ c)$$
$$\Phi_{IV}(g,u) \equiv \frac{1}{\sqrt{2}}\left[\phi_g(1)\phi_u(2) - \phi_g(2)\phi_u(1)\right] \quad \mathbf{ug}^{\mathbf{ant}} \ \ d) \qquad (8.26)$$

In view of the four spin wavefunctions χ^{ant} and χ^{symm}, in principle 16 spin-molecular orbitals could be constructed. Due to the Pauli principle, in the H$_2$ molecule one finds only 6 states, the ones of antisymmetric character.

The ground MO state $(\sigma_g 1s)(\sigma_g 1s)\chi^{ant}$ can be detailed by referring to the LCAO specialization, so that the complete spin-MO is

$$\phi_{TOT}^{MO}(1,2) = \chi_{S=0}^{ant}\left[\phi_{VB} + \phi_A(1)\phi_A(2) + \phi_B(1)\phi_B(2)\right] \qquad (8.27)$$

namely the VB form with the "ionic" states, as already mentioned.

To find the excited MO state corresponding to the $(VB)^-$ wavefunction, in Eqs. 8.26 one can look for the one that without the ionic states does correspond to 8.21 ϕ_- without the ionic states. From Eq. 8.26d) with the LCAO specialization it is found that

$$\Phi_{IV} = \frac{1}{\sqrt{2}}\left[\phi_A(1)\phi_B(2) - \phi_A(2)\phi_B(1)\right] \qquad (8.28)$$

is the same as ϕ_{VB}^-.

From another point of view, now one understands why the $^3\Sigma_u$ state is unstable: it corresponds to have one electron in the g bonding MO orbital and

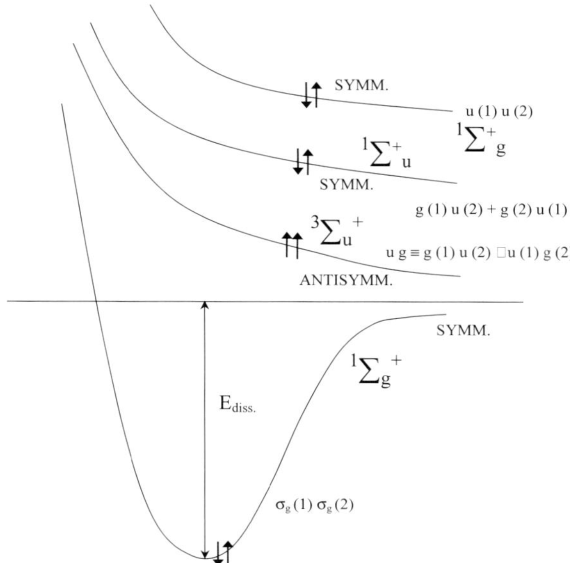

Fig. 8.8. Schematic energy curves for H$_2$ corresponding to the wavefunctions in Eq. 8.26. More accurate forms of the energies for the $^1\Sigma^+_g$ and $^3\Sigma^+_u$ states are reported in Fig. 8.9, in comparison with the VB eigenvalues.

one in the u antibonding MO (see § 8.2), this latter being strongly repulsive. In Fig. 8.8 the lowest energy levels in H$_2$ corresponding to 8.26 are sketched.

For a more quantitative comparison of the MO and the VB descriptions in H$_2$, let us look at the values for the dissociation energies and the equilibrium distances (see also Fig. 8.9) in the ground state:

$$\phi_{VB} \qquad E_{diss} \simeq 3.14\,eV \qquad R^{eq.}_{AB} = 1.7\,a_0$$

$$\phi_{MO} \qquad E_{diss} \simeq 2.7\,eV \qquad R^{eq.}_{AB} = 1.7\,a_0$$

$$Experimental \qquad E_{diss} \simeq 4.75\,eV \qquad R^{eq.}_{AB} = 1.4\,a_0$$

One should remark that the VB orbital does not include the ionic states while the MO-LCAO overestimates their weight. In fact, the energy to remove the electron from the Hydrogen atom (13.56 eV) is much higher than the energy gain Δ in setting it on the configuration H$^-$. The energy gain Δ (sometimes called **electron affinity**) in principle could be estimated from

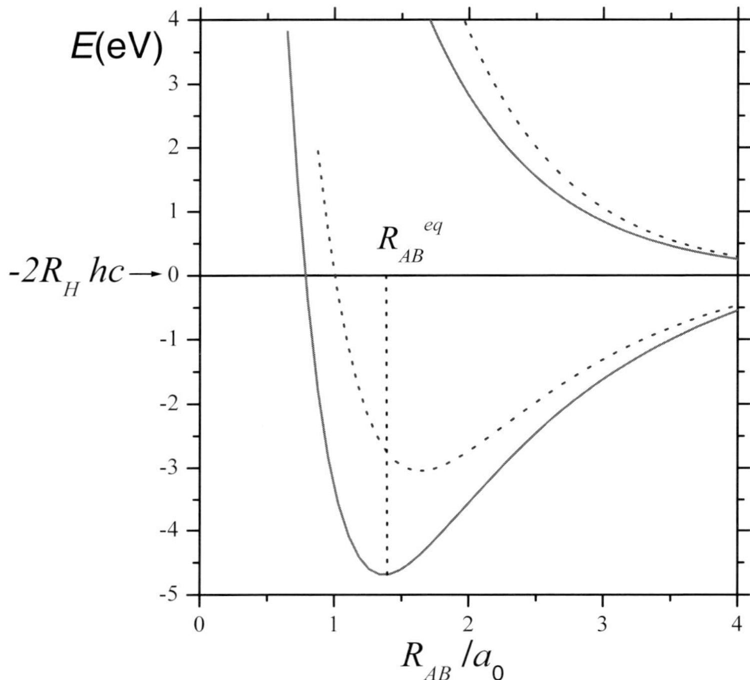

Fig. 8.9. Energy curves for the lowest energy states in H_2: dotted lines, within the VB approach; solid lines, more accurate evaluations for the $^1\Sigma_g^+$ and the $^3\Sigma_u^+$ states according to the procedure outlined in the text.

the Coulomb integral in Helium atom (§2.2), with $Z=1$ for the nuclear charge (however, see Problem II.2.4). From accurate estimates one actually would find $\Delta = 0.75\,\text{eV}$. Therefore the ionic states cannot be weighted as much as they are MO-LCAO orbital. This observation suggests a tentative wavefunction of the form

$$\phi_{VB} + \lambda\phi_{ionic} \; , \tag{8.29}$$

namely a mixture of the covalent VB and of ionic states with a coefficient λ, for instance to be estimated variationally. From the derivative of the energy function $E(\lambda)$ one could find that the minimum corresponds to $\lambda = 0.25$. Therefore, from the normalization of the wavefunction the weight of the ionic states is given by $\lambda^2/(\lambda^2 + 1)$, about 6 percent.

How could the MO description of the ground state in H_2 be improved? Since the wavefunctions 8.26 involve the ionic states with different coefficients, it is conceivable that a better approximation is obtained if a proper combination of the wavefunctions correspondent to different configurations is attempted. This procedure is an example of the approach called **configuration interaction (CI)**. In the combination one has to take into account that the mixture must involve states with the same symmetry properties and same spin. Thus one should combine the **gg** state with the **uu** one, both coupled to χ^{ant}:

$$\phi_{CI}(1,2) = \phi_I + k\phi_{II} \tag{8.30}$$

From this wavefunction, as usual, one generates two energy levels, one of them having energy $E < E_+$, E_+ being the energy for ϕ_I. In this way one could find a dissociation energy and equilibrium distance close to the experimental values. Furthermore those quantities are found to **coincide** with the ones associated with the VB wavefunction with addition of the ionic states! This is not by chance. In fact, by collecting the various terms involving the atomic orbitals, one can rewrite Eq. 8.30 in the form

$$\phi_{CI}(1,2) = (1-k)\phi_{VB} + (1+k)\phi_{ionic}$$

and by defining $\lambda = (1+k)/(1-k)$ one sees that it coincides with Eq. 8.29.

This is an example of a more general issue: *the MO-LCAO method with interaction of the configurations is equivalent to the VB approach with addition of the ionic states to the covalent wavefunctions.*

8.5 Heteronuclear molecules and the electric dipole moment

In the following we shall recall some novel aspects present in diatomic molecules when the two atoms are different.

First of all one remarks that the inversion symmetry, with the Hamiltonian $\mathcal{H}(\mathbf{r})$ equal to $\mathcal{H}(-\mathbf{r})$, no longer holds. Therefore, within the separated atoms scheme one cannot longer classify the states as g or u and the one-electron states become (see Fig. 7.6) $\sigma 1s_A$, $\sigma 1s_B$, $\sigma 2s_A$, $\sigma 2s_B$,...

Within the MO-LCAO scheme the one-electron orbital is written

$$\phi = c_A\phi_A + c_B\phi_B$$

with $c_A \neq c_B$. Equivalently, in the normalized form

$$\phi_{MO}^{LCAO} = \frac{1}{(1 + \lambda^{*2} + 2\lambda^* S_{AB})^{\frac{1}{2}}} \left[\phi_A + \lambda^*\phi_B\right] . \tag{8.31}$$

Here λ^* can vary from $-\infty$ to $+\infty$ and it characterizes the **polarization** of the orbital, namely measuring the **electronic charge transfer** from one atom to the other. As illustrative example in Fig. 8.10 the molecular orbital for the HCl molecule is sketched.

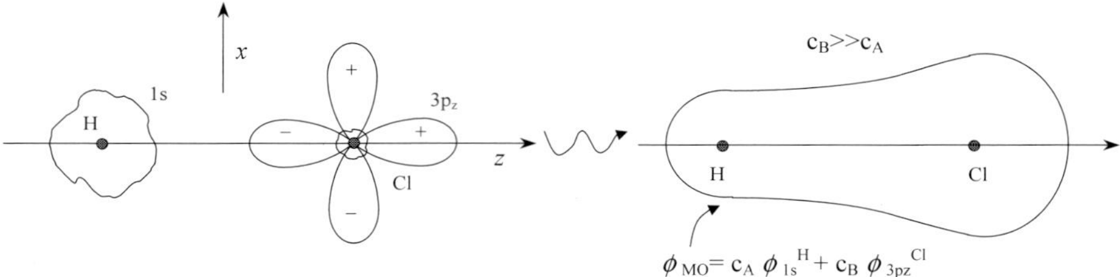

Fig. 8.10. Sketchy illustration of the polarized MO-LCAO orbital in HCl. The p_x and p_y atomic orbitals are scarcely involved in the formation of the molecule since they imply small overlap integral S_{AB} and resonance integral H_{AB} (see text at § 8.2).

In the VB description the only way to account for the charge transfer is to add the ionic states in the molecular orbital, no longer with the same weight as for the homonuclear molecules (see Eq. 8.29). In practice only the ionic configuration favoured by the polarity of the molecule can be included. Then

$$\phi_{VB}^{het} = \phi_{VB} + \lambda\phi_{ionic}$$

The parameters λ in the above definition and λ^* in Eq. 8.31 are difficult to evaluate from first principles. They have been empirically related to the electronegativity of the atoms or to the difference between the ionization energy with respect to the one pertaining to the purely covalent configuration.

An illustrative relationship of λ and λ^* to molecular properties is the one involving the electric dipole moment μ_e. By referring to the sketched schematization for a given molecular orbital with two electrons (pag. 277),

the dipole moment is written $\mu_e = 2e < z >$, with $< z >$ the expectation value of the coordinate, corresponding to the first moment of the electronic charge distribution.

For an MO-LCAO orbital as in Eq. 8.31, one has

$$< z >= \frac{1}{(1 + \lambda^{*2} + 2\lambda^* S_{AB})} \int z \left[|\phi_A|^2 + \lambda^{*2}|\phi_B|^2 + 2\lambda^* \phi_A\phi_B \right] d\tau \quad (8.32)$$

The mixed term $< A|z|B >$ is usually negligible. By assuming for simplicity $S_{AB} \ll 1$ one obtains

$$< z >= \frac{1}{(1 + \lambda^{*2})} \left[-\frac{1}{2}R_{AB} + \frac{1}{2}R_{AB}\lambda^{*2} \right] \quad (8.33)$$

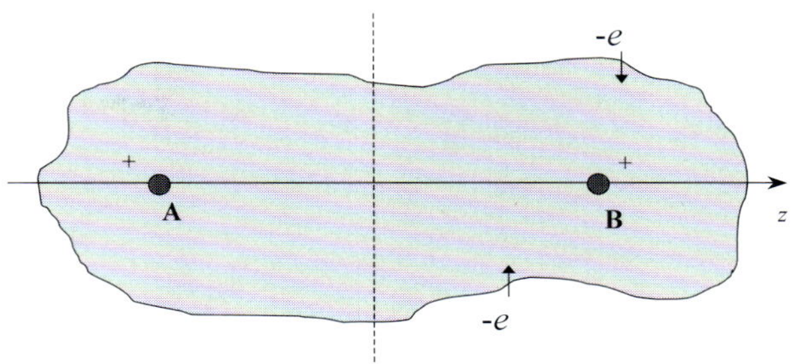

By defining $g = \mu_e/eR_{AB}$ as **degree of ionicity** (g being the unit for total charge transfer and dipole moment $\mu_e^{max} = eR_{AB}$), λ^* can be expressed in terms of a relevant property of the molecule:

$$g = \frac{\lambda^{*2} - 1}{\lambda^{*2} + 1} \tag{8.34}$$

In analogous way in the VB framework, where g is evidently given by the weight of the ionic structure, one has

$$g = \frac{\lambda^2}{\lambda^2 + 1} \quad . \tag{8.35}$$

As for the homonuclear molecules, the energy curve $E(R_{AB})$ in principle could be evaluated in terms of the overlap and resonance integrals.

Direct understanding of the mechanism leading to the bonded state can easily be achieved by referring to a model of **totally ionic molecule**, i.e. $\phi_{MO} = \phi_B$ (or configuration $A^+ B^-$) and in the assumption of Coulombic interaction between point charge ions. This is an oversimplified way to derive the eigenvalue as a function of the interatomic distance, still allowing one to grasp the main source of the bonding.

For numerical clarity let us refer to the NaCl molecule (Fig. 8.11). One observes that for distances R_{AB} above about 10 Å the energy of the neutral atoms is below the one for ions. When the distance is smaller than the R_{AB}^* for which $e^2/R_{AB}^* \simeq (E_I - E_A)$, the ionic configuration is favoured and the system reduces the energy by decreasing the interatomic distance.

At short distance a repulsive term is acting. Its phenomenological form can be written

$$E_{rep} \sim B \, exp\,[-R_{AB}/\rho], \tag{8.36}$$

an expression known as **Born–Mayer** repulsion. Thus the energy curve depicted as solid line in the Figure 8.11 is generated.

The dissociation energy $E(R_{AB}^{min})$ can be evaluated by estimating the distance where the energy minimum occurs. A detailed calculation of this type will be used for the cohesive energy in ionic crystals (§13.2.1).

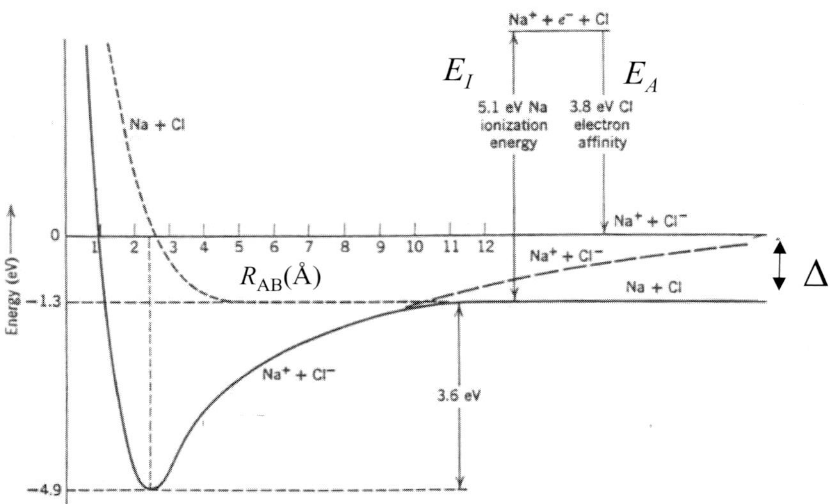

Fig. 8.11. Energies of the neutral atoms and of the ionic configuration in the NaCl molecule. E_I is the ionization energy of Na, about $5.14\,\text{eV}$ while $E_A = 3.82\,\text{eV}$ is the electron affinity in Cl and it corresponds to the energy to remove an electron from Cl^-.

Finally, for some polar diatomic molecules the electric dipole moment μ_e, the degree of ionicity and the value of λ^* according to Eqs. 8.32 are reported below (having used for S_{AB} a value around 0.3).

	μ_e (Debye)	μ_e/eR_{AB}	λ^*
HF	1.91	0.43	1.88
HCl	1.08	0.17	1.28
HBr	0.78	0.11	1.19
KF	8.6	0.67	2.93
KCl	10.27	0.77	3.36

(1 Debye $= 10^{-18}$ u.e.s cm). In H_2O, $\mu_e = 1.86$ Debye.

Problems VIII.5

Problem VIII.5.1 Write the ground state configuration for the molecules CO, LiH, HBr, CN, NO.
  Solution:

$$
\begin{array}{ccc}
\text{CO} & & {}^{1}\Sigma^{+} \\
\text{LiH} & & {}^{1}\Sigma^{+},\ {}^{3}\Sigma^{+} \\
\text{HBr}\quad(\text{Br atom in }{}^{2}P\text{ state}) & & {}^{1}\Sigma^{+},\ {}^{3}\Sigma^{+},\ {}^{1}\Pi,\ {}^{3}\Pi \\
\text{CN}\quad(\text{C atom in }{}^{3}P,\ \text{N atom in }{}^{4}S) & & {}^{2}\Sigma^{+},\ {}^{4}\Sigma^{+},\ {}^{6}\Sigma^{+},\ {}^{2}\Pi,\ {}^{4}\Pi,\ {}^{6}\Pi \\
\text{NO} & & {}^{2}\Pi
\end{array}
$$

Problem VIII.5.2 In the ionic bond approximation assume for the eigenvalue in the NaCl molecule the expression

$$
E(R_{AB}) = -\frac{e^2}{R_{AB}} + \frac{A}{R_{AB}^n}\,.
$$

From the equilibrium interatomic distance $R_{AB}^{eq} = 2.51\,\text{Å}$ and knowing that the vibrational frequency is $1.14 \cdot 10^{13}$ Hz, obtain A and n and estimate the dissociation energy.
  Solution:
  At the minimum

$$
\left(\frac{dE}{dR_{AB}}\right)_{R_{AB}=R_{AB}^{eq.}\equiv R_e} = \frac{e^2}{R_e^2} - n\,A\,\frac{1}{R_e^{n+1}} = 0
$$

thus

$$
\frac{e^2}{n\,A} = \frac{1}{R_e^{n-1}}\,.
$$

The elastic constant is (see Problem VIII.1.3)

$$
k = \left(\frac{d^2 E}{dR_{AB}^2}\right)_{R_{AB}=R_e} = -\frac{2e^2}{R_e^3} + A\,n(n+1)\,\frac{1}{R_e^{n+2}}
$$

Then

$$
k = \frac{e^2}{R_e^3}(n-1)\,.
$$

For the reduced mass

$$\mu = 2.3 \cdot 10^{-23}\, g$$

the elastic constant takes the value

$$k = 4\pi^2 \mu \nu_0^2 = 1.18 \cdot 10^5 \ \text{dyne/cm}$$

(see §10.3).
Then

$$n - 1 = \frac{k\, R_e^3}{e^2} \simeq 8.$$

From

$$A = \frac{e^2 R_e^{n-1}}{n}$$

the energy at R_e is

$$E_{min} = -\frac{e^2}{R_e} + \frac{A}{R_e^n} = -\frac{e^2}{R_e}\left(1 - \frac{1}{n}\right) = -5.1\, eV\,.$$

and then the dissociation energy turns out

$$E_{diss} = -\left[E_{min} + \frac{1}{2}\, h\nu_0\right] \simeq 5\, eV\,.$$

Problems F.VIII

Problem F.VIII.1 The first ionization energy in the K atom is 4.34 eV while the electron affinity for Cl is 3.82 eV. The interatomic equilibrium distance in the KCl molecule is 2.79 Å. Assume for the characteristic constant in the Born-Mayer repulsive term $\rho = 0.28$ Å. In the approximation of point-charge ionic bond, derive the energy required to dissociate the molecule in neutral atoms.

Solution:
From

$$V(R) = -\frac{e^2}{R} + B\, e^{-\frac{R}{\rho}}$$

and the equilibrium condition

$$\left(\frac{dV}{dR}\right)_{R=R_e} = 0 = \frac{e^2}{R_e^2} - \frac{B}{\rho}\, e^{-\frac{R_e}{\rho}}$$

one obtains

$$V(R_e) = -\frac{e^2}{R_e}\left(1 - \frac{\rho}{R_e}\right) \simeq -4.66\, eV.$$

The energy for the ionic configuration $K^+ + Cl^-$ is $(4.34 - 3.82)\text{eV} = 0.52\text{eV}$ above the one for neutral atoms. Then the energy required to dissociate the molecule is

$$E_{diss} = (+4.66 - 0.52)\,\text{eV} \simeq 4.12\,\text{eV}.$$

Problem F.VIII.2 In the molecule KF the interatomic equilibrium distance is 2.67 Å and the bonding energy is $0.5\,\text{eV}$ smaller than the attractive energy of purely Coulomb character. By knowing that the electron affinity of Fluorine is $4.07\,\text{eV}$ and that the first ionization potential for potassium is $4.34\,\text{V}$, derive the energy required to dissociate the molecule in neutral atoms.

Solution:

Since

$$E_{Coulomb} = \frac{e^2}{R_e} = 8.6 \cdot 10^{-12}\,\text{erg} = 5.39\,\text{eV}$$

the energy required for the dissociation in ions is $E_i = (5.39 - 0.5) = 4.89\,\text{eV}$. For the dissociation in neutral atoms $E_a = E_i + A_f - P_{ion} = 4.89 + 4.07 - 4.34\,\text{eV} = 4.62\,\text{eV}$.

Problem F.VIII.3 Report the ground state configuration and the first excited configuration of the molecules H_2, Li_2, N_2, LiH and CH.

Solution:

Molecule	Ground state configuration	
H_2	$(\sigma_g 1s)2$	$^1\Sigma_g^+$
Li_2	$KK(\sigma_g 2s)2$	$^1\Sigma_g^+$
N_2	$KK(\sigma_g 2s)2(\sigma_u 2s)2(\pi_u 2p)4(\sigma_g 2p)2$	$^1\Sigma_g^+$
LiH	$K(2s\sigma)2$	$^1\Sigma^+$
CH	$K(2s\sigma)2(2s\sigma)22p\pi$	$^2\Pi$

Molecule	First excited configuration	
H_2	$(\sigma_g 1s)\sigma_u 1s$	$^1\Sigma_u^+,\ ^3\Sigma_u^+$
Li_2	$KK(\sigma_g 2s)\sigma_u 2s$	$^1\Sigma_u^+,\ ^3\Sigma_u^+$
N_2	$KK(\sigma_g 2s)2(\sigma_u 2s)2(\pi_u 2p)4(\sigma_g 2p)1\pi_g 2p$	$^1\Pi_g,\ ^1\Pi_u$
LiH	$K(2s\sigma)2p\sigma$	$^1\Sigma^+,\ ^3\Sigma^+$
CH	$K(2s\sigma)2(2p\sigma)(2p\pi)2$	$^4\Sigma^-,\ ^2\Delta,\ ^2\Sigma^+,\ ^2\Sigma^-$

Problem F.VIII.4 Derive the structure of the hyperfine magnetic states for the ground-state of the Hydrogen molecular ion. Then numerically evaluate their energy separation in the assumption of $\sigma_g 1s$ molecular orbital in the form of linear combination of $1s$ atomic orbitals (the interatomic equilibrium distance can be assumed $2a_0$).

Solution:

From the extension of Eq. 5.3

$$H_{mag}^{hyp} = A_{H_2^+}(\mathbf{I}_A + \mathbf{I}_B) \cdot \mathbf{s},$$

with $A_{H_2^+}$ the hyperfine coupling constant. From $\mathbf{I} = \mathbf{I}_A + \mathbf{I}_B$, $I = 0$ or $I = 1$, namely states with $F = 1/2, 3/2$ are obtained.

Since $\mathbf{I}.\mathbf{s} = (1/2)[F(F+1) - I(I+1) - s(s+1)]$

the $F = 3/2$ and $F = 1/2$ levels are separated by $\Delta E = (3/2)A_{H_2^+}$.

$A_{H_2^+}$ can be obtained from

$$\phi_{\sigma_g 1s} = \frac{1}{\sqrt{2(1 + S_{AB})}} \frac{1}{\sqrt{\pi}a_0^{3/2}} \left[e^{-r_A/a_0} + e^{-r_B/a_0} \right]$$

considering $r_A = 0$ and $r_B = R_{AB}$. Then

$$|\phi_{\sigma_g 1s}(0)|^2 = \frac{1}{\pi a_0^3} \frac{\left[1 + e^{-R_{AB}/a_0} \right]^2}{2(1 + S_{AB})} \simeq \frac{0.41}{\pi a_0^3}$$

for $S_{AB} \approx 0.58$ (see Eq. 8.9).

In atomic Hydrogen where $|\phi_{1s}(0)|^2 = 1/\pi a_0^3$ the separation between the $F = 1$ and $F = 0$ hyperfine levels is $A_H/h = 1421.8$ MHz. Then in H_2^+ one deduces $\Delta E/h = (3/2)0.41A_H \simeq 810$ MHz. (For the difference between the ortho-states at $I=1$ and the para-state at $I=0$ read §10.9).

Problem F.VIII.5 In the assumption that an electric field $\mathcal{E}$ applied along the molecular axis of H_2^+ can be considered as a perturbation, evaluate the electronic contribution to the electric polarizability (for rigid molecule and for molecular orbital LCAO).

Solution:

$$\mathcal{H}_P = -ez\mathcal{E}$$

At first order $< g|\mathcal{H}_P|g >= 0$ (where $|g >= (1/\sqrt{2(1 + S_{AB})})(\phi_A + \phi_B)$), since it corresponds to the first moment of the electronic distribution, evidently zero for a homonuclear molecule (see §8.5).

At second order, involving only the first excited state $|u >= (1/\sqrt{2(1 - S_{AB})})(\phi_A - \phi_B)$, one has

$$E^{(2)} = e^2 \mathcal{E}^2 \frac{|<u|z|g>|^2}{E_+ - E_-}$$

From

$$<u|z|g> = \frac{1}{2} \frac{1}{\sqrt{1 + S_{AB}}} \frac{1}{\sqrt{1 - S_{AB}}} \int (\phi_A + \phi_B) z (\phi_A - \phi_B) d\tau$$

$$= \frac{1}{\sqrt{1 - S_{AB}^2}} \left[-\frac{R_{AB}}{2} - \frac{R_{AB}}{2} \right],$$

$$E^{(2)} = -\frac{e^2 \mathcal{E}^2 R_{AB}^2}{4(1 - S_{AB}^2)(E_- - E_+)}$$

and then

$$\alpha_{H_2^+} = -\frac{1}{\mathcal{E}} \frac{\partial E^{(2)}}{\partial \mathcal{E}} = \frac{e^2 R_{AB}^2}{2(1 - S_{AB}^2)(E_- - E_+)}$$

From $R_{AB}^{eq} \simeq 2a_0$, $S_{AB}^2 \ll 1$ and $(E_- - E_+) \simeq 0.1 e^2/a_0$ one has $\alpha_{H_2^+} \simeq (R_{AB}^{eq})^3/0.4$, of the expected order of magnitude (see Problem X.3.4).

Problem F.VIII.6 For two atoms A and B with $J = S = 1/2$, in the initial spin state $\alpha_A \beta_B$, **spin-exchange collision** is the process by which at large distances (no molecule is formed) they interact and end up in the final spin state $\beta_A \alpha_B$ (This process is often used in atomic optical spectroscopy to induce polarization and **optical pumping**). From the extension of the VB description of H_2 (§8.3) one can assume a spin-dependent interaction $\mathcal{H} = -2K(R)\mathbf{S}_A \cdot \mathbf{S}_B$, where $K(R)$ is the **negative**, R-dependent exchange integral favouring the $S = 0$ ground-state.

Discuss the condition for the spin-exchange process by making reasonable assumptions for the collision time R_c/v, R_c being an average interaction distance and v the relative velocity of the two atoms.

Solution:
In the singlet ground-state the interaction is

$$E(R) = -2K(R) \left[\frac{S^2 - S_A^2 - S_B^2}{2} \right] = \frac{6K(R)}{4}$$

An approximate estimate of the time required to shift from $\alpha_A \beta_B$ to $\beta_A \alpha_B$ can be obtained by referring to the Rabi equation (App.I.2), in a way somewhat analogous to the exchange of the electron discussed at §8.1.2 for the H_2^+ molecule. Here the Rabi frequency has to be written $\Gamma \approx E(R)/\hbar$, for R around R_c. Then the time for spin exchange is of the order of $(\pi/3)(\hbar/|K(R_c)|)$ while the time for interaction is $\tau_c = R_c/v$ (v can be considered the thermal velocity at room temperature in atomic vapors, i.e. $\approx 3 \cdot 10^4$ cm/s).

Thus one derives $-3K(R_c)R_c \simeq \pi \hbar v$.

For an order of magnitude estimate one can assume that at large distance $K(R_c)$ is in the range $10^{-3} - 10^{-4}$eV thus yielding R_c in the range $6 - 60$ Å.

These crude estimates for the spin-exchange process and the limits of validity are better discussed at Chapter 5 of the book by **Budker, Kimball** and **De Mille** quoted in the Preface, where a more rigorous analysis of this problem can be found.

9

Electronic states in selected polyatomic molecules

Topics

Polyatomic molecules as formed by bonds between pairs of atoms
Hybrid atomic orbitals
Geometry of some molecules
Bonds for carbon atom
Electron delocalization and the benzene molecule

In this Chapter a few general aspects of electronic states in polyatomic molecules shall be discussed. Some details will be given for typical molecules involving novel phenomena that do not occur in diatomic molecules.

The electronic structure in polyatomic molecules is based on the same principles described for diatomic molecules. As already mentioned, a general theory somewhat equivalent to the Slater theory for many-electron atoms can be developed. The steps of that approach are the following. Molecular orbitals of the form $\phi(\mathbf{r}_i, \mathbf{s}_i) = \sum_p c_p^{(i)} \phi_p(\mathbf{r}_i) \chi_{spin}^{(i)}$ are assumed as a basis, in terms of linear combination of atomic spin-orbitals centered at the various sites with unknown coefficients $c_p^{(i)}$. The determinantal wavefunction for all the electrons is then built up and the energy function is constructed from the full Hamiltonian $(T + V_{ne} + V_{ee})$ (see § 7.1). The Hartree-Fock variational procedure is then carried out in order to derive the coefficients $c_p^{(i)}$. This approach is known as **MO.LCAO.SCF** (self-consistent field). Advanced computational methods are required and the one developed by **Roothaan** is the most popular. More recently the **density functional theory** is often applied in *ab-initio* procedures, based on the idea that the energy can be written in terms of electron probability density, thus becoming a **functional** of the charge distribution, while the **local density approximation** is used to account for the exchange-correlation corrections. Configuration interaction (see § 8.4) is usually taken into account. We will not deal with these advanced topics, essentially belonging to the realm of computational quantum chemistry.

We shall see how qualitative aspects can be understood simply in terms of the idealization of **independent bonds**, by considering the molecule as resulting from pairs of atoms, each pair corresponding to a given bond. In this way the main aspects worked out in diatomic molecules (Chapter 8) can be extended to polyatomic molecules. Typical illustrative example is the NH_3 molecule.

At § 9.2 we will discuss the molecular bonds involving **hybrid atomic orbitals** and giving rise to particular geometries of the molecules, typically the ones related to the variety of bonds involving the carbon atom. In § 9.3 the **delocalization** of the electrons will be addressed, with reference to the typical case of the benzene molecule.

9.1 Qualitative aspects of NH_3 and H_2O molecules

In the spirit of the simplified picture of localized orbitals and independent bonds, by considering the molecule as due to bonds between pairs of atoms, one can sketch the formation of the NH_3 molecule as resulting from three mutually perpendicular σ MO orbitals involving LCAO combination of $2p$ N atomic orbitals and $1s$ Hydrogen orbital (see Fig. 9.1).

Similar qualitative picture can be given for the H_2O molecule (Fig. 9.2).

From those examples one can understand how the geometry of the molecules, with certain angles between bonds, is a consequence of the maxima for the probability of presence of the electrons controlled by atomic orbitals, coupled with the criterium of **strong overlap**, in order to maximize the resonance integral.

However it should be remarked that the above picture is incomplete. In fact the angles between bonds are far from being $90°$, in general. For instance in H_2O the angle between the two OH bonds is about $105°$. As we shall see in the next Section, the geometry of the molecules is consistent with the assumption of **hybrid atomic orbitals** involved in the formation of the MO's.

9.2 Bonds due to hybrid atomic orbitals

By naively referring to the electrons available to form bonds by sharing the molecular orbitals, Be, B and C atoms would be characterized by valence numbers $n_V = 0, 1$ and 2, correspondent to the electrons outside the closed shells. The experimental findings ($n_V = 2, 3$ and 4, respectively) could qualitatively be understood by assuming that when molecules are formed, one electron in those atoms is promoted to an excited state. The increase in the number of bonds, with the related decrease of the total energy upon bonding, would account for the energy required to promote the electron to the excited

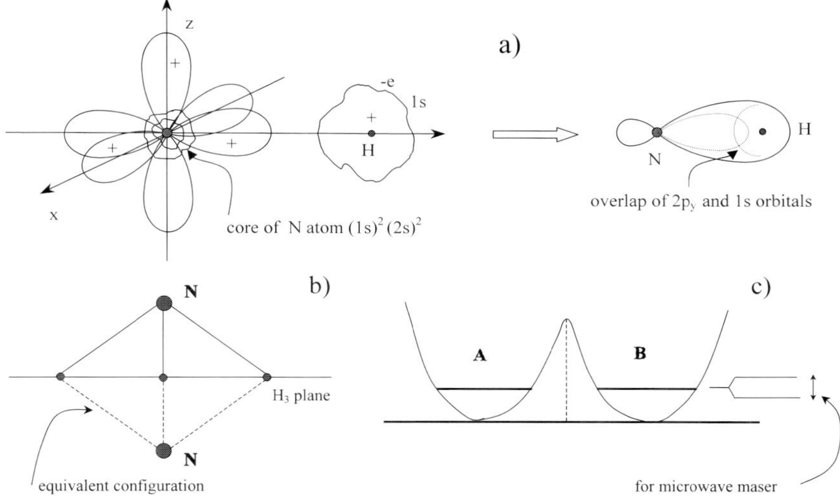

Fig. 9.1. Pictorial view of the formation of the NH_3 molecule in terms of combination of localized $1s$ Hydrogen and of $2p_{x,y,z}$ N atomic orbitals, with the criterium of the **maximum overlap** to grant the largest contribution to the bonding energy from each bond (a). The equivalent configuration is shown in part (b), where the molecule can be thought to result from the approach of the H atoms along the opposite directions of the coordinate axes.
The evolution of the two level states is sketched in part (c) with the **inversion doublet** resulting from the removal of the degeneracy. The separation energy of the doublet is related to the exchange integral. These two states were used to obtain the first **maser** operation (see Appendix IX.1).

state. This argument by itself cannot justify the experimental evidence. At the sake of clarity, let us refer to the CH_4 molecule: its structure, with four equivalent C-H bonds, with angles $109°28'$ in between, can hardly be justified by assuming for the carbon atom one electron in each $2s, 2p_x, 2p_y$ and $2p_z$ atomic orbitals. A related consideration, claiming for an explanation of the molecular geometry, is the one aforementioned for the angles between bonds in the H_2O molecule.

Again referring to CH_4 and in the light of the equivalence of the four C-H bonds, one can still keep the criterium of the **maximum overlap** provided that atomic orbitals, not corresponding to states of definite angular momentum, are supposed to occur in the atom when the molecule is being formed. These atomic orbitals are called **hybrid**.

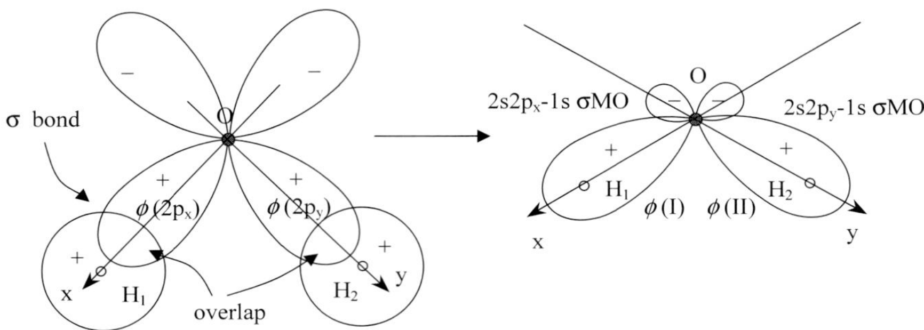

Fig. 9.2. Schematic view of the H_2O molecule as resulting from two σ MO's involving $2p$ O and $1s$ H atomic orbitals, with strong overlap of the wavefunctions when the Hydrogen atoms approach the Oxygen along the directions of the x and y axes.

To account for the geometry of the bonds in CH_4 we have to generate hybrid orbitals with maxima in the probability of presence along the directions of the **tetrahedral** environment, as sketched below

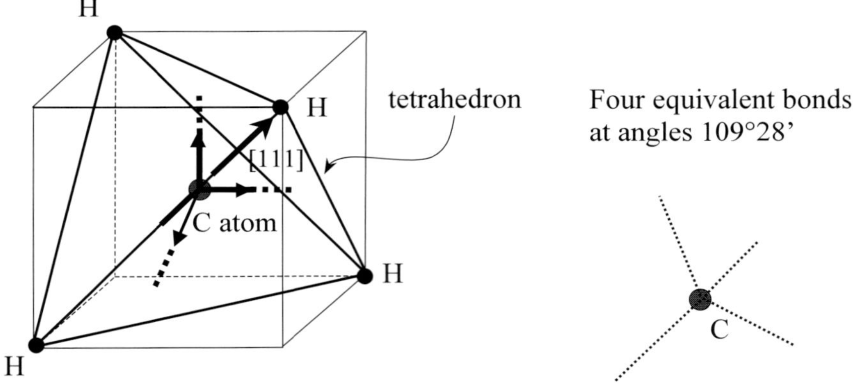

From the linear combination

$$\phi_C = a\phi_{2s} + b\phi_{2p_x} + c\phi_{2p_y} + d\phi_{2p_z}$$

by resorting to the orthonormality condition, to the requirement of electronic charge displaced along the tetrahedral directions and by considering that for symmetry reasons the s electron has to be equally distributed on the four hybrid orbitals, one can figure out that the coefficients must be

$$a^2 = 1/4 \qquad b^2 + c^2 + d^2 = 3/4 \qquad b^2 = c^2 = d^2,$$

yielding maxima for the probability of presence along the directions

$$(1, 1, 1) \quad , \quad (1, -1, -1) \quad , \quad (-1, 1, -1) \quad , \quad (-1, -1, 1) \ .$$

Thus the hybrid orbitals of the C atom are

$$\phi_I = \frac{1}{2}\left[2s + 2p_x + 2p_y + 2p_z\right]$$

$$\phi_{II} = \frac{1}{2}\left[2s + 2p_x - 2p_y - 2p_z\right]$$

$$\phi_{III} = \frac{1}{2}\left[2s - 2p_x + 2p_y - 2p_z\right]$$

$$\phi_{IV} = \frac{1}{2}\left[2s - 2p_x - 2p_y + 2p_z\right]$$

The individual bonds with the H atoms can then be thought to result from σ MO, given by linear combinations of the C hybrids and of the $1s$ H atomic orbitals, as sketched below

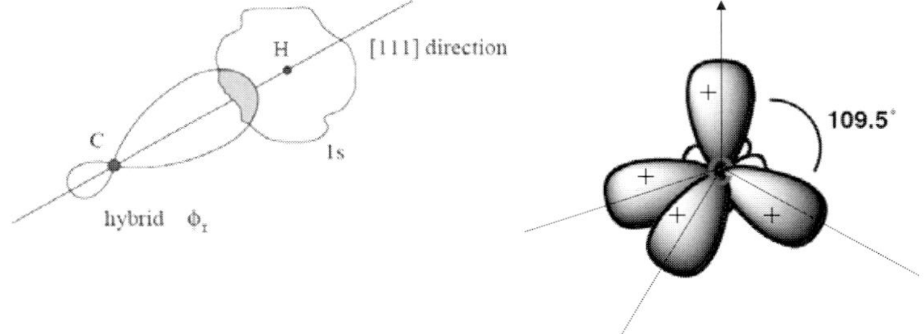

That type of hybridization is called (sp^3) or **tetragonal**. Besides the methane molecule, is the one that can be thought to occur in the molecular-like bonding in some crystals, primarily in diamond (C) and in semiconductors such as Ge, Si and others (see Chapter 11).

Another type of hybridization involving the carbon atom is (sp^2) or **trigonal** one, giving rise to planar geometry of the molecule, with three equivalent bonds forming angles of 120° between them, such as in the ethylene molecule, C_2H_4. The hybrid orbital can be derived in a way similar to the tetragonal hybridization, from a linear combination of $2s, 2p_x$ and $2p_y$. By taking into account that the coefficients b and c are proportional to the cosine of the related angles and that $a^2 = 1/3$, $b^2 + c^2 = 2/3$, one has the following picture

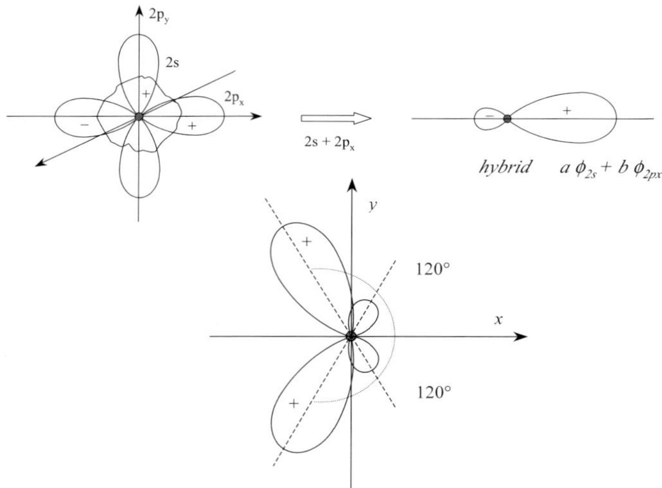

corresponding to C hybrid orbitals

$$\phi_I = \frac{1}{\sqrt{3}}\,\phi_{2s} + \frac{\sqrt{2}}{\sqrt{3}}\,\phi_{2px}$$

$$\phi_{II} = \frac{1}{\sqrt{3}}\,\phi_{2s} - \frac{1}{\sqrt{6}}\,\phi_{2px} + \frac{1}{\sqrt{2}}\,\phi_{2py}$$

$$\phi_{III} = \frac{1}{\sqrt{3}}\,\phi_{2s} - \frac{1}{\sqrt{6}}\,\phi_{2px} - \frac{1}{\sqrt{2}}\,\phi_{2py}$$

Therefore the σ MO bonds are generated from the linear combination of the $1s$ H orbitals or of the equivalent hybrid orbital of the other carbon atom, as sketched below

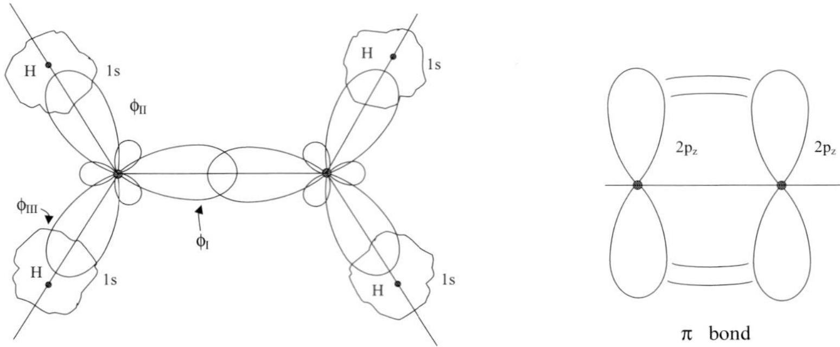

The $2p$ electron described by the atomic orbital $2p_z$, perpendicular to the plane of the molecule, is not involved in the hybrid and therefore it can form a π C-C MO of the type already seen in diatomic molecules, leading to an additional **weak bond** (see the N_2 molecule at §8.2).

Another interesting hybrid orbital for the carbon atom and leading to linear molecule, such as acetylene (C_2H_2) is the **digonal** (sp) hybrid. It mixes the s electron and one p electron only. The electronic configuration sketched below is derived

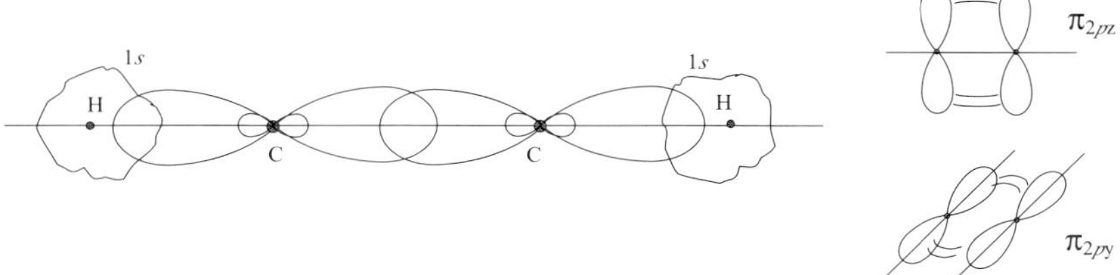

The C-C bond here is a triple one (one strong bond and two weak bonds).

More complex hybrid orbitals are generated in other multi-atoms molecules, with particular geometries. For its importance and at the sake of illustration we mention the (d^2sp^3) atomic orbitals, occurring in atoms with incomplete d shells. The hybridization implies **six bonds**, along the positive and negative directions of the Cartesian axes. By combining with $2p$ oxygen orbitals the octahedral structure depicted in Fig. 9.3 originates, for example for $BaTiO_3$.

We shall come back to this relevant atomic configuration, characteristic of the perovskite-type ferroelectric titanates such as $BaTiO_3$, at §13.3 when dealing with the **CuO_6 octahedron**, which is the structural core of high-temperature superconductors.

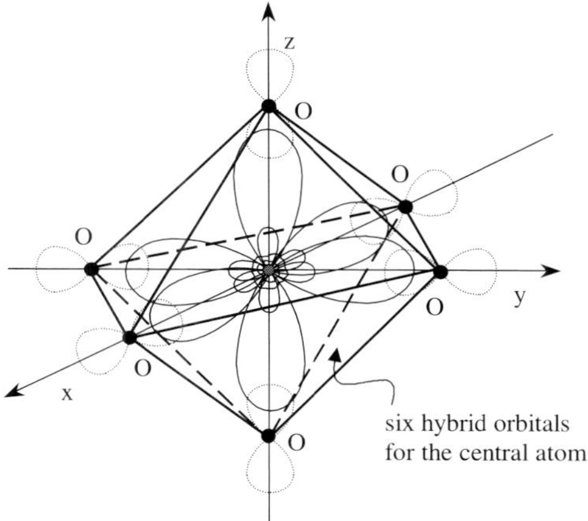

Fig. 9.3. The configuration of σ bonds involving the six atomic orbitals of the central atom (for example Titanium) associated with the d^2sp^3 hybridization. This atomic configuration is the one occurring, for example in the $(TiO_3)^{2-}$ molecular-like unit in the $BaTiO_3$ crystal (oxygens are shared by two units) (see §11.4).

9.3 Delocalization and the benzene molecule

Experimental evidences, such as X-ray (in the solid state) and roto-vibrational spectra (see Chapter 10) indicate that the benzene molecule, C_6H_6, is characterized by planar hexagonal structure, with the carbon atoms at the vertices of the hexagon. The C-H bonds form $120°$ angles with the adjacent pair of C-C bonds. According to this atomic configuration one understands that the Carbon atom is in the sp^2 trigonal hybridization, as the one discussed for the C_2H_4 molecule (§9.2). The remaining $2p_z$ electrons of the Carbon atoms, not involved in the hybrids, can form a πMO between adjacent C atoms, yielding three double bonds. However, all the C-C bonds are equivalent and the distances C-C are the same. This is one of the evidences that the simplified picture of localized electrons, with "independent" bonds between pairs of atoms, in some circumstances has to be abandoned. We shall see that the structure of the benzene molecule, as well as of other molecules with π-bonded atoms like the polyenes, can be justified only by **delocalizing** the $2p_z$ electrons all along the carbon ring. The delocalization process is a further mechanism of bonding, since the total energy is decreased. At Chapter 12 we shall see that the electronic states in crystals can be described as related to the delocalization of the electrons. Thus, for certain aspects the benzene molecule can also be regarded as a prototype for the electronic states in crystals.

By extending to the six Carbon atoms in the benzene ring the MO.LCAO description, the one-electron orbital is written

$$\phi_{MO}(i) = \sum_r c_r \phi_{2p_z}^{(r)} \tag{9.1}$$

where $\phi_{2p_z}^{(r)}$ are C $2p$ orbitals centered at the r-th site of the hexagon (r runs from 1 to 6). Then the energy function, by referring only to the Hamiltonian $\mathcal{H}$ for the $2p_z$ C electrons, is

$$E(c_r) = \frac{\int \phi_{MO}^* \,\mathcal{H}\, \phi_{MO}\, d\tau}{\int \phi_{MO}^* \,\phi_{MO}\, d\tau} \tag{9.2}$$

By resorting to the concepts already used in diatomic molecules (§8.1) we label $\beta_{rs} = \int \phi_{2p_z}^{*(s)} \,\mathcal{H}\, \phi_{2p_z}^{(r)}\, d\tau$ as resonance integral, while $S_{rs} = \int \phi_{2p_z}^{*(s)} \,\phi_{2p_z}^{(r)}\, d\tau$ is the overlap integral. One has $\int \phi_{2p_z}^{*(r)} \,\mathcal{H}\, \phi_{2p_z}^{(r)}\, d\tau = E_0$, energy of $2p$ electron in the C atom and $\int \phi_{2p_z}^{*(r)} \,\phi_{2p_z}^{(r)}\, d\tau = 1$.

It is conceivable to assume $S_{rs} = 0$ for $r \neq s$ and to take into account the resonance integral only between adjacent C atoms: $\beta_{rs} = \beta$ for $r = s \pm 1$ and zero otherwise.

Then the secular Equation for the energy function $E(c_r)$ reads

$$\begin{pmatrix} (E_0 - E) & \beta & 0 & 0 & 0 & \beta \\ \beta & (E_0 - E) & \beta & 0 & 0 & 0 \\ 0 & \beta & (E_0 - E) & \beta & 0 & 0 \\ 0 & 0 & \beta & (E_0 - E) & \beta & 0 \\ 0 & 0 & 0 & \beta & (E_0 - E) & \beta \\ \beta & 0 & 0 & 0 & \beta & (E_0 - E) \end{pmatrix} = 0 \ .$$

The roots are $E_0 + 2\beta, \quad E_0 \pm \beta$ (twice), $\quad E_0 - 2\beta \quad$ (note that $\beta < 0$).

The lowest energy delocalized π orbital, correspondent to the eigenvalue $E_0 + 2\beta$, can accommodate two electrons, while on the state at energy $E_0 + \beta$ one can place four electrons, as sketched below

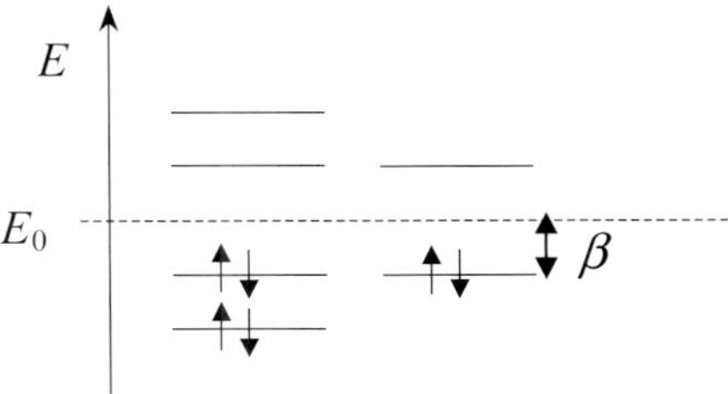

The bonding energy turns out $2(2\beta) + 4\beta = 8\beta$, lower than the energy 6β that one would obtain for localized electrons. The energy 2β can be considered the contribution to the ground state energy due to the delocalization.

In correspondence to the root $(E_0 + 2\beta)$ the coefficients c_r in Eq. 9.1 are equal. The normalization yields $c_r = 1/\sqrt{6}$ and therefore the molecular orbital is

$$\phi_{MO}(\mathbf{r}_i) = \frac{1}{\sqrt{6}} \left[\phi_{2p_z}(\mathbf{r}_i - \mathbf{l}_1) + ... + \phi_{2p_z}(\mathbf{r}_i - \mathbf{l}_6) \right] \tag{9.3}$$

where $\mathbf{l}_r$ indicate $n_r \mathbf{a}$ and specifies the position of the Carbon atom along the ring of step $\mathbf{a}$. The wavefunction 9.3 is sketched in Fig. 9.4.

In correspondence to the root $(E_0 + \beta)$ different choices for the coefficients c_r are possible (see Problem F.IX.2 for similar situation). One choice is

$$\phi_{MO}(\mathbf{r}_i) = \frac{1}{\sqrt{12}} \left[2\phi_{2p_z}(\mathbf{r}_i - \mathbf{l}_1) + \phi_{2p_z}(\mathbf{r}_i - \mathbf{l}_2) - \phi_{2p_z}(\mathbf{r}_i - \mathbf{l}_3) - 2\phi_{2p_z}(\mathbf{r}_i - \mathbf{l}_4) - \right.$$

$$\left. - \phi_{2p_z}(\mathbf{r}_i - \mathbf{l}_5) + \phi_{2p_z}(\mathbf{r}_i - \mathbf{l}_6) \right] \tag{9.4}$$

The eigenvalues can be written in the form

$$E_p = E_o + 2\beta cos[(2\pi/6a)pa] \ , \tag{9.5}$$

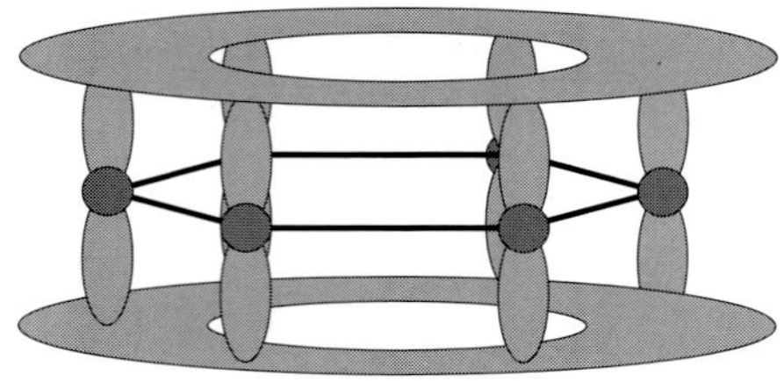

Fig. 9.4. Pictorial view of the πMO delocalized orbital correspondent to Eq. 9.3 (eigenvalue $E_0 + 2\beta$).

while for the coefficients

$$c_p = (e^{2\pi i p/6})/(6)^{1/2} \ , \tag{9.6}$$

where $p = 0, \pm 1, \pm 2, 3$.

The benzene ring can be considered the cyclic repetition of a "crystal" of six Carbon atoms. The eigenvalues and the coefficients in the forms 9.5 and 9.6 are somewhat equivalent to the band states in a one-dimensional crystal (see Chapter 12).

The quantitative evaluation (by means of numerical methods or by resorting to approximate radial parts of the wavefunctions) of the electronic eigenvalue as a function of the interatomic distance a yields a minimum for $a \simeq 1.4\,\text{Å}$, in between the values $a' = 1.34\,\text{Å}$ and $a'' = 1.54\,\text{Å}$ pertaining to double and to simple C-C bond, respectively.

The structural anisotropy of the molecule is reflected, for instance, in the strong dependence of the diamagnetic susceptibility χ_{dia} on the orientation. In fact, by extending the arguments discussed for atoms (§4.5) one can expect $\chi_{dia} \propto \sum_i < r_i^2 sin^2\theta_i >$, with θ angle between the magnetic field and the positional vector of a given electron. Then, in the benzene molecule, $\chi_{dia}^{\parallel} << \chi_{dia}^{\perp}$ (with $\parallel$ and $\perp$ to the plane of the hexagon). (See Problem F.IX.3).

Appendix IX.1 Ammonia molecule in electric field and the Ammonia maser

According to Fig. 9.1 the Ammonia molecule can be found in two equivalent configurations, depending on the position of the N atom above (state $|1 >$) or below (state $|2 >$) the xy plane of the H atoms. By considering the molecule in its ground electronic state and neglecting all other degrees of freedom, let us discuss the problem of the position of the N atom along the z direction perpendicular to the xy plane, therefore involving the vibrational motion in which N oscillates against the three coplanar H atoms (for details on the vibrational motions see §10.3 and §10.6).

The potential energy $V(z)$, that in the framework of the Born-Oppenheimer separation (§7.1) controls the nuclear motions and that is the counterpart of the energy $E(R_{AB})$ in diatomic molecules, has the shape sketched below

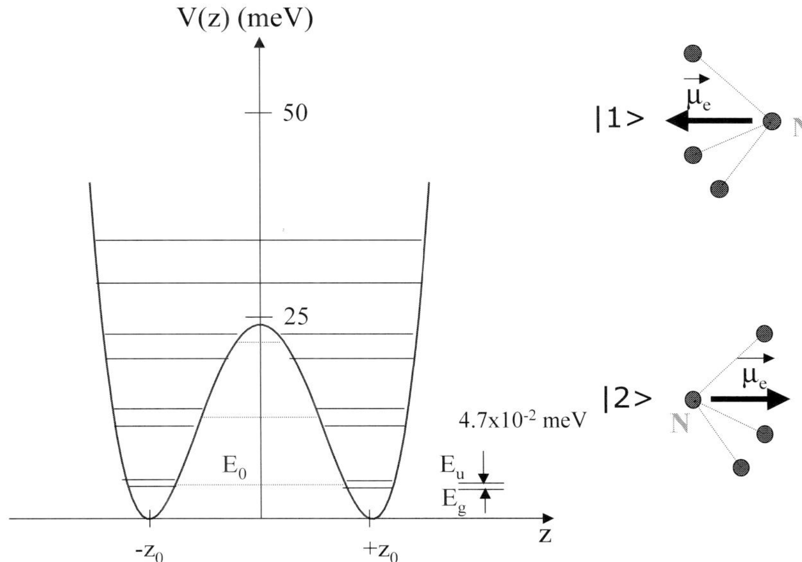

The distance of the N atom from the xy plane corresponding to the minima in $V(z)$ is $z_o = 0.38$ Å, while the height of the potential energy for $z = 0$ is $V_o \simeq 25$meV. In the state $|1>$ the molecule has an electric dipole moment $\boldsymbol{\mu}_e$ along the negative z direction, while in the state $|2>$ the dipole moment is parallel to the reference z-axis. Within each state the N atoms vibrate around $+z_o$ or $-z_o$. As for any molecular oscillator the ground state has a zero-point energy different from zero, that we label E_o (correspondent to the two levels A and B sketched in Fig. 9.1.c). The vibrational eigenfunction in the ground state is a Gaussian one, centered at $\pm z_o$ (see §10.3). The effective mass of the molecular oscillator is $\mu = 3M_H M_N/(3M_H + M_N)$.

Thus the system is formally similar to the H_2^+ molecule discussed at §8.1, the $|1>, |2>$ states corresponding to the electron hydrogenic states $1s_A$ and $1s_B$, while the vibration zero-point energy corresponds to $-R_H hc$. Therefore, the generic state of the system is written

$$|\psi> = c_1|1> + c_2|2> \qquad (A.IX.1)$$

with coefficients c_i obeying to Eqs. 8.13. Here $H_{12} = -A$ is the probability amplitude that because of the quantum tunneling the N atom jumps from $|1>$ to $|2>$ and *viceversa*, in spite of the fact that $E_o \ll V_o$. Two stationary states are generated, say $|g>$ and $|u>$, with eigenvalues $E_o - A$ and $E_o + A$, respectively. The correspondent eigenfunctions are linear combinations of the Gaussian functions describing the oscillator in its ground state (see §10.3):

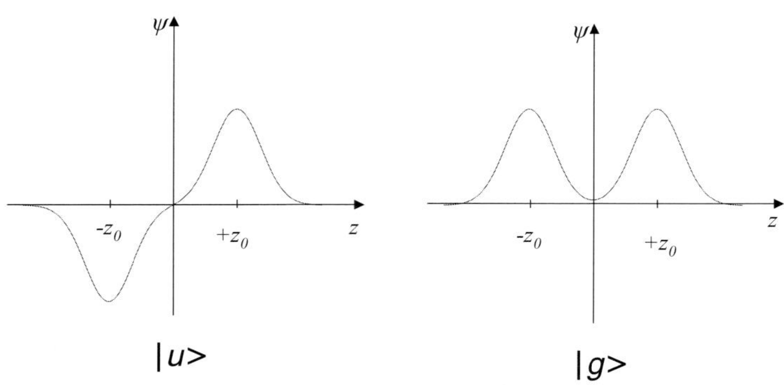

The degeneracy of the original states is thus removed and the vibrational levels are in form of doublets (**inversion doublets**). For the ground-state the splitting $E_g - E_u = 2A$ corresponds to 0.793 cm^{-1}, while it increases in the excited vibrational states, owing to the increase of H_{12}. For the first excited state $2A' =36.5$ cm^{-1} and for the second excited state $2A'' =312.5$ cm^{-1}. It can be remarked that the vibrational frequency (see §10.3) of N around the minimum in one of the wells is about 950 cm^{-1}.

The inversion splitting are drastically reduced in the deuterated Ammonia molecule ND_3 where for the ground-state $2A =0.053$ cm^{-1}. Thus the tunneling frequency, besides being strongly dependent on the height of the effective potential barrier V_o, is very sensitive to the reduced mass μ. For instance, in the AsH_3 molecule, the time required for a complete tunneling cycle of the As atom is estimated to be about two years. These marked dependences on V_o and μ explains why in most molecules the inversion doublet is too small to be observed.

In NH_3 the so-called **inversion spectrum** was first observed (**Cleeton** and **Williams**, 1934) as a direct absorption peak at a wavelength around 1.25 cm, by means of microwave techniques. This experiment opened the field presently known as **microwave spectroscopy**.

The typical experimental setup is schematically shown below

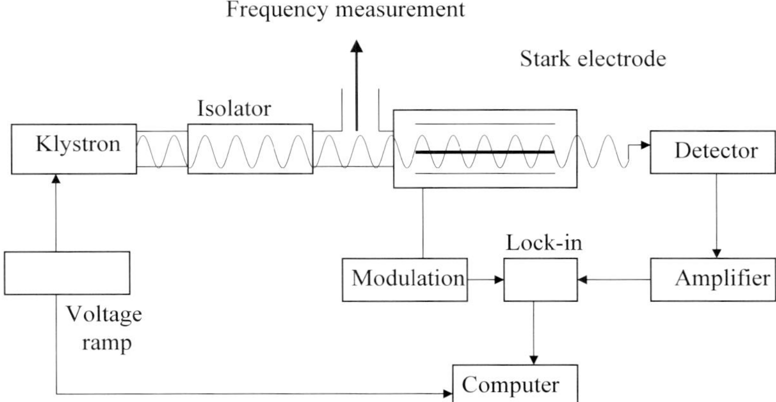

Finally it should be remarked that the rotational motions of the molecule (§10.2), as well as the magnetic and quadrupolar interactions (Chapter 5), in general cause fine and hyperfine structures in the inversion spectra.

As already mentioned the $|g>$ and $|u>$ states of the inversion doublet in NH_3 have been used in the first experiment (**Townes** and collaborators) of **microwave amplification** by **stimulated emission of radiation** (see Problem F.I.1). The **maser** action requires that the statistical population N_u is maintained larger than N_g while a certain number of transitions from $|u>$ to $|g>$ take place.

Now we are going to discuss how the Ammonia molecule behaves in a static electric field. Then we show how by applying an electric field gradient **(quadrupolar electric lens)** one can select the Ammonia molecule in the upper energy state.

In the presence of a field $\mathcal{E}$ along z the eigenvalue for the states $|1>$ and $|2>$ become

$$H_{11} = E_o + \mu_e\mathcal{E} \quad \text{and} \quad H_{22} = E_o - \mu_e\mathcal{E}$$

The rate of exchange can be assumed approximately the same as in absence of the field, namely $H_{12} = -A$. The analogous of Eqs. 8.13 for the coefficients c_i in A.IX.1 are then modified in

$$i\hbar\frac{dc_1}{dt} = (E_o + \mu_e\mathcal{E})c_1 - Ac_2 \qquad\qquad \text{(A.IX.2)}$$

$$i\hbar\frac{dc_2}{dt} = (E_o - \mu_e\mathcal{E})c_2 - Ac_1 \qquad\qquad \text{(A.IX.3)}$$

The solutions of these equations must be of the form $c_i = a_i exp(-iEt/\hbar)$, with E the unknown eigenvalue. The resulting Eqs. for a_i are

$$(E - E_o - \mu_e \mathcal{E})a_1 + A a_2 = 0$$
$$A a_1 + (E - E_o + \mu_e \mathcal{E})a_2 = 0$$

and the solubility condition yields

$$E_\pm = \frac{H_{11} + H_{22}}{2} \pm \sqrt{\frac{(H_{11} - H_{22})^2}{4} + A^2} = E_o \pm \sqrt{A^2 + \mu_e^2 \mathcal{E}^2} \quad \text{(A.IX.4)}$$

(representing a particular case of the perturbation effects described in Appendix I.2 (Eqs. A.I.2.4)). When the perturbation is not too strong compared to the inversion splitting, Eq. A.IX.4 can be approximated in the form

$$E_\pm = E_o \pm A \pm \frac{\mu_e^2 \mathcal{E}^2}{2A} . \quad \text{(A.IX.5)}$$

$E_\pm$ are reported below as a function of the field.

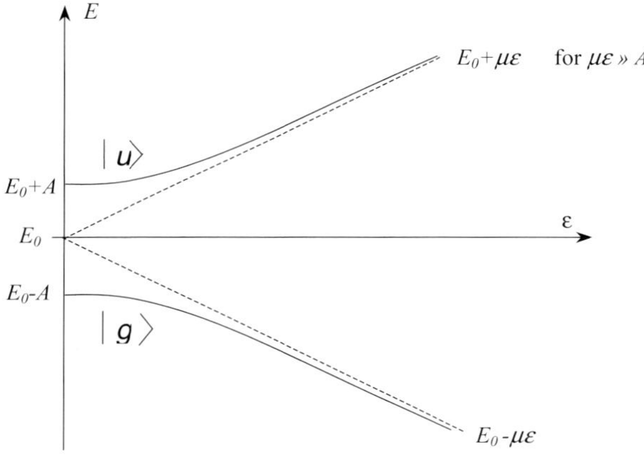

Eq. A.IX.5 can be read in terms of induced dipole moments $\mu_{ind}^\pm = -dE_\pm/d\mathcal{E} = \mp \mu_e^2 \mathcal{E}/A$. Therefore, if a collimated beam of molecules passes in a region with an electric field gradient across the beam itself, molecules in the $|u >$ and $|g >$ states will be deflected along **opposite directions** (this effect is analogous to the one observed in the Rabi experiment at §6.2). In particular, the molecules in the $|g >$ state will be deflected towards the region of stronger $\mathcal{E}^2$, owing to the force $-\boldsymbol{\nabla}[-(\mu_e \mathcal{E})^2/2A]$.

In practice, to obtain a beam with molecules in the upper energy state one uses quadrupole electric lenses, providing a radial gradient of $\mathcal{E}^2$. The square of the electric field varies across the beam. Passing through the lens the beam is enriched in molecules in the excited state and once they enter the microwave cavity the maser action becomes possible. The experimental setup of the Ammonia maser is sketched in the following Figure.

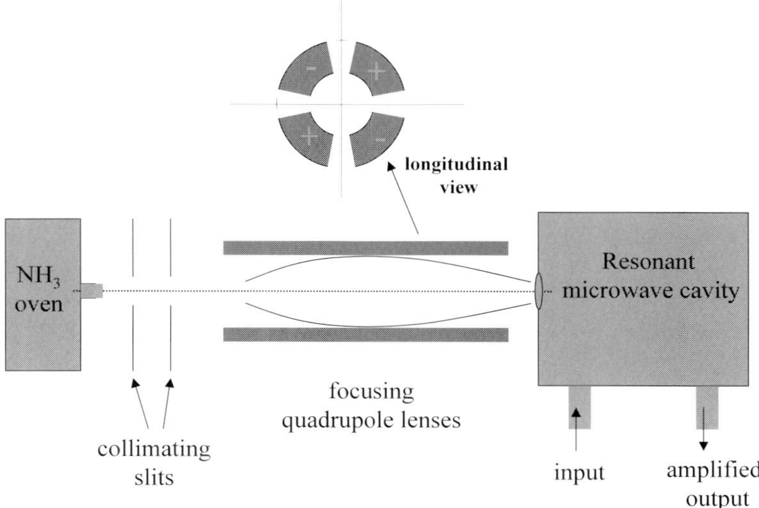

The basic principles outlined above for the Ammonia maser are also at work in other type of atomic or solid-state masers. In the Hydrogen or Cesium atomic maser the stimulated transition involves the hyperfine atomic levels (see Chapter 5). For the line at 1420 MHz, for instance, the selection of the atoms in the upper hyperfine state with $F = 1$ is obtained by a magnetic multipolar lens. Then the atomic beam enters a microwave cavity tuned at the resonance frequency. The resolution (ratio between the linewidth and 1420 MHz) can be improved up to 10^{-10}, since the atoms can be kept in the cavity up to a time of the order of a second. The experimental value of the frequency of the $F = 1 \rightarrow 0$ transition in Hydrogen is presently known to be (1420405751.781 ± 0.016 Hz), while for ^{133}Cesium the $F = 4 \rightarrow 3$ transition is estimated 9192631770 Hz, which is the frequency used to calibrate the unit of time.

Solid state masers are usually based on crystals with a certain number of paramagnetic transition ions, kept in a magnetic field and at low temperature, in order to increase the spin-lattice relaxation time T_1 and to reduce the linewidth associated with the life-time broadening (see Chapter 6) (as well as to reduce the spontaneous emission acting against the populations inversion). A typical solid state maser involves ruby, a single crystal of Al_2O_3 with diluted Cr^{3+} ions (electronic configuration $3d^3$). The crystal field removes the degeneracy of the $3d$ levels (details will be given at §13.3) and the magnetic field causes the splitting of the $M_J = \pm 3/2, \pm 1/2$ levels. The population inversion between these levels is obtained by microwave irradiation of proper polarization.

Here we have presented only a few aspects of the operational principles of masers, which nowadays have a wide range of applications, due to their resolution (which can be increased up to 10^{-12}) and sensitivity (it can be recalled that maser signals reflected on the surface of Venus have been detected).

Problems F.IX

Problem F.IX.1 Under certain circumstances the cyclobutadiene molecule can be formed in a configuration of four C atoms at the vertices of a square. In the MO.LCAO picture of delocalized $2p_z$ electrons derive the eigenvalues and the spin molteplicity of the ground state (within the same approximations used for C_6H_6).

Solution:

The secular equation is

$$\begin{vmatrix} \alpha - E & \beta & 0 & \beta \\ \beta & \alpha - E & \beta & 0 \\ 0 & \beta & \alpha - E & \beta \\ \beta & 0 & \beta & \alpha - E \end{vmatrix} = 0.$$

By setting $\alpha - E = x$, one has $x^4 - 4\beta^2 x^2 = 0$ and then $E_1 = -2|\beta| + \alpha,\qquad E_{2,3} = \alpha,\qquad E_4 = 2|\beta| + \alpha.$

Ground state: $4\alpha - 4|\beta|,$ triplet .

Problem F.IX.2 Refer to the C_3H_3 molecule, with carbon atoms at the vertices of an equilateral triangle. Repeat the treatment given for C_6H_6, deriving eigenfunctions and the energy of the ground-state. Then release the assumption of zero overlap integral among orbitals centered at different sites and repeat the derivation. Estimate, for the ground-state configuration, the average electronic charge per C atom.

Solution:

For $S_{ij} = 0$ for $i \neq j$, the secular equation is

$$\begin{vmatrix} E_0 - E & \beta & \beta \\ \beta & E_0 - E & \beta \\ \beta & \beta & E_0 - E \end{vmatrix} = 0$$

so that

$$E_I = E_0 + 2\beta \qquad\qquad E_{II,III} = E_0 - \beta$$

and the ground-state energy is

$$E_g = 3E_0 + 4\beta - \beta = 3E_0 + 3\beta$$

The eigenfunctions turn out

$$\phi_I = \frac{1}{\sqrt{3}}[\phi_1 + \phi_2 + \phi_3] \equiv \frac{A}{\sqrt{3}}$$

$$\phi_{II} = \frac{1}{\sqrt{2}}[\phi_1 - \phi_3] \equiv \frac{B}{\sqrt{2}}$$

$$\phi_{III} = \frac{1}{\sqrt{6}}[-\phi_1 + 2\phi_2 - \phi_3] \equiv \frac{C}{\sqrt{6}}$$

The total amount of electronic charge on a given atom (e.g. atom 1) is given by the sum of the squares of the coefficient pertaining to ϕ_1 in $\phi_{I,II,III}$:

$$q = 2(\frac{1}{\sqrt{3}})^2 + \frac{1}{2}[(\frac{1}{\sqrt{2}})^2 + (\frac{1}{\sqrt{6}})^2] = 1$$

(having taken the average of the two degenerate states).

For $S_{ij} \equiv S \neq 0$, the secular equation becomes

$$\begin{vmatrix} E_0 - E & \beta - SE & \beta - SE \\ \beta - SE & E_0 - E & \beta - SE \\ \beta - SE & \beta - SE & E_0 - E \end{vmatrix} = 0$$

and the eigenvalues are

$$E_I = \frac{E_0 + 2\beta}{1 + 2S} \qquad\qquad E_{II,III} = \frac{E_0 - \beta}{1 - S}$$

The ground-state energy is

$$E_g = 2E_I + E_{II} = 3\frac{E_0 + \beta(1 - 2S)}{(1 + 2S)(1 - S)}$$

with normalized eigenfunctions

$$\phi_I = \frac{A}{\sqrt{3(1 + 2S)}}$$

$$\phi_{II} = \frac{B}{\sqrt{2(1 - S)}}$$

$$\phi_{III} = \frac{C}{\sqrt{6(1 - S)}}$$

Again, by estimating the squares of the coefficients the charge at a given atom turns out

$$q' = \frac{1}{(1-S)(1+2S)} \; .$$

The charge in the region "in between" two atoms (e.g. atoms 1 and 2) is obtained by evaluating the sum of the coefficients $c_1 c_2$ (for ϕ_1 and ϕ_2) in $\phi_{I,II,III}$, multiplied by the overlap integral. Thus

$$q" = \frac{S(1-2S)}{(1-S)(1+2S)} \; .$$

Problem F.IX.3 Estimate the order of magnitude of the diamagnetic contributions to the susceptibility in benzene, for magnetic field perpendicular to the molecular plane.

Solution:
The diamagnetic susceptibility (per molecule) can approximately be written

$$\chi_\psi = -\frac{n_\psi e^2}{6mc^2} < r^2 >_\psi ,$$

where n_ψ is the number of electrons in a molecular state ψ and $< r^2 >_\psi$ is the mean square distance.

In benzene there are 12 $1s$ electrons of C, with $< r^2 >_{1s} \simeq a_0^2/Z^2$ $(Z = 6)$. Then there are 24 electrons in σ bonds for which, approximately,

$$< r^2 >_\sigma \simeq \int_{-\frac{L}{2}}^{\frac{L}{2}} \frac{dx}{L} x^2 = \frac{L^2}{12} ,$$

the length of the σ bond being $L = 1.4 \; \overset{\circ}{A}$.

Finally there are 6 electrons in the delocalized bond π_z, where one can assume $< r^2 >_{\pi_z} \simeq L^2$. The diamagnetic correction at the center of the molecule is dominated by the delocalized electrons and one can crudely estimate $\chi_\pi \approx -6 \cdot 10^{-29} \; cm^3$.

10

Nuclear motions in molecules and related properties

Topics

Rotations and vibrations in diatomic molecules
How the rotational and vibrational states are studied
The normal modes in polyatomic molecules
Basic principles of rotational, vibrational and Raman spectroscopies
Nuclear spin statistics and symmetry-related effects

10.1 Generalities and introductory aspects for diatomic molecules

In the framework of the Born-Oppenheimer separation ($\S\,7.1$), once that the electronic state has been described and the eigenvalue $E_e(\mathbf{R})$ and wavefunction $\phi_e(\mathbf{r}, \mathbf{R})$ have been found, then the motions of the nuclei are described by a function $\phi_\nu^{(g)}(\mathbf{R})$, where g represents the quantum numbers for the electrons and ν are the quantum numbers (to be found) for the nuclei. This wavefunction is solution of the Equation

$$\{-\sum_\alpha \frac{\hbar^2}{2M_\alpha}\nabla_\alpha^2 + E_e(\mathbf{R})\}\,\phi_\nu^{(g)}(\mathbf{R}) = E_{g,\nu}\,\phi_\nu^{(g)}(\mathbf{R}) \tag{10.1}$$

(note that V_{nn} in Eq. 7.3 and 7.5 has been included in $E_e(\mathbf{R})$, see for example Eq. 8.1).

Let us refer to a diatomic molecule in the ground electronic state, for which we assume $\Lambda = 0$ and $S = 0$ ($^1\Sigma$ state) and let us indicate the effective potential energy, resulting from the electronic eigenvalue and the nucleus-nucleus repulsion, with $V(R)$, R being the interatomic distance (previously often indicated by R_{AB}). It is reminded that $V(R)$ has the form sketched below

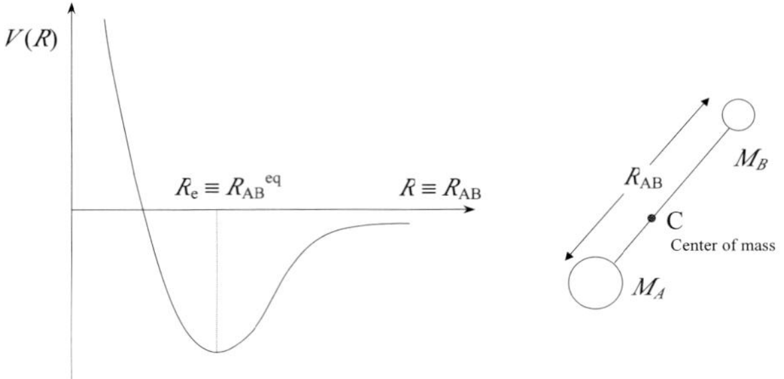

By introducing the reduced mass $\mu = M_A M_B/(M_A + M_B)$ the molecule becomes equivalent to a single particle. By recalling the treatment used for the Hydrogen atom, Eq. 10.1 is rewritten

$$\left\{-\frac{\hbar^2}{2\mu}\left[\frac{1}{R^2}\frac{\partial}{\partial R}\left(R^2\frac{\partial}{\partial R}\right)\right] + T_\theta + T_\varphi + V(R)\right\}\phi(R,\theta,\varphi) = E\,\phi(R,\theta,\varphi) \quad (10.2)$$

where the polar coordinates R, θ and φ have been introduced and $(T_\theta + T_\varphi)$ involves the angular momentum operator $\mathbf{L}^2$. The only difference with respect to the radial part of the Schrödinger equation for Hydrogen is in the potential energy $V(R)$, obviously different from the Coulomb form. Thus the factorization of the wavefunction follows:

$$\phi(R,\theta,\varphi) = \mathcal{R}(R)Y(\theta,\varphi) \;, \quad (10.3)$$

$Y_{KM}(\theta,\varphi)$ being the spherical harmonics characterized by quantum numbers K and M (the analogous of l and m in the H atom), related to the eigenvalues for $\mathbf{L}^2$ and L_z.

The radial part $\mathcal{R}(R)$ obeys the Equation

$$T_R\mathcal{R} + \left[V(R) + \frac{K(K+1)\hbar^2}{2\mu R^2}\right]\mathcal{R} = E\mathcal{R} \quad (10.4)$$

and corresponds to the one-dimensional probability of presence along a given direction under a potential energy including the centrifugal term, as sketched below

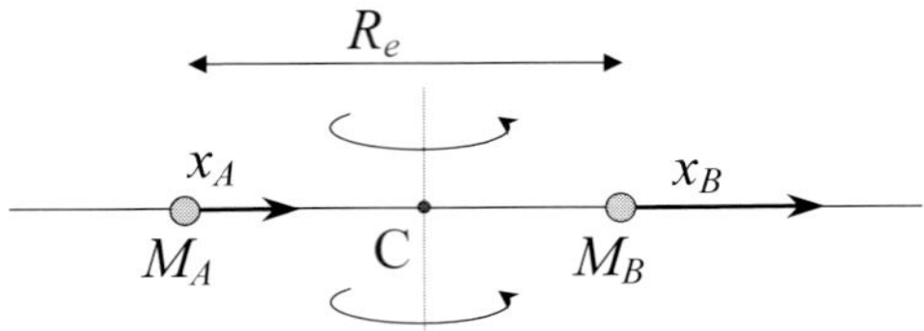

By indicating with Q the internuclear distance R with respect to the equilibrium distance R_e, in terms of the local displacements x_A and x_B one has

$$Q = R - R_e = x_B + R_e - x_A - R_e \equiv x_B - x_A. \tag{10.5}$$

Thus Q is a **non-local** coordinate (we shall return to this point when discussing the vibrational motions in polyatomic molecules, §10.6). Then the centrifugal term in Eq. 10.4 can be written

$$\frac{K(K+1)\hbar^2}{2\mu R_e^2} \left[\frac{1}{1+\frac{Q}{R_e}} \right]^2 \simeq \frac{K(K+1)\hbar^2}{2\mu R_e^2} \left(1 - \frac{2Q}{R_e} \right) \tag{10.6}$$

having taken into account that $Q/R_e \ll 1$.

In Eq. 10.6 the term $2Q/R_e$ couples the vibrational and the rotational motions. In a first approximation this term can be considered as perturbation and one can deal with the rotational part of the Schrodinger equation only. After the analysis of the vibrational part and the derivation of the correspondent wavefunction $\mathcal{R}(R) \equiv \phi_{vib}(R)$, it will be possible to take into account the roto-vibrational coupling by referring to unperturbed states described by

$$\phi(R, \theta, \varphi) = \phi_{vib}(R)\, \phi_{rot}(\theta, \varphi) \tag{10.7}$$

with $\phi_{rot}(\theta, \varphi) \equiv Y_{KM}(\theta, \varphi)$.

10.2 Rotational motions

10.2.1 Eigenfunctions and eigenvalues

From §10.1 it follows that the contribution to the energy of the molecule from the rotational motion is

$$E_{rot} = K\,(K+1)\,\hbar^2/2\,\mu\,R_e^2 \tag{10.8}$$

This result can be thought to derive directly from the quantization of the angular momentum $\mathbf{P}$ in the classical expression of the rotational energy $P^2/2I$, I being the moment of inertia $I = R_e^2 \mu$.

The eigenfunctions $\phi_{rot}(\theta, \varphi)$, so that $\phi_{rot}^* \phi_{rot}\, d\Omega$ yields the probability that the molecular axis is found inside the elemental solid angle $d\Omega = sin\theta\, d\theta\, d\varphi$, coincide with the spherical harmonics. In the light of the classical relation $|\mathbf{P}| = I\omega$, to a given quantum state with eigenvalue $|\mathbf{P}| = [K(K+1)]^{1/2}\,\hbar$, one can associate a frequency of rotation $\nu_{rot} = (h/4\pi^2 I)\,[K(K+1)]^{1/2}$.

A fundamental **rotational frequency**

$$\nu_{rot} = \hbar/2\,\pi\,\mu\,R_e^2 \tag{10.9}$$

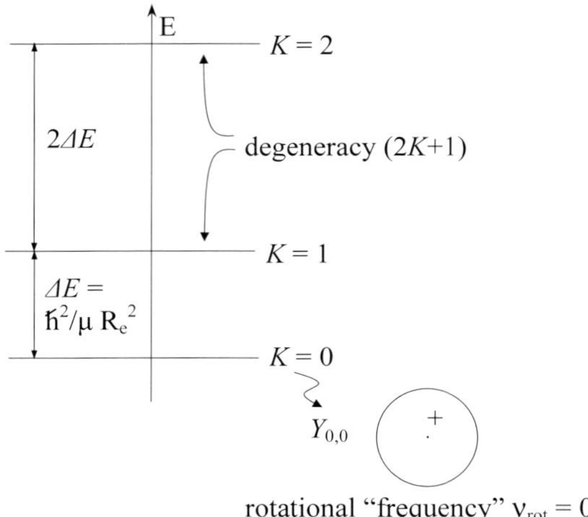

Fig. 10.1. Levels diagram for the lowest-energy rotational states.

or equivalently a **rotational constant** (in cm^{-1}) $B = (\hbar/4\pi I c)$, is usually defined $(B \equiv (\hbar^2/2\mu R_e^2)/hc)$.

The energy diagram for the first rotational states is reported in Fig. 10.1.

The probability distribution of the molecular axis involves $Y_{KM}^* Y_{KM}$. Since the φ-dependence of the spherical harmonics goes as $exp(\pm i M \varphi)$, the distribution of the molecular axes is characterized by rotational symmetry with respect to a given z direction. For $M = K$ and large value of the quantum numbers the distribution tends to the classical one, as expected from the **correspondence principle**.

10.2.2 Principles of rotational spectroscopy

The rotational states are experimentally studied by means of spectroscopic techniques involving the microwave range (typically $10^{-1} \div 10 \, \text{cm}^{-1}$) or the far infrared range $(10 \div 500 \, \text{cm}^{-1})$ of the electromagnetic spectrum (see Appendix I.1). Usually the sample is a gas at reduced pressure, since frequent collisions would prevent the definition of a precise quantum state (which is hard to be defined in the liquid state, for instance).

The generators of the radiation are often metals at high temperature or arc lamps while the detectors are semiconductor devices (for wavenumbers typically larger than 10 cm^{-1}). When low frequencies are required (say below 150 GHz), klystrons, magnetrons or Gunn diodes (usually fabricated with GaAs) are the microwave sources. Wave guides, resonant cells and again semiconductor detectors are commonly used.

Without going into details of technical character, we shall devote attention to the selection rules for electric dipole transitions between states K', M' and K'', M''.

The electric dipole matrix element reads

$$\mathbf{R}_{1 \to 2} = \int Y^{*}_{K'' M''}(\theta, \varphi)\, \boldsymbol{\mu}_e\, Y_{K' M'}(\theta, \varphi)\, sin\theta \, d\theta \, d\varphi \qquad (10.10)$$

where $\boldsymbol{\mu}_e$ is the dipole moment of the molecule. Therefore only **heteronuclear polar molecules**, where $|\boldsymbol{\mu}_e| \neq 0$ can be driven into transitions between different rotational states. Homonuclear molecules cannot interact with the radiation. From the matrix element in Eq. 10.10, in a way similar to the deduction of the selection rules for atomic transitions (see § 3.5 and App. I.3), the selection rules for electric dipole transitions between rotational states in polar molecules are

$$\Delta K = \pm 1 \qquad \Delta M = 0, \pm 1 \ , \qquad (10.11)$$

the latter being relevant when a static electric field is applied (see §10.2.4).

The energy difference between the states K and $(K+1)$ is

$$\Delta E_{K+1,\, K} = \frac{\hbar^2}{2I}\left[(K+2)(K+1) - K(K+1)\right] = \frac{\hbar^2}{I}\,(K+1) \ . \qquad (10.12)$$

Then in principle one expects rotational transitions at frequencies $\nu = n\,\nu_{rot}$ (Eq. 10.9), with n integer.

The intensities of the lines, to a good approximation, are controlled by the statistical populations of the rotational levels. The number of molecules on a given K state, at the thermal equilibrium, is

$$N_K = A\, g_K\, e^{-E_K/k_B\,T} \qquad (10.13)$$

with $g_K = (2K+1)$. The normalization constant can be expressed in terms of the population on the $K = 0$ level and thus one writes

$$N_K(T) = N_{K=0}(T)\,(2K+1)\, e^{-\frac{K(K+1)\hbar^2}{2Ik_B T}} \ , \qquad (10.14)$$

a function of the "variable" K of the form sketched below

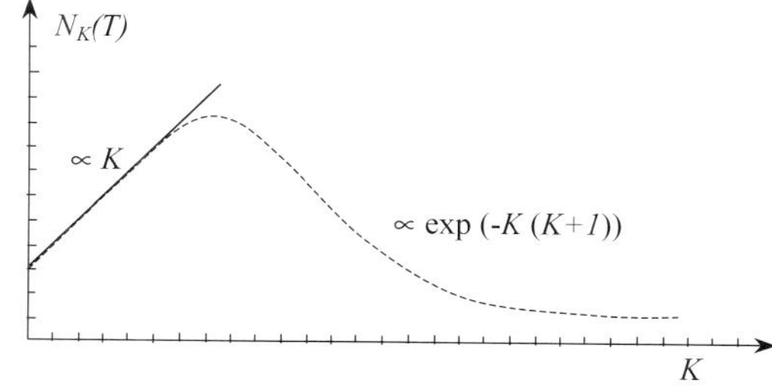

and implying typical absorption spectrum of the form shown in Figs. 10.2 and 10.3.

The fundamental rotational constants $B = \hbar/4\pi Ic$ for some diatomic molecules are reported below

Molecule	$B(\mathrm{cm}^{-1})$
H_2	60.8
N_2	2.01
O_2	1.45
HCl	10.6
NaCl	0.19

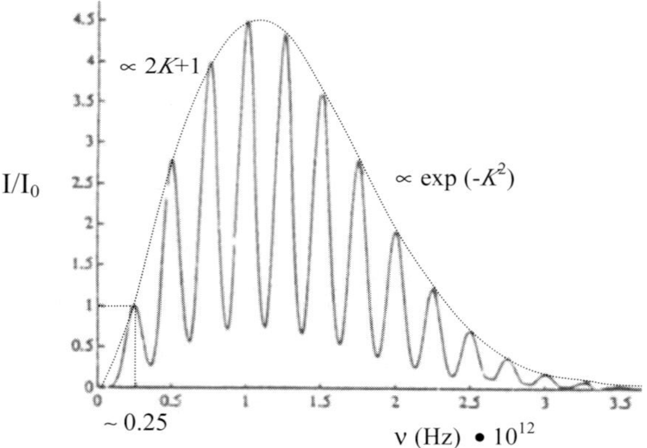

Fig. 10.2. Sketch of the expected absorption rotational spectrum for the DBr molecule on the basis of Eqs. 10.14 and 10.9. The intensities of the lines are normalized to the one of the $K = 0 \to K = 1$ line. The rotational frequency $\bar{\nu} = \nu_{rot}/c$ is around $8\,cm^{-1}$, corresponding to **rotational temperature** $h\nu_{rot}/k_B \approx 12\,\mathrm{K}$. The separation between adjacent lines is $2B$ and θ_{rot} is often defined as $\theta_{rot} = \hbar^2/2Ik_B = Bhc/k_B$ (see §10.2.1).

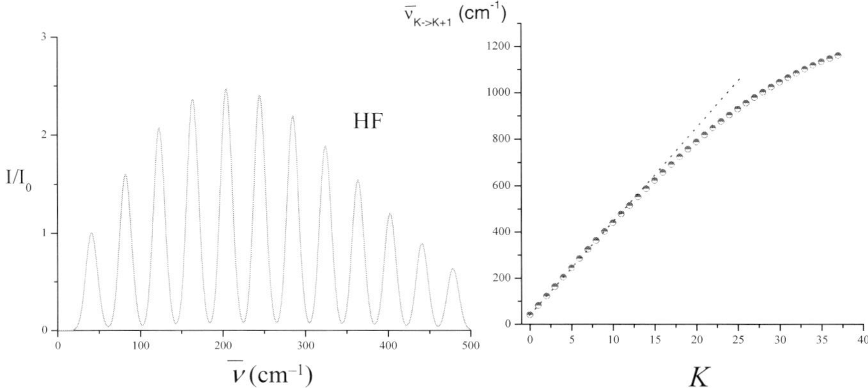

Fig. 10.3. (left) Absorption rotational spectrum of the HF molecule. The intensities of the lines are normalized to the $K = 0 \rightarrow K = 1$ line. In the right panel the wavenumbers associated with the $K \rightarrow K + 1$ transitions are reported as a function of K. A departure from the interval rule is observed at high K, owing to the increased strength of the coupling between rotational and vibrational motions (see §10.5).

10.2.3 Thermodynamical energy from rotational motions

Once the structure of the quantum levels is known, from the statistical distribution function it is possible to derive the thermodynamical energy U_{rot} and the specific heat. One has

$$U_{rot} = \sum_{K} N_K E_K \tag{10.15}$$

with N_K given by Eq. 10.14, where $N_{K=0}(T)$ can be written

$$N_{K=0} = N/Z_{rot} \qquad (N \text{ total number of molecules})$$

The rotational partition function Z_{rot} is

$$Z_{rot} = \sum_{K} (2K + 1)\, e^{-E_K/k_B T} \tag{10.16}$$

It is noted that the energy in terms of Z_{rot} (see Problem VI.1.4) is

$$U_{rot} = k_B T^2 \frac{d}{dT} \ln Z_{rot} \qquad .$$

In the high temperature limit $T \gg \theta_{rot} = \hbar^2/2Ik_B$ ($\theta_{rot} \sim 5 \div 10\,K$ for most molecules, with the exception of H_2 where $\theta_{rot} \simeq 87\,K$), the sum over K can be transformed to an integral:

$$Z_{rot} \approx \int_0^\infty 2K\, e^{-K^2\,\theta_{rot}/T}\, dK = \frac{T}{\theta_{rot}} \ . \qquad (10.17)$$

For one mole $(N = N_A)$, $U_{rot} \approx N_A\, k_B\, T$ and the specific heat turns out $C_V \approx R$, as expected from classical statistics.

The temperature behavior of the specific heat (see Problem F.VI.1) is schematically reported below

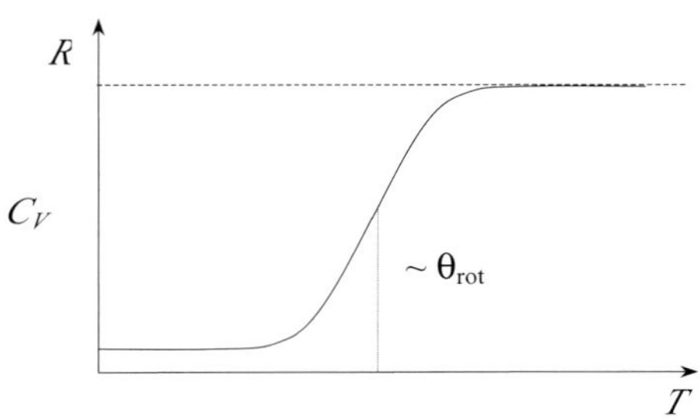

10.2.4 Orientational electric polarizability

Let us outline how to describe the effect of a static electric field on a system of dipolar diatomic molecules and how the polarizability is evaluated. The perturbation to the rotational states is given by the Hamiltonian $\mathcal{H}_p = -\boldsymbol{\mu}_e \cdot \boldsymbol{\mathcal{E}} \equiv -|\boldsymbol{\mu}_e|\,\mathcal{E}\,cos\theta$, with $\mathcal{E}$ electric field along the z direction. From parity argument one notes that the first order contribution to the energy is zero:

$$\langle K', M' \,|\, \mathcal{H}_p \,|\, K', M' \rangle = 0$$

Thus a correction term to the eigenvalues of the form $\Delta E \propto \mathcal{E}^2$ and $(\mu_{eff})_z \propto \mathcal{E}$ is expected, implying an **effective induced dipole moment** and therefore **positive** polarizability, somewhat similar to the paramagnetic susceptibility derived at §4.4.

For the ground-state at $K = M = 0$ the second-order perturbation correction

$$\Delta E_0^{(2)} = \sum_{K,M \neq 0} \frac{<0|\mathcal{H}_p|K,M><K,M|\mathcal{H}_p|0>}{E_0 - E_{K,M}}$$

reduces to

$$\Delta E_0^{(2)} = -\frac{I}{\hbar^2}\mu_e^2\mathcal{E}^2\Big|\int sin\theta d\theta d\phi \frac{1}{\sqrt{4\pi}} cos\theta \frac{\sqrt{3}}{\sqrt{4\pi}} cos\theta\Big|^2 = -\frac{1}{3}\mu_e^2\mathcal{E}^2\frac{I}{\hbar^2} \ ,$$

$$(10.18)$$

having taken into account that the only matrix element different from zero is the one connecting the state $K = 0$ to the state $K = 1, M = 0$, with eigenfunctions $1/\sqrt{4\pi}$ and $\sqrt{3/4\pi}cos\theta$, respectively. Then

$$\alpha(0,0) = -\frac{1}{\mathcal{E}}\frac{\partial \Delta E_0^{(2)}}{\partial \mathcal{E}} = \frac{2\mu_e^2 I}{3\hbar^2} \tag{10.19}$$

For the states at $K \neq 0$ we report the result of the estimate similar to the one given above:

$$E(K, M, \mathcal{E}) = E_K^0 + \frac{\mu_e^2 \, \mathcal{E}^2 \, I}{\hbar^2}\left[\frac{K(K+1) - 3\,M^2}{K\,(K+1)\,(2K-1)\,(2K+3)}\right], \tag{10.20}$$

for $K \neq 0$. From the sum over M in a given state $|K, M >$ (in first approximation the energy can be considered to depend only on K) and from Eq. 10.20, one deduces

$$\alpha(K) \simeq \sum_M \alpha(K, M) = 0 \quad \text{for} \quad K \neq 0$$

(note that $\sum_K M^2 = K(K+1)(2K+1)/3$).

Then only the ground state $(K = M = 0)$ contributes to the orientational **effective** electric moment along the field. The polarizability is temperature dependent, since the population of this state is affected by the temperature.

The thermal statistical average reads

$$\langle \alpha \rangle_T = \frac{\alpha(0,0)}{Z_{rot}} \tag{10.21}$$

When for Z_{rot} the sum over K can be transformed into an integral (see Eq. 10.17), by taking into account Eq. 10.18 with $\theta_{rot} = \hbar^2/2Ik_B$, the single-molecule polarizability becomes

$$\langle \alpha \rangle_T = \frac{\mu_e^2}{3\,k_B\,T} \;, \tag{10.22}$$

similar to the classical Langevin form

$$\langle \mu_e \rangle_z = \mu_e \, \langle cos\theta \rangle$$

$$\langle cos\theta \rangle = \frac{\int e^{\mu\,\mathcal{E}\,cos\theta/k_B\,T}\,cos\theta\,d\Omega}{\int e^{\mu\,\mathcal{E}\,cos\theta/k_B\,T}\,d\Omega} = ctnh\,x - \frac{1}{x} \equiv L(x)$$

with $L(x)$ Langevin function, that for $x = \mu\mathcal{E}/k_B T \ll 1$ becomes $L(x \ll 1) \simeq x/3$.

10.2.5 Extension to polyatomic molecules and effect of the electronic motion in diatomic molecules

In the following we sketch how the rotational eigenvalues can be obtained also in polyatomic molecules when particular symmetries allows one to extend the quantum rules for the angular momentum.

The classical rotational Hamiltonian reads

$$\mathcal{H}_{rot} = \frac{P_A^2}{2\,I_A} + \frac{P_B^2}{2\,I_B} + \frac{P_C^2}{2\,I_C} \tag{10.23}$$

where $I_{A,B,C}$ are the moments of inertia with respect to the principal axes of the tensor of inertia (conventionally $I_A < I_B < I_C$).

When the molecule is a **prolate** rotator, namely $I_A < I_B = I_C$, as for instance in CH_3F sketched below

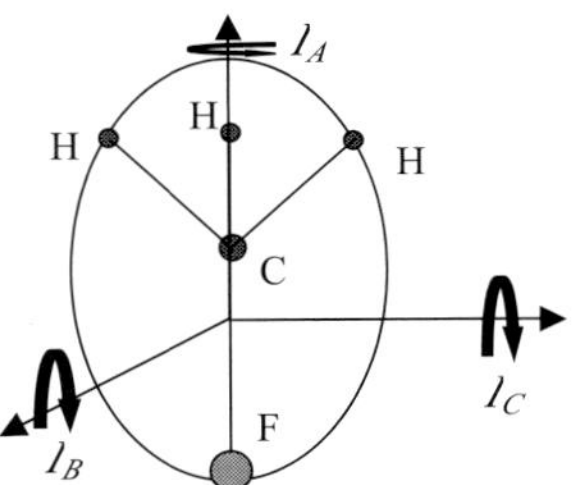

then the Hamiltonian can be rewritten

$$\mathcal{H}_{rot} = \frac{P_A^2}{2\,I_A} + \frac{P_A^2}{2\,I_B} - \frac{P_A^2}{2\,I_B} + \frac{P_B^2 + P_C^2}{2\,I_B} = \frac{P^2}{2\,I_B} + P_A^2 \left(\frac{1}{2\,I_A} - \frac{1}{2\,I_B}\right) \tag{10.24}$$

Therefore from the quantization rules

$$P^2 = K(K+1)\,\hbar^2 \qquad P_A = M\hbar$$

the eigenvalues of the rotational energy turn out

$$E(K,M) = \frac{K(K+1)\,\hbar^2}{2\,I_B} + M^2\,\hbar^2 \left(\frac{1}{2\,I_A} - \frac{1}{2\,I_B}\right) , \tag{10.25}$$

where now M refers to the component along the molecular axis A.

Equivalently, for an **oblate** rotator, where $I_A = I_B < I_C$ (as for instance in C_6H_6 (see §9.3)), one has a similar result.

In the general case, when $I_A \neq I_B \neq I_C$, no simple expressions can be derived for the eigenvalues and therefore reference to limit situations is usually made.

Up to now, in discussing the rotational motions, the electronic motions have been disregarded. In fact, for diatomic molecules it has been assumed the most common case of $^1\Sigma$ ground-state where the components of the orbital and spin moments along the molecular axis are zero.

The derivation of the rotational eigenvalues carried out for the prolate polyatomic rotator leading to Eq. 10.25 can be used to include the effect of the electron motion for diatomic molecules in electronic state $\Lambda \neq 0$. In a simplified picture, in fact, the electronic clouds can be regarded as a rigid charge distribution rotating around the molecular z-axis. Thus the diatomic molecule can be considered somewhat equivalent to a prolate rotator, with moments of inertia $I_A \equiv I_{elec}$ and $I_B = I_C = I_{nucl.}$, with $I_A \ll I_{B,C}$. Then, from extension of Eq. 10.25, at variance with Eq. 10.8 the rotational eigenvalues for diatomic molecules in an electronic state different from Σ turn out

$$E_{rot}(K,\Lambda) = \frac{\hbar^2}{2\mu R_e^2}[K(K+1) - \Lambda^2] + \frac{\hbar^2}{2I_{elec}}\Lambda^2 \,, \qquad (10.26)$$

Λ being the quantum number for L_z (or for J_z in the case of strong spin-orbit coupling). The last term in Eq. 10.26, much larger than the first one and independent from the rotational levels, is the one involved in the electron kinetic energy. When the molecule is in a state at $\Lambda \neq 0$ the roto-vibrational structure in the spectra involving electronic states display an **extra line** correspondent to a transition at $\Delta K = 0$, called **Q-branch**. This line is frequently observed in the electronic lines of band spectra (§10.8) or in Raman spectroscopy (§10.7), when transitions between different electronic states are involved (see for example Figs. 10.8 and 10.9).

Problems X.2

Problem X.2.1 As sketched in the following scheme the emission spectrum in the far infrared region from HBr molecules displays a series of lines regularly shifted by about $15\,\mathrm{cm}^{-1}$.

Derive the statistical populations of the rotational levels for $T = 12\,K$, $36\,K$ and $120\,K$. Estimate the interatomic equilibrium distance and obtain the relationship between temperature and rotational number K_{max} corresponding to the line of maximum intensity.

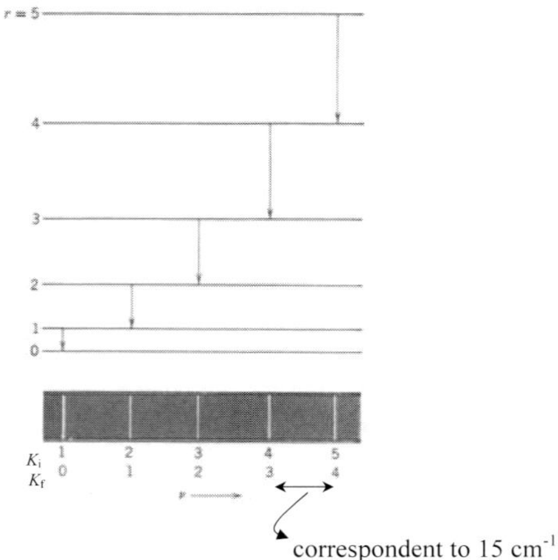

Solution:

The separation among adjacent lines is $2Bhc$, then $\hbar^2/2I = 1.06$ meV and $R_e = 1.41$ Å . The maximum intensity implies $(\partial N(K)/\partial K)_{K_{max}} = 0$. Then, from Eq. 10.14, $T = \hbar^2(2K_{max} + 1)^2/4k_B I$.

The statistical populations as a function of K are reported below:

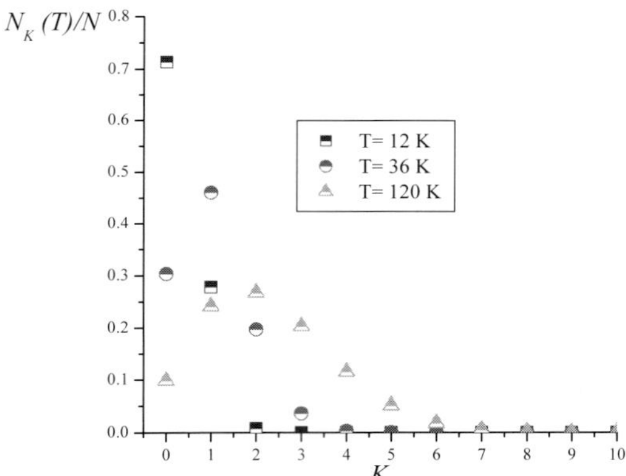

Problem X.2.2 For an ensemble of diatomic molecules at the thermal equilibrium write the contribution from rotational motions to the free energy, to the entropy, to the thermodynamical energy and to the specific heat, in the limit $T \gg \theta_{rot}$.

Solution:

The partition function is

$$Z = \sum_K (2K+1)e^{-\frac{K(K+1)\hbar^2}{2Ik_BT}} \ .$$

By substituting the sum with an integral

$$Z \simeq \int_0^\infty dK(2K+1)e^{-K(K+1)\Theta_{rot}/T} = \frac{T}{\Theta_{rot}}$$

one derives $F = -k_BT \ln Z$, $U = k_BT^2 \frac{\partial}{\partial T} \ln Z$, $S = \frac{U-F}{T} = -\frac{\partial F}{\partial T}$, $C_V = \frac{\partial U}{\partial T}$.

Problem X.2.3 By referring to the three rotational levels depicted in Fig. 10.1, plot the splittings induced by a static electric field (**Stark effect**). Indicate the transition that can be observed in rotational spectroscopy and estimate the order of magnitude of the resolution required to evidence the splittings, for external field $\mathcal{E} = 10^3 \, Volt/cm$.

Solution:
From Eqs. 10.18 and 10.20

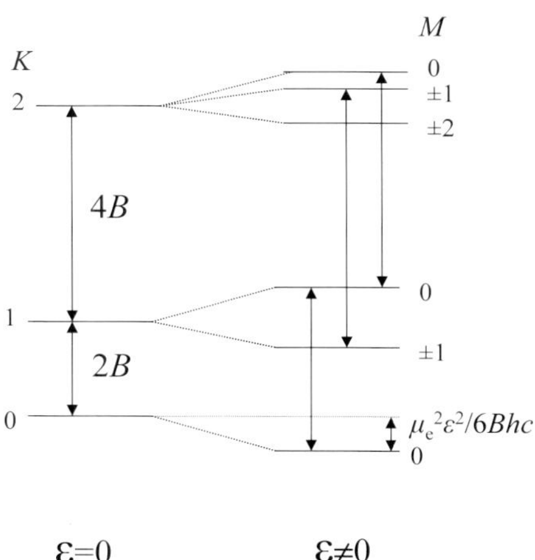

and for $\mu_e \sim 10^{-18}$ u.e.s.cm and $B \sim 10$ cm^{-1}, $\Delta\nu/\nu \sim 10^{-7}$.

Problem X.2.4 In the rotational spectrum of $H^{35}Cl$ two lines are detected with the same strength at $106\,cm^{-1}$ and at $233.2\,cm^{-1}$. Derive the temperature of the gas.

Solution:

From

$$\bar{\nu} = 2B(K+1),$$

with $B = 10.6\,cm^{-1}$, one has for $\bar{\nu}_1 = 106.0\,cm^{-1}$ $K_1 = 4$ and for $\bar{\nu}_2 = 233.2\,cm^{-1}$ $K_2 = 10$. The intensity is proportional to the population of the rotational levels. Then

$$(2K_1 + 1)e^{-\frac{hcBK_1(K_1+1)}{k_BT}} = (2K_2 + 1)e^{-\frac{hcBK_2(K_2+1)}{k_BT}}$$

and

$$T = \frac{90\,hcB}{k_B\,\ln\frac{7}{3}} \simeq 1620\,K \ .$$

10.3 Vibrational motions

10.3.1 Eigenfunctions and eigenvalues

Going back to the radial part of the Schrödinger equation (Eq. 10.4), again disregarding the term $-2Q/R_e$ and without including the rotational terms $K(K+1)\,\hbar^2/2I$ which does not depend on $(R - R_e)$, the function $U(R) = R\mathcal{R}(R)$ is introduced, so that the equation becomes

$$\frac{d^2U}{dR^2} + \frac{2\mu}{\hbar^2}\left[E - V(R)\right] = 0 \tag{10.27}$$

While the full expression of the potential energy $V(R)$ is unknown, the vibration involves small displacements around the equilibrium position R_e and then one can write

$$V(R) = V(R_e) + \left(\frac{dV}{dR}\right)_{R_e}(R - R_e) + \frac{1}{2}\left(\frac{d^2V}{dR^2}\right)_{R_e}(R - R_e)^2 + \dots \tag{10.28}$$

Since $\left(\frac{dV}{dR}\right)_{R_e}$ is zero, by omitting the constant $V(R_e)$ (namely the energy E has to be added to the electronic energy at the equilibrium position, see Fig. 7.2), in terms of the non-local coordinate $Q = x_A - x_B$ (Eq. 10.5), one has

$$V(R) = \frac{1}{2}kQ^2 + \dots, \tag{10.29}$$

where k is the curvature of $V(R)$ around the equilibrium position. In the **harmonic approximation** higher order terms in the expansion 10.28 are neglected. The equation for the vibration of the nuclei around the equilibrium position is thus written

$$-\frac{\hbar^2}{2\mu}\frac{d^2U}{dQ^2} + \frac{1}{2}kQ^2U = EU \ , \tag{10.30}$$

the well known form for the harmonic oscillator (with $-R_e \leq Q < \infty$). Then the eigenfunctions are related to the **Hermite polynomials** and the eigenvalues are

$$E_v = (v + 1/2)h\nu_o \tag{10.31}$$

with quantum number $v = 0, 1, 2, 3...$, while $\nu_o = (1/2\pi)\sqrt{k/\mu}$ corresponds to the frequency of the classical oscillator with same mass and elastic constant.

The eigenvalues and eigenfunctions for low energy states are depicted in Fig. 10.4. The ground state ($v = 0$) is described by the wavefunction

$$U(Q) \propto e^{\frac{-Q^2 b}{2}}, \quad \text{with} \ \ b = \frac{2\pi\nu_o\mu}{\hbar} \tag{10.32}$$

implying a behavior significantly different from the one expected for classical oscillator for which the maxima of probability of presence are at the boundaries of the motion. Other relevant differences with respect to classical oscillator are the extension of the "motion" outside the extreme elongations and the occurrence of zero-point energy $E_{v=0} = (1/2)h\nu_o$.

The mean square displacement from the equilibrium position reads

$$< Q^2 >_v = \int U_v^* \, Q^2 \, U_v \, dQ \tag{10.33}$$

and from the expressions of the Hermite polynomials one finds

$$< Q^2 >_v = \frac{\hbar}{\sqrt{\mu k}}(v + 1/2) = E_v/k \tag{10.34}$$

implying $E_v = k < Q^2 >_v$, the same relation holding for the classical oscillator.

To give a few representative examples, in the HCl molecule the vibrational frequency is $\nu_0 = 8.658 \times 10^{13}$ Hz, corresponding to a force constant $k = 4.76 \times 10^5$ dyne/cm, while in CO $\nu_0 = 6.51 \times 10^{13}$ Hz, corresponding to a force constant $k = 18.65 \times 10^5$ dyne/cm.

Some vibrational constants $\bar{\nu}_0$ for homonuclear diatomic molecules are reported below:

Molecule	(cm^{-1})
H_2	4159 (see caption in Fig. 10.4)
N_2	2330
O_2	1556 (see caption in Fig. 10.4)
Li_2	246
Na_2	158
Cs_2	42

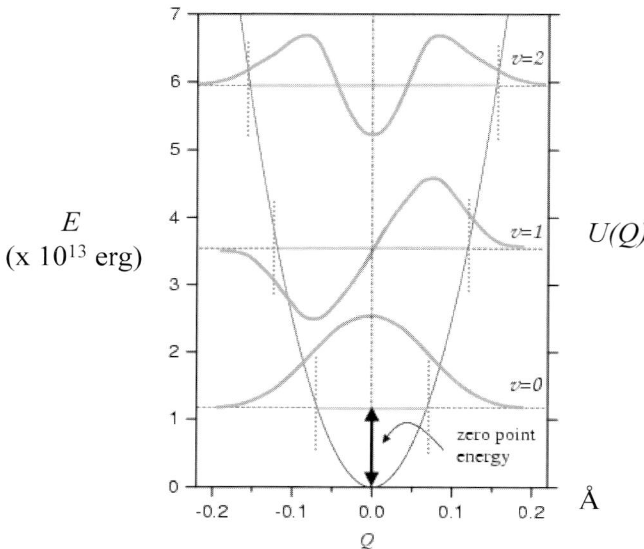

Fig. 10.4. First vibrational states in a diatomic molecule, having assumed $k = 5 \times 10^5$ dyne/cm and effective mass $\mu = 10^{-23}$ g, corresponding to vibrational frequency $\nu_0 = 3.6 \cdot 10^{13}$ Hz. The dotted lines correspond to the maxima elongations according to the classical oscillator in the parabolic potential energy indicated by the solid line.

Typical values for the force constants are i) in H_2 $k \simeq 5 \times 10^5$ dyne/cm; ii) in O_2 (where a double bond is present) $k \simeq 11 \times 10^5$ dyne/cm; iii) in N_2 (triple bond) $k \simeq 23 \times 10^5$ dyne/cm; iv) in NaCl (ionic bond) $k \simeq 1.2 \times 10^5$ dyne/cm.

10.3.2 Principles of vibrational spectroscopy and anharmonicity effects

In regards of the main aspects, the spectroscopic studies of the vibrational states in molecules are similar to the ones in optical atomic spectroscopy. The spectral range typically is in the range $100 - 4000$ cm^{-1} (see Appendix I.1) and the devices are no longer based on glasses but rather use alkali halides, to reduce the absorption of the **infrared radiation**. The diffraction gratings grant a better resolution and the detectors are usually semiconductor devices. Details of technical character can be found in the exhaustive book by **Svanberg**, quoted in the Preface.

As for rotational spectroscopy, being more interested into the fundamental aspects, we turn our attention to the transition probability due to the electric dipole mechanism. For two vibrational states at quantum numbers v' and v'', the component along the molecular axis of the electric dipole matrix element reads

$$(\mathbf{R}_{v' \to v''})_z = \int U_{v''}^* \mu_e U_{v'} dQ \tag{10.35}$$

where $\mu_e(Q)$ is a complicate function of the interatomic distance R.

The sketch of a plausible dependence of μ_e with R is given below

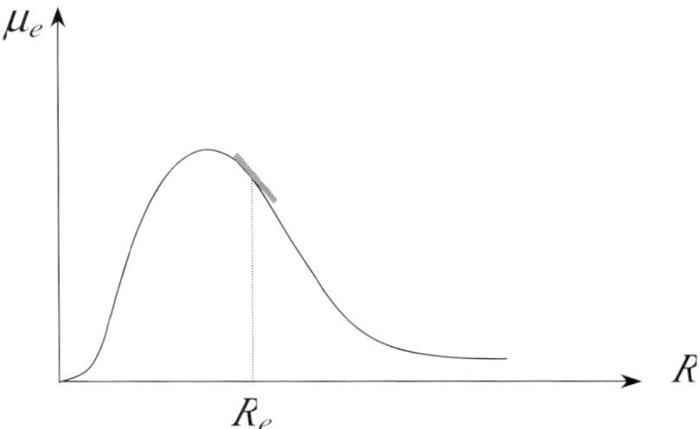

One can expand μ_e around R_e in terms of Q:

$$\mu_e = \mu_e(0) + \left(\frac{d\mu_e}{dQ}\right)_0 Q + \ldots \,, \tag{10.36}$$

where the first term is involved in the rotational selection rules, while the expansion has been limited to the linear term in Q (often called **linear electric approximation**).

From Eq. 10.35 and 10.36 one concludes that only **heteronuclear molecules**, with $\mu_e \neq 0$, can be driven to transitions among vibrational states. Furthermore, from the term linear in Q one deduces that only states of different parity imply a matrix element different from zero. Indeed, as it can be seen from inspection to the Hermite polynomials, only transitions between adjacent states are allowed : $\Delta v = \pm 1$. One can also remark that the frequency emitted or absorbed is the one expected for a classical Lorentz-like oscillator.

Thus in the harmonic approximation (Eq. 10.29) and in the linear dipole approximation (Eq. 10.36) (sometimes called **electrical harmonicity**), one expects a **single absorption line**, at the frequency ν_o. The line yields the curvature of the energy of the molecule at the equilibrium interatomic distance. The intensity of each component, to a large extent, is controlled by the statistical population on the vibrational levels:

$$N_v(T) = Ae^{\frac{-(v+1/2)h\nu_o}{k_B T}} \equiv N_0(T)e^{\frac{-vh\nu_o}{k_B T}} \tag{10.37}$$

with

$$N_0(T) = (N/Z_{vib})e^{-h\nu_o/2k_B T}\,, \tag{10.38}$$

and N total number of molecules.

For $k_B T \ll h\nu_o$, as it is often the case, the ground state is by large the most populated and therefore the absorption line is related practically to the transition $v = 0 \to v = 1$.

Now a brief discussion of the **anharmonicity** effects is in order. The electrical anharmonicity originates from the term in Q^2, neglected in the expansion 10.36. According to the correspondent matrix element in Eq. 10.35, because of parity characters of the operator and of the Hermite polynomials, that term implies transitions between states at the same parity. Therefore the selection rule $\Delta v = \pm 2$ results, for states pertaining to the mechanical harmonic approximation (Eq. 10.29).

The qualitative effect of the terms proportional to Q^3 and Q^4 in the expansion 10.28 (**mechanical anharmonicity**) is to cause a progressive reduction in the separation between the states at high quantum number v, as sketched below

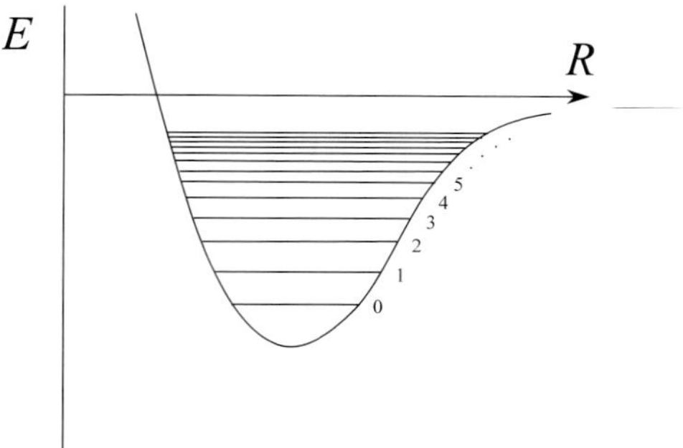

Then from the matrix element of the form correspondent to Eq. 10.35, transitions at frequencies different from ν_o have to be expected.

The anharmonic terms can be analyzed as perturbation of the vibrational states described by the wavefunctions $U_v(Q)$. The term in Q^3 must be considered up to the second order, its expectation value being zero for unperturbed states. Thus the eigenvalues turn out of the form

$$E_v = (v + 1/2)h\nu_o - a(v + 1/2)^2 h\nu_o \tag{10.39}$$

where the constant a, much smaller than the unit, is related to the ratio $[(d^3V/dR^3)_{R_e}]^2/k^{5/2}$. To give an idea, for the hydrogen molecule H_2, $a = 0.027567$. In $H^{35}Cl$, $\nu_0/c = 2990.95$ cm^{-1} and $a = 0.0176$.

For an heuristic potential $V(R)$ in the Morse-like form (see §10.4) one derives $a = h\nu_0/4D_e$, with $D_e \equiv -V(R_e)$, as we shall discuss in a later Section (see Eq. 10.41 and 10.47).

Problems X.3

Problem X.3.1 Consider the H_2 molecule in the vibrational ground state and in the first excited rotational state $(K = 1)$. Evaluate the number of oscillations occurring during one rotation.

Solution:

From

$$\nu_{rot} = \frac{[K(K+1)]^{1/2}\,\hbar}{2\pi I} = \frac{\sqrt{2}\hbar}{2\pi\mu R_e^2}$$

and $\nu_{vib} = (1/2\pi)\sqrt{k/\mu}$ the number of oscillations is

$$n_{osc} = \frac{1}{2\pi}\sqrt{\frac{k}{\mu}} \cdot \frac{2\pi\mu R_e^2}{\sqrt{2}\hbar} = \frac{\sqrt{k\,\mu}R_e^2}{\sqrt{2}\hbar} \simeq 25\,.$$

Problem X.3.2 The dissociation energy in the D_2 molecule is increased by 0.08 eV with respect to the one in H_2 (4.46 eV). Estimate the zero point energy for both molecules.

Solution:

From

$$E = -A + \hbar\omega\left(v + \frac{1}{2}\right)$$

the dissociation energy is given by $E_d = +A - \frac{1}{2}\hbar\omega$, for $v = 0$. Since $A(H) = A(D)$,

$$E_d(D) - E_d(H) = -\frac{1}{2}\hbar[\omega(D) - \omega(H)].$$

Hence

$$\frac{\hbar\omega(H)[1 - \omega(D)/\omega(H)]}{2} = \frac{\hbar\omega(H)[1 - 1/\sqrt{2}]}{2} = 0.08 \text{ eV},$$

and the zero-point energy of H_2 turns out $\hbar\omega(H)/2 = 0.27$ eV while for D_2 $0.27/\sqrt{2}\,\text{eV} = 0.19$ eV.

Problem X.3.3 The infrared spectrum of a gas of diatomic molecules is characterized by lines equally spaced by the amount of about 10^{11} Hz. A static electric field of $3kV/cm$ is applied. The lowest frequency line, with intensity 2.7 times smaller than the adjacent one, splits in a doublet, with 1 MHz of separation between the lines. Derive the molar polarizability.

Solution:
From the ratio of the intensities $I(1)/I(0) = 2.7 = 3exp(-h\nu_{\mathrm{rot}}/k_B T)$, with $\nu_{\mathrm{rot}} = 10^{11}$ Hz, the temperature is deduced: $T \simeq 45.6$ K.

Since $k_B T \gg h\nu_{\mathrm{rot}}$ the molar polarizability reads (see Eq. 10.22)
$\alpha = N_A \mu_e^2 / 3k_B T$.

The electric field partially removes the degeneracy of the $K = 1$ level. The separation between levels at $M_{K=0}$ e $M_{K=\pm 1}$ turns out (see Eq. 10.20)

$$\Delta \nu = 10^6 = \frac{\mu_e^2 \mathcal{E}^2}{h^2 \nu_{\mathrm{rot}}} \frac{3}{10}$$

Then $\mu_e^2 = 14.6 \times 10^{-38}$ u.e.s.^{2}cm^2 and $\alpha(T = 45.6K) = 4.7$ emu/mole.

Problem X.3.4 Evaluate the order of magnitude of the electronic, rotational and vibrational polarizabilities for the HCl molecule at $T = 1000$ K (the elastic constant is $k = 4.76 \times 10^5$ dyne/cm and the internuclear distance $R_e = 1.27$ Å). From the Clausius-Mossotti relation estimate the dielectric constant of the gas, at ambient pressure.

Solution:
For order of magnitude estimates one writes (see Problem F.VIII.5 and Eq. 10.22)

$$\alpha_{el} \simeq 8a_0^3 \sim 10^{-24} \text{ cm}^3,$$

$$\alpha_{rot} = \frac{e^2 R_e^2}{3k_B T} \simeq 10^{-22} \text{ cm}^3.$$

For the vibrational contribution (see Prob. IV.2.3 and Prob. X.5.6)

$$\alpha_v \simeq \frac{e^2}{k} \simeq 5 * 10^{-25} \text{ cm}^3.$$

From the equation of state $PV = RT$, by taking into account that at ambient pressure and temperature the molar volume is $V = 2.24 * 10^4$ cm^3, at $T = 1000°$ K one finds $V \sim 8.2 * 10^4$ cm^3, corresponding to the density $N \sim 10^{19}$ molecules cm^{-3}.

From $\Delta \epsilon = (4\pi N \alpha_{rot})/[1 - 4\pi N \alpha_{rot}/3] \simeq 4\pi N \alpha_{rot}$, one derives $\epsilon \approx 1.01$.

10.4 Morse potential

An heuristic energy curve for diatomic molecules that can be assumed as potential energy for the vibrational motion of the nuclei (Eq. 10.27) is the one suggested by **Morse**:

$$V_M = D_e \left[1 - exp[-\beta(R - R_e)]\right]^2, \qquad (10.40)$$

of the form sketched below.

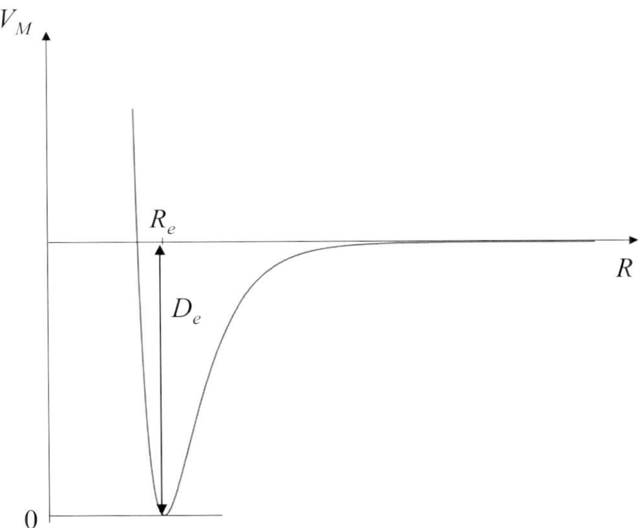

This expression retains a satisfactory validity for R around the interatomic equilibrium distance R_e. D_e corresponds to the energy of the molecule for $R = R_e$ (the real dissociation energy being D_e minus the zero-point vibrational energy $(1/2)h\nu_0$), while β is a characteristic constant.

It is noted that for R close to R_e Eq. 10.40 yields $V_M \simeq D_e Q^2 \beta^2$, namely the harmonic potential with elastic constant $k = 2D_e\beta^2$, with $\nu_0 = (\beta/\sqrt{2\pi})\sqrt{D_e/\mu}$.

The Morse potential, often useful for approximate expression of the electronic eigenvalue $E(R_{AB})$ in diatomic molecules, has the advantage that the Schrödinger equation for the vibrational motion (Eq. 10.27) can be solved analytically, although with cumbersome calculations. The eigenvalues turn out

$$E_M = h\nu_0[(v + 1/2) - a(v + 1/2)^2] \qquad (10.41)$$

with $a = (h\nu_0/4D_e)$ (see Eq. 10.39). The eigenfunctions are no longer even or odd functions for v even or odd, respectively, at variance with the ones derived in the harmonic approximation. Therefore one has transitions at $\Delta v \neq \pm 1$ without having to invoke electrical anharmonicity.

Problems X.4

Problem X.4.1 From the approximate expression for the energy of a diatomic molecule

$$V(R) = A\left(1 - exp[-B(R - C)]\right)^2$$

with $A = 6.4$ eV, $B = 0.7 \times 10^8$ cm^{-1} and $C = 10^{-8}$ cm, derive the main properties of the rotational and vibrational motions.
Sketch the qualitative temperature dependence of the specific heat. Assume for the reduced mass the one pertaining to HF molecule.
   Solution:
The elastic constant is $k = 2AB^2 = 10^5$ dyne/cm . For the reduced mass $\mu = 0.95M$ (with M the proton mass), the fundamental vibrational frequency is $\nu_0 = 2 \times 10^{13}$ Hz, corresponding to the vibrational temperature $\Theta_v = h\nu_0/k_B \simeq 960$ K.
   For an equilibrium distance $R_e = C$, the moment of inertia is $I = 1.577 \times 10^{-40}$ g cm^2. The separation between adjacent lines in the rotational spectrum is $\Delta\bar{\nu} = 35.6$ cm^{-1} and therefore $\Theta_r = Bhc/k_B = 25.6$ K.The qualitative temperature dependence of the molar specific heat is sketched below.

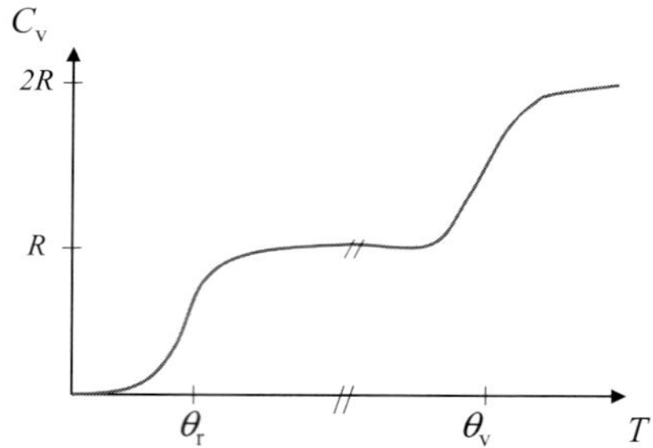

Problem X.4.2 In the RbH molecule ($R_e = 2.36$ Å) the fundamental vibrational frequency is $\nu_0 = 936.8$ cm^{-1} and the dissociation energy $D_e = 15505$ cm^{-1}. Derive the Morse potential and the correction due to the rotational term for $K = 40$ and $K = 100$. Discuss the influence of the rotation on the dissociation energy.

Solution:

The parameter $\beta = 2\pi\nu_0\sqrt{\mu/2D_e}$ and from $\mu = 1.65 \cdot 10^{-24}$ g one has $\beta = 9.14 \cdot 10^7$ cm^{-1}.

The rotational constant B at the equilibrium distance R_e is

$$B = \frac{h}{8\pi^2 c \mu R_e^2} \simeq 3 \text{ cm}^{-1}.$$

To account for the rotational contribution the energy has to be written in terms of the R-dependent rotational constant in Eq. 10.4. Therefore

$$E_{\rm rot}(R) = hcBK(K+1) \cdot \frac{R_e^2}{R^2} \quad,$$

so that the effective potential becomes

$$V_{\rm eff}(R) = V_M(R) + E_{\rm rot}(R),$$

plotted below.

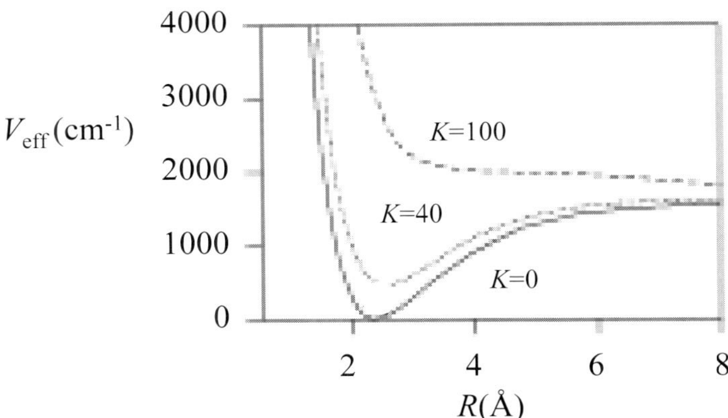

On increasing the rotational number the molecular bond is weakened.

10.5 Roto-vibrational eigenvalues and coupling effects

In high resolution an absorption line involving transitions between vibrational
states evidences the fine structure related to simultaneous transitions between
rotational states (See Fig. 10.5).

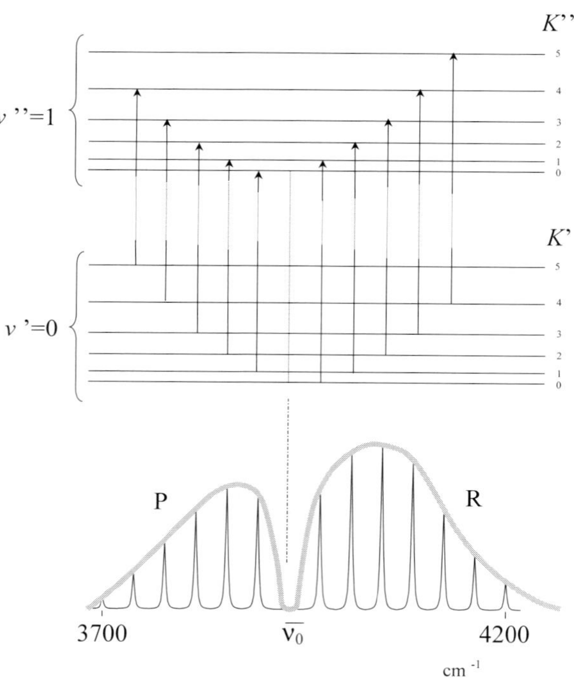

Fig. 10.5. Rotational structure in the vibrational spectral line of the HF molecule
and illustration of the transitions generating the P and R branches.

Still assuming weak roto-vibrational coupling, the wavefunction is $\phi_{rot}^{K,M}\,\mathcal{R}_v(R)$
and the eigenvalues are

$$E_{K,v} = (v + \frac{1}{2})h\nu_o + \frac{\hbar^2 K(K+1)}{2\mu R_e^2} \qquad (10.42)$$

The electric dipole matrix element connecting two states K', v' and K'', v''
is

$$\mathbf{R}_{K',v'\to K'',v''} \propto \int \phi_{rot}^{K'',M''*}\mathcal{R}_{v''}(R)^* \cdot (\boldsymbol{\mu}_e + c\mathbf{j}Q)\phi_{rot}^{K',M'}\mathcal{R}_{v'}(R)sin\theta d\theta d\phi dQ$$

$$(10.43)$$

where $\mathbf{j}$ is a unitary vector along the molecular axis. The term involving $\boldsymbol{\mu}_e$ drives the purely rotational transitions while

$$c\int \phi_{rot}^{K'',M''}\mathbf{j}\phi_{rot}^{K',M'}sin\theta d\theta d\phi \int \mathcal{R}_{v''}^*(R)Q\mathcal{R}_{v'}(R)dQ \qquad (10.44)$$

implies transitions with the selection rules $\Delta K = \pm1$ and $\Delta v = \pm1$ (see Eq. 10.11 and §10.3.2). When in the $v' \to v''$ transition the quantum number K increases then the correspondent line is found at a frequency $\nu > \nu_o$ (**branch R** in Fig. 10.5) while when K decreases one has $\nu < \nu_0$ (**branch P**).

It is noted that the line at $\nu = \nu_0$ is no longer present. When electronic states at $\Lambda \neq 0$ are involved in a transition a component at ν_0 can be observed (called **Q** branch), usually in form of a broad line (see §10.2.5, Eq. 10.26 and examples at Figs. 10.8 and 10.9).

From Eqs 10.42 and 10.44 the wavenumbers associated with the roto-vibrational transitions are

$$\bar{\nu}_R = \bar{\nu}_o + B_{v''}(K + 1)(K + 2) - B_{v'}K(K + 1) \qquad (10.45)$$
$$\bar{\nu}_P = \bar{\nu}_o + B_{v''}K(K - 1) - B_{v'}K(K + 1) \qquad (10.46)$$

Since $B_{v'} \simeq B_{v''} \simeq B_v$ the separation between the adjacent lines turns out about $2B_v$, as shown in Fig. 10.5.

Now a brief comment on the role of the terms coupling the rotational and vibrational motions is in order. In the framework of the perturbative approach, with unperturbed eigenfunctions $Y_{K,M}(\theta, \phi)\mathcal{R}(Q)$, according to Eq. 10.6 the perturbing Hamiltonian is

$$\mathcal{H}_P = \frac{\hbar^2 K(K + 1)}{2\mu_e R_e}\left(-2\frac{Q}{R_e}\right).$$

No correction terms to the unperturbed eigenvalues are expected at the first order. Thus the evaluation of the roto-vibrational coupling has to be carried out at the second order in $\mathcal{H}_P$.

The final result for the second order correction is

$$\Delta E^{(2)} = -a_1 h\nu_0(v + 1/2)^2 + a_2(v + 1/2)K(K + 1) - a_3 K^2(K + 1)^2 \quad (10.47)$$

The term in a_1 is the one due to mechanical anharmonicity, already discussed (Eq. 10.39). The term in a_2 is related to the effect on the elastic constant produced by the centrifugal potential and by the contribution in Q and Q^3. Finally the term in a_3 in Eq. 10.47 reflects the increase of the moment of inertia due to the rotation of the molecule (**centrifugal distortion**).

The detailed expressions for the coefficients a_i in Eq. 10.47 are $a_1 = (\hbar/384\,\pi\,\mu\,k\,\nu_0^3)\,(5\,\alpha^2 - 3\,k\,\beta)$, $a_2 = (\hbar^3/k^2\,R_e^2)\,(R_e\,\alpha + 3\,k)$ and $a_3 = \hbar^4/2\,k\,\mu^2\,R_e^6$

(as it can be obtained also classically by writing $kQ = \mu \omega^2 R$ and then evaluating the rotational energy). Here α and β are the coefficients of the terms in Q^3 and in Q^4 in the perturbative Hamiltonian resulting in the expansion in Eq. 10.6.

Problems X.5

Problem X.5.1 The rotovibrational transmission spectrum for the HCl molecule is shown below.

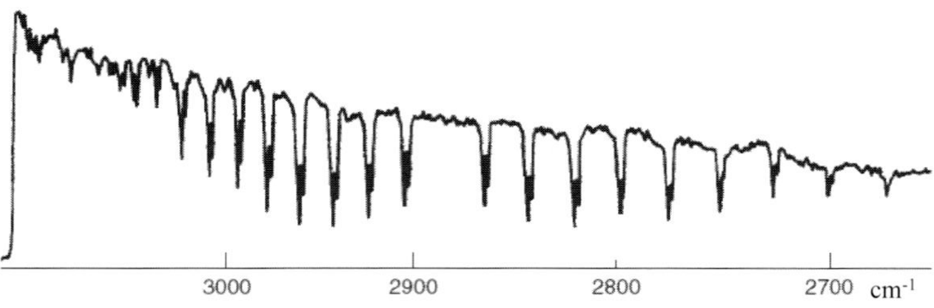

Derive the equilibrium distance and the elastic constant for the molecule. Which is the origin of the doublets observed at each peak? How does the spacing between adjacent lines change for the deuterated molecule?

Solution:
From the spacing among adjacent lines $\bar{\Delta\nu} \simeq 21\mathrm{cm}^{-1} = 2Bhc$, the moment of inertia being $I = 2.67 \times 10^{-40}$ g cm^2, the equilibrium distance turns out $R_e = 1.27$ Å. The vibrational frequency is about $\nu_0 = 2890\mathrm{cm}^{-1}c \simeq 8.67 \times 10^{13}$ Hz, implying an elastic constant $k = 4\pi^2\nu_0^2\mu = 4.56 \times 10^5$ dyne/cm.
The doublet arises from the spectrum of the $H^{35}Cl$ and $H^{37}Cl$ molecules.
Deuteration implies an increase of the reduced mass by about a factor 2, leading to a spacing of the lines about $\bar{\Delta\nu} \simeq 10.5\mathrm{cm}^{-1}$. The vibrational frequency is reduced by a factor close to $\sqrt{2}$.

Problem X.5.2 The fundamental vibrational frequency of the NaCl molecule is $\nu_0 = 1.14 \cdot 10^{13}$ Hz. Report in a plot the temperature dependence of mean-square displacement from the equilibrium interatomic distance.

Solution:
From Eqs. 10.34 and 10.37 (see Problem F.X.1 for the average energy):

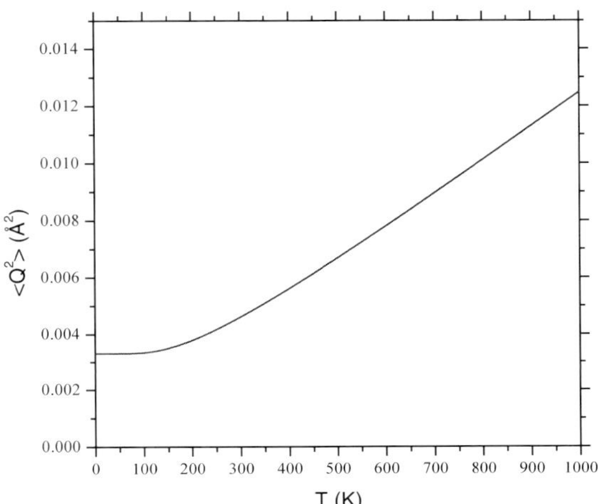

Problem X.5.3 Derive the equilibrium distance and the vibrational frequency of a diatomic molecule in the assumption of interatomic effective potential $V(R) = 4U((a/R)^{12} - (a/R)^6)$, with $a = 3.98$ Å and $U = 0.02$ eV, for reduced mass $\mu = 10^{-22}$ g.

Solution:
From $\partial V/\partial R\big|_{R=R_{eq}} = 0$ one has $R_{eq} = a \cdot (2)^{\frac{1}{6}} = 4.47$ Å , while

$$V(R_{eq}) = 4U\left[\frac{1}{4} - \frac{1}{2}\right] = -U = -0.02 \text{ eV}.$$

By deriving $V(R)$ twice, one finds

$$k = \frac{4U}{a^2}\left(12 \cdot 13 \cdot 2^{-7/2} - 6 \cdot 7 \cdot 2^{-4/3}\right) = 57.144\,\frac{U}{a^2} = 11.558 \cdot 10^2 \text{ dyne/cm}.$$

Then

$$\nu = \frac{1}{2\pi}\sqrt{\frac{k}{\mu}} = 5.18 \cdot 10^{11} \text{ Hz}.$$

Problem X.5.4 In a diatomic molecule the eigenvalue $E(R)$ for the electronic ground state is approximated in the form

$$E(R) = -2V_0 \left[\frac{1}{\rho} - \frac{1}{2\rho^2} \right]$$

(with $\rho = R/a$ and a characteristic length). Derive the rotational, vibrational and roto-vibrational energy levels in the harmonic approximation.

Solution:
The equivalent of Eq. 10.4 is

$$-\frac{\hbar^2}{2\mu}\frac{d^2\mathcal{R}}{dR^2} + \left[-2V_0 \left(\frac{1}{\rho} - \frac{1}{2\rho^2} \right) + \frac{K(K+1)\hbar^2}{2\mu a^2 \rho^2} \right] \mathcal{R} = \mathcal{E}\mathcal{R},$$

where μ is the reduced mass. The effective potential has the minimum for

$$\rho = \rho_0 \equiv 1 + \frac{K(K+1)\hbar^2}{2\mu a^2 V_0} \equiv 1 + B$$

For $V(\rho) = -V_0 (1+B)^{-1} + V_0 (1+B)^{-3} (\rho - \rho_0)^2$ the Schrödinger equation takes the form for the harmonic oscillator. Then

$$E + V_0 (1+B)^{-1} = \hbar \sqrt{\frac{2V_0}{\mu a^2} (1+B)^{-3}} \left(v + \frac{1}{2} \right)$$

.

For $B \ll 1$,

$$E = -V_0 + \frac{K(K+1)\hbar^2}{2\mu a^2} + h\nu_0 \left(v + \frac{1}{2} \right) - \frac{3}{2} \frac{\hbar^3 K(K+1)\left(v + \frac{1}{2}\right)}{\mu^2 a^4 2\pi\nu_0}$$

where $\nu_0 = (1/2\pi)\sqrt{2V_0/\mu a^2}$.

Problem X.5.5 From the data reported in the Figure at Problem X.5.1 for the HCl molecule, estimate the vibrational contribution to the molar specific heat, at room temperature.

Solution:
From the thermodynamical energy $<E> = \sum_v E_v N_v(T)$ (see Eq. 10.37) the molar specific heat turns out

$$C_V = R (T_D/T)^2 \frac{e^{T_D/T}}{\left(e^{T_D/T} - 1\right)^2},$$

where T_D is the **vibrational temperature** $T_D = h\nu_0/k_B$, given by $T_D = 4.15 \cdot 10^3 K$ for $\nu_0 = 0.87 \cdot 10^{14}$ Hertz.

At room temperature $T \ll T_D$ and $C_V \simeq R(T_D/T)^2 exp(-T_D/T) = 1.5 \cdot 10^4$ erg/K mole.

Problem X.5.6 A static and homogeneous electric field $\mathcal{E}$ is applied along the molecular axis of an heteronuclear diatomic molecule. In the harmonic approximation for the vibrational motion, by assuming an effective mass μ and an effective charge $-ef$ (with $0 < f \le 1$, see §8.5) derive the contribution to the electrical polarizability, in the perturbative approach.

Prove that the result derived in this way is actually the **exact one**.

Solution:

For

$$\mathcal{H}_p = fez\mathcal{E}$$

the first order correction is $< v|z|v >= 0$. The matrix elements $\mathcal{H}_{vv'}$ for z are

$$\mathcal{H}_{vv'} =< v|z|v' >= \left(\frac{v+1}{2\alpha}\right)^{1/2} \qquad \text{for} \quad v' = v+1$$

and

$$\left(\frac{v}{2\alpha}\right)^{1/2} \qquad \text{for} \quad v' = v-1 \,,$$

where $\alpha = \frac{\mu 2\pi\nu_0}{\hbar}$ and $\nu_0 = \frac{1}{2\pi}\sqrt{k/\mu}$, with k force constant.

The second order correction to the energy E_v^0 turns out

$$E_v^{(2)} = (fe\mathcal{E})^2 \left\{ \frac{|\mathcal{H}_{v,\,v+1}|^2}{-h\nu_0} + \frac{|\mathcal{H}_{v,\,v-1}|^2}{h\nu_0} \right\} =$$

$$= (fe\mathcal{E})^2 \left\{ \frac{\frac{v+1}{2\alpha}}{-h\nu_0} + \frac{\frac{v}{2\alpha}}{h\nu_0} \right\} = - (fe\mathcal{E})^2 \frac{1}{8\pi^2\mu\nu_0^2} \,.$$

Then the electric polarizability is

$$\chi = N\alpha = N\frac{1}{\mathcal{E}} \left(-\frac{\partial E_v^{(2)}}{\partial \mathcal{E}} \right) = N (fe)^2 \frac{1}{k}$$

independent of the state of the oscillator (N number of molecule for unit volume).

The result for the single molecule polarizability α is the exact one. In fact, going back to the Hamiltonian of the oscillator in the presence of the field

$$\mathcal{H} = -\frac{\hbar^2}{2\mu}\frac{d^2}{dz^2} + \frac{1}{2}kz^2 + fe\mathcal{E}z$$

by the substitution $z = z' - (fe\mathcal{E}/k)$ it becomes

$$\mathcal{H}' = -\frac{\hbar^2}{2\mu}\frac{d^2}{dz'^2} + \frac{1}{2}k(z')^2 - \frac{1}{2k}(fe\mathcal{E})^2$$

implying the eigenvalues

$$E'_v = E^0_v - \frac{(fe\mathcal{E})^2}{2k}$$

and therefore $\alpha = (fe)^2/k$.

10.6 Polyatomic molecules: normal modes

In a polyatomic molecule with S atoms, $(3S - 6)$ degrees of freedom involve the oscillations of the nuclei around the equilibrium positions. If q_i indicate local coordinates expressing the displacement of a given atom, as sketched below,

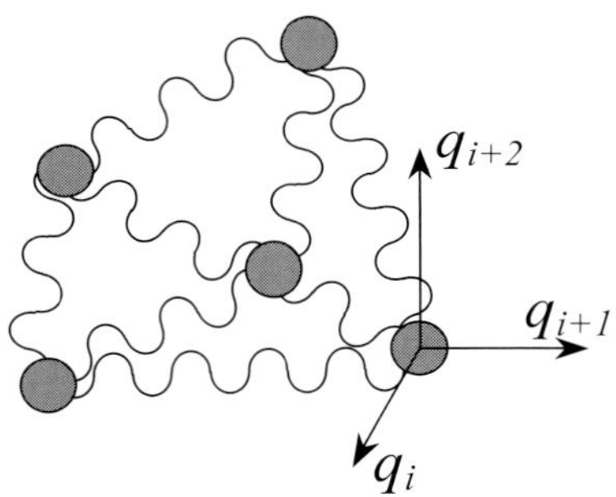

for small displacements the potential energy, in the harmonic approximation, can be written

$$V = V_o + \sum_i \left(\frac{\partial V}{\partial q_i}\right)_o q_i + \frac{1}{2}\sum_{i,j}\left(\frac{\partial^2 V}{\partial q_i \partial q_j}\right)_o q_i q_j \simeq \sum_{i,j} b_{ij} q_i q_j \qquad (10.48)$$

Similarly, the kinetic energy is $T = \sum_{i,j} a_{ij}\dot{q}_i\dot{q}_j$. The classical equations of motion become

$$\frac{d}{dt}\frac{\partial \mathcal{L}}{\partial \dot{q}_i} - \frac{\partial \mathcal{L}}{\partial q_i} = \sum_j a_{ij}\ddot{q}_j + \sum_j b_{ij} q_j = 0 \ , \qquad (10.49)$$

namely $(3S - 6)$ coupled equations, corresponding to complex motions that can hardly be formally described. Before moving to the quantum mechanical formulation it is necessary to introduce a new group of coordinates $Q = \sum_j h_j q_j$ (and a group of constants c_1, c_2, etc...) so that, by multiplying the first Eq. 10.48 by c_1, the second by c_2, etc... and adding up, one obtains equations of the form $d^2 Q / dt^2 + \lambda Q = 0$. This is the classical approach to describe small displacements around the equilibrium positions. The conditions to achieve such a new system of equations are

$$\sum_i c_i a_{ij} = h_j \tag{10.50}$$

$$\sum_i c_i b_{ij} = \lambda h_j \ . \tag{10.51}$$

Therefore, in terms of the constants c_i

$$\sum_i c_i (\lambda a_{ij} - b_{ij}) = 0 \ , \tag{10.52}$$

implying
$$|\lambda a_{ij} - b_{ij}| = 0 \tag{10.53}$$

This secular equation yields the roots $\lambda^{(1)}$, $\lambda^{(2)}$, ... corresponding to the conditions allowing one to find h_j so that the equations of motions become

$$\frac{d^2}{dt^2} Q_i + \lambda^{(i)} Q_i = 0 \tag{10.54}$$

These equations in terms of the **non-local, collective** coordinates

$$Q_i = \sum_j h_{ij} q_j \tag{10.55}$$

correspond to an Hamiltonian in the normal form

$$\frac{1}{2} \sum_i \dot{Q}_i^2 + \frac{1}{2} \sum_i \lambda^{(i)} Q_i^2, \tag{10.56}$$

where
$$\lambda^{(i)} \equiv (\frac{\partial^2 V}{\partial Q_i^2})_o \tag{10.57}$$

The Q's are called **normal coordinates**. The normal form of the Hamiltonian will allow one to achieve a direct quantum mechanical description of the vibrational motions in polyatomic molecules and in crystals (see Chapter 14).

A few illustrative comments about the role of the normal coordinates are in order. From the inverse transformation the local coordinates are written

$$q_i = \sum_j g_{ij} Q_j \tag{10.58}$$

and therefore, from Eq. 10.54,

$$q_i = \sum_j g_{ij} A_j \cos\left[\sqrt{\lambda^{(j)}}\, t + \varphi_j\right] \tag{10.59}$$

Thus the **local motion is the superposition of normal modes** of vibration. Each normal mode corresponds to an harmonic motion of the full system, with all the S atoms moving with the same frequency $\sqrt{\lambda^{(j)}}$ and the same phase. The amplitudes of the local oscillations change from atom to atom, in general.

Taking a look back to a diatomic molecule and considering the vibration along the molecular axis (see § 10.3) it is now realized that the normal coordinate is $Q = (x_A - x_B)$. The root of the secular equation analogous to Eq. 10.53 yields the frequency $\omega = \sqrt{k/\mu}$ and the (single) normal mode implies the harmonic oscillation, in phase opposition, of each atom, with relationship in the amplitudes given by $x_A = -x_B(M_B/M_A)$.

The formal derivation of the normal modes in polyatomic molecules in most cases is far from being trivial and the symmetry operations are often used to find the detailed form of the normal coordinates. For a linear molecule with three atoms, as CO_2, the description of the longitudinal vibrational motions in the harmonic approximation is straightforward (see for an illustrative example Prob. F.X.5). Fig. 10.6 provides the illustration of the four normal modes.

The coupled character of the motions is hidden in the collective frequency $\lambda^{(i)} \equiv \left(\frac{\partial^2 V}{\partial Q_i^2}\right)_o$, namely in the curvature of the potential energy under the variation of the i-th normal coordinate.

It is noted that the **stability** of the system is related to the **sign** of λ. Structural and ferroelectric **phase transitions** in crystals, for instance, are associated with the temperature dependence of the frequency of a normal mode, so that at a given temperature the structure becomes unstable ($\lambda^{(i)}$ is approaching zero) and a transition to a new phase, restoring large and positive $\lambda^{(i)}$, is driven.

Once that the vibrational motions are described by normal coordinates Q_i, the quantum formulation is straightforward. In fact, in view of the form of the classical Hamiltonian, the eigenfunction $\Phi(Q_1, Q_2, ...)$ is the solution of the equation

$$\sum_i \left(-\frac{\hbar^2}{2}\frac{\partial^2}{\partial Q_i^2} + \frac{1}{2}\lambda^{(i)} Q_i^2\right)\Phi(Q_1, Q_2, ...) = E\Phi(Q_1, Q_2, ...) \tag{10.60}$$

(the nuclear masses are included in mass-weighted coordinates Q's). Therefore the wavefunctions and the eigenvalues are

$$\Phi(Q_1, Q_2, ...) = \prod_i \Phi(Q_i) \tag{10.61}$$

$$E = \sum_i (n_i + 1/2)\hbar\sqrt{\lambda^{(i)}}, \qquad n_i = 0, 1, ... \tag{10.62}$$

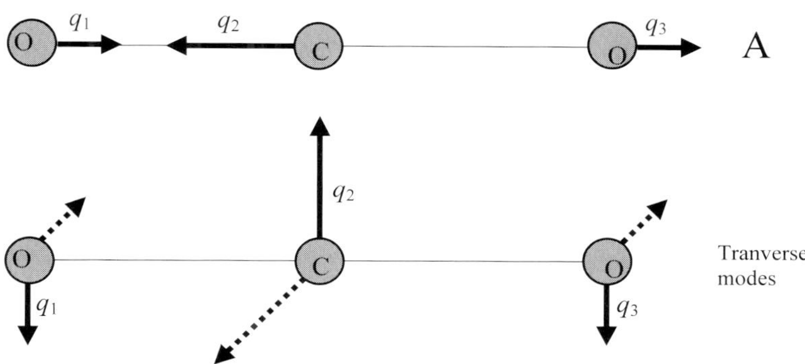

Fig. 10.6. Normal modes (longitudinal and transverse) in the CO_2 molecule. The symmetric mode S is not **active** in the infrared absorption spectroscopy, the selection rule requiring that the normal mode causes the variation of the **electric dipole moment** (see Eq. 10.66), while the antisymmetric mode A is active. The inverse proposition holds for Raman spectroscopy (see § 10.7) where the variation of the **polarizability** rather than of the dipole moment is required to allow one to detect the normal mode of vibration.

where $\Phi(Q_i)$ and $E_i = (n_i + 1/2)\hbar\sqrt{\lambda^{(i)}}$ are the single normal oscillator eigenfunctions and eigenvalues. Thus, by recalling the results for the diatomic molecule (§ 10.3), the vibrational state is described by a **set of numbers** $n_1, n_2, ...$, labelling the state of each normal mode.

Now we are going to show that within the harmonic approximation any normal oscillator interacts individually with the electromagnetic radiation, in other words the normal modes are **spectroscopically independent**.

The electric dipole matrix element for a transition from a given initial state to a final one, reads

$$\mathbf{R}_{in \to f} \propto \int \Phi^*_{n_1^f}(Q_1)\Phi^*_{n_2^f}(Q_2)...\mu_e(Q_1, Q_2, ...)\Phi_{n_1^{in}}(Q_1)\Phi_{n_2^{in}}(Q_2)... \ . \quad (10.63)$$

In the approximation of electrical harmonicity (see § 10.3)

$$\mu_{ex,y,z} \propto \sum_i \left(\frac{\partial \mu_{ex,y,z}}{\partial Q_i}\right)_o Q_i \quad (10.64)$$

Eq. 10.63 involves a sum of terms of the form

$$\left(\frac{\partial \mu_{e\,x,y,z}}{\partial Q_i}\right)_o \int \Phi^*_{n^f_1} \Phi_{n^{in}_1} dQ_1 \int \cdots \int \Phi^*_{n^f_i} Q_i \Phi_{n^{in}_i} dQ_i \cdots \tag{10.65}$$

This term is different from zero when

$$\left(\frac{\partial \mu_{e\,x,y,z}}{\partial Q_i}\right)_o \neq 0 \ , \tag{10.66}$$

meaning that the i-th normal mode must imply a variation of the electric dipole moment of the molecule. At the same time it is necessary that

$$n^f_j = n^{in}_j, \qquad \text{for } j \neq i$$
$$n^f_i = n^{in}_i \pm 1 \tag{10.67}$$

Therefore each normal oscillator interacts with the electromagnetic radiation independently from the others, with absorption spectrum displaying lines in correspondence to the eigenfrequencies of the various modes.

When the selection rule in Eq. 10.66 is verified the mode is said to be **infrared active**. As a consequence of this condition, one can infer that in the CO_2 molecule only the antisymmetric longitudinal mode can interact with the radiation while the symmetric one is silent (see Fig. 10.6).

Finally we just mention that in the harmonic approximation the contribution to the thermodynamic energy in polyatomic molecules is obtained by adding the contributions expected from each mode, of the form derived in diatomic molecules (see Problem X.5.5).

10.7 Principles of Raman spectroscopy

As it has been remarked, by means of infrared or microwave absorption spectroscopies some rotational or vibrational motions cannot be directly studied. This is the case of rotations and vibrations in homonuclear molecules or for normal modes which do not comply with the selection rule given by Eq. 10.66. Motions of those types can often be investigated by means of a spectroscopic technique based on the analysis of **diffuse radiation**: the **Raman** spectroscopy.

Phenomenologically the Raman effect can be described by referring to the experimental set-up schematically reported below

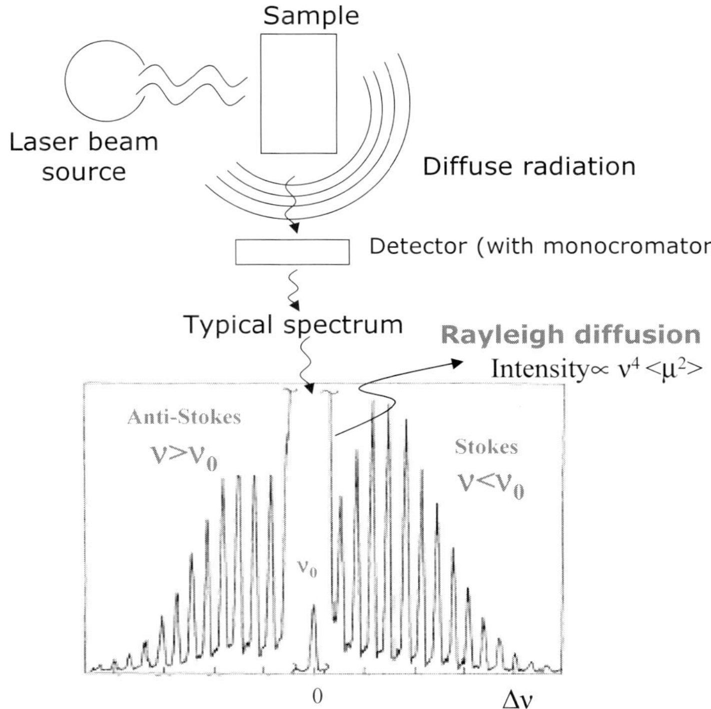

The classical explanation for the occurrence of Stokes and of anti-Stokes lines in the diffuse radiation, although not appropriate in some respects, still it enlightens the physical basis of the phenomenon. A normal mode of vibration can be thought to cause a time dependence of the molecular polarizability:

$$\alpha = \alpha_o + \alpha^* cos(\omega_i t) \tag{10.68}$$

Therefore the electric component of the radiation $\mathbf{E}(t) = \mathbf{E}_o cos(\omega_o t)$ (the wavelength is much larger than the molecular size) induces an electric dipole moment

$$\boldsymbol{\mu}_{ind} = \mathbf{E}(t)\alpha(t) = \alpha_o \mathbf{E}_o cos(\omega_o t) + \frac{1}{2}\alpha^* \mathbf{E}_o cos(\omega_o - \omega_i)t + \frac{1}{2}\alpha^* \mathbf{E}_o cos(\omega_o + \omega_i)t \tag{10.69}$$

From the phenomenological picture of *oscillating dipoles* as source of radiation one can realize that components of the diffuse light at frequencies $\omega_o \pm \omega_i$ have to be expected.

The inadequacy of the classical description can be emphasized by observing that the experimental findings indicate that the anti-Stokes lines, in general, are less intense than the Stokes lines. The interpretation based on the oscillating dipole as in Eq. 10.69, would predict intensities proportional to the fourth power of the frequency and then the anti-Stokes lines should be more intense than the Stokes ones.

The quantum description of the Raman effect is based on the process of **scattering of photons** and provides a satisfactory description of all the aspects of the phenomenon. The intensity of the lines, in fact, are controlled by the statistical populations on the ground-state and on excited vibrational states, as it can be grasped from the sketch of the inelastic scattering of the photon $(h\nu_i)$ given below:

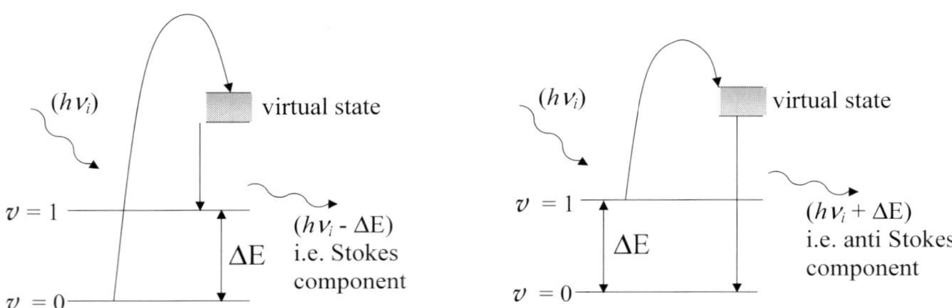

The basic aspects of the Raman radiation can be realized by extending the idea of electric dipole moment associated with a pair of states (already used in a variety of cases) to include the **field induced dipole moment**. Then in

$$\mathbf{R}_{n \to m}(t) = \left[\int \Phi_m^* \boldsymbol{\mu}_e \Phi_n d\tau \right] e^{\frac{i(E_m - E_n)t}{\hbar}}$$

the dipole moment $\alpha E_0 cos\omega_o t$, induced by the electric field of the incident radiation, is included:

$$\mathbf{R}_{n \to m}(t) = \left[\mathbf{E}_o \int \Phi_m^* \alpha \Phi_n d\tau \right] e^{-i(\omega_o - \frac{E_m - E_n}{\hbar})t} \; . \qquad (10.70)$$

Again interpreting this expression as a kind of microscopic source of radiation somewhat equivalent to irradiating dipoles, one sees that lines at the frequencies $\omega_o \pm \omega_{mn}$ have to be expected.

The amplitude of the matrix element of the **polarizability** in Eq. 10.70 controls the strength of the Raman components and therefore the selection rules. By referring for simplicity to scalar polarizability, in a first order approximation the analogous of Eq. 10.64 can be written

$$\alpha = \alpha_o + \sum_i (\frac{\partial \alpha}{\partial Q_i})_o Q_i \qquad (10.71)$$

Thus to have Raman radiation the conditions

$$\left(\frac{\partial \alpha}{\partial Q_i}\right)_o \neq 0 \tag{10.72}$$

$$v_i^m = v_i^n \pm 1 \tag{10.73}$$

must be fulfilled (see Eqs. 10.66 and 10.67).

Going back to Fig. 10.6 now one realizes that the S mode, which is not active in direct infrared absorption, can give Raman diffusion and *vice versa*[1].

Raman spectroscopy can also be used to study the rotational motions. In this case the fundamental aspect to pay attention to is the tensorial character of the molecular polarizability. The rotation of the molecule implies the rotation of the frame of reference Σ^P of the principal axes of the polarizability tensor $\bar{\bar{\alpha}}$, thus modulating the component along the direction of the electric field in the laboratory frame Σ^L, as sketched below

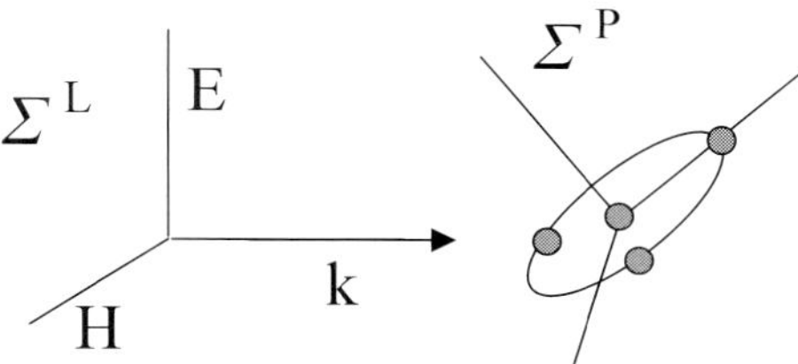

Therefore the incident radiation interacts with a time-dependent molecular polarizability, "modulated" at a frequency $2\nu_{rot}$ (the tensor being symmetric). For a molecule to be active in rotational Raman spectroscopy is not required to have a dipole moment. Any molecule not spherically symmetric and thus having **anisotropic polarizability**, is Raman active, in principle. The selection rule in terms of the quantum number K is $\Delta K = 0, \pm 2$, according to parity arguments, at variance with the selection rules 10.11 for the direct electric dipole transition between rotational states.

---

[1] This statement regarding the alternative role of symmetric and antisymmetric modes in Raman and infrared activity is a general one, holding in any molecule with **inversion symmetry**. It is related to the fact that the polarizability upon inversion transforms as a second order tensor while the dipole moment is a vector.

Problems X.7

Problem X.7.1 For a gas of diatomic molecules, with non-identical nuclei the roto-vibrational energy diagram is sketched below

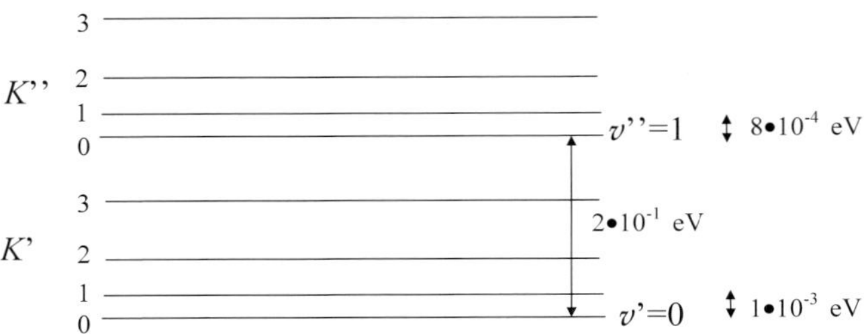

Figure out which lines can be detected in infrared and in Raman spectroscopies.

What do you expect if the molecule has two identical nuclei, with spin $I = 0$ or $I = 1/2$?

Solution:

The solution follows directly from Fig. 10.5 and from the selection rules $\Delta K = \pm 1, \Delta v = \pm 1$ for infrared absorption and from $\Delta K = \pm 2$ for Raman lines. For identical nuclei read §10.9.

10.8 Franck - Condon principle

The details of the **band spectra** (usually in the optical or in the UV ranges of the electromagnetic spectrum (see Appendix I.1)) associated with simultaneous transitions between electronic and roto-vibrational states in molecules involve rather complex selection rules.

In diatomic molecules one can easily illustrate a relevant and general aspect: **the Franck-Condon principle**. In Fig. 10.7 the typical energy curves for the ground and the first excited states are sketched and some transitions involving the vibrational states are indicated.

The classical description of the principle (given by Franck) was based on the following arguments. The nuclei-electron coupling is weak, the electronic transitions occur in very short times (typically $10^{-15} \div 10^{-16}$ seconds in comparison to the typical periods, around 10^{-13} s, of the vibrational motions). Therefore the interatomic distance can hardly change while the electrons are carried from one electronic state to the other. Since for the classical oscillator the probability to find the atoms at a given distance is large in correspondence to the maxima elongations, it is conceivable to expect a certain prevalence of the end-to-end transitions, as the one indicated in Fig. 10.7 by the arrow on the right side.

The basic aspect of the quantum description is outlined hereafter. The transition probability is controlled by the matrix element

$$\mathbf{R}_{g_1 \to g_2, v_1 \to v_2, K_1 \to K_2} = \int \phi_e^{g_2*} \phi_{vib}^{v_2*} \phi_{rot}^{K_2*} \left[\boldsymbol{\mu}_e + \boldsymbol{\mu}_N \right] \phi_e^{g_1} \phi_{vib}^{v_1} \phi_{rot}^{K_1} d\tau_e d\tau_N$$

$$(10.73)$$

with $\boldsymbol{\mu}_e = -e \sum_i \mathbf{r}_i$ and $\boldsymbol{\mu}_N = e \sum_\alpha Z_\alpha \mathbf{R}_\alpha$. The rotational part of the wavefunction involves only the angles θ and ϕ and therefore it can be considered separately. Thus one is left with

$$\mathbf{R}_{g_1 \to g_2, v_1 \to v_2} = \int \phi_e^{g_2*} \phi_{vib}^{v_2*} \left[\boldsymbol{\mu}_e + \boldsymbol{\mu}_N \right] \phi_e^{g_1} \phi_{vib}^{v_1} d\tau_e d\tau_N \qquad (10.74)$$

This term can be separated in two, the one involving $\boldsymbol{\mu}_N$ being zero since the electronic wavefunctions for g_1 and g_2 are orthogonal. Then only the term

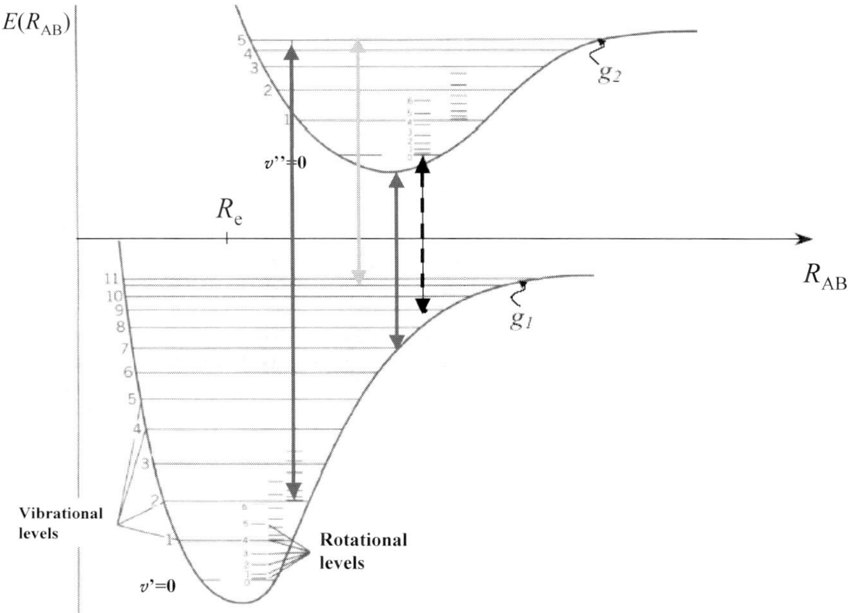

Fig. 10.7. Energy curves for the electronic ground state g_1 and the first excited state g_2 and sketches of the transitions involving the vibrational v' and v'' levels. The solid lines indicate transitions with large Franck-Condon factors, at variance with the classical prediction (dashed arrow on the right). The grey line refers to a transition with small Franck-Condon factor.

involving $\boldsymbol{\mu}_e$ has to be considered and by assuming that the electronic wavefunctions are only slightly modified when the interatomic distance is varied, the matrix element is written

$$\mathbf{R}_{g_1 \rightarrow g_2, v_1 \rightarrow v_2} = \int \phi_{vib}^{v_2 *} \phi_{vib}^{v_1} d\tau_N \int \phi_e^{g_2 *} \left[\boldsymbol{\mu}_e \right] \phi_e^{g_1} d\tau_e = S_{FC} \int \phi_e^{g_2 *} \left[\boldsymbol{\mu}_e \right] \phi_e^{g_1} d\tau_e.$$
(10.75)

Thus the matrix element appears as the usual electronic term multiplied by the **Franck-Condon factor** S_{FC}. For $v_1 \neq v_2$ S_{FC} can be different from zero since two different electronic states are involved in correspondence to the two vibrational levels.

The Franck-Condon factor is a kind of **overlap integral** and now it can be realized why for large quantum vibrational numbers the empirical formulation of the principle is again attained. The intensity of the transition line, proportional to the square of the transition dipole moment given by Eq. 10.75, is controlled by the factor $|S_{FC}|^2$ (see Problem X.8.1).

Problems X.8

Problem X.8.1 Evaluate the Franck-Condon term $|S_{FC}|^2$ involving two $v = 0$ vibrational states for electronic states g_1 and g_2 (see Fig. 10.7) having the same curvature at the equilibrium distances, one at R_e and the other at $R_e + \Delta R_e$.

Solution:
The vibrational wavefunctions are

$$\phi_0^{(1)} = \left(\frac{b}{\pi} \right)^{1/4} e^{-bQ^2/2}, \quad \phi_0^{(2)} = \left(\frac{b}{\pi} \right)^{1/4} e^{-b(Q-\Delta R_e)^2/2}$$

where $b = \mu \omega_0 / \hbar$ (see Eq. 10.32). The overlap integral is

$$S_{FC}(0,0) = \left(\frac{b}{\pi} \right)^{1/2} \int_{-\infty}^{+\infty} e^{-b(Q^2/2) - [b(Q-\Delta R_e)^2/2]} dQ =$$

$$= \left(\frac{b}{\pi} \right)^{1/2} e^{-b(\Delta R_e/2)^2} \int_{-\infty}^{+\infty} e^{-b(Q-\Delta R_e/2)^2} dQ = e^{-b(\Delta R_e/2)^2}$$

and

$$|S_{FC}|^2 = e^{-b(\Delta R_e)^2/2},$$

as plotted below:

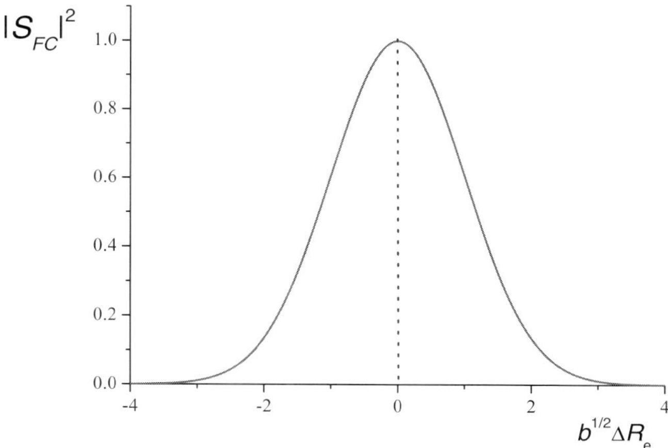

10.9 Effects of nuclear spin statistics in homonuclear diatomic molecules

Now we turn to an impressive demonstration of quantum principles, without any classical counterpart: the influence of the nuclear spins on the statistics, the related selection of molecular states and the occurrence of **zero-temperature rotations**.

In a diatomic homonuclear molecule, let us consider the behavior of the total wavefunction

$$\phi_T = \phi_e\,\phi_{vib}\,\phi_{rot}\,\chi_{spin} \tag{10.76}$$

upon exchange of the nuclei, each having nuclear spin I. The total number of spin wavefunctions is $(2I + 1)^2$. $(2I + 1)$ of them are symmetric, since the magnetic quantum numbers m_I are the same for both nuclei. Half of the remaining wavefunctions are symmetric and half antisymmetric. Thus $[(2I + 1)^2 - (2I + 1)]/2 + (2I + 1) = (I + 1)(2I + 1)$ are **symmetric** and the remaining $I(2I + 1)$ **antisymmetric**. Therefore the ratio of the **ortho** (symmetric) to **para** (antisymmetric) molecules is $(N_{para}/N_{ortho}) = I/(I+1)$. For example, for Hydrogen 75% of the molecules belong to orthohydrogen type and 25% to parahydrogen.

Let P indicate the operator exchanging spatial and spin coordinates of the nucleus A with the ones of the nucleus B. One has $P\phi_T = +\phi_T$ for nuclei with integer I (**bosons**) while $P\phi_T = -\phi_T$ for nuclei with half integer spin (**fermions**).

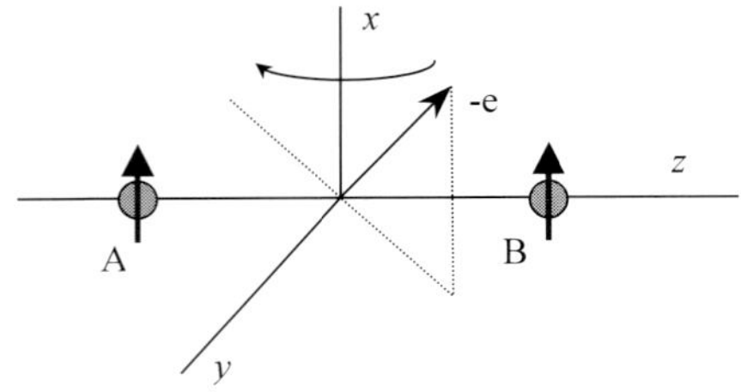

For the electronic wavefunction ϕ_e the exchange of the nuclei is equivalent to:

i) rotation by 180 degrees around the x axis;

ii) inversion of the electronic coordinates with respect to the origin;

iii) reflection with respect to the yz plane.

For the most frequent case of electronic ground state Σ_g^+ one concludes

$$P\phi_e = +\phi_e \tag{10.77}$$

The vibrational wavefunction is evidently symmetrical, i.e. $P\phi_{vib} = +\phi_{vib}$, since it depends only on $(R - R_e)$. For the rotational wavefunctions one has

$$P\phi_{rot} = (-1)^K \phi_{rot} \tag{10.78}$$

namely they are symmetrical (positive parity) upon rotation when the number K is even while are antisymmetric (negative parity) the ones having odd rotational numbers K. By taking into account Eqs. 10.76-10.78 one deduces the requirement that for half integer nuclear spin (total wavefunction antisymmetric upon exchange of the nuclei) **ortho molecules** (having symmetric spin functions) can be found only in rotational states with odd K. On the contrary **para molecules** (having antisymmetric spin functions) can be found only in rotational states with K even. For integer nuclear spins the propositions are inverted. It is noted that ortho to para transitions are hardly possible, for the same argument used to discuss the (almost) lack of transitions from singlet to triplet states in the Helium atom (see § 2.2).

A relevant spectroscopic consequence of the symmetry properties in diatomic molecules, for instance, is the fact that in Raman spectra in H_2 the lines associated to transitions starting from rotational states at K odd (see the illustrative plot in the following Figure) are approximately three time stronger than the ones involving the states at K even, once that thermal equilibrium is established between the two species ortho and para (see Problem X.9.1). For D_2 an opposite alternation in the intensities occurs (by a factor of two), the nuclear spin of deuterium being $I = 1$.

For O_2, the electronic ground state being $^3\Sigma_g^-$, the nuclear spin for ^{16}O is zero and χ_{spin} is necessarily symmetric. Then only odd K states are allowed. Thus only the rotational lines corresponding to $\Delta K = \pm 2$ and involving odd

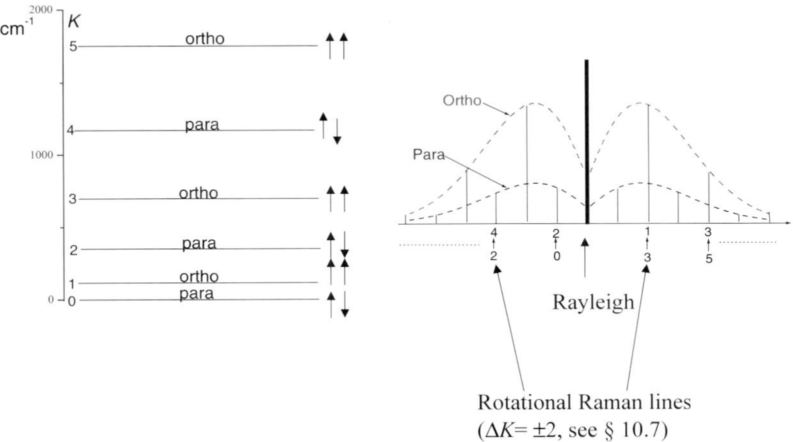

Rotational Raman lines
($\Delta K = \pm 2$, see § 10.7)

K states are observed in Raman spectroscopy (see Fig. 10.8). As soon as one of the nuclei is substituted by its isotope ^{17}O, all rotational lines are detected.

For N_2, the nuclear spin being $I = 1$, the roto-vibrational structure shows the same alternation in the intensities expected for D_2 (see Fig. 10.9).

Analogous spectroscopic effects are observed in polyatomic molecules having inversion symmetry, such as CO_2 or C_2H_2.

It should be stressed that for Raman spectroscopy, where virtual electronic states are involved, the remarks given above imply that these states retain the symmetry properties of the ground state. For optical and UV transitions between different electronic states these considerations can be applied to the roto-vibrational fine structure.

As a final remark one should observe that for ortho-Hydrogen molecules the lowest accessible rotational state in practice is the one at $K = 1$, unless one waits for the thermodynamical equilibrium for very long times. Thus even at the lowest temperature the molecules (in solid hydrogen) are still rotating. This is an example of the so called **quantum rotators**.

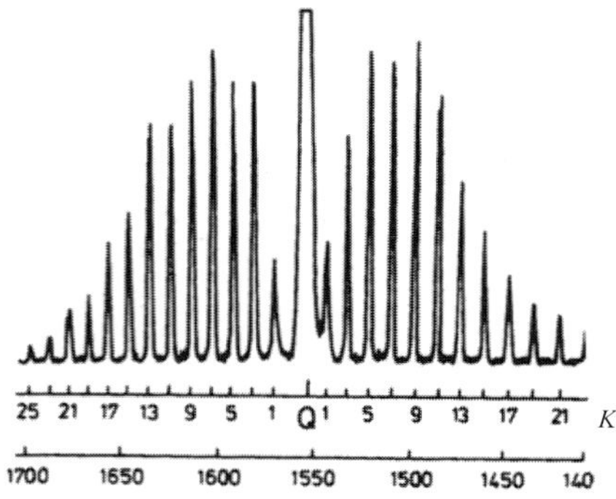

Fig. 10.8. A Stokes Raman component in $^{16}O_2$ displaying the rotational structure. At $\bar{\nu} \simeq 1556$ cm^{-1} the Q branch (§10.5), for $\Delta K = 0$, is observed (broad line). The lines with even K are missing (experimental spectrum reported in the book by Haken and Wolf (2004) quoted in the preface).

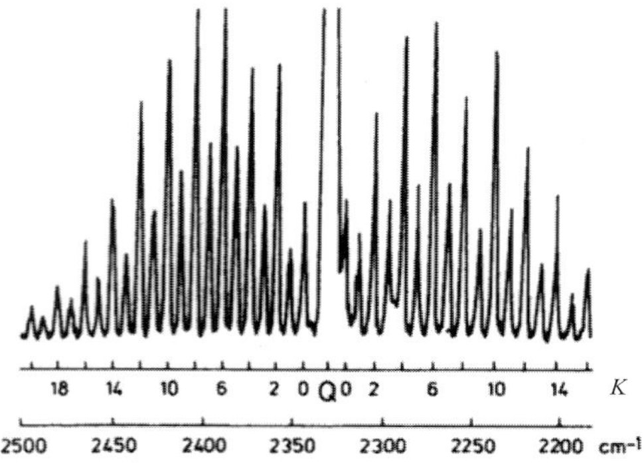

Fig. 10.9. Roto-vibrational Raman spectrum in N$_2$ (for ^{14}N -^{14}N). The alternation in the line intensities is in the ratio 1:2 (experimental spectrum reported in the book by Haken and Wolf (2004) quoted in the preface).

It can be mentioned that the bosonic character of ^{14}N nucleus has been claimed for the first time by **Heitler** and **Herzberg** in 1929 just from the alternation in the line intensities, before the discovery of the neutron which later on explained why $I = 1$.

Problems X.9

Problems X.9.1 For the molecules ^{3}He$_2$ and ^{4}He$_2$ (existing in excited states, assumed of Σ_g^+ character), derive the rotational quantum numbers that are allowed. By assuming thermal equilibrium obtain the ratio of the intensities of the absorption lines in the roto-vibrational spectra, at high temperature.

Solution:

The nuclear spin for ^{3}He is $I = 1/2$ while for ^{4}He is $I = 0$. Therefore in ^{3}He$_2$ only states at K even are possible for total spin 0 and only states at K odd for total spin 1. The intensity of the lines for $E_K \ll k_B T$ is proportional to the degeneracy. Therefore, by remembering that the rotational degeneracy is $(2K + 1)$, while $(2I + 1)(I + 1)$ spin states are symmetric and $(2I + 1)I$ are antisymmetric, one can expect for the ratio of the intensities

$$\frac{\text{Intensity of transitions from } 2K}{\text{Intensity of transitions from } (2K - 1)} = \frac{I(4K + 1)}{(I + 1)(4K - 1)}.$$

For large rotational numbers the ratio reduces to $I/(I + 1)$, namely 1:3, as discussed at §10.9.

For ^{4}He$_2$, I being 0, no antisymmetric nuclear spin functions are possible, only the rotational states at $K = 0, 2, 4...$ are allowed and every other line is absent.

Problem X.9.2 In the assumption that in the low temperature range only the rotational states at $K \leq 2$ contribute to the rotational energy of the H$_2$ molecule, derive the contribution to the molar specific heat.

Solution:

Since the ortho-molecules (on the $K = 1$ state) cannot contribute to the increase or the internal energy U_{rot} upon a temperature stimulus, only the partition function Z_{rot}^{para} of the para-molecules (on the $K = 0$ and $K = 2$ rotational states) have to be considered in

$$U_{rot} = N k_B T^2 \frac{d}{dT} \ln Z_{rot}.$$

From $\sum_K (2K + 1) \exp[K(K + 1)\theta_{rot}/T]$ (with $\theta_{rot} \equiv \hbar^2/2I k_B \simeq 87$ K in H$_2$), $Z_{rot}^{para} \simeq 1 + 5 \exp[-6\theta_{rot}/T]$. Then $U_{rot} \simeq N k_B 30 \theta_{rot} \cdot \exp[-6\theta_{rot}/T]$. Since the number of para-molecules in a mole can be considered $N_A/4$,

$$C_V \simeq N_A/4 \cdot 180(\theta_{rot}/T)^2 \exp[-6\theta_{rot}/T].$$

Problem X.9.3 Estimate the fraction of para-Hydrogen molecules in a gas of H_2 at temperature around the rotational temperature $\theta_{rot} = 87$ K and at $T \simeq 300$ K, in the assumption that thermal equilibrium has been attained. (Note that after a thermal jump it may take very long times to attain the equilibrium, see text). Then evaluate the fraction of para molecules in D_2.

Solution:
For $T \simeq 300K$ the fraction of para molecules is controlled by the spin statistical weights. Thus, for H_2, $f_{\text{para}} \simeq 1/(1+3) \simeq 0.25$. Around the rotational temperature one writes

$$f_{para} = \frac{\sum_{K\,even}(2K+1)\exp[-K(K+1)\theta_{rot}/T]}{Z_{para} + Z_{ortho}} \simeq$$

$$\simeq \frac{1 + 5e^{-6} + \dots}{1 + 5e^{-6} + \dots + 3[3e^{-2} + 7e^{-12} + \dots]} \simeq 0.49$$

For D_2 the rotational temperature is lowered by a factor 2 and the spin statistical weights are $I + 1 = 2$ for K even and $I = 1$ for states at K odd. At $T \simeq 300K \quad f_{para} = 0.33$. At $T \simeq 87K$ one writes

$$f_{para} = \frac{\sum_{K\ odd}(2K+1)\exp[-K(K+1)\theta_{rot}/2T]}{2\Sigma_{K\ even}\dots + 1\sum_{K\ odd}\dots} \simeq \frac{3\exp[-\theta_{rot}/T]}{2 + 1[3e^{-1}]} \simeq 0.35$$

Problems F.X

Problem F.X.1 Derive the temperature dependence of the mean square amplitude $< (R - R_e)^2 >\equiv< Q^2 >$ of the vibrational motion in a diatomic molecule of reduced mass μ and effective elastic constant k. Then evaluate the mean square amplitude of the vibrational motion for the $^1H^{35}Cl$ molecule at room temperature, knowing that the fundamental absorption frequency is $\nu_0 = 2990\ cm^{-1}$.

Solution:
From the virial theorem $< E >= 2 < V >$ and then
$< E >= 2 \cdot (k/2) < Q^2 > \equiv \mu\omega^2 < Q^2 >$ (see also Eq. 10.34).
Since (see Problem F.I.2), for the thermal average

$$< \overline{E} >= \hbar\omega\left(\frac{1}{2} + < n >\right), \qquad \text{with} \quad < n >= 1/(e^{\frac{\hbar\omega}{k_B T}} - 1)$$

one writes

$$< \overline{Q^2} >= \frac{\hbar}{\mu\omega}\left(\frac{1}{2} + \frac{1}{e^{\hbar\omega/k_B T} - 1}\right).$$

For $k_B T >> \hbar\omega$,

$$< \overline{Q^2} >\simeq const + \frac{k_B T}{\mu\omega^2}.$$

(see plot in Prob. X.5.2).

For $h\nu_0 \gg k_B T$ $<\overline{Q^2}> \simeq (1/2)h\nu_0/k$ and $\sqrt{<\overline{Q^2}>} \simeq \sqrt{h/8\pi^2\nu_0\mu} \simeq$ 0.076 Å.

Problem F.X.2 At room temperature and at thermal equilibrium condition the most populated rotational level for the CO_2 molecule is found to correspond to the rotational quantum number $K_M = 21$. Estimate the rotational constant B.

Solution:

From Eq. 10.14, by deriving with respect to K one finds

$$2K_M + 1 = \left(\frac{2k_B T}{Bhc}\right)^{1/2}$$

Then $Bhc = 2k_B T/(2K_M + 1)^2 = 4.48 \cdot 10^{-17}$ erg.

Problem F.X.3 When a homogeneous and static electric field $\mathcal{E} = 1070$ V cm^{-1} is applied to a gas of the linear molecule OCS the rotational line at 24325 MHz splits in a doublet, with frequency separation $\Delta\nu = 3.33$ MHz. Evaluate the rotational eigenvalues for $K = 1$ and $K = 2$ in the presence of the field, single out the transitions originating the doublet and derive the electric dipole moment of the molecule.

Solution:

From Eq. 10.20

$$E\left(1, M, \mathcal{E}\right) = 2Bhc + \mu^2\mathcal{E}^2(2 - 3M^2)/20Bhc$$
$$E\left(2, M, \mathcal{E}\right) = 6Bhc + \mu^2\mathcal{E}^2(2 - M^2)/84Bhc$$

The transitions at $\Delta K = \pm 1$ and $\Delta M = 0$ yield the frequencies $\nu = \nu_0 + \delta\nu(M)$, where $\nu_0 = 4Bc$ while the correction due to the field is

$$\delta\nu(M) = \frac{\mu^2\mathcal{E}^2}{Bh^2c}\frac{(29M^2 - 16)}{210}$$

Then $\delta\nu(0) = -(8/105)(\mu^2\mathcal{E}^2)/(Bh^2c)$, $\delta\nu(1) = (13/210)(\mu^2\mathcal{E}^2)/(Bh^2c)$ and the separation between the lines is

$$\Delta\nu = \delta\nu(1) - \delta\nu(0) = \frac{29}{210}\frac{\mu^2\varepsilon^2}{Bh^2c} = \frac{58}{105}\frac{\mu^2\mathcal{E}^2}{h^2\nu_0}$$

The dipole moment of the molecule turns out

$$\mu_e = \frac{h}{\mathcal{E}}\sqrt{\frac{105}{58}\nu_0\Delta\nu} = 2.37\cdot 10^{-21} \ \text{erg cm V}^{-1} = 0.71 \ \text{Debye}$$

Problem F.X.4 Derive an approximate expression, valid in the low temperature range, for the rotational contribution to the specific heat of a gas of HCl molecules.

Solution:

At low temperature only the first two rotational levels E_0 and E_1 can be considered and the rotational partition function is written

$$Z \simeq 1 + 3e^{-\frac{E_1}{k_B T}} \ .$$

The energy is $U_{rot} = -\partial lnZ/\partial\beta$, $(\beta = 1/k_B T)$ and then

$$U_{rot}(T \to 0) = 3E_1 \, e^{-E_1/k_B T}$$

and the specific heat (per molecule) becomes

$$(C_V)_{T\to 0} = \frac{3E_1^2}{k_B T^2} e^{-E_1/k_B T} \ .$$

This expression can actually be used only for

$$k_B T \ll \frac{\hbar^2}{\mu R_e^2} \equiv 2Bhc, \qquad \text{where} \quad B = 10.6\,cm^{-1}$$

(see §10.2.2), i.e. for

$$T_{val} \ll 30K \ .$$

Problem F.X.5 Derive the longitudinal normal modes for the system sketched below, assuming that the force constant of the spring in between the two masses is twice the ones for the lateral springs, that are stuck at fixed points (the springs have negligible mass).

Solution:

In terms of local coordinates (see §10.6)

$$T = \frac{M}{2}(\dot{q}_1^2 + \dot{q}_2^2), \qquad V = \frac{k}{2}q_1^2 + \frac{2k}{2}(q_2 - q_1)^2 + \frac{k}{2}q_2^2$$

The equations of motion in Lagrangian form are

$$M\ddot{q}_1 + 3kq_1 + 2kq_2 = 0 \ \text{ and } \ M\ddot{q}_2 + 3kq_2 - 2kq_1 = 0$$

By multiplying by $c_{1,2}$ and summing

$$M(c_1\ddot{q}_1 + c_2\ddot{q}_2) + q_1(3kc_1 - 2kc_2) + q_2(-2kc_1 + 3kc_2) = 0$$

The normal form in terms of the coordinates Q_i is obtained for

$$Mc_1 = \frac{1}{\lambda}(3kc_1 - 2kc_2) = h_1 \ \text{ and } \ Mc_2 = \frac{1}{\lambda}(-2kc_1 + 3kc_2) = h_2$$

yielding the secular equation

$$\begin{vmatrix} \lambda M - 3k & 2k \\ 2k & \lambda M - 3k \end{vmatrix} = 0\,.$$

with roots $\lambda_1 = 5k/M$ (implying $c_1 = -c_2$) and $\lambda_2 = k/M$ ($c_1 = c_2$), so that

$$Q_1 = (Mq_1 - Mq_2) \ \text{ and } \ Q_2 = (Mq_1 + Mq_2)$$

The equations of motion in the normal form are:

$$\ddot{Q}_1 + \frac{5k}{M}Q_1 = 0 \ , \ \ \ddot{Q}_2 + \frac{k}{M}Q_2 = 0$$

One normal mode corresponds to $q_1 = q_2$, and frequency $\omega_2 = \sqrt{k/M}$, while the second one corresponds to $q_1 = -q_2$ and frequency $\omega_1 = \sqrt{5k/M}$.

Crystal structures

Topics

Elementary crystallography
Translational invariance
Reciprocal lattice
The Bragg law
Brillouin zone
Typical crystal structures

In this Chapter and in the following three Chapters we shall be concerned with the general aspects of the solid state of the matter, namely the atomic arrangements where the interatomic interactions are strong enough to keep the atoms bound at well defined positions. We will address the bonding mechanisms leading to the formation of the crystals, the electronic structure and the vibrational dynamics of the atoms. The liquid and solid states are similar in many respects, for instance in regards of the density, short range structure and interactions. The difference between these two states of the matter relies on the fact that in the former the thermal energy is larger than the cohesive energy and the atoms cannot keep definite equilibrium positions.

Before the advent of quantum mechanics the solid-state physics was practically limited to phenomenological descriptions of macroscopic character, thus involving quantities like the compressibility, electrical resistivity or other mechanical, dielectric, magnetic and thermal constants. After the application of quantum mechanics to a model system of spatially ordered ions (the **crystal lattice**, indicated by **Laue X-ray diffraction** experiments) quantitative studies of the microscopic properties of solids began.

During the last forty years the study of the condensed matter has allowed one to develop the **transistors**, the **solid state lasers**, novel devices for **opto-electronics**, the **SQUID**, **superconducting magnets** based on new materials, *etc...* . As regards the development of the theory, solid state physics has triggered monumental achievements for **many-body systems**, such as the theories for **superconductivity** or of **quantum magnetism** for strongly correlated electrons, as well as the explanation of the **fractional quantum Hall effect**.

Besides the spatially ordered crystalline structures there are other types of solids, as polymers, amorphous and glassy materials, Fibonacci-type quasi-crystals, which are not characterized by regular arrangement of the atoms. Our attention shall be devoted to the simplest model, the **ideally perfect crystal**, with no defects and/or surfaces, where the atoms occupy spatially regular positions granting **translational invariance**. In the first Chapter we shall present some aspects of elementary crystallographic character in order to describe the crystal structures and to provide the support for the quantum description of the fundamental properties. Many solid-state physics books (and in particular the texts by **Burns**, by **Kittel** and by **Aschcroft** and **Mermin** recalled in the foreword) report in the introductory Chapters more complete treatments of **crystallography**, the "geometrical" science of crystals.

11.1 Translational invariance, Bravais lattices and Wigner-Seitz cell

In an ideal crystal the physical properties found at the position **r**

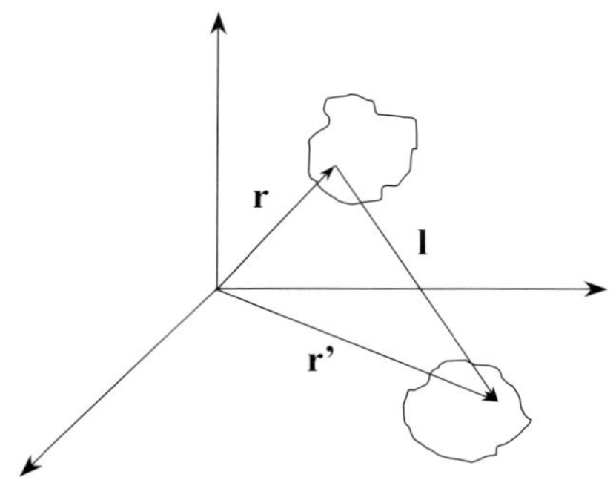

are also found at the position $\mathbf{r}' = \mathbf{r} + \mathbf{l}$, where

$$\mathbf{l} \equiv m\mathbf{a} + n\mathbf{b} + p\mathbf{c} \tag{11.1}$$

with m, n, p integers and $\mathbf{a}, \mathbf{b}, \mathbf{c}$ **fundamental translational vectors** which characterize the crystal structure.

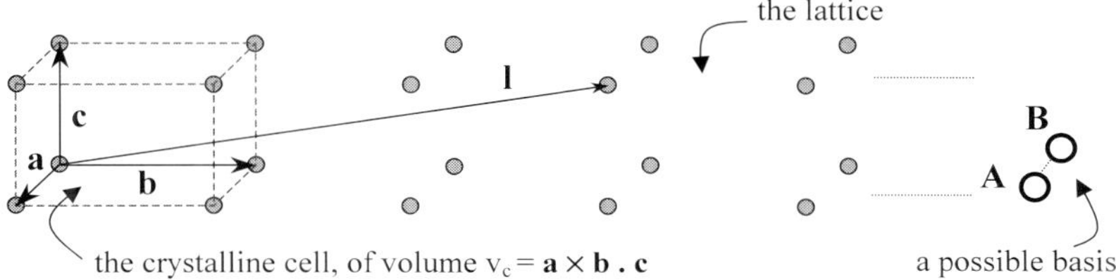

the crystalline cell, of volume $v_c = \mathbf{a} \times \mathbf{b} \cdot \mathbf{c}$ a possible basis

This property is called **translational symmetry** or **translational invariance**. As we shall see in Chapter 12, it is a symmetry property analogous to the ones utilized for the electronic states in atoms and molecules.

The extremes of the vectors $\mathbf{l}$, when the numbers m, n, p in Eq. 11.1 are running, identify the points of a geometrical network in the space, called **lattice**. By placing at each lattice point an atom or an identical group of atoms, called the **basis**, the real crystal is obtained. Thus one can ideally write **crystal$\equiv$ lattice + basis**.

The lattice and the fundamental translational vectors $\mathbf{a}, \mathbf{b}, \mathbf{c}$ are called **primitive** when Eq. 11.1 holds for any arbitrary pair of lattice points. Accordingly, in this case one has the maximum density of lattice points and the basis contains the minimum number of atoms, as it can be realized from the sketchy example reported below for a two-dimensional lattice:

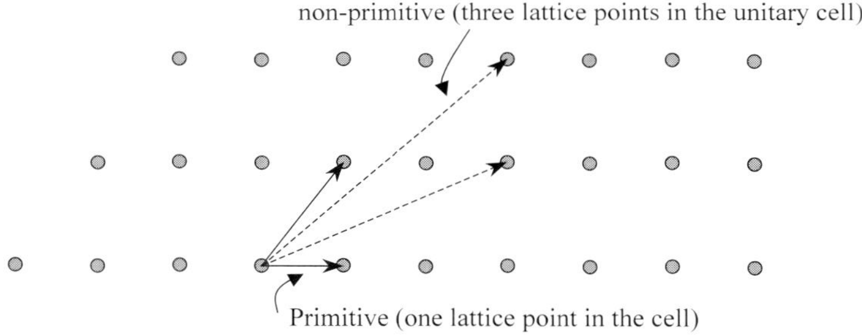

The geometrical figure resulting from vectors $\mathbf{a}, \mathbf{b}, \mathbf{c}$ is called the **crystalline cell**. The lattice originates from the repetition in space of this fundamental **unitary cell** when the numbers m, n and p run. The unitary cell is called **primitive** when it is generated by the primitive translational vectors. The primitive cell has the smallest volume among all possible unitary cells and it contains just **one lattice point**. Therefore it can host one basis only.

Instead of referring to the cell resulting from the vectors $\mathbf{a}, \mathbf{b}, \mathbf{c}$ one can equivalently describe the structural properties of the crystal by referring to the **Wigner-Seitz (WS) cell**. The WS cell is given by the region included within the planes bisecting the vectors connecting a lattice point to its neighbors, as in the example sketched below.

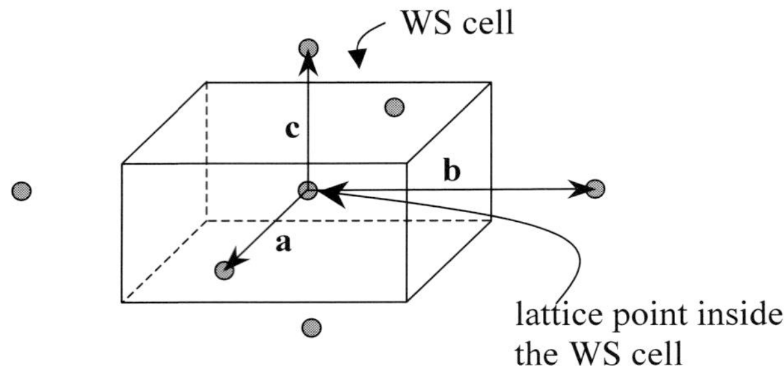

The lattice points are then at the center of the WS cells.

The translation of the WS cell by all the vectors **l** belonging to the group **T** of the **translational operations** (see Eq. 11.1) generates the whole lattice.

A few statements of geometrical character are the following:

i) The orientation of a plane of lattice points is defined by the **Miller indexes** (hkl), namely by the set of integers without common factors, inversely proportional to the intercepts of the plane with the crystal axes. The reason of such a definition will be clear after the discussion of the properties of the reciprocal lattice (§11.2).

ii) A direction in the crystal is defined by the smallest integers $[hkl]$ having the same ratio of its components along the crystal axes. For example, in a crystal with a cubic unitary cell the diagonal is identified by $[111]$. One should observe that the direction $[hkl]$ is perpendicular to the plane having Miller indexes (hkl) (see Problem F.XI.1).

iii) The position of a lattice point, or of an atom, within the cell is usually expressed in terms of fractions of the axial lengths a, b and c.

The **symmetry operations** are the ones which bring the lattice into itself, while leaving a particular lattice point fixed. The collection of the symmetry operations is called **point group** (of the lattice or of the crystal). When also the translational operations through the lattice vectors are taken into account, one speaks of **space group**. For non-monoatomic basis the spatial group also involves the symmetry properties of the basis itself. The point groups are groups in the mathematical sense and are at the basis of an elegant theory (the **group theory**) which can predict most symmetry-related properties of crystal just from the geometrical arrangement of the atoms.

Crystal System	Bravais Lattice	Unit Cell Dimensions
Triclinic	Primitive (P)	$a \neq b \neq c$ $\alpha \neq \beta \neq \gamma \neq 90°$
Monoclinic	Primitive (P) Base-Centered (C)	$a \neq b \neq c$ $\alpha = \gamma = 90° \neq \beta$
Orthorhombic	Primitive (P) Base-Centered (C) Body-Centered (I) Face-Centered (F)	$a \neq b \neq c$ $\alpha = \beta = \gamma = 90°$
Trigonal	R-Centered (R)	$a = b \neq c$ $\alpha = \beta = \gamma \neq 90° < 120°$
Hexagonal	Primitive (P)	$a = b \neq c$ $\alpha = \beta = 90° \; \gamma = 120°$
Tetragonal	Primitive (P) Body-Centered (I)	$a = b \neq c$ $\alpha = \beta = \gamma = 90°$
Cubic	Primitive (P) Body-Centered (I) Face-Centered (F)	$a = b = c$ $\alpha = \beta = \gamma = 90°$

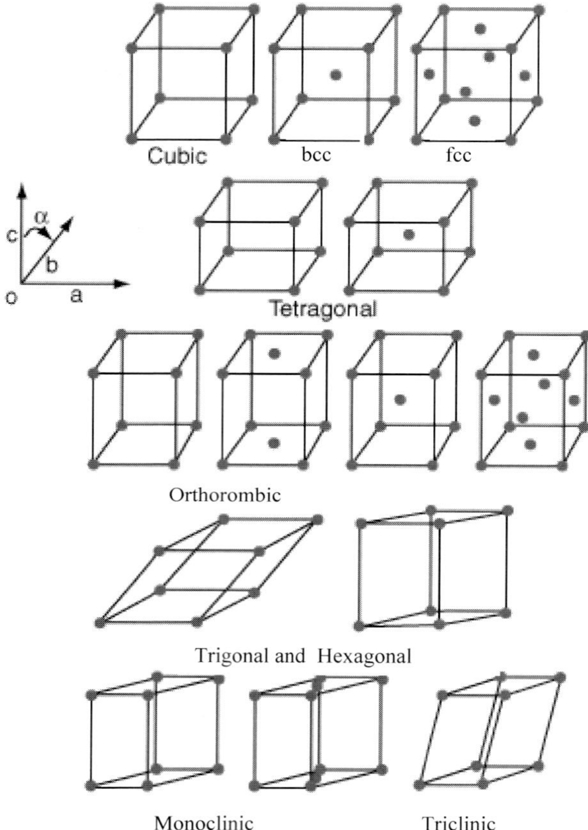

Fig. 11.1. Bravais crystal lattices with the conventional unitary cells, with the relations among the lattice lengths and among the characteristic angles.

The crucial point is that the requirement of **translational invariance limits the number of symmetry operations** that can be envisaged to define the crystal structures. To illustrate this restriction it is customary to recall that in a plane the unitary cell cannot be a pentagon (which is characterized by a rotational invariance after a rotation by an angle $2\pi/5$) since in that case one cannot achieve translational invariance.

In three dimensions (3D) there are **32 point groups** and **230 space groups** collecting all the symmetry operations compatible with translational invariance and with the symmetry of the basis. These groups define **14 fundamental lattices**, called the **Bravais lattices**. These lattices are shown in Fig. 11.1, where the **unitary conventional cell** generally used is indicated. It is noted that some cells might appear non-primitive, since there is more than one lattice point within them (see for instance the **bcc** lattice). However, one can easily identify the fundamental lattice vectors defining the primitive cell of the body-centered-cubic (**bcc**) Bravais lattice, in terms of the more frequently used non-primitive cubic lattice vectors $\mathbf{a}, \mathbf{b}, \mathbf{c}$ shown in the Figure. For the analogous case of the **fcc** (face-centered cubic) lattice, see Fig. 11.4 and Prob. F.XI.4.

11.2 Reciprocal lattice and Brillouin cell

As a consequence of the translational invariance in the ideal crystal, any local function $f(\mathbf{r})$ of physical interest (for instance, the energy or the probability of presence of electrons) must be spatially periodic, in other words invariant under the translation $\mathbf{T_l}$ by a vector belonging to the translational group:

$$\mathbf{T_l} f(\mathbf{r}) = f(\mathbf{r} + \mathbf{l}) = 1.f(\mathbf{r}). \tag{11.2}$$

Then one can abide by the Fourier expansion of $f(\mathbf{r})$ and by referring for simplicity to a crystal with orthogonal axes $\mathbf{a}, \mathbf{b}$ and $\mathbf{c}$ and choosing x, y and z along these axes, one writes

$$f(\mathbf{r}) = \sum_{-\infty}^{+\infty} {}_{n_x} A_{n_x}(y, z) e^{[i n_x x (2\pi/a)]} = \sum_{-\infty}^{+\infty} {}_{n_x} A_{g_x} e^{[i g_x x]} ,$$

where n_x is an integer and $g_x = n_x(2\pi/a)$ are reciprocal lattice lengths. The coefficients A_{n_x} can be Fourier-expanded along y and z and so one can put the function $f(\mathbf{r})$ in the form

$$f(\mathbf{r}) = \sum_{\mathbf{g}} A_{\mathbf{g}} e^{i \mathbf{g} \cdot \mathbf{r}} \tag{11.3}$$

where

$$A_{\mathbf{g}} = \frac{1}{v_c} \int_{-\infty}^{+\infty} f(\mathbf{r}) e^{-i \mathbf{g} \cdot \mathbf{r}} d\mathbf{r}, \tag{11.4}$$

v_c being the volume of the unitary cell. $\mathbf{g}$ is a **reciprocal lattice vector** built up by linear combination, with integer numbers $n_{x,y,z}$, of the **fundamental reciprocal vectors**, i.e.

$$\mathbf{g} = n_x(2\pi/a)\hat{x} + n_y(2\pi/b)\hat{y} + n_z(2\pi/c)\hat{z}. \tag{11.5}$$

It follows that for any reciprocal lattice vector $\mathbf{g}$ and for any translational vector $\mathbf{l}$, given by Eq. 11.1, one has

$$e^{i\mathbf{g}\cdot\mathbf{l}} = 1 \; , \tag{11.6}$$

corresponding to the necessary and sufficient condition to allow the Fourier expansion of local functions.

The above arguments can be generalized for non-orthogonal crystal axes by defining the fundamental reciprocal vectors $\mathbf{a}^*, \mathbf{b}^*$ and $\mathbf{c}^*$ in the form

$$\mathbf{a}^* = \frac{2\pi}{(\mathbf{a} \times \mathbf{b}.\mathbf{c})}(\mathbf{b} \times \mathbf{c}) = \frac{2\pi}{v_c}(\mathbf{b} \times \mathbf{c}),$$

$$\mathbf{b}^* = \frac{2\pi}{v_c}(\mathbf{c} \times \mathbf{a}),$$

$$\mathbf{c}^* = \frac{2\pi}{v_c}(\mathbf{a} \times \mathbf{b}). \tag{11.7}$$

The set of points, in the reciprocal space, reached by the vectors

$$\mathbf{g} = h\mathbf{a}^* + k\mathbf{b}^* + l\mathbf{c}^* \tag{11.8}$$

with h, k and l integers, defines the **reciprocal lattice**:

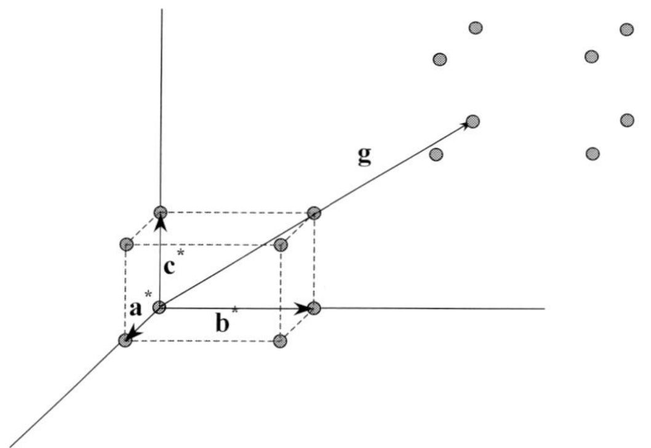

Instead of referring to the reciprocal lattice cell defined by $\mathbf{a}^*, \mathbf{b}^*$ and $\mathbf{c}^*$, it is often convenient to use its Wigner-Seitz equivalent, having a reciprocal lattice point at the center. This cell is called the **Brillouin cell** and it is shown schematically below for orthogonal axes:

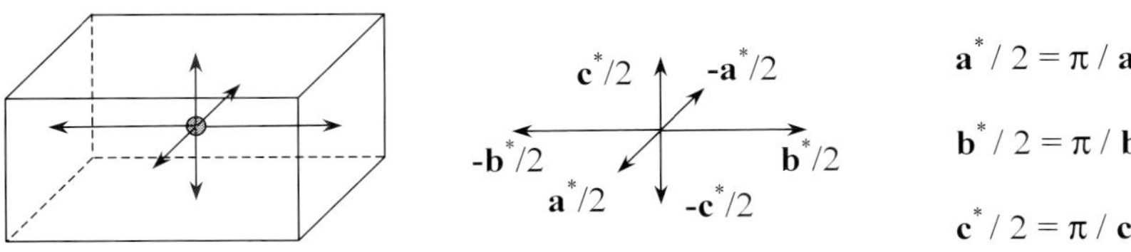

For instance, the Brillouin cell for the fcc lattice is obtained by taking eight reciprocal lattice vectors (bcc lattice, see Problem F.XI.4) bisected by planes perpendicular to such vectors and when the six next-shortest reciprocal lattice vectors are also bisected. This Brillouin cell is depicted in Fig. 11.2.

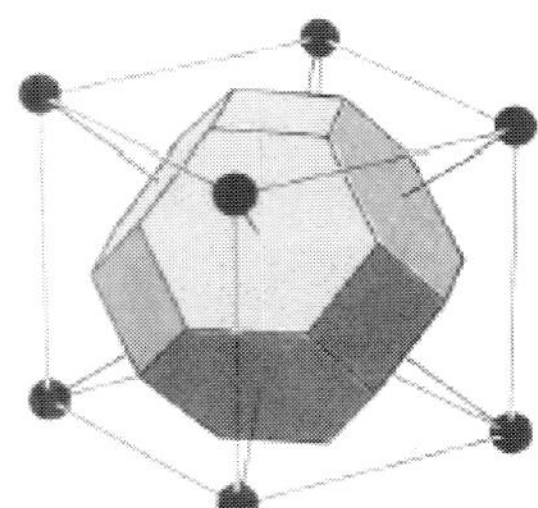

Fig. 11.2. Brillouin cell for fcc lattice

From the definitions of reciprocal lattice and of fundamental reciprocal vectors, one can derive the following properties (see Problem F.XI.1):

i) $\mathbf{g}(h, k, l)$ is perpendicular to the planes with Miller indexes (hkl);

ii) $|\mathbf{g}|$ is inversely proportional to the distance among the lattice planes (hkl).

The reciprocal lattice plays a relevant role in solid state physics. Its importance was first evidenced in diffraction experiments when it was noticed that each point of the reciprocal lattice corresponds to a diffraction spot. When the momentum of the electromagnetic wave (or of the De Broglie neutron wave) as a consequence of the scattering process changes by any reciprocal lattice vector, then the wave does not propagate through the crystal but undergoes Bragg reflection, as sketched below:

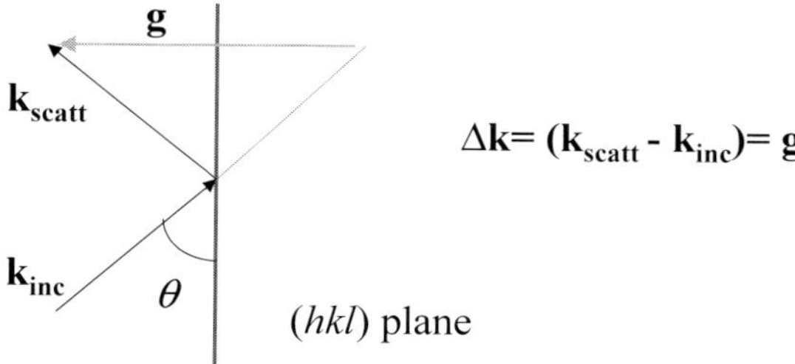

This condition corresponds to the Bragg law in the form

$$n\lambda = 2d\sin\theta \tag{11.9}$$

for the constructive interference of the radiation diffused by adjacent planes (d separation between the planes, n= 1,2,3..., X-ray beam incident at the angle θ the planes). In fact $\Delta k = g$ is equivalent to $2\pi/|\Delta k| = d(hkl)$, while $|k_{inc}| = |k_{scatt}| = 2\pi/\lambda$ (for elastic scattering) and $\Delta k = (4\pi/\lambda)\sin\theta$.

Furthermore, as we shall see at Chapter 12, the generators of the Brillouin cell, cut in a way related to the number of the cells in a reference volume, define the generators of a three-dimensional network in the reciprocal space. These vectors correspond to the wave-vectors of the excitations that can propagate through the crystal. Meantime they set the quantum numbers of the electron states.

11.3 Typical crystal structures

CsCl is the prototype of a family of cubic primitive (P) crystals with the basis formed by two atoms, one at position (0,0,0) and the other at (1/2,1/2,1/2). As sketched below the coordination number, i.e. the number of nearest neighbors around the Cs (or Cl) atoms is 8:

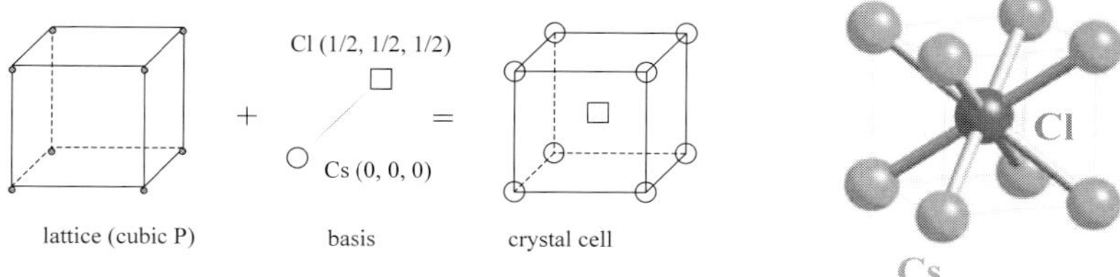

Other diatomic crystals with the same structure are TlBr, TlI, AgMg, AlNi and BeCu. Elements having the simple cubic (the basis being formed by one atom) Bravais lattice are P and Mn.

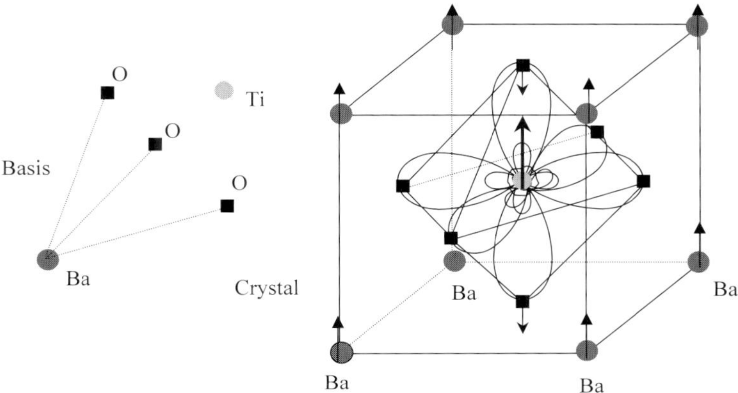

Fig. 11.3. Sketch of the crystal cell in $BaTiO_3$ (in the cubic phase). At $T_c \simeq 120$ C a displacive phase transition occurs, to a structure of tetragonal symmetry. The arrows indicate the directions of the displacements of the ions, having taken the oxygen ions at $c/2$ fixed (also a slight shrinkage in the ab plane occurs). The displacement of the positive and negative ions in opposite directions are responsible for the **spontaneous polarization** arising as a consequence of the transition from the cubic to the tetragonal phase (**ferroelectric state**).

A group of interesting crystals having a P cubic lattice with a more complex basis are the perovskite-type titanates and niobates, such as $BaTiO_3$, $NaNbO_3$, $KNbO_3$. At high temperature ($T \geq 120$ C for $BaTiO_3$) the atomic arrangement is the one reported in Fig. 11.3. The oxygen octahedra having the Ti (or Nb) atom at the center result from the d^2sp^3 hybrid orbitals (see Fig. 9.3). These octahedra are directly involved in the structural transitions driven by the softening of the $q = 0$ or of the zone-boundary vibrational modes (see §10.6 for a comment and Chapter 14). The distortion of the cubic cell is the microscopic source of the **ferroelectric transition** and of the **electro-optical properties** which characterize that crystal family. For all the crystal lattices described above the reciprocal lattice is cubic and the Brillouin cell is also cubic.

NaCl crystal is a typical example of face-centered cubic (fcc) lattice. The non-primitive, conventional, unitary cell and the primitive cell are shown in Fig. 11.4. The basis is formed by two atoms at the positions (0,0,0) and (1/2,1/2,1/2). The coordination number is 6. The fcc lattice characterizes

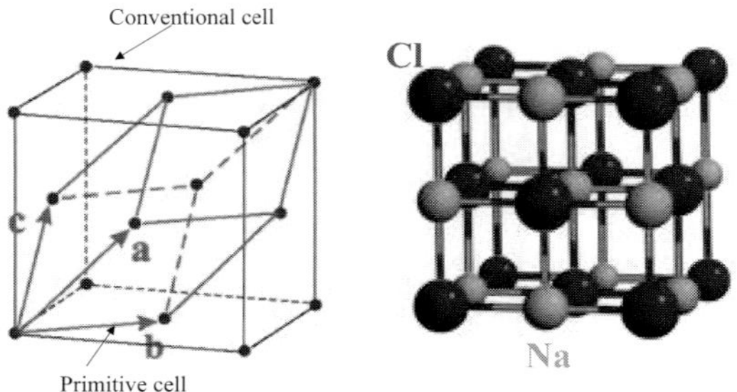

Fig. 11.4. Conventional and primitive cells for NaCl.

also the structure of KBr, AgBr and LiH and of several metal elements such as Al, Ca, Cu, Au, Pb, Ni, Ag and Sr.

The fcc lattice also characterizes the **diamond** (C) and the semiconductors Si, Ge, GaAs and InSb. In these cases the basis is given by two atoms (both C for diamond, Si and Ge) at the positions $(0,0,0)$ and $(1/4,1/4,1/4)$. Each atom has a tetrahedral coordination that may be thought to result from the formation of sp^3 hybrid atomic orbitals (§9.2), as sketched below:

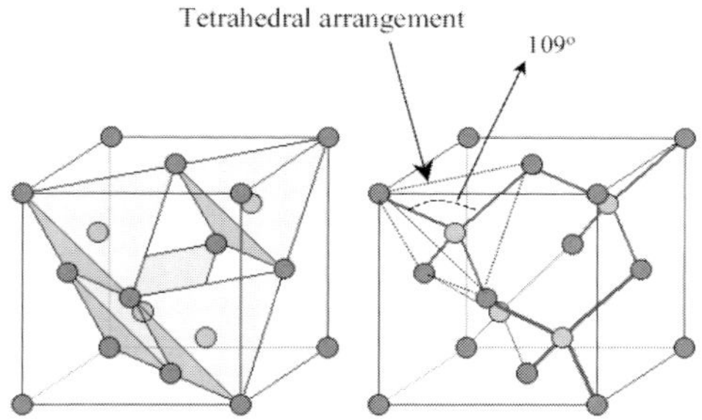

Carbon is known to crystallize also in the form of **graphite**, where the sp^2 hybridization of the C atomic orbitals yields a planar (2D) atomic arrangement. The 2D lattice is formed by two interpenetrating triangular lattices (see Fig. 11.5).

It should be mentioned that carbon can also crystallize in other forms, as for example in the fcc **fullerene**, where at each fcc lattice site there is a C_{60} molecule, with the shape of truncated icosahedron (a cage of hexagons and pentagons).

Another relevant crystalline form is the one having the **hexagonal close-packed** lattice, with the densest packing of hard spheres placed at the lattice

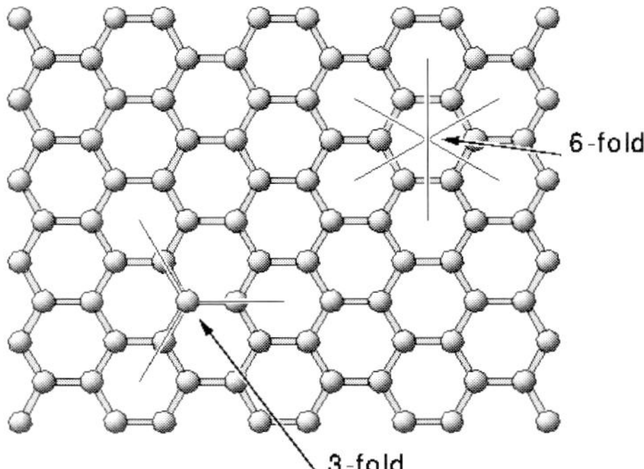

Fig. 11.5. In-plane atomic arrangement of C atoms in graphite.

points. The arrangement is obtained by placing the atoms at the vertexes of planar hexagons and then creating a second layer with "spheres" superimposed in contact with the three spheres of the underlying layer. The crystal lattice is the P hexagonal and the basis is given by two atoms placed at $(0,0,0)$ and at $(2/3,1/3,1/2)$.

In the hard sphere model 74 % of the volume is occupied and the ratio c/a is 1.633. In real crystals with this structure one has values of c/a slightly different, as 1.85 for Zn and 1.62 for Mg.

Problems F.XI

Problem F.XI.1 From geometrical considerations derive the relationships between the reciprocal lattice vector $\mathbf{g}(hkl)$ and the lattice planes with Miller indexes (hkl).

Solution:
For

$$\mathbf{g} = h\mathbf{a}^* + k\mathbf{b}^* + l\mathbf{c}^*.$$

let us take a plane perpendicular, containing the lattice points $m\mathbf{a}$, $n\mathbf{b}$ and $p\mathbf{c}$. Then, since $m\mathbf{a} - n\mathbf{b}$, $m\mathbf{a} - p\mathbf{c}$ and $n\mathbf{b} - p\mathbf{c}$ lie in this plane, one has

$$\mathbf{g}.(m\mathbf{a} - n\mathbf{b}) = \mathbf{g}.(m\mathbf{a} - p\mathbf{c}) = \mathbf{g}.(n\mathbf{b} - p\mathbf{c}) = 0.$$

Then $hm - kn = 0$, $mh = pl$ and $nk = pl$, yielding $m = 1/h$, $n = 1/k$ and $p = 1/l$.

From the definition of the Miller indexes one finds that the plane perpendicular to $\mathbf{g}$, passing through the lattice points $m\mathbf{a}$, $n\mathbf{b}$ and $p\mathbf{c}$ is the one characterized by (hkl).

Now it is proved that the distance $d(hkl)$ between adjacent (hkl) planes is $2\pi/|\mathbf{g}(hkl)|$. Let us consider a generic vector $\mathbf{r}$ connecting the lattice points of two adjacent (hkl) planes. Since $\mathbf{g}(hkl)$ is perpendicular to these planes one has $\mathbf{r}.\hat{g}(hkl) = d(hkl)$. One can arbitrarily choose $\mathbf{r} = \mathbf{a}/h$. Then $\mathbf{a}.\mathbf{g}(hkl) = 2\pi h$ and since $\hat{g} = \mathbf{g}/|\mathbf{g}|$ one has $\mathbf{r}.\hat{g} = 2\pi/|g|$. Therefore

$$d(hkl) = \frac{2\pi}{|\mathbf{g}(hkl)|}$$

Problem F.XI.2 Derive the density of the following compounds from their crystal structure and lattice constants:

Iron (bcc, $a = 2.86$ Å), Lithium (bcc, $a = 3.50$ Å), Palladium (fcc, $a = 3.88$ Å), Copper (fcc, $a = 3.61$ Å), Tungsten (bcc, $a = 3.16$ Å).

Solution:

$$\text{Fe} : \qquad \rho = \frac{\text{atomic mass} \cdot 2}{v_c} = 7.93 \text{ g cm}^{-3} .$$

$$\text{Li} : \qquad \rho = \frac{2 \cdot 1.660 \cdot 10^{-24} \cdot 6.939}{(3.5 \cdot 10^{-8})^3} = 0.537 \text{ g cm}^{-3} .$$

$$\text{Pd} : \qquad \rho = 12.095 \text{ g cm}^{-3} .$$

$$\text{Cu} : \qquad \rho = 8.968 \text{ g cm}^{-3} .$$

$$\text{W} : \qquad \rho = 19.344 \text{ g cm}^{-3} .$$

Problem F.XI.3 Estimate the order of magnitude of the kinetic energy of the neutrons used in diffraction experiments to obtain the crystal structures. By assuming that the neutron beam arises from a gas, estimate the order of magnitude of the temperature required to have diffraction.

Solution:
The neutron wavelength has to be of the order of the lattice spacing, i.e. of the order of 1 Å. Then $E_{kin} = h^2/2M_n\lambda^2 \simeq 80$ meV. The corresponding velocity is around 4×10^5 cm/s. Since $E_{kin} = 3k_BT/2$, one has $T \simeq 630$ K.

Problem F.XI.4 Show that the reciprocal lattice for the fcc lattice is a bcc lattice and *vice-versa*.

Solution:

In terms of the side a of the conventional cubic cell the **primitive** lattice vectors of the fcc structure are (Fig. 11.4):

$$\mathbf{a}_1 = \frac{a}{2}(\mathbf{i} + \mathbf{j})$$

$$\mathbf{a}_2 = \frac{a}{2}(\mathbf{i} + \mathbf{k})$$

$$\mathbf{a}_3 = \frac{a}{2}(\mathbf{j} + \mathbf{k})$$

($\mathbf{i}, \mathbf{j}, \mathbf{k}$ orthogonal unit vectors parallel to the cube edges). Then the primitive vectors of the reciprocal lattice are

$$\mathbf{a}_1^* = \frac{2\pi \mathbf{a}_2 \times \mathbf{a}_3}{a^3/4}$$

and similar expressions for $\mathbf{a}_2^*$ and $\mathbf{a}_3^*$ (Eq. 11.7) (in the unit cube of volume a^3 there are four lattice points). Thus

$$\mathbf{a}_1^* = \frac{2\pi}{a}(-\mathbf{i} - \mathbf{j} + \mathbf{k})$$

$$\mathbf{a}_2^* = \frac{2\pi}{a}(-\mathbf{i} + \mathbf{j} - \mathbf{k})$$

$$\mathbf{a}_3^* = \frac{2\pi}{a}(\mathbf{i} - \mathbf{j} - \mathbf{k})$$

The shortest (non-zero) reciprocal lattice vectors are given by the eight vectors $(2\pi/a)(\pm\mathbf{i} \pm \mathbf{j} \pm \mathbf{k})$ which generate the bcc (reciprocal) lattice.

A similar procedure applied to the primitive translational vectors of the bcc lattice

$$\mathbf{a}_1 = \frac{a}{2}(\mathbf{i} + \mathbf{j} + \mathbf{k})$$

$$\mathbf{a}_2 = \frac{a}{2}(-\mathbf{i} + \mathbf{j} + \mathbf{k})$$

$$\mathbf{a}_3 = \frac{a}{2}(-\mathbf{i} - \mathbf{j} + \mathbf{k})$$

(yielding the volume $a^3/2$ for the primitive cell) implies

$$\mathbf{a}_1^* = \frac{2\pi}{a}(\mathbf{i} + \mathbf{k})$$

$$\mathbf{a}_2^* = \frac{2\pi}{a}(-\mathbf{i} + \mathbf{j})$$

$$\mathbf{a}_3^* = \frac{2\pi}{a}(-\mathbf{j} + \mathbf{k})$$

as primitive vectors of fcc lattice.

The Brillouin cell of the bcc lattice is shown below (compared to the one in Fig. 11.2).

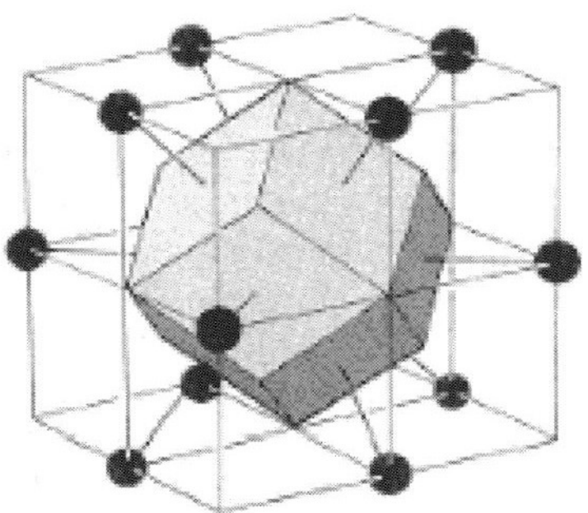

12

Electron states in crystals

Topics

The band of energy levels
Bloch orbital and crystal momentum
Effective electron mass
Density of states
Free-electron model and properties of metals
Perturbative effects on free-electron states and energy gap
Tight-binding model

12.1 Introductory aspects and the band concept

A fundamental issue in solid state physics is the structure of the electronic states. Transport, magnetic and optical properties, as well as the very nature (**metal**, **insulator** or **semiconductor**) of the crystal, are indeed controlled by the arrangement of the energy levels.

The complete form of the Schrödinger equation for electrons and nuclei can hardly be solved, even by computational approaches. Therefore to describe the electron states in a crystal it is necessary to rely on adequate approximate models. Usually the crystal is ideally separated into ions (the atoms with the core electrons practically keeping their atomic properties) and the **valence** electrons, which are affected by the crystalline arrangement. The Born-Oppenheimer separation (§7.2) is usually the starting point, often in the adiabatic approximation [1]. From the many-body problems for the electrons, by means of Hartree-Fock description, one can devise the one-electron effective potential that takes into account the interaction with the positive ions, the Coulomb-like repulsion among the electrons as well as the generalized exchange integrals. We shall not derive the potential energy in detail on the basis of that approach. Rather, similarly to atoms and molecules, we shall address the main aspects of the electronic structure in crystals on the basis of the fundamental symmetry property, namely the **translational invariance** for the potential energy:

$$V(\mathbf{r} + \mathbf{l}) = V(\mathbf{r}) \tag{12.1}$$

with $\mathbf{l}$ lattice vector (Eq. 11.1 and §11.2).

First we shall derive the general properties and a suitable classification of the electronic states. Then a deeper description will be made on the basis of particular models, at the sake of illustration of the generalities, meantime describing typical groups of solids.

Henceforth, by extending the molecular orbital approach (§8.1) in the LCAO form, one can express the one-electron wave function as **Bloch orbital**. This is somewhat equivalent to the delocalized MO introduced for the benzene molecule (§9.3).

Generalizing the concept used for Hydrogen molecule (§8.2) and referring to an ideal crystal formed by a chain of N one-electron atoms,

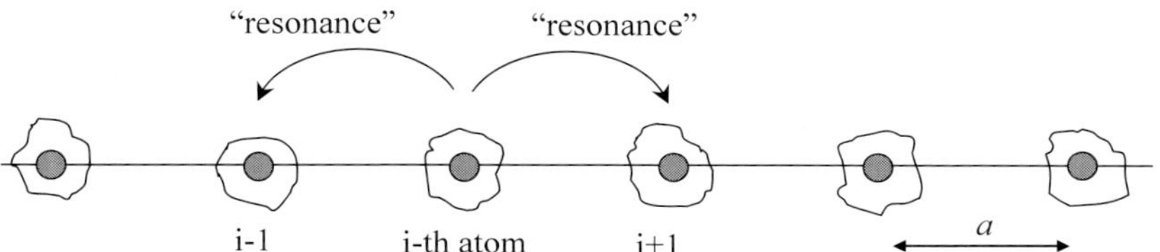

the formation of the **band** of electron levels can be understood as resulting from the removal of the degeneracy of the atomic levels. In fact, by taking into account the resonance of one electron among neighboring atoms (see sketch above), the wave function of the electron centered at i-th site is written

$$i\hbar \frac{d\psi_i}{dt} = E_o \psi_i + A\psi_{i-1} + A\psi_{i+1}, \tag{12.2}$$

where $A < 0$ is the **resonance integral between adjacent sites** (equivalent to H_{AB} in §8.1). From what has been learned for the H_2^+ molecule, we look for a solution of Eq. 12.2 in the form

---

[1] As already mentioned (§7.1) several relevant phenomena, for instance electrical resistivity and superconductivity, require to go beyond the adiabatic approximation.

$$\psi_i = \phi_i e^{-iEt/\hbar} \tag{12.3}$$

where E is the unknown eigenvalue, while ϕ_i is the electron eigenfunction for the atom centered at the site i. Then

$$E\phi_i = E_o\phi_i + A(\phi_{i-1} + \phi_{i+1}) \ , \tag{12.4}$$

with $\phi_i = \phi(x_i)$ and $\phi_{i\pm1} = \phi(x_i \pm a)$. By looking for a solution of the form $exp(ikx_i)$, typical of the difference equations and already used for the benzene molecule, Eq. 12.4 is rewritten in the form

$$Ee^{ikx_i} = E_o e^{ikx_i} + A\left[e^{ik(x_i+a)} + e^{ik(x_i-a)}\right], \tag{12.5}$$

yielding

$$E = E_o + 2A\cos(ka) \tag{12.6}$$

The formation of a **band** of electronic levels, each level labelled by k, as a consequence of the removal of the degeneracy existing for non-interacting atoms, is illustrated below

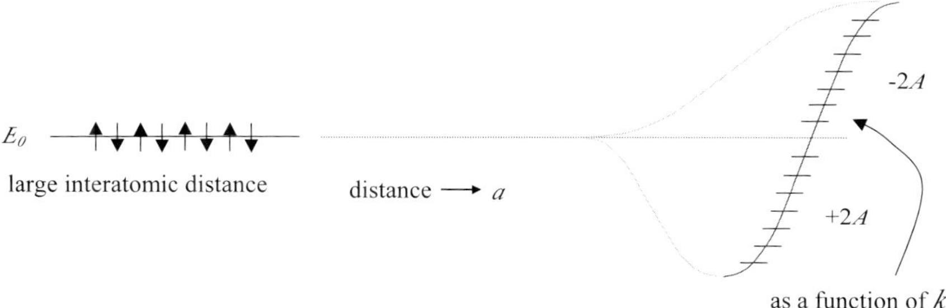

The band of N electron levels is the generalization of the g and u levels in the H_2 molecule or of the four levels in the C_6H_6 molecule. The energy interval between two adjacent bands, related to different atomic eigenvalues E_o, will be called **gap**. We shall come back to the problem of labelling the electron states and to the mechanisms leading to the appearance of a gap, after the discussion of the crystal models (§12.7).

12.2 Translational invariance and the Bloch orbital

In ideal crystals, with no defects and without surfaces, the translation operator $\mathbf{T_1}$ (see Eq. 11.2) commutes with the Hamiltonian:

$$\mathbf{T_1}\mathcal{H}(\mathbf{r})\phi(\mathbf{r}) = \mathcal{H}(\mathbf{r} + \mathbf{l})\phi(\mathbf{r} + \mathbf{l}) = \mathcal{H}(\mathbf{r})\mathbf{T_1}\phi(\mathbf{r})$$

Then the one-electron eigenfunction ϕ must be eigenfunction of $\mathbf{T}_1$ also, with eigenvalues c_1 satisfying the condition $|c_1|^2 = 1$, since

$$|\phi(\mathbf{r}+\mathbf{l})|^2 \equiv |\mathbf{T}_l\phi(\mathbf{r})|^2 = |\phi(\mathbf{r})|^2.$$

On the other hand, two translations $\mathbf{T}_{l_1}\mathbf{T}_{l_2} \equiv \mathbf{T}_{l_1+l_2}$, must yield the same result of the translation by $l_1 + l_2$:

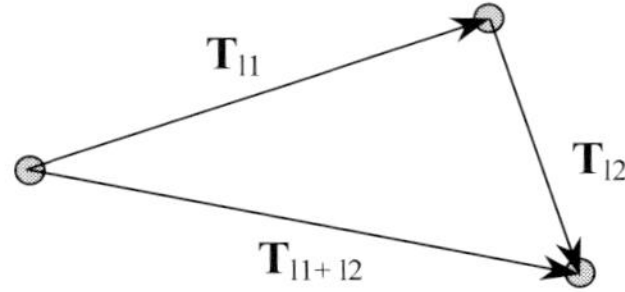

This suggests for the eigenvalue the form $c_1 = exp(i\lambda_1)$, so that

$$\mathbf{T}_{l_1+l_2}\phi = \mathbf{T}_{l_1}e^{i\lambda_2}\phi = e^{i\lambda_2}e^{i\lambda_1}\phi = e^{i(\lambda_1+\lambda_2)}\phi \ ,$$

with λ_l **real number.**

For any translation vector $\mathbf{l}$ one can find in the reciprocal space a vector $\mathbf{k}$ so that $\lambda_l = \mathbf{k}.\mathbf{l}$. Therefore one writes

$$\mathbf{T}_l\phi(\mathbf{r}) = \phi(\mathbf{r}+\mathbf{l}) = e^{i\mathbf{k}.\mathbf{l}}\phi(\mathbf{r}),$$

and by multiplying by $e^{-i\mathbf{k}.\mathbf{r}}$

$$e^{-i\mathbf{k}.\mathbf{r}}\phi(\mathbf{r}) = e^{-i\mathbf{k}.(\mathbf{r}+\mathbf{l})}\phi(\mathbf{r}+\mathbf{l}).$$

This condition shows that the function $u_\mathbf{k}(\mathbf{r}) = exp(-i\mathbf{k}.\mathbf{r})\phi(\mathbf{r})$ has the **periodicity of the lattice.**

Then the one-electron wave function can be written as **Bloch orbital**, i.e.

$$\phi_\mathbf{k}(\mathbf{r}) = u_\mathbf{k}(\mathbf{r})e^{i\mathbf{k}.\mathbf{r}}$$
$$u_\mathbf{k}(\mathbf{r}+\mathbf{l}) = u_\mathbf{k}(\mathbf{r}) \ , \tag{12.7}$$

which couples the free-electron wave function $exp(i\mathbf{k}.\mathbf{r})$ (characteristic of the **empty lattice,** namely in the limit $V(\mathbf{r}) \to 0$) with an unknown wave function $u_\mathbf{k}(\mathbf{r})$ having the **lattice periodicity.**

It can be remarked that up to now $\mathbf{k}$ in the Bloch orbital is just a vector in the reciprocal space used to label the one-electron states in a periodic potential. In the next Section we shall discuss the role and the physical properties of $\mathbf{k}$.

In order to illustrate the Bloch orbital we will take into consideration a particular form for the function $u_\mathbf{k}(\mathbf{r})$. $u_\mathbf{k}(\mathbf{r})$ can be found from the one-electron Schrödinger equation $\mathcal{H}\phi(\mathbf{r}) = E\phi(\mathbf{r})$ by writing for $\phi(\mathbf{r})$ the Bloch orbital according to Eq. 12.7:

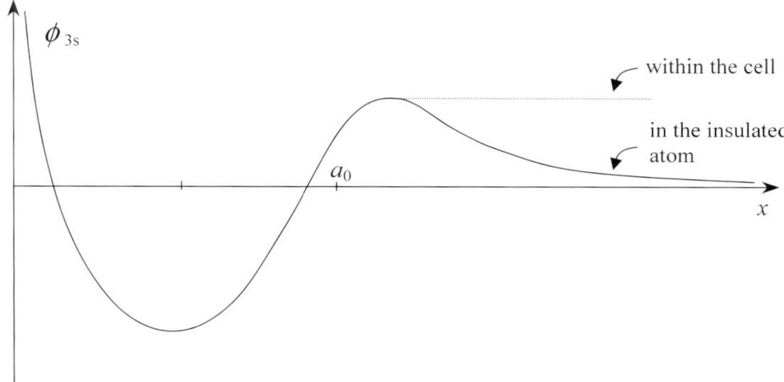

Fig. 12.1. Sketch of $u_{\mathbf{k}=0}(\mathbf{r})$ for $3s$ electron in the Na crystal, derived by Wigner and Seitz by means of the cellular method (by approximating the WS cell to a sphere; see also the book by Slater quoted in the preface).

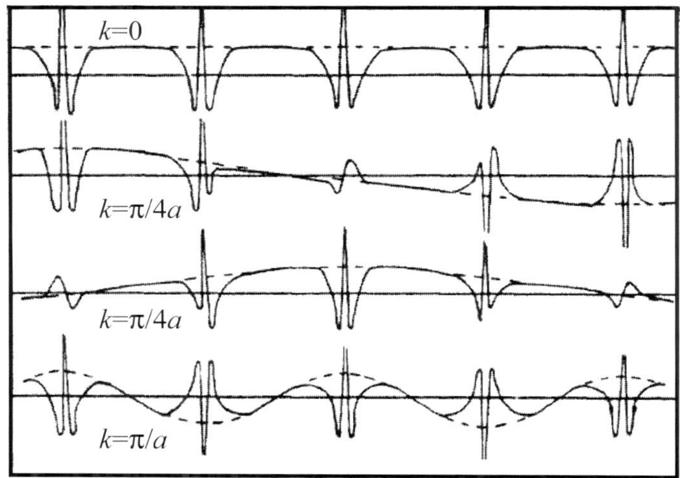

Fig. 12.2. Sketchy illustration of the Bloch orbital in the Na crystal along the [111] direction for different values of k, with a interatomic distance.

$$\left[\frac{-\hbar^2}{2m}(\boldsymbol{\nabla} + i\mathbf{k})^2 + V(\mathbf{r})\right]u_{\mathbf{k}}(\mathbf{r}) = E_{\mathbf{k}}u_{\mathbf{k}}(\mathbf{r}) \tag{12.8}$$

Under the assumption that $u_{\mathbf{k}}(\mathbf{r})$ is weakly $\mathbf{k}$-dependent, i.e. $u_{\mathbf{k}}(\mathbf{r}) \simeq u_{\mathbf{k}=0}(\mathbf{r})$, one can write $\phi_{\mathbf{k}}(\mathbf{r}) = u_{\mathbf{k}=0}(\mathbf{r})e^{i\mathbf{k}\cdot\mathbf{r}}$. From Eq. 12.8 one sees that for $\mathbf{k} = 0$ $u_{\mathbf{k}=0}(\mathbf{r})$ is the solution of the atomic-type Schrödinger equation. The only difference is in the boundary conditions which impose the continuity at the border of the Wigner-Seitz cell (see §11.1). In Fig. 12.1 the function $u_{\mathbf{k}=0}(\mathbf{r})$ for the $3s$ electron, in the Na crystal derived under these constraints (this procedure is the core of the so-called **cellular method**), is sketched. The corresponding Bloch orbitals are schematically depicted in Fig. 12.2.

12.3 Role and properties of k

The reciprocal space vector $\mathbf{k}$, labelling the eigenvalues of the translational operator, is a **constant of motion**: its components k_x, k_y and k_z have to be considered as good quantum numbers for the one-electron states. Hence, as far as the translational invariance condition holds, the electron remains in a given state $\mathbf{k}$. [2]

A first illustration of the role of $\mathbf{k}$ can be provided by considering the limiting case of vanishing potential energy $V(\mathbf{r})$, often called the **empty lattice** condition, as already mentioned. Then the eigenfunctions are

$$\phi_{\mathbf{k}}(\mathbf{r}) \propto e^{i\mathbf{k}\cdot\mathbf{r}} \tag{12.9}$$

with eigenvalues

$$E_{\mathbf{k}} = \frac{\hbar^2 k^2}{2m}. \tag{12.10}$$

Therefore for the empty lattice, $\mathbf{k}$ represents the **momentum of the electron**, in $\hbar$ units.

When $V(\mathbf{r}) \neq 0$ $\hbar\mathbf{k}$ is no longer the momentum of the electron (it is not the eigenvalue of $-i\hbar\boldsymbol{\nabla}$). In fact, by referring for simplicity to the x direction, one sees that

$$-i\hbar\frac{\partial}{\partial x}u_{k_x}(x)e^{ik_x x} \neq \hbar k_x u_{k_x}(x)e^{ik_x x}.$$

The expectation value of the momentum is given by

$$-i\hbar\int u_{k_x}^* e^{-ik_x x}\frac{\partial}{\partial x}u_{k_x}e^{ik_x x}dx = \hbar k_x + (-i\hbar)\int u_{k_x}^*\frac{\partial}{\partial x}u_{k_x}\;, \tag{12.11}$$

where the second term can be considered as an "average momentum" transferred to the lattice. Nevertheless, even for $V(\mathbf{r}) \neq 0$, $\mathbf{k}$ continues to be a constant of motion and labels the state. Furthermore $\mathbf{k}$ plays the role of an **electron momentum in regards of external forces**.

---

[2] The translational invariance can be broken by defects, free surfaces or by the vibrational motions of the ions. In this respect, it should be observed that, at variance with the states in molecules, here the $\mathbf{k}$-electron states are very close in energy and the vibrational motions of the ions may cause variation of the electron state. These processes contribute to the electrical **resistivity** (§13.4 for qualitative remarks).

A semiclassical way to prove this role of $\mathbf{k}$ is to consider the elemental work δL made by an external force $\mathbf{F}_e$ (e.g. the one due to an external electric field). Since

$$\delta L = \mathbf{F}_e.\mathbf{v}_g\delta t$$

with the group velocity $\mathbf{v}_g = (1/\hbar)\partial E_\mathbf{k}/\partial \mathbf{k}$, one has

$$\delta L = \mathbf{F}_e.\frac{1}{\hbar}\frac{\partial E_\mathbf{k}}{\partial \mathbf{k}}\delta t \ .$$

By equating the elemental work δL to $\delta E = (\partial E_\mathbf{k}/\partial \mathbf{k}).\delta \mathbf{k}$, one derives

$$\frac{\delta \mathbf{k}}{\delta t}\hbar = \hbar\dot{\mathbf{k}} = \mathbf{F}_e. \tag{12.12}$$

One sees that $\hbar\mathbf{k}$ behaves as a momentum and thus it can be defined as **pseudo-momentum** or **crystal momentum**.

Up to now $\mathbf{k}$ is a continuous vector in the reciprocal space. As already seen in atoms and in molecules, the boundary conditions determine discrete values for $\mathbf{k}$. In this respect one possibility would be to fix the nodes of the wavefunctions at the surface of the crystal. The same quantum conditions found for a particle in a box would be obtained. However, this procedure would imply the transformation of the wavefunctions from running waves to stationary waves and surface effects would arise. It is often more convenient to impose **periodic boundary conditions (Born-Von Karman** procedure), as we shall see in the next Section.

Problems XII.3

Problem XII.3.1 For k-dependence of the electron eigenvalues given by

$$E(k) = Ak^2 - Bk^4$$

derive the eigenvalue $E(k^*)$ for which phase and group velocities of the electrons are the same. Give the proper order of magnitude and units for the coefficients A and B.

Solution:
From $v_{ph} = \omega/k = (Ak - Bk^3)/\hbar$ and $v_g = \partial\omega/\partial k = (2Ak - 4Bk^3)/\hbar$, one has $2A - 4Bk^{*2} = A - Bk^{*2}$, yielding

$$k^* = \left(\frac{A}{3B}\right)^{\frac{1}{2}}$$

and $E^* = k^{*2}(A - Bk^{*2}) = 2A^2/9B$. The orders of magnitude of A and B are $A \sim$ eV Å^2 and $B \sim$ eV Å^4.

Problem XII.3.2 Discuss the trajectory of an electron under the Lorentz force due to an external magnetic field along the z-direction, for energy eigenvalues $E_k = \alpha k_x^2 + \beta k_y^2$.

Solution:

According to the extension of Eq. 12.12, from

$$\hbar \frac{d\mathbf{k}}{dt} = \frac{-e\mathbf{v}_g}{c} \times \mathbf{H} \ ,$$

with $\mathbf{v}_g$ the group velocity, one has

$$\frac{d\mathbf{k}}{dt} = -\frac{e}{\hbar^2 c}(\boldsymbol{\nabla}_k E_k \times \mathbf{H}) \ .$$

For magnetic field along the z-direction one finds

$$\frac{d\mathbf{k}}{dt} = \frac{2eH}{\hbar^2 c}(\alpha k_x \mathbf{j} - \beta k_y \mathbf{i})$$

or

$$\dot{k}_x = -\frac{2eH}{\hbar^2 c}\beta k_y \qquad \dot{k}_y = -\frac{2eH}{\hbar^2 c}\alpha k_x$$

yielding

$$k_x = k_{x0}\cos(\omega t + \phi), \quad k_y = k_{y0}\sin(\omega t + \phi) \ ,$$

where

$$\omega = \frac{2eH}{\hbar^2 c}(\alpha\beta)^{1/2}$$

The trajectory in the k plane is an ellipse, as well as the one in the real space. The motion induced by the magnetic field is called **cyclotron motion** (see App.XIII.1 for details).

Problem XII.3.3 In a cubic crystal the **k**-dependence of the electron eigenvalues is

$$E(\mathbf{k}) = C - 2V_1[\cos k_x a + \cos k_y a + \cos k_z a]$$

(a form that can be obtained in the framework of the tight-binding model, see §12.7.3). Derive the acceleration of an electron due to an electric field.

Solution:

From the time derivative of the group velocity $\mathbf{v}_g$, by considering that $\dot{\mathbf{k}} = \mathbf{F}_e/\hbar$ the tensor describing the relationship between the electric field $\mathcal{E}$ and the acceleration $\dot{\mathbf{v}}_g$ turns out

$$\begin{pmatrix} A\cos k_x a & 0 & 0 \\ 0 & A\cos k_y a & 0 \\ 0 & 0 & A\cos k_z a \end{pmatrix},$$

with $\quad A = \frac{2V_1 a^2}{\hbar^2}$. Then, for $\mathcal{E} = \mathcal{E}_x \mathbf{i} + \mathcal{E}_y \mathbf{j} + \mathcal{E}_z \mathbf{k}$ the acceleration is

$$\dot{\mathbf{v}}_g = \left(\frac{-2V_1 a^2 e\mathcal{E}_x}{\hbar^2} \cos k_x a \right) \mathbf{i} + \left(\frac{-2V_1 a^2 e\mathcal{E}_y}{\hbar^2} \cos k_y a \right) \mathbf{j} + \left(\frac{-2V_1 a^2 e\mathcal{E}_z}{\hbar^2} \cos k_z a \right) \mathbf{k}.$$

Since no off-diagonal elements of the tensor are present, the acceleration is along the same direction of the field. The ratio between the external force and the acceleration leads to the concept of effective mass (see §12.6).

12.4 Periodic boundary conditions and reduction to the first Brillouin zone

Let us refer to a region of macroscopic size in an ideal crystal containing N cells, N_1 along the $\mathbf{a}$ direction, N_2 along $\mathbf{b}$ and N_3 along $\mathbf{c}$. The reference volume is $N v_c$, with $v_c = (\mathbf{a} \times \mathbf{b}).\mathbf{c}$. The electron wavefunctions $\phi_{\mathbf{k}}$ have to be identical in equivalent points of that region and of a replica region. By assuming for simplicity that the crystal axes are perpendicular and considering the vector $\mathbf{L} = N_1 \mathbf{a} + N_2 \mathbf{b} + N_3 \mathbf{c}$, according to Eq. 12.7 one has to write

$$e^{i\mathbf{k}.\mathbf{r}} = e^{i\mathbf{k}.(\mathbf{r}+\mathbf{L})} , \tag{12.13}$$

the equality of $u_{\mathbf{k}}(\mathbf{r})$ being obviously granted. Then

$$k_x = n_1 \frac{2\pi}{aN_1} \quad , \quad k_y = n_2 \frac{2\pi}{bN_2} \quad , \quad k_z = n_3 \frac{2\pi}{cN_3} \tag{12.14}$$

with n_i integers. By referring to the reciprocal lattice vectors (§11.3) $\mathbf{a}^*, \mathbf{b}^*, \mathbf{c}^*$, thus extending the above arguments to non-perpendicular crystal axes, the periodic boundary conditions yield

$$\mathbf{k} = n_1 \frac{\mathbf{a}^*}{N_1} + n_2 \frac{\mathbf{b}^*}{N_2} + n_3 \frac{\mathbf{c}^*}{N_3} . \tag{12.15}$$

It should be noticed that $\mathbf{k}$ can be outside of the Brillouin cell. However, as we shall see in the following, an electron state $\mathbf{k}$ outside the Brillouin cell (or first Brillouin cell (BZ)) is equivalent to a state within the cell. Therefore, one can classify the states by means of the set of discrete N vectors $\mathbf{k}$ given by Eq. 12.15, with n_i such that $\mathbf{k}$ lies within the **Brillouin zone**. This statement can be understood with the aid of the planar reciprocal lattice given below:

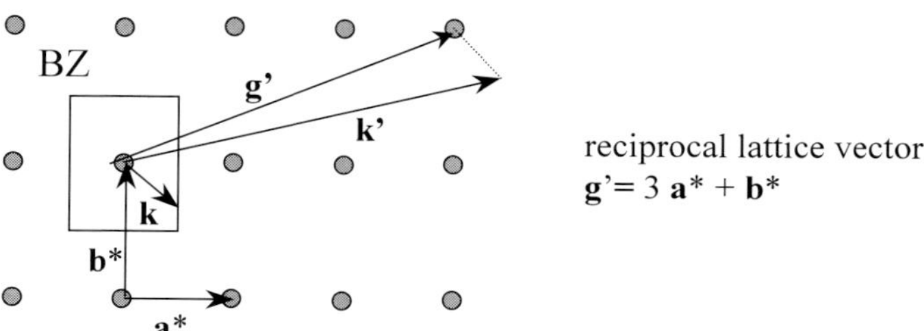

$\mathbf{g}'$ is a reciprocal lattice vector that brings from $\mathbf{k}'$ outside the BZ to a point inside it. Thus $\mathbf{k} = \mathbf{k}' - \mathbf{g}'$ and the wavefunction $\phi_{\mathbf{k}'}$ can be written

$$\phi_{\mathbf{k}'} = u_{\mathbf{k}'}(\mathbf{r})e^{i\mathbf{k}'\cdot\mathbf{r}} = e^{i\mathbf{k}\cdot\mathbf{r}}e^{i\mathbf{g}'\cdot\mathbf{r}}u_{\mathbf{k}'}(\mathbf{r}) \ . \tag{12.16}$$

One has to observe that $e^{i\mathbf{g}'\cdot\mathbf{r}}u_{\mathbf{k}'}(\mathbf{r})$ has the lattice periodicity since, according to Eq. 11.6, $e^{i\mathbf{g}'\cdot\mathbf{l}} = 1$. Hence $e^{i\mathbf{g}'\cdot\mathbf{r}}u_{\mathbf{k}'}(\mathbf{r}) = u_{\mathbf{k}}(\mathbf{r})$ is the function which makes $\phi_{\mathbf{k}}$ a Bloch orbital. Then

$$\phi_{\mathbf{k}'} = \phi_{\mathbf{k}} \quad \text{and} \quad E_{\mathbf{k}'} = E_{\mathbf{k}} \tag{12.17}$$

and the electron states **can be classified by means of N vectors k inside the BZ**. The states $\mathbf{k}'$ outside this zone merely correspond to equivalent states, in a representation called **extended zone representation**. This representation has to be compared to the **reduced zone representation** where all states are reported inside the BZ. The detail of the electron states in the framework of specific models (see §12.7) will better clarify this aspect.

For a one-dimensional (1D) crystal one has the illustrative plots reported below, for a band of the form as in Eq. 12.6.

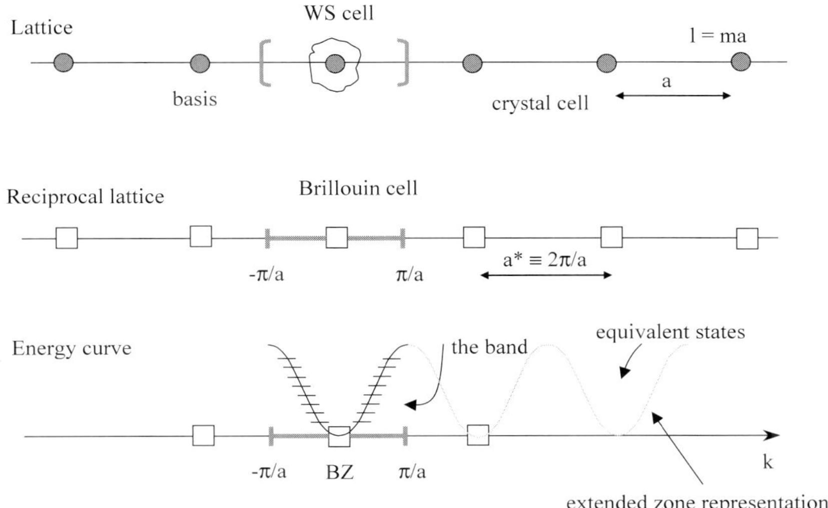

12.5 Density of states, dispersion relations and critical points

As discussed in the previous Section, the electron states can be described by referring to the reciprocal space and in particular to the first Brillouin zone. The state of the whole crystal can be thought to result from the assignment of two electrons, with opposite spins, to each state **k**, in a way similar to the *aufbau* principle used in atoms and in molecules. For the moment we shall refer for simplicity to the condition of zero temperature, so that one can disregard the thermal excitations to higher energy states.

One can sketch the situation as below,

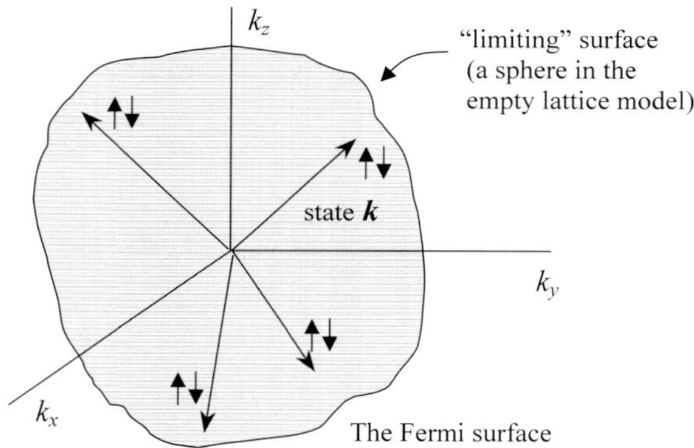

with a limit surface in the reciprocal space including all electron states. This surface, corresponding to a sphere in the empty lattice model (Eq. 12.10), is called **Fermi surface**.

The following points should be remarked:

i) if one increases the reference volume Nv_c, by increasing the number of crystal cells, the total number of **k** states increases;

ii) if the crystal cell is expanded (v_c increases) the BZ volume decreases;

iii) for monoatomic crystals, with the basis formed by a single atom with one valence electron, the BZ is half filled by occupied states;

iv) again for monoatomic crystal, when each atom contributes with two valence electrons, the BZ is fully occupied (the surface of the Brillouin cell not necessarily coincides with the Fermi surface).

The density of **k** states $D(\mathbf{k})$ can be derived once it is noticed that within the BZ there are N states, equally spaced in the reciprocal volume. Then, the BZ volume being $v_c^* = 8\pi^3/v_c$, one has

$$D(\mathbf{k}) = \frac{N}{v_c^*} = \frac{Nv_c}{8\pi^3}. \tag{12.18}$$

The reference volume is often assumed 1 cm^3. Since for such a volume the number of states within the BZ is around 10^{22}, although **k** in principle is a discrete variable, in practice it is often convenient to treat it as a continuous variable, so that

$$\sum_{\mathbf{k}'} \rightarrow \int D(\mathbf{k})d\mathbf{k} \equiv \frac{Nv_c}{8\pi^3} \int d\mathbf{k} \tag{12.19}$$

The sequence of energy levels $E(\mathbf{k})$ is the **band**, that we have already introduced qualitatively in §12.1. In analogy to wave optics, the **k**-dependence of the eigenvalues is called **dispersion relation**.

An important quantity characterizing the structure of the energy levels is the density of energy states $D(E)$ (**density of states**), namely the number of electronic states within a unitary interval of energy around $E = E(\mathbf{k})$. $D(E)$ is related both to $D(\mathbf{k})$ and to the dispersion relation. A general expression for $D(E)$ can be obtained by estimating the number of states lying between the two surfaces, in the reciprocal space, having constant energy given by E and $E + dE$, respectively (see sketch below).

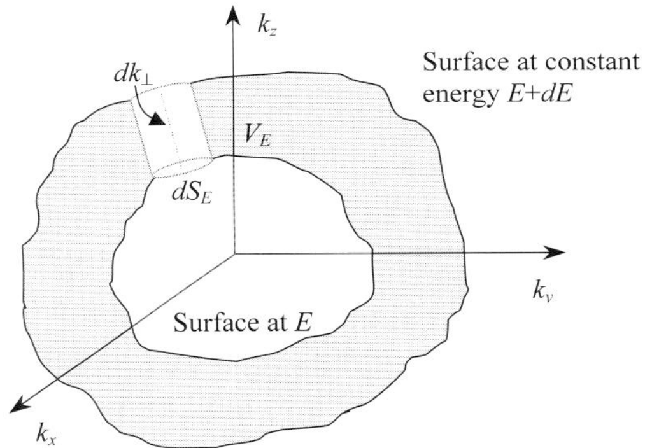

The number of states in the volume V_E is

$$D(E)dE = \frac{N v_c}{8\pi^3}.2.V_E \ ,$$

the factor 2 accounting for the spin degeneracy. For the volume V_E in between the two surfaces one has

$$V_E = \int_S dS_E dk_\perp = \int_S dS_E \frac{dE}{|\nabla_\mathbf{k} E(\mathbf{k})|}$$

since $dk_\perp = dE/|\partial E/\partial \mathbf{k}|$. Therefore

$$D(E) = \frac{N v_c}{4\pi^3} \int_S dS_E \frac{1}{|\nabla_\mathbf{k} E(\mathbf{k})|}. \tag{12.20}$$

From the above expression it is evident that $D(E)$ has singularities (**Van Hove singularities**) whenever the gradient of $E(\mathbf{k})$ in the reciprocal space vanishes. The points, in the reciprocal space, where this condition is fulfilled are called **critical points**. These critical points are particularly relevant for the optical and transport properties since they imply a marked denseness of states. As it will be shown in the next Section, electrons around a critical point behave as if they had particular **effective masses**.

12.6 The effective electron mass

As shown in Section 12.3 the **k**-dependence of the energy controls the behavior of the electron under external forces. In fact $\hbar\dot{\mathbf{k}} = \mathbf{F}_e$, while the group velocity is $\mathbf{v}_g = (1/\hbar)(\partial E(\mathbf{k})/\partial \mathbf{k})$.

By differentiating $\mathbf{v}_g$ one has

$$\mathbf{a} = \frac{d\mathbf{v}_g}{dt} = \frac{1}{\hbar}\frac{\partial^2 E(\mathbf{k})}{\partial \mathbf{k}^2}\frac{\partial \mathbf{k}}{\partial t} = \frac{1}{\hbar^2}\frac{\partial^2 E(\mathbf{k})}{\partial \mathbf{k}^2}\mathbf{F}_e \qquad (12.21)$$

On the basis of the classical analogy, the relationship between the force and the acceleration points out that the electron reacts to the external force as if it had a mass

$$\widetilde{m^*} = \hbar^2\left(\frac{\partial^2 E(\mathbf{k})}{\partial \mathbf{k}^2}\right)^{-1}. \qquad (12.22)$$

In the empty lattice limit, or free electron model, the **effective mass** coincides with the real electron mass: $m^* = \hbar^2/[\partial^2(\hbar^2 k^2/2m)/\partial k^2] \equiv m$.

In order to illustrate the concept of effective mass let us refer to the dispersion curve derived in Section 12.1 by applying to a linear chain the idea of resonance among adjacent atoms: $E(k) = 2A\cos(ka)$, with k along x axis and $A < 0$. In this case, from Eq. 12.22, the effective mass turns out

$$m^* = -\frac{\hbar^2}{2Aa^2}\frac{1}{\cos(ka)} \qquad (12.23)$$

(Fig. 12.3).

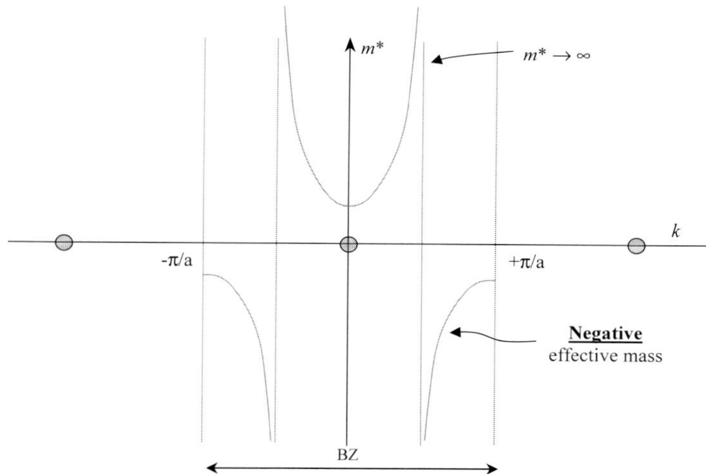

Fig. 12.3. Behavior of the effective mass m^* as a function of k for a 1D model crystal, in correspondence to the dispersion relation $E(k) = 2A\cos(ka)$, with $A < 0$.

Finally it should be remarked that in general the effective mass $\widetilde{m}^*$ is a tensor. Its components are

$$m^*_{\alpha\beta} = \hbar^2 \left(\frac{\partial^2 E(\mathbf{k})}{\partial k_\alpha \partial k_\beta}\right)^{-1}. \tag{12.24}$$

(see Problem XII.3.3).

Problems XII.6

Problem XII.6.1 In a one-dimensional crystal the dispersion relation is

$$E(k) = E_1 + (E_2 - E_1) \sin^2\left(\frac{ka}{2}\right),$$

with lattice step $a = 1$ Å. By assuming that a single electron is present and by neglecting any scattering process (with defects, boundaries or impurities) derive the effective mass, the velocity and the motion of the electron in the real space, under the action of a constant electric field $\mathcal{E}$. For $\mathcal{E} = 100 \,\mathrm{V/m}$ and $(E_2 - E_1) = 1$ eV, obtain the period and the amplitude of the oscillatory motion.

Solution:

The group velocity is $v_g = [a(E_2 - E_1)/2\hbar]\sin(ak_x)$. The effective mass is $m^* = \hbar^2[d^2 E/dk_x^2]^{-1} = m_0 \sec(ak_x)$, where $m_0 = [2\hbar^2/a^2(E_2 - E_1)]$ is the mass at the bottom of the band. m^* becomes infinite for $k_x = \pm\pi/2a$ (see plots).

For a single non-scattered electron in a time-independent electric field $\mathcal{E}_x$ the force implies $dk_x/dt = (-e\mathcal{E}_x/\hbar)$. Then k_x scans repetitively through the Brillouin zone, with period $t^* = (2\pi\hbar/ae\mathcal{E}_x)$.

In the assumption that at $t = 0$ $E = E_1$, $m^* = m_o$, $k_x = 0$, the electron has finite positive mass for some time, becoming infinite at $t = t^*/4$.

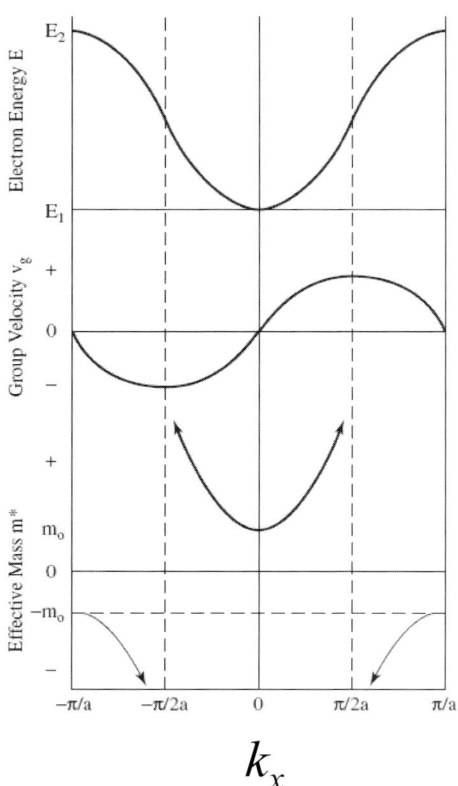

At the time $t = t^*/2$ the electron arrives at $k_x = -(\pi/a)$. The equivalence of this state with the one at $k_x = +\pi/a$ corresponds to the return into the BZ from this point (this process corresponds to the Bragg reflection of the De Broglie wave, see also §12.7.2). Then k_x decreases again and the mass divergence is reached at $t = (3/4)t^*$.

From the velocity

$$v_g = [a(E_2 - E_1)/2\hbar] \sin(-2\pi t/t^*) = [a(E_2 - E_1)/2\hbar] \sin(-ae\mathcal{E}_x t/\hbar)$$

it is found that in the real space an oscillatory motion occurs:

$$x(t) = \int v_g dt = [(E_2 - E_1)/2e\mathcal{E}_x] \cos(-ae\mathcal{E}_x t/\hbar)$$

For $a = 1\text{Å}$ and $\mathcal{E}_x = 10^2$ V/m, $t^* \simeq 4 \times 10^{-7}$ s and the distance covered is about 1 cm.

For the case of a sinusoidally modulated electric field, see the problem 3.30 in the book by **Blakemore** quoted in the preface.

12.7 Models of crystals

Now we are going to apply the general formulation given in previous Sections to particular models of crystals. This should allow one to achieve a better understanding of the physical concepts. Meantime the models to be referred to, to a good approximation correspond to particular groups of solids.

12.7.1 Electrons in empty lattice

The condition of potential energy $V(\mathbf{r})$ going to zero has already been occasionally addressed. Now we shall explore in more detail this ideal situation and derive some finite-temperature properties which reflect the thermal excitations and the statistical effects.

When $V(\mathbf{r}) \to 0$ the electrons delocalize in the reference volume Nv_c and are described by Bloch orbitals (Eq. 12.7) with constant $u_{\mathbf{k}}(\mathbf{r})$. According to the one-electron Schrödinger equation one has

$$\phi_{\mathbf{k}} = \frac{1}{\sqrt{Nv_c}} e^{i\mathbf{k}.\mathbf{r}} \tag{12.25}$$

and

$$E(\mathbf{k}) = \frac{\hbar^2 k^2}{2m}. \tag{12.26}$$

The valence electrons can be thought to move freely in the reference volume and they become responsible for the electric conduction. This model is suited to describe the **metals**.

The theory of metals in the framework of the **free electron model** was actually developed before the advent of quantum mechanics. Significant successes were achieved, as the derivation of **Ohm law** and of the relationship between thermal and electrical conductivity (**Wiedemann-Franz law**). At variance, the behaviour of other quantities, such as the heat capacity and the magnetic susceptibility, requiring in the derivation the use of Fermi-Dirac distribution, could hardly be explained in the early theories. On the other hand, in spite of the successful quantum mechanical description, the limits of the free electron model become obvious when one recalls the huge change in the electrical conductivity from metals to insulators or the existence of semiconductors. In these compounds the role played by a non-zero lattice potential is crucial (see next Section).

The dispersion curve for electrons in empty lattice (Eq. 12.26) is reported in Fig. 12.4 in the **extended, reduced** and **repeated zone** representations, along a reciprocal space axis.

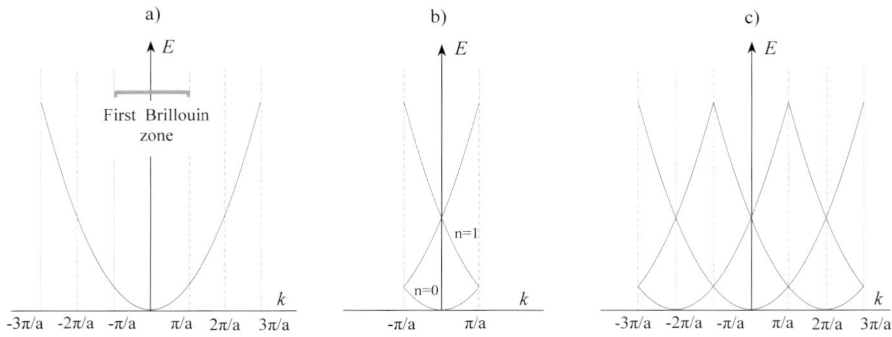

Fig. 12.4. Dispersion curves for the empty lattice model, in a crystal of lattice step a, within: a) the extended zone scheme, b) the reduced zone scheme, c) the repeated zone scheme. The indexes (in b)) indicate the number of reciprocal lattice vectors **a*** required for the reduction to the first BZ.

The constant energy surfaces in **k** space are spherical, as sketched below

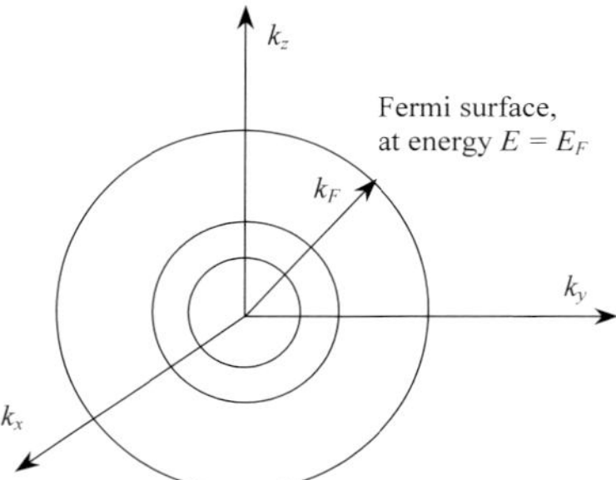

At $T = 0$ the electrons fill all the states up to a given wavevector of modulus k_F, called the **Fermi wavevector**, which corresponds to the radius of the **Fermi surface** (see §12.6). In a crystal with N cells, each containing Z electrons, k_F can be directly derived by considering the volume of the Fermi sphere, the density of states $D(\mathbf{k})$ (Eq. 12.18) and the spin variable for each **k** state:

$$ZN = \frac{Nv_c}{8\pi^3} \cdot 2 \cdot \frac{4\pi}{3} k_F^3 \tag{12.27}$$

The Fermi energy $E_F = \hbar^2 k_F^2 / 2m$ turns out

$$E_F = \frac{\hbar^2}{2m}(3\pi^2 \frac{Z}{v_c})^{2/3} \propto n_d^{2/3} , \tag{12.28}$$

where $n_d = Z/v_c$ is the number density of electrons (per cubic centimeter).

The Fermi wavevector k_F is of the order of 10^8 cm^{-1}, the correspondent velocity is of the order of 10^8 cm/s, while the Fermi energy is of the order of $1-10$ eV.

The density of states can be derived starting from Eq. 12.20:

$$D(E) = \frac{N v_c}{4\pi^3} \int_S dS_E \frac{1}{|\boldsymbol{\nabla}_\mathbf{k} E(\mathbf{k})|} = \frac{N v_c}{2\pi^2}\left(\frac{2m}{\hbar^2}\right)^{3/2} E^{1/2} = \frac{3NZ}{2} \frac{E^{1/2}}{E_F^{3/2}}.$$

The density of states per unit cell is reported below:

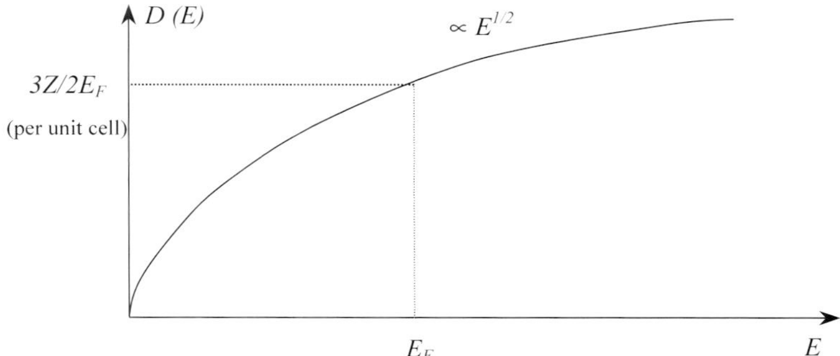

$D(E)$ is often defined per unit volume or per atom.

Now we briefly discuss the situation occurring at finite temperature, when the statistical excitation of the electrons above the Fermi level has to be taken into account. The probability of occupation of the level at energy E is given by the Fermi function

$$f(E) = \frac{1}{e^{\frac{E-\mu}{k_B T}} + 1}, \tag{12.29}$$

where the chemical potential μ can be considered to coincide with the Fermi energy E_F, for temperatures much lower than the Fermi temperature $T_F = E_F/k_B$, (of the order of 10^4 K).

Then the distribution function and the density of occupied states take the forms plotted below:

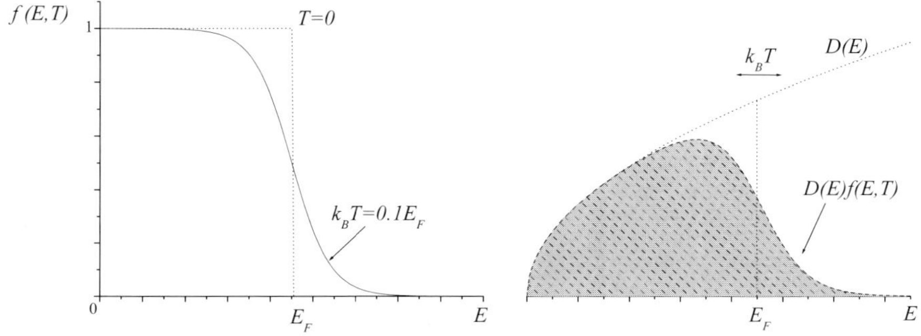

The average energy is

$$< E >= \int ED(E)f(E)dE,$$

and for $T \rightarrow 0$

$$< E >= \int_0^{E_F} ED(E)dE = \frac{3}{5}NZE_F \ , \tag{12.30}$$

while at finite temperatures (Prob. F.XII.1) it turns out

$$< E >=\simeq \frac{3}{5}NZE_F + \frac{\pi^2}{4}NZk_BT\frac{T}{T_F} \ . \tag{12.31}$$

It is noted that the contribution at $T \neq 0$ takes a form of the classical energy $3k_BT/2$ (per electron) times the "fraction" $\sim T/T_F$ of electrons in the neighborhood of the Fermi level.

The specific heat C_V and the magnetic susceptibility χ_P can be derived as illustrated in the Problems F.XII.1 and XII.7.6.

A simple way to estimate the order of magnitude of C_V and χ_P is to consider that only a fraction T/T_F of all the electrons can be thermally or magnetically excited. In fact, the states at $E \ll E_F$ are all occupied and the Pauli principle prevents double occupancies. Then, from the classical expressions for Boltzmann statistics one can approximately write

$$C_V \simeq \frac{\partial}{\partial T}(\frac{3}{2}n_dk_BT)\frac{T}{T_F} = \gamma T \tag{12.32}$$

and

$$\chi_P \simeq \frac{n_d\mu_B^2}{3k_BT}\frac{T}{T_F} = \frac{n_d\mu_B^2}{3k_BT_F} \ . \tag{12.33}$$

12.7.2 Weakly bound electrons

As already mentioned the free electron model cannot account for the properties of crystals different from metals, as for instance the semiconductors, not even at a qualitative level. In order to explain the basic aspects of those solids one has to take into account, at least in the perturbative limit, the effects of the lattice potential in the so called **nearly free electron approximation**. Even a weak perturbation causes relevant modifications with respect to the empty lattice situation and yields the appearance of the **gap**, namely the energy interval where no electron states can exist. In particular, the gap arises for the electrons at De Broglie wavelength (of the order of the inverse of $|\mathbf{k}|$) close to the lattice step, in analogy with the diffraction phenomenon in optics.

The simplest way to account for the effect of the lattice potential $V(\mathbf{r})$ in modifying the electron dispersion curve $E(\mathbf{k})$ is to consider the perturbative correction to empty-lattice states $E^o(\mathbf{k})$:

$$E(\mathbf{k}) = E^o(\mathbf{k}) + <\mathbf{k}|V(\mathbf{r})|\mathbf{k}> + \sum_{\mathbf{k}' \neq \mathbf{k}} \frac{|<\mathbf{k}|V(\mathbf{r})|\mathbf{k}'>|^2}{E^o(\mathbf{k}) - E^o(\mathbf{k}')} \; , \tag{12.34}$$

where

$$|<\mathbf{k}|V(\mathbf{r})|\mathbf{k}'> \equiv \int e^{-i(\mathbf{k}-\mathbf{k}')\cdot \mathbf{r}} V(\mathbf{r}) d\mathbf{r}. \tag{12.35}$$

For $\mathbf{k} - \mathbf{k}' \neq \mathbf{g}$, with $\mathbf{g}$ reciprocal lattice vectors, the integral vanishes due to the fast oscillations with $\mathbf{r}$ of the function $e^{-i(\mathbf{k}-\mathbf{k}')\cdot\mathbf{r}}$. Whereas for $\mathbf{k} - \mathbf{k}' = \mathbf{g}$ the matrix element reads

$$|<\mathbf{k}|V(\mathbf{r})|\mathbf{k}-\mathbf{g}> \equiv \int e^{-i\mathbf{g}\cdot\mathbf{r}} V(\mathbf{r}) d\mathbf{r} = V_{\mathbf{g}} \; , \tag{12.36}$$

which is non zero since it corresponds to the coefficent $V_{\mathbf{g}}$ of the Fourier expansion of the periodic lattice potential:

$$V(\mathbf{r}) = \sum_{\mathbf{g}} V_{\mathbf{g}} e^{i\mathbf{g}\cdot\mathbf{r}}. \tag{12.37}$$

It can be remarked that for degenerate states, where $E^o(\mathbf{k}) = E^o(\mathbf{k}')$ at the denominator in Eq. 12.34, one should rely on the perturbation theory for degenerate states and still $<\mathbf{k}|V(\mathbf{r})|\mathbf{k}'> = 0$ for $\mathbf{k}' \neq \mathbf{k} + \mathbf{g}$.
Therefore Eq. 12.34 is rewritten

$$E(\mathbf{k}) = E^o(\mathbf{k}) + V^o + \sum_{\mathbf{g} \neq 0} \frac{|V_{\mathbf{g}}|^2}{E^o(\mathbf{k}) - E^o(\mathbf{k}-\mathbf{g})}, \qquad V^o = \int V(\mathbf{r}) d\mathbf{r} \tag{12.38}$$

which modifies the dispersion curve for free electrons at the second order. The validity of Eq. 12.38 requires the convergence of $|V(\mathbf{g})|^2$, which should be granted by the choice of a plausible lattice potential (often a **pseudo-potential**). In addition it requires that

$$E^o(\mathbf{k}) \neq E^o(\mathbf{k} - \mathbf{g})$$

which corresponds to avoid the wavevectors $\mathbf{k}$ at the BZ boundary. In fact, recalling that $E^o(\mathbf{k}) = \hbar^2 k^2/2m$, the condition $E^o(\mathbf{k}) = E^o(\mathbf{k} - \mathbf{g})$ implies $(\mathbf{k})^2 = (\mathbf{k} - \mathbf{g})^2$ and then

$$\mathbf{k}.\mathbf{g} = \frac{g^2}{2}, \tag{12.39}$$

corresponding to $\mathbf{k}$ at the BZ boundary, as depicted below for a 2D lattice.

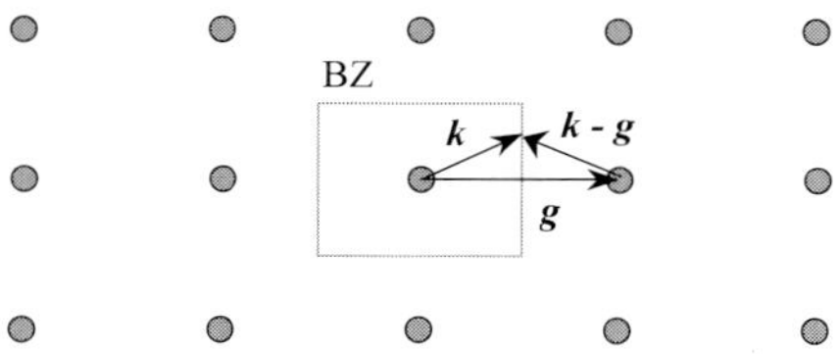

Thus at the BZ boundaries, where the states $\phi_{\mathbf{k}}$ and $\phi_{\mathbf{k}-\mathbf{g}}$ have the same energy, the perturbative approach leading to Eq. 12.38 breaks down.

The situation arising at the zone boundaries can be deduced by mean of arguments essentially based on the perturbation theory for degenerate states. An illustrative example is easily carried out for a **one-dimensional** lattice, with perturbative periodic potential of the form sketched below:

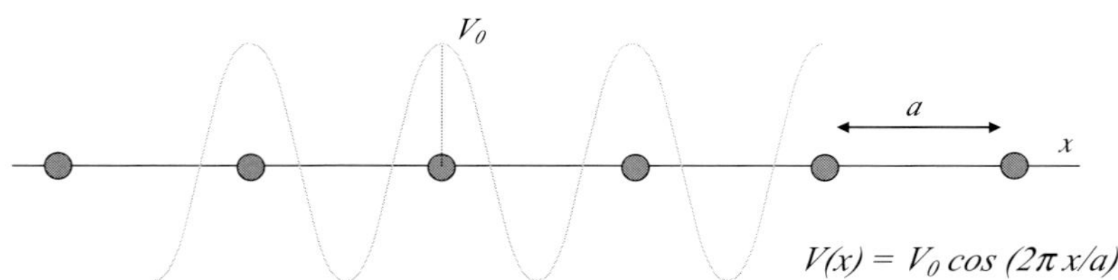

The zero-order wave function is

$$\phi_{\mathbf{k}}^{(1)} = c_1 \phi_{\mathbf{k}} + c_2 \phi_{\mathbf{k}-\mathbf{g}}$$

and the secular equation becomes

$$\begin{pmatrix} <\mathbf{k}|V(\mathbf{r})|\mathbf{k}> -\varepsilon & <\mathbf{k}|V(\mathbf{r})|\mathbf{k}-\mathbf{g}> \\ <\mathbf{k}-\mathbf{g}|V(\mathbf{r})|\mathbf{k}> & <\mathbf{k}-\mathbf{g}|V(\mathbf{r})|\mathbf{k}-\mathbf{g}> -\varepsilon \end{pmatrix} = 0$$

The choice of the potential implies $< \mathbf{k}|V(\mathbf{r})|\mathbf{k} >=< \mathbf{k}-\mathbf{g}|V(\mathbf{r})|\mathbf{k}-\mathbf{g} >= 0$ and

$$< \mathbf{k}|V(\mathbf{r})|\mathbf{k} - \mathbf{g} >= \frac{1}{2a} \int_0^a e^{-igx} V_o(e^{\frac{i2\pi x}{a}} + e^{\frac{-i2\pi x}{a}})dx = \frac{1}{2}V_o. \qquad (12.40)$$

Thus the correction to the unperturbed eigenvalues turns out $\varepsilon_\pm = \pm V_o/2$, implying a **gap** for the states around the BZ boundaries, as schematically shown in Fig. 12.5 (to be compared to Fig. 12.4).

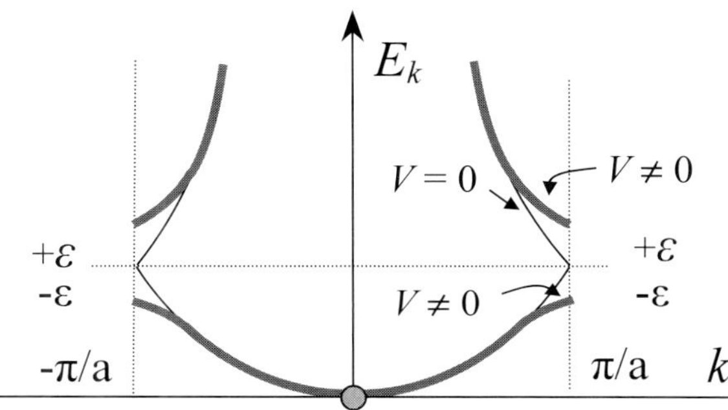

Fig. 12.5. Schematic representation of the dispersion curve for 1D crystal, in the nearly free electron approximation, by taking into account that for **k** far from the BZ boundaries Eq. 12.38 is a good approximation, while approaching the BZ boundaries the correction given by Eq. 12.40 has to be considered.

The gap can be thought to arise from the **Bragg reflection** occurring when the De Broglie wavelength is $\lambda = 2a$. In fact, in this case (see Eq. 11.9) the Bragg reflected wave, travelling in opposite direction, induces standing waves, as sketched below:

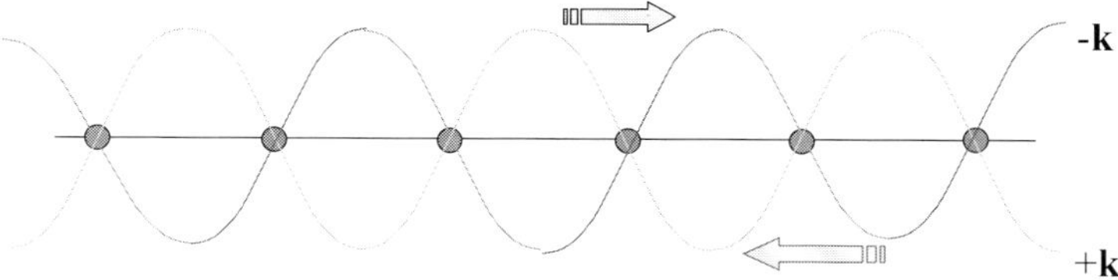

The cosine and sine standing waves formed by the $\pm$ linear combination of $\exp[(\pm ikx)]$, with $k = \pi/a$, yield different distributions of probability density.

Thus the electron charge $\rho(x)$ around the lattice sites imply different energies:

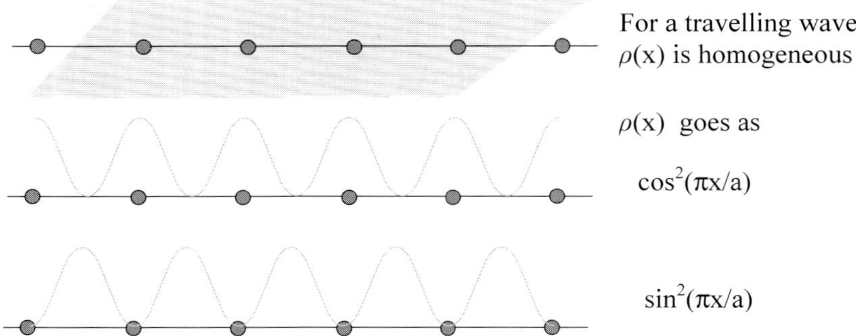

The characteristic feature of the gap generation for weakly perturbed electrons can be derived by constructing the complete **k**-dependence of the eigenvalues in a periodic square-well potential, in one dimension. The potential energy in the Schrödinger equation is assumed $V(x) = 0$ for $0 < x \leq a$ and $V(x) = V_0$ for $a < x \leq a + b$, the lattice parameter being $(a + b)$. **Kronig** and **Penney** solved this artificial model and derived the k-dependent eigenvalues. In the limit where $V(x)$ is characterized by Dirac δ functions separated by distance a (the product $V_0 b$ remaining finite) the dispersion curve in the extended zone scheme has the form sketched below:

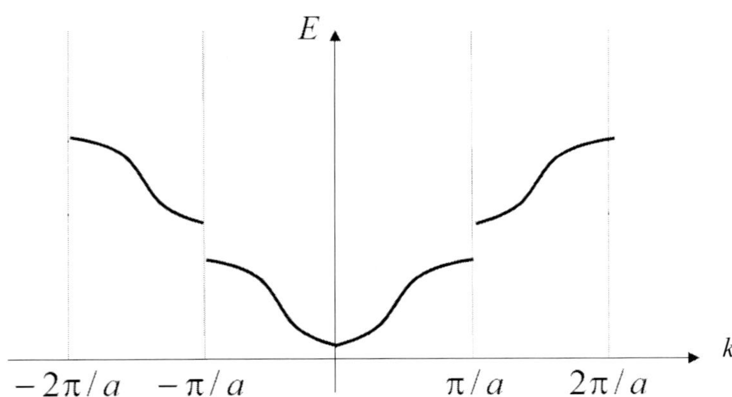

12.7.3 Tightly bound electrons

In this model the electrons are assumed to keep, to a large extent, the properties they have in the neighborhood of the atoms. Only in the region in between the atoms sizeable effects occur and the atomic levels are thus spread in a band. The model allows one to understand how the Bloch orbitals are related to the atomic states, in a way similar to the case discussed for the benzene molecule (§9.3).

Let us refer to the lattice potential reported in Fig. 12.6, along a given direction in the crystal.

By extending the idea of the molecular orbital used for the delocalization of the $2p$ electrons along the C_6H_6 ring, we shall assume a one-electron wavefunction of the form

$$\phi_\mathbf{k} = \sum_\mathbf{l} e^{i\mathbf{k}.\mathbf{l}}\phi_a(\mathbf{r} - \mathbf{l}) \ . \tag{12.41}$$

$\phi_a(\mathbf{r} - \mathbf{l})$ is an atomic wavefunction centered at the l-th site and an eigenfunction of the equation

$$\left\{ -\frac{\hbar^2}{2m}\nabla^2 + V_a(\mathbf{r} - \mathbf{l}) \right\} \phi_a(\mathbf{r} - \mathbf{l}) = E_a\phi_a(\mathbf{r} - \mathbf{l}). \tag{12.42}$$

To show that $\phi_\mathbf{k}$ in Eq. 12.41 is a Bloch orbital, one multiplies by $exp(i\mathbf{k}.\mathbf{r}).exp(-i\mathbf{k}.\mathbf{r})$:

$$\phi_\mathbf{k}(\mathbf{r}) = e^{i\mathbf{k}.\mathbf{r}} \sum_\mathbf{l} e^{-i\mathbf{k}.(\mathbf{r}-\mathbf{l})}\phi_a(\mathbf{r} - \mathbf{l}) \ .$$

Then it can be observed that the term multiplying the plane wave function has the lattice periodicity and plays the role of $u_\mathbf{k}(\mathbf{r})$ in Eq. 12.7, as requested. One also notices that $\phi_\mathbf{k}$ in the form 12.41 is a combination of localized atomic

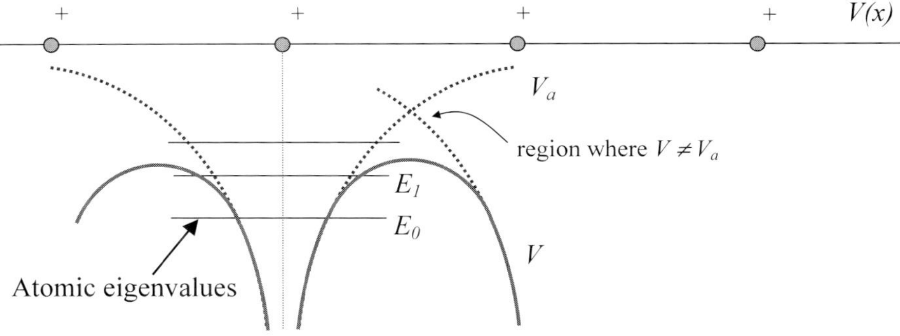

Fig. 12.6. Schematic form of the potential energy for tightly bound electrons, along a given direction in the crystal.

orbitals and in the neighborhood of an atom the orbital behaves in a way similar to the one for insulated atoms. The phase factor $exp(i\mathbf{k}.\mathbf{l})$ modifies the orbital from site to site while $|\phi_{\mathbf{k}}|^2$ is unaffected.

To obtain the eigenvalues $E_{\mathbf{k}}$, the eigenfunction in Eq. 12.41 is inserted in the one-electron Schrödinger equation $(-\hbar^2\nabla^2/2m + V)\phi_{\mathbf{k}} = E_{\mathbf{k}}\phi_{\mathbf{k}}$. By recalling Eq. 12.42 one obtains

$$(E_{\mathbf{k}} - E_a)\sum_{\mathbf{l}} e^{i\mathbf{k}.\mathbf{l}}\phi_a(\mathbf{r} - \mathbf{l}) = \sum_{\mathbf{l}}(V - V_a)e^{i\mathbf{k}.\mathbf{l}}\phi_a(\mathbf{r} - \mathbf{l}). \qquad (12.43)$$

By multiplying both sides of this equation by $\phi_a^*(\mathbf{r} - \mathbf{l})$ and integrating, one has

$$(E_{\mathbf{k}}-E_a)\sum_{\mathbf{l}} e^{i\mathbf{k}.\mathbf{l}}\int \phi_a^*(\mathbf{r}-\mathbf{l}')\phi_a(\mathbf{r}-\mathbf{l})d\mathbf{r} = \sum_{\mathbf{l}}\int \phi_a^*(\mathbf{r}-\mathbf{l}')(V-V_a)e^{i\mathbf{k}.\mathbf{l}}\phi_a(\mathbf{r}-\mathbf{l})d\mathbf{r}.$$

$$(12.44)$$

When the orthogonality condition for $\mathbf{l} \neq \mathbf{l}'$ is assumed

$$\int \phi_a^*(\mathbf{r} - \mathbf{l}')\phi_a(\mathbf{r} - \mathbf{l})d\mathbf{r} = 0 \ , \qquad (12.45)$$

by taking into account that the sum in Eq. 12.44 only depends on the difference $\mathbf{h} = \mathbf{l} - \mathbf{l}'$, one finds

$$E_{\mathbf{k}} = E_a + \sum_{\mathbf{h}} e^{i\mathbf{k}.\mathbf{h}}\int \phi_a^*(\mathbf{r} + \mathbf{h})V_1\phi_a(\mathbf{r})d\mathbf{r}. \qquad (12.46)$$

In the matrix element in this equation, somewhat analogous to the resonance integral (§8.1.2), V_1 is the difference between the local $V(\mathbf{r})$ and the atomic potential energy V_a (see Fig.12.6). The matrix element is **negative**.

For cubic crystal, with atoms of the same species,

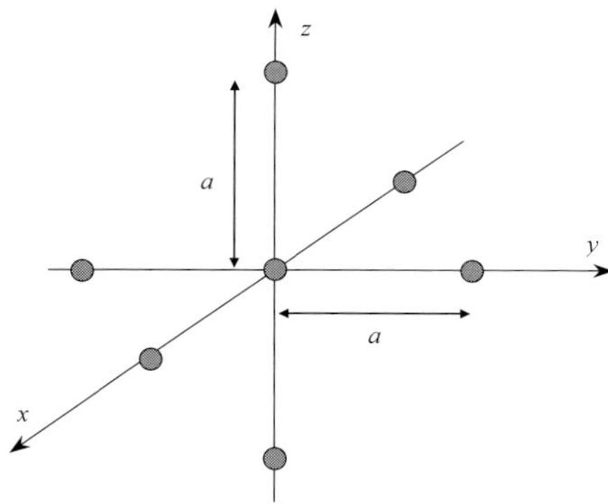

assuming $V_1 \neq 0$ only when nearest neighbors are involved, Eq. 12.46 takes the form

$$E_{\mathbf{k}} = E_a + V_o + 2V_1 \left[cos(k_x a) + cos(k_y a) + cos(k_z a) \right],$$ (12.47)

depicted in Fig.12.7, along the k_x direction.

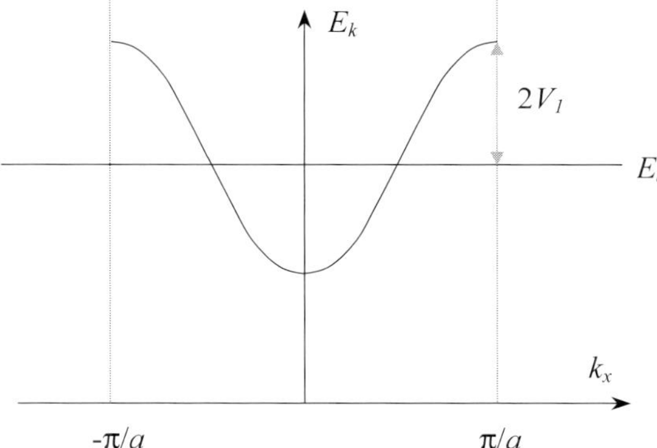

Fig. 12.7. Dispersion relation $E(\mathbf{k})$ for $\mathbf{k}$ along one of the reciprocal lattice axis in a cubic monoatomic crystal, according to Eq. 12.47. V_0 is usually negligible.

The band $E(\mathbf{k})$ results from the spread of the atomic energy level when the interatomic distance in the crystal is reduced. The gap is the direct consequence of the discrete character of the atomic eigenvalues E_a's. One also realizes that the number of states in a single band is $N(2l+1)$, for N atoms in the reference volume of the crystal (l quantum number for the atomic orbital momentum). The band width is proportional to V_1 and, therefore, to the **overlap**, in a way somewhat equivalent to the molecules (see §8.1). This explains why the internal bands are narrow and why the cores states are little affected by the formation of the crystal, as sketched below:

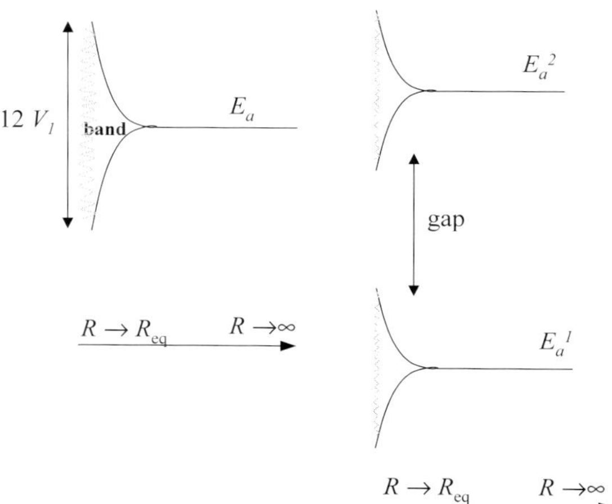

In the framework of the tight binding model the **effective mass** of the electron can be derived from Eq. 12.47. For small $\mathbf{k}$, by expanding E_k, one obtains

$$E_{\mathbf{k}} = E_o + V_o - 6V_1 - V_1 a^2 k^2 \ ,$$

yielding

$$m^* = \frac{-\hbar^2}{2a^2 V_1} > 0,$$

For $k_x, k_y, k_z \to \pi/a$ one has

$$m^* = \frac{\hbar^2}{2a^2 V_1} < 0,$$

and the electron responds to external forces as a positive charge.

When the spread of the atomic levels leads to the superposition of adjacent bands related to different states, one has a **degenerate band** that can be thought to result from hybrid atomic orbitals:

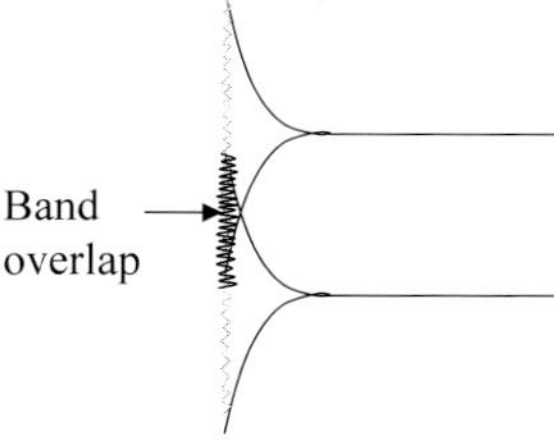

This happens, for instance, in the case of diamond, Si and Ge, as shown in Fig. 12.8.

The energy bands are usually labelled by referring to the atomic orbitals which lead to their formation. Furthermore, since the $\mathbf{k}$-dependence of the energy in the reciprocal space reflects all the symmetry properties of the point group, one could classify the electron states in a crystal on the basis of the symmetry properties.

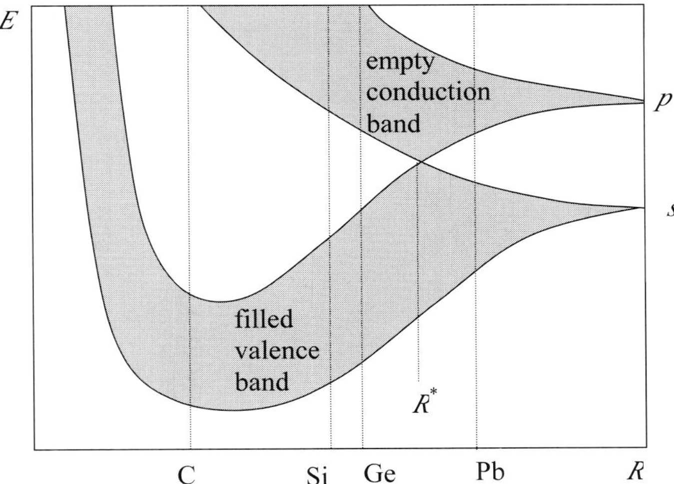

Fig. 12.8. Sketchy picture of the energy bands for $(ns)^2(np)^2$ electronic configurations, showing how below a certain interatomic distance R the p and s bands
overlap, changing the electronic structure of the crystal. While for $R > R^*$ one has
a partially filled p band and the possibility of charge transport (see Chapter 13)
(this is the case of Pb, in regards of the $6p$ and $6s$ electrons), for $R < R^*$ one has
an entirely filled valence band and therefore an insulator.
Note that for those elements there are two atoms for each unitary cell. Thus for
$R < R^*$ when the zones overlap, the lower zone system is exactly filled by eight
electrons per unit cell. When the gap to the upper band (which is empty at $T = 0$)
is comparable to the thermal energy $k_B T$, then the electrons can be promoted to the
upper conducting band. In this case one can have a **semiconductor**, as it happens
for Si and Ge, in terms of the $n = 3$ and $n = 4$ electrons (see §13.1).

Problems XII.7

Problem XII.7.1 The Fourier components of a perturbative one-dimensional
potential energy are V_G. Evaluate the behavior of the effective mass m^* for
$k = 0$ in terms of the lattice step a.

Reformulate the evaluation for $V(x) = 2V_1 \cos(2\pi x/a)$.

Solution:

From (see Eq. 12.38)

$$E_k = E_k^0 + V_0 + \sum_{G \neq 0} \frac{|V_G|^2}{(E_k^0 - E_{k-G}^0)},$$

with $E_k^0 = \hbar^2 k^2 / 2m$, one has

$$E_k = \frac{\hbar^2 k^2}{2m} + V_0 - \frac{2m}{\hbar^2} \sum_{G \neq 0} \frac{|V_G|^2}{G(G - 2k)} .$$

The effective mass is given by

$$\frac{1}{m^*} = \frac{1}{\hbar^2} \frac{\partial^2 E}{\partial k^2} = \frac{1}{m} - \frac{8m}{\hbar^4} \sum_{G \neq 0} \frac{|V_G|^2}{G(G - 2k)^3} .$$

For $k = 0$ one finds

$$\frac{1}{m^*} = \frac{1}{m} - \frac{ma^4}{\hbar^4 \pi^4} \sum_{n=1}^{\infty} \frac{|V_{G_n}|^2}{n^4} .$$

For $V(x) = 2V_1 \cos(2\pi x/a)$ only V_G for $n = 1$ is non-zero and then

$$m^* = m \left(1 - \frac{m^2 a^4 V_1^2}{\hbar^4 \pi^4} \right)^{-1} .$$

Problem XII.7.2 For a 1D crystal of lattice step a, generalize the result obtained at §12.7.2 at the zone boundary in order to obtain the energy $E(k)$ when k is close to π/a.

Solution:

Near $k = \pi/a$ the wavefunction is written as the linear combination of the two degenerate unperturbed eigenfunctions at the two boundaries of the BZ, as it was done in order to derive Eq. 12.40:

$$\psi = c_1 e^{ikx} + c_2 e^{i(k - 2\pi/a)x} .$$

By substituting that tentative wavefunction into the Schrödinger equation

$$-\frac{\hbar^2}{2m} \frac{d^2 \psi}{dx^2} + V\psi = \varepsilon \psi ,$$

first multiply by e^{-ikx} and integrate over all space. Then multiply by $e^{-i(k - 2\pi/a)x}$ and integrate over all space. From the secular equation for c_1 and c_2 (see the equivalent at §12.7.2), the eigenvalues turns out

$$E = \frac{\hbar^2 k^2}{2m} + \frac{\hbar^2 \pi}{ma} \left\{ \left(\frac{\pi}{a} - k \right) \pm \left[\left(\frac{\pi}{a} - k \right)^2 + \left(\frac{amV_0}{2\pi\hbar^2} \right)^2 \right]^{1/2} \right\} .$$

The energy E reduces to the free electron eigenvalues for k well away from the zone boundary, while for $k = \pi/a$ implies the gap V_0, as in §12.7.2.

Problem XII.7.3 Consider a metal with one electron per unit cell in geometric dimension $n = 1, 2$ and 3 and derive the density of states $D(E)$ as a function of n. Then give a general expression for $D(E)$ in terms of the Fermi energy.

Finally derive the chemical potential μ (Hint: write the total number of electrons in terms of the Fermi distribution and use the identity $\int_{-\infty}^{+\infty} f(t)\frac{e^t dt}{(1+e^t)^2} = f(0) + \frac{\pi^2}{6} f''(0)$).

Solution:

From $E_F = \hbar^2 k_F^2/2m$, since

$$N = 2D(\mathbf{k})\frac{4\pi k_F^3}{3} \qquad \text{for } n = 3$$

$$N = 2D(\mathbf{k})\pi k_F^2 \qquad \text{for } n = 2$$

$$N = 2D(\mathbf{k})2\pi k_F \qquad \text{for } n = 1$$

one finds $E_F = (\hbar^2/2m)(3\pi^2 N/v_c)^{2/3}$ for $n = 3$, $E_F = \pi\hbar^2 N/ma_c$ for $n = 2$ and $E_F = (\hbar^2/2m)(N/2l_c)^2$ for $n = 1$.

One can write

$$D(\mathbf{k})d^n k = 2(\frac{\sqrt{2m}}{2\pi\hbar})^n d^n x$$

with $x = \sqrt{E}$. From $D(E)dE = \int D(\mathbf{k})d^n k$ one finds

$$D(E) = \frac{3N}{2}\frac{E^{1/2}}{E_F^{3/2}} \qquad \text{for } n = 3$$

$$D(E) = \frac{m}{\pi\hbar^2} \qquad \text{for } n = 2$$

$$D(E) = \frac{N}{2}\frac{E^{-1/2}}{E_F^{1/2}} \qquad \text{for } n = 1 \ .$$

In general $D(E)dE = Nd(E/E_F)^{n/2}$.

The total number of electrons is

$$N = \int_0^\infty \frac{D(E)}{e^{\beta(E-\mu)} + 1}dE = N \int_0^\infty \frac{1}{e^{\beta(E-\mu)} + 1}d(\frac{E}{E_F})^{n/2} =$$

$$= N \int_{-\beta\mu}^\infty \frac{1}{e^t + 1}d(\frac{\mu + (t/\beta)}{E_F})^{n/2} = N \int_{-\beta\mu}^\infty \frac{e^t}{(e^t + 1)^2}(\frac{\mu + (t/\beta)}{E_F})^{n/2}dt$$

For $\beta \to \infty$ (i.e. $T \to 0$) one finds

$$1 = (\frac{\mu}{E_F})^{n/2} + \frac{\pi^2}{6}\frac{n(n - 2)}{4}(\frac{T}{T_F})^2(\frac{\mu}{E_F})^{(n/2)-2} + \ldots$$

yielding

$$\mu = E_F(1 - \frac{\pi^2}{12}(n-2)(\frac{T}{T_F})^2 + ...)$$

Problem XII.7.4 The specific mass (density) of alluminum is $d = 2.75\,\mathrm{g/cm^3}$. Evaluate the Fermi energy, the Fermi velocity, the average velocity of the conduction electrons and the quantum pressure (for $T \to 0$).
  Solution:
  The number of atoms per cubic cm is

$$N = 0.54 \cdot 10^{23}$$

Thus, for three free-electrons per atom

$$E_F = \frac{\hbar^2}{2m}(3\pi^2 N_e/V)^{2/3} = 11.7\,\mathrm{eV}\,,$$

(with $N_e = 3N$) and

$$v_F = \sqrt{\frac{2E_F}{m}} = 2.03 \cdot 10^8\,\mathrm{cm/s}\,,$$

The distribution function for the velocity is

$$p(v)dv = D(E)dE = N_e d(\frac{v}{v_F})^3 = 3N_e\frac{v^2}{v_F^3}dv\,,$$

then

$$< v >= \int_0^{v_F} v p(v)dv = \frac{3}{4}\,v_F = 1.44 \cdot 10^8\,\mathrm{cm/s}\,.$$

From Eq. 12.30 and

$$P = -\partial < E > /\partial V = \frac{2}{3}\frac{3}{5}N_e E_F = 1.2 \cdot 10^{12}\,\mathrm{dyne/cm^2}\,.$$

Problem XII.7.5 In the assumption that the electrons in a metal can be described as a classical free-electron gas, show that no magnetic susceptibility would arise from the orbital motion.
  Solution:
  For the classical free-electron gas the partition function is

$$Z = \int_{-\infty}^{\infty} dq_x dq_y dq_z \int_{-\infty}^{\infty} dp_x dp_y dp_z\, exp[-\frac{\mathbf{p}^2}{2mk_B T}]$$

Neglecting the electron magnetic moment, in the presence of a magnetic field the partition function becomes

$$Z = \int_{-\infty}^{\infty} dq_x dq_y dq_z \int_{-\infty}^{\infty} dp_x dp_y dp_z \, exp[-\frac{(\mathbf{p} + \frac{e}{c}\mathbf{A})^2}{2mk_B T}]$$

with $\mathbf{A}$ the vector potential. Therefore the only effect of $\mathbf{A}$ is to shift the origin of the integration, yielding no effect on the integral. The partition function and then also the free energy are field independent and the magnetization is zero.

Problem XII.7.6 Derive the paramagnetic susceptibility due to the free electrons in a metal (**Pauli susceptibility**).

Solution:

In the absence of a magnetic field the number of electrons with spin up N_+ is equal to the number of electrons with spin down N_- and the total magnetization of the metal $M = \mu_B(N_+ - N_-)$ is zero.

In a field H the energy of spins up is lowered by an amount $\mu_B H$, while the one of the spins down is increased by the same amount. The unbalance in the populations yields a non-zero magnetization. The number of spins up is

$$N_+ = \int_{-\mu_B H}^{\infty} f(E,T)\frac{D(E + \mu_B H)}{2}dE,$$

(the factor $1/2$ in the density of states $D(E)$ takes into account that we are considering the electrons with spin up only). Introducing $E' = E + \mu_B H$ one can write

$$N_+ = \int_{0}^{\infty} f(E' - \mu_B H, T)\frac{D(E')}{2}dE'.$$

In the weak-field limit $f(E' - \mu_B H, T) \simeq f(E',T) - \mu_B H(\partial f/\partial E)_{E'}$ and

$$N_+ = \frac{1}{2}\int_{0}^{\infty} f(E',T)D(E')dE' - \frac{1}{2}\int_{0}^{\infty} \mu_B H(\frac{\partial f}{\partial E})_{E'}D(E')dE'.$$

In the same way

$$N_- = \frac{1}{2}\int_{0}^{\infty} f(E',T)D(E')dE' + \frac{1}{2}\int_{0}^{\infty} \mu_B H(\frac{\partial f}{\partial E})_{E'}D(E')dE'$$

For $H \to 0$ $\chi_P = M/H$ and therefore from $M = \mu_B(N_+ - N_-)$,

$$\chi_P = \mu_B^2 \int_{0}^{\infty} (\frac{-\partial f}{\partial E})_{E'}D(E')dE'$$

For $E_F \gg k_B T$ and provided that the density of states varies smoothly around E_F, one writes $(-\partial f/\partial E)_{E'} \simeq \delta(E' - E_F)$, so that

$$\chi_P = \mu_B^2 D(E_F)$$

(see Eq. 12.33). ($D(E)$ density of states per unit volume and χ_P is dimensionless).

Problem XII.7.7 For a metal with bcc structure and lattice step $a = 5$ Å and electron density 2×10^{-2} electrons per cell, evaluate the temperature at which the electron gas can be considered degenerate and write the approximate form for the specific heat above that temperature.

Solution:

The electron density is

$$n = \frac{2 \times 10^{-2}}{a^3} = 1.6 \times 10^{20} \ \text{e/cm}^3$$

and the average spacing among the electrons is $d \simeq (3/4\pi n)^{1/3} \simeq 11$ Å. The electron gas can be considered degenerate when $d \leq \lambda_{DB}$, the **De Broglie** wavelength. Since $\lambda_{DB} = h/\sqrt{3mk_BT}$, the gas can be considered degenerate for $T < h^2/(3d^2 mk_B) \simeq 8600$ K. Above that temperature the gas is practically a classical one and the specific heat (per particle) is $C_V \simeq (3/2)k_B$.

Problem XII.7.8 The bulk modulus $B = -V(\partial P/\partial V)_T$ of potassium crystal at low temperature is $B = 0.38 \times 10^{11}$ dyne/cm^2. Discuss this result in the assumption that B is entirely due to the electron gas.

Solution:

The pressure of the electron Fermi gas is $P = (2/5)nE_F$, with n electron density. Then

$$B = -V\frac{\partial P}{\partial V} = \frac{2}{3}nE_F$$

For potassium, at density 0.86 g/cm^3, the electron density is 1.4×10^{22} e/cm^3 and the Fermi energy is $E_F = 2.1$ eV. Then $B \simeq 0.32 \times 10^{11}$ dyne/cm^2, in rather good agreement with the experimental finding.

Problem XII.7.9 Prove that in a semiconductor at thermal equilibrium the concentration of holes and electrons is given by

$$n = N_c e^{-(E_c - E_F)/k_BT} \ , \ \ p = N_v e^{-(E_F - E_v)/K_BT}$$

where

$$N_c = 2\left(\frac{2\pi m_e k_BT}{h^2}\right)^{3/2} \ \ N_v = 2\left(\frac{2\pi m_h k_BT}{h^2}\right)^{3/2} \ ,$$

E_F is the Fermi level (in the middle of the gap), E_c the bottom of the conduction band and E_v the top of the valence band (m_e and m_h are the effective masses of electrons and holes). Assume parabolic bands, going as $(k - k_c)^2$ and $(k - k_v)^2$.

Then evaluate

(a) the value of N_c for $m_c = m$ ($m \equiv$ electron mass) and $T = 300°$K ;

(b) the carriers concentration in Si, at $T = 300°$K, assuming $m_c = m_h = m$, and a gap of 1.14 eV.

Solution:

At thermal equilibrium the concentration of electrons is given by

$$n = \int_{E_{min}}^{E_{max}} D_c(E) f(E) dE$$

the density of states $D_c(E)$ per unit volume being

$$D_c(E) = (4\pi/h^3)(2m_c)^{3/2}(E - E_c)^{1/2}.$$

The Fermi-Dirac distribution for $(E - E_F) \gg k_B T$ can be approximated as

$$f(E) = \frac{1}{1 + e^{(E-E_F)/k_B T}} \simeq e^{-(E-E_F)/k_B T}$$

Then

$$n = \frac{4\pi}{h^3}(2m_c)^{3/2} e^{-(E_c - E_F)/k_B T} \int_0^{E_{max} - E_c} x^{1/2} e^{-x/k_B T} dx$$

$$\simeq \frac{4\pi}{h^3}(2m_c)^{3/2} e^{-(E_c - E_F)/k_B T} \int_0^{\infty} x^{1/2} e^{-x/k_B T} dx$$

$$= \frac{4\pi}{h^3}(2m_c)^{3/2} [e^{-(E_c - E_F)/k_B T}] \frac{1}{2}\pi^{1/2}(k_B T)^{3/2}$$

yielding

$$n = 2 \left(\frac{2\pi m_e k_B T}{h^2} \right)^{3/2} e^{-(E_c - E_F)/k_B T}$$

In an analogous way the hole concentration can be derived.

For intrinsic Si at $300°$K one has $N_c = 2.51 \times 10^{19} \text{cm}^{-3}$ and the electron and hole concentration are $n = p = 3.14 \times 10^9 \text{cm}^{-3}$.

Problems F.XII

Problem F.XII.1 Derive the contribution to the specific heat associated with the conduction electrons in a metal for temperature small compared to E_F/k_B.

Solution:

In a way analogous to the derivation of the Pauli susceptibility (see Problem XII.7.6) the increase of the electron energy when the temperature is brought from 0 to T is written in the form

$$U(T) = \int_{E_F}^{\infty} (E - E_F)f(E)D(E)dE + \int_0^{E_F} (E_F - E)(1 - f(E))D(E)dE \ .$$

In the second integral $(1 - f(E))$ gives the probability that an electron is removed from a state at energy below E_F. Then

$$C_V = \int_0^{\infty} (E - E_F)\frac{\partial f}{\partial T}D(E)dE \simeq D(E_F) \int_0^{\infty} (E - E_F)\frac{\partial f}{\partial T}dE \ .$$

Since

$$\frac{\partial f}{\partial T} = \frac{(E - E_F)}{k_B T^2} \frac{e^{(E-E_F)/k_B T}}{[e^{(E-E_F)/k_B T} + 1]^2}$$

by utilizing $\int_{-\infty}^{\infty} x^2 e^x dx/(e^x + 1)^2 = (\pi^2/3)$, one obtains

$$C_V = \frac{\pi^2}{3}D(E_F)k_B^2 T$$

This result can be read as the derivative of the product $k_B T$ times the fraction T/T_F of the electrons in the energy range $k_B T$ around E_F.

Problem F.XII.2 Derive the equation of state (relation between P, V and T) for the Fermi gas, in the limit $T \to 0$.

Solution:

From the energy (see Eq. 12.31)

$$U = (3/5)NE_F \left(1 + \frac{5\pi^2}{12}\left(\frac{k_B T}{E_F}\right)^2 + \dots\right)$$

with

$$E_F = (\hbar^2/2m)\left(3\pi^2\frac{N}{V}\right)^{2/3}$$

$$P = -\frac{\partial U}{\partial V} = \frac{2}{5}\frac{NE_F}{V}\left(1 - \frac{5\pi^2}{18}\left(\frac{k_B T}{E_F}\right)^2 + ...\right)$$

namely

$$PV = (2/5)NE_F\left(1 - \frac{5\pi^2}{18}\left(\frac{T}{T_F}\right)^2 + ...\right)$$

(see Prob. XII.7.3).

Problem F.XII.3 The temperature dependence of the specific heat in Gallium is reported in the Figure

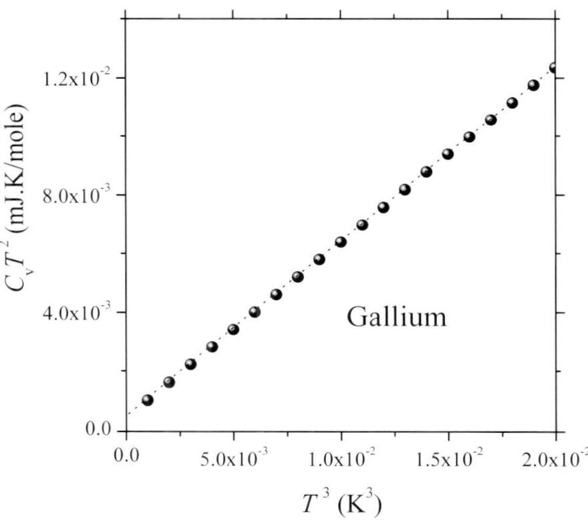

Derive the Fermi energy and the electric field gradient at the nucleus (assuming for simplicity that ^{69}Ga with $I = 3/2$ and $Q = 0.168$ barn is the only isotope and noticing that at low temperature the contribution to the specific heat due to lattice vibrations can be neglected).

Solution:
From the measurements one deduces the straight line $C_v T^2 = (a + bT^3)$ with $a \simeq 4 \cdot 10^{-4}$mJ $\cdot$ K/mole and $b \simeq 0.6$mJ/moleK^2, the second contribution being associated with conduction electrons. From $C_v = (\pi^2/3) \cdot k_B{}^2 T D(E_F)$, one derives $E_F \simeq 6$ eV.

The first term for C_V, going as $1/T^2$, is the high-temperature tail of the Schottky-like specific heat C_v^Q associated with the hyperfine split by quadrupolar interaction, with energy separation E. Since for $k_B T \gg E$, $C_v^Q = N_A k_B (E/k_B T)^2$, one finds $E \simeq 6.6 \times 10^{-20}$ erg. For $I = 3/2$ the splitting between the $M_I = \pm 1/2$ and $M_I = \pm 3/2$ levels due to quadrupole interaction is $E = 2eQV_{zz}$, with V_{zz} the principal component of the electric field gradient (see §5.3). Then one obtains $V_{zz} = 4 \times 10^{14}$ u.e.s./cm^3.

Problem F.XII.4 Consider two cubic clusters of Lithium (lattice step $a = 3.5$ Å and bcc structure) formed by 1.6×10^7 and 16000 atoms, respectively. Evaluate the Fermi energy for each cluster and estimate the separation among the electronic levels in proximity of the center of the Brillouin zone.

Solution:

The electron density is $n = 2/a^3 = 4.6 \times 10^{22}$ cm^{-3} and the Fermi energy $E_F = 4.7$ eV, size independent. The size affects the spacing among k states. The first cluster is a cube of size $L_1 = 200a$, while the second one of size $L_2 = 20a$. Then the separation among the lowest energy levels is

$$\Delta E_{1,2} = \frac{\hbar^2}{2m}\left(\frac{2\pi}{L_{1,2}}\right)^2 ,$$

namely $\Delta E_1 = 1.6 \times 10^{-15}$ erg and $\Delta E_2 = 1.6 \times 10^{-14}$ erg.

Problem F.XII.5 The density of Lithium is 0.53 g/cm^3. Evaluate the contribution to the bulk modulus due to electrons, in the low temperature range. Compare the estimated value with the experimental result $B \simeq 0.12 \times 10^{12}$ dyne/cm^2.

Solution:

Following Problem XII.7.8

$$B = -V\frac{\partial P}{\partial V} = \frac{2}{3}nE_F$$

The electron density is $n = 4.6 \times 10^{22}$ e/cm^3 and the Fermi energy turns out $E_F = 4.7$ eV. Then $B = 2.4 \times 10^{11}$ dyne/cm^2, not far form the experimental result.

Problem F.XII.6 In semiconductors the concentration of itinerant electrons is low and one can expect that the Pauli susceptibility turns to a Curie-like susceptibility characteristic of localized electrons. Discuss the derivation of the Pauli susceptibility for semiconductors (neglect the electron-electron Coulomb interaction).

Solution:

The Pauli susceptibility is (see Problem XII.7.6)

$$\chi_P = \mu_B^2 \int_0^\infty \left(\frac{-\partial f}{\partial E}\right)_{E'} D(E')dE'$$

(see Problem XII.7.6).

For diluted Fermi gas, at room temperature the statistical distribution function can be written
$f(E) \simeq e^{-(E-E_F)/k_B T}$ (see Problem XII.7.9).

Thus $-\partial f/\partial E = f/k_B T$ and the susceptibility turns out

$$\chi_P = \frac{\mu_B^2}{k_B T} \int_0^\infty f(E')D(E')dE' = n\frac{\mu_B^2}{k_B T}$$

with n concentration of conduction electrons.

13

Miscellaneous aspects related to the electronic structure

Topics

Covalent, metallic, ionic and molecular crystals
Cohesive energies and bonding mechanisms
Lennard-Jones potential
Crystal-field effects in magnetic ions
Electric current flow
Magnetic properties of itinerant electrons

13.1 Typology of crystals

In the light of the main aspects involving the electronic properties, a classification of crystalline solids can be devised. This can be done either in a valence bond scenario by looking at the bonding mechanisms or by referring to the electric conduction and to the band structure.

In the first case the crystals can be divided in **covalent**, **metallic**, **ionic** and **molecular**. In **covalent crystals** the bonding mechanism and the strength of bonds are similar to the ones in covalent molecules. In other words, the crystal can be conceived as a "macroscopic" molecule with marked directional bonds between pairs of atoms where spin-paired electrons can be placed. Therefore, covalent crystals are stiff, scarcely plastic and fragile. Illustrative examples can be found in carbon-based crystals, such as diamond and graphite. Diamond, as well as the isostructural Ge, Si, Sn and Pb crystals, result from an ideally infinite network of sp^3 hybrid orbitals (§9.2). On the other hand, in graphite the sp^2 hybridization yields a planar atomic arrangement, with weak interaction among adjacent planes (see §11.3).

Metallic crystals are somewhat equivalent to large molecules with electrons delocalized through all the volume, an extension of what discussed in benzene (§9.3). The description of these systems in a VB-like framework would require the superposition of a large number of equivalent configurations. It is evident that a Bloch-like approach is more convenient for the metallic crystals. A suitable way to describe these solids is to refer to a model of positive ions at the lattice sites embedded in a *sea* of electrons, with a nearly uniform charge distribution. In general the bonds are not saturated. For instance, in Li metal (bcc structure) each ion has 8 nearest neighbors and in a molecular-like picture one can think that there is 1/4 of electron on each orbital.

In **ionic crystals** the electrons are characterized by molecular-like orbitals centered at the atoms having larger electronegativity, as in the case of strongly heteronuclear molecules (see §8.5). The attractive interaction may often be approximated to the one for point charge ions and in a crude approximation the ions can be assumed to have the closed shells configurations. For example, in LiF crystal, the $(1s)^2$ shell for Li^+ and the $(2p)^6$ shell for F^-. From the X-ray diffraction peaks one can estimate the actual number of electrons at a given site. For instance, in NaCl it turns out that there are 17.85 electrons at Cl site. Thus the order of magnitude of the bond energy **per pair** is $-(0.85e)^2/R$, with R interatomic distance.

The hydrogen bond O-H-O typical of hydrides, of the ferroelectric KDP (potassium dihydrogen phosphate) and of other organic compounds, can be considered as a type of ionic bond. The hydrogen atom can be thought in a local double-well potential. Several electric and elastic properties of these crystals are rather well explained within this simple model.

In molecular-like scenarios one can hardly devise any bonding mechanism for neutral molecules at high ionization energy or for closed shell atoms, such as inert gases. In these cases the aggregation into a solid state can occur because of an interaction that we have not directly considered in molecules: the **Van der Waals** forces, associated with fluctuating electric dipoles. This mechanism yields a weak attractive potential decreasing as R^{-6} and leads to the formation of **molecular crystals** (a mention has been given in Problem VIII.3.1 and that interaction shall be described in some detail at §13.2.2).

The classification scheme based on the bonding mechanisms is not very suited to describe the properties involving the electrical transport. This aim is better achieved by referring to the band scheme, in the framework of Bloch orbitals. Let us remind that a band arising from s atomic states can be occupied by $2N$ electrons (N the number of atoms in a reference volume) while in the p band this number is $3 \times 2N$. As already mentioned, the electrical conductivity originates from the "acceleration" (namely from the change of state) induced by the electric field, for a single electron described by the equation $\hbar d\mathbf{k}/dt = -e\mathbf{E}$ (see §12.3). A few observations can be made in regards of the current flow (for some more detail see §13.4). In a fully filled band each $\mathbf{k}$ state is occupied by two electrons and a neat flow of current is not possible (unless the electric field is so strong to alter the unperturbed bands) and one has an **insulator**. For a partially filled band electrical transport can occur and one has a **conductor**. When the gap between full **valence band** and an empty **conduction band** is of the order of 0.1-1 eV, then one has an intrinsic **semiconductor**. These crystals are insulators for $T \to 0$, while progressive increase in the conductivity with increasing temperature occurs, as a consequence of the partial filling up of the conduction band. At variance, in a metal the conductivity decreases with increasing temperature due to the increase in the scattering rate between the electrons and the ionic vibrational modes.

In such a scenario one can predict that alkali crystals Li, Na, Rb, etc... and transition metals as Cu, Ag and Au are metallic conductors, since they have an odd number of electrons per unit cell. This rule however, is not quite valid and often one has to pay attention to other details. For instance, although As, Sb and Bi atoms convey five electrons in the conduction band, they generate a crystal which is essentially an insulator. Without going into the real aspects of the electronic band structure, we only mention that the reason for the quasi-insulating character is related to the generation of five bands that are completely filled by the valence electrons of the two atoms present in the unit cell.

In some crystals there is also the possibility of a tiny superposition of bands, giving a small metallic character and causing electrical conduction with a particular temperature dependence: these are the **semimetals**.

Strong **band overlap** can drastically change the simplified picture given above. For instance, according to the previous statements Be crystals (atomic configuration $(1s)^2(2s)^2$) should be insulators. This is not the case: the overlap between s and p orbitals generates a partially occupied **hybrid band**, a situation similar to the one in diamond (see Fig. 12.8). The overlap of these bands yields a fully occupied valence band and an empty conduction band in this latter crystal. At the equilibrium distance characteristic of Si and Ge the gap between the two band diminishes and a semiconducting behavior can be observed. On the other hand, Sn can undergo a transition from metallic to semiconductor, in view of the proximity to the overlap condition. Finally Pb is a metal, since the $6p$ band is only partially filled (Fig. 12.8).

Semiconducting behavior can be expected for a class of materials with tetrahedral structure generated by sp^3 hybridization. The so-called III-V semiconductors are the crystals in which the basis, instead of being formed by the same two atoms at $(0,0,0)$ and $(1/4,1/4,1/4)$ in the fcc structure (as in C, Si, Ge and Sn) involves one element of the third group (Ga for instance) and one of the fifth group (As, for example). The covalent "transfer" of one electron from As to Ga gives rise to the s^2p^2 configuration in both atoms, as in Ge or Si, thus triggering the sp^3 hybrid bands and the semiconducting behavior. The band gaps in Ge (0.75 eV) and in Si (1.17 eV) (indirect gaps, the maximum of the valence band and the minimum of the conduction band occurring at different points of the Brillouin zone) are of the same order of magnitude in GaAs (1.52 eV) and in GaSb (0.81 eV).

In ionic crystals the gap between the fully occupied valence band and the empty conduction band can be larger than about 5 eV, thus explaining their insulating behavior.

13.2 Bonding mechanisms and cohesive energies

The **cohesive energy** is defined as the difference between the energies of the atoms for interatomic distance $R \rightarrow \infty$ and the one for $R = R_e$, the interatomic equilibrium distance. From thermodynamical and spectroscopic measurements the order of magnitude of the cohesive energies turn out:

 i) around 5 eV/atom in covalent crystals (e.g. 7.36 eV for diamond);
 ii) around 1 eV/atom in metallic alkali crystals;
 iii) around 5-10 eV/(pair of atoms) in ionic crystals;
 iv) from 10^{-2} to 10^{-1} eV in molecular crystals, with a sizeable increase in the binding energy with increasing atomic number for inert atoms crystals.

Quantitative estimates of the cohesive energy are evidently difficult, since in principle they correspond to the derivation of the eigenvalues in the Schrödinger equation for the electronic states. It is possible to achieve satisfactory descriptions of the relevant aspects of the binding mechanisms and to obtain rather good estimates of the cohesive energies by referring to limit ideal situations. For instance, one usually refers to molecular-like scenarios or to ionic atomic configurations. In covalent crystals, where the bonds are similar to the ones in molecules, the cohesive energy per molecule is expected around the one described at Chapter 8. The bonding mechanism in metals can be considered as due to the attractive term related to the electron delocalization (favored by the band overlap) and the repulsive term arising from the increase of the Fermi energy when the electron density increases (see Eq. 12.28).

More quantitative descriptions of the binding energies for ionic and molecular crystals shall be given in the subsequent Subsections.

13.2.1 Ionic crystals

Let us refer to a crystal with N positive and N negative ions, per cubic cm. In the point-charge approximation the interaction between two ions is written

$$V_{ij} = \pm \frac{e^2}{R_{ij}} + B e^{-R_{ij}/\rho}, \tag{13.1}$$

where the sign of the first term depends on the signs of the charges at the i-th and j-th ions. The second term has the **Born-Mayer** form used to take into account the short-range repulsion in heteronuclear molecules (Eq. 8.36).

For a given i-th ion the energy is $V_i = \sum_j' V_{ij}$ and by writing the distance $R_{ij} = p_{ij} R$ (R being the nearest neighbour distance) one has

$$V_i = \sum_j' V_{ij} = \frac{e^2}{R} \sum_j' \frac{(\pm 1)}{p_{ij}} + zBe^{-R/\rho}, \qquad (13.2)$$

where in view of its short-range character the repulsive term has been limited to the z nearest neighbours. Then the total energy becomes

$$V_T = NV_i = -\frac{Ne^2}{R}\alpha + NzBe^{-R/\rho}, \qquad (13.3)$$

with

$$\alpha = \sum_j \frac{(\pm 1)}{p_{ij}}, \qquad (13.4)$$

the **Madelung constant**.

From Eq. 13.3 one realizes that α has to be positive, in order to grant the aggregation of the ions to form a crystal. At the equilibrium interatomic distance

$$\left(\frac{dV}{dR}\right)_{R=R_e} = 0 = \frac{N\alpha e^2}{R_e^2} - \frac{Nz}{\rho}Be^{-R_e/\rho} \qquad (13.5)$$

and then $(\rho\alpha e^2/zB) = R_e^2 e^{-R_e/\rho}$, thus giving for the total energy

$$V_T^{eq} = -\frac{N\alpha e^2}{R_e}\left[1 - \frac{\rho}{R_e}\right]. \qquad (13.6)$$

The characteristic constant $\rho \ll R_e$ can be estimated from the crystal compressibility (see Problem XIII.2.1) and usually turns out of the order of $0.1R_e$. Thus Eq. 13.6 shows that the cohesive energy is of the order of the dissociation energy of the ideal molecule formed by positive and negative ions and that it is largely controlled by the Madelung constant.

The estimate of α is not trivial due to the slow convergence of the sum in Eq. 13.4. It has to be noticed that the series of the positive and of the negative terms, taken separately, diverge in 3D crystals. Numerical methods to grant fast convergence for α have been devised a long ago, based on the choice of reference regions where the monopole contribution vanishes (**Ewald procedure**). The remaining dipole or quadrupole contributions converge with increasing distance faster than the Coulomb terms.

Typical values for the Madelung constants are $\alpha = 1.7475$ for crystals with NaCl-type structure and $\alpha = 1.7626$ for crystals with CsCl-type structure. In the simple case of a chain with alternating positive and negative ions the evaluation of α is straightforward:

$$\alpha = 2[1 - \frac{1}{2} + \frac{1}{3} - \frac{1}{4} + ...] = 2ln2. \qquad (13.7)$$

13.2.2 Lennard-Jones interaction and molecular crystals

In order to describe the bonding mechanism in molecular crystals let us first derive the form of the attractive interaction among the atoms when no molecular-like mechanisms (as the ones described in Chapters 8 and 9) are active. The mechanism we shall consider originates from **fluctuating electric dipoles**, first described as **Van der Waals interaction** and later on known in the quantum mechanical scenario as **London interaction**.

To derive the London interaction let us refer to two hydrogen atoms along the x direction:

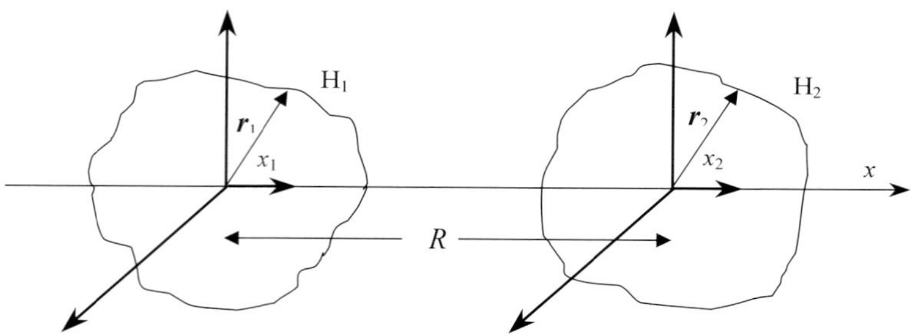

The distance R is larger than the one at which the bonding mechanisms leading to the Hydrogen molecule would become relevant (in other words R is a distance where the overlap, resonance or exchange integrals can be neglected; see Problem VIII.3.1). Then the unperturbed wavefunction is

$$\phi^o(1,2) = \phi_{n_1,l_1}(\mathbf{r_1})\phi_{n_2,l_2}(\mathbf{r_2}) \ , \tag{13.8}$$

with eigenvalue $E^o = E^o_{n1,l1} + E^o_{n2,l2}$. The perturbation Hamiltonian is the dipolar one

$$\mathcal{H}_{dip} = \frac{e^2}{R^3}\left[\mathbf{r_1}.\mathbf{r_2} - 3(\mathbf{r_1}.\hat{x})(\mathbf{r_2}.\hat{x})\right] \ ,$$

that is rewritten in the form

$$\mathcal{H}_{dip} = -\frac{e^2}{R^3}\left[2x_1x_2 - y_1y_2 - z_1z_2\right] \ . \tag{13.9}$$

From second order perturbation theory the ground-state energy turns out

$$E(R) = 2E^o_{1s} + <0|\mathcal{H}_{dip}|0> + \sum_{k\neq0}\frac{<0|\mathcal{H}_{dip}|k><k|\mathcal{H}_{dip}|0>}{E^o_0 - E^o_k} \tag{13.10}$$

$\mathcal{H}_{dip}$ is an odd function and $<0|\mathcal{H}_{dip}|0> = 0$.

By resorting to arguments already used in the derivation of the atomic polarizability (§4.2) and noticing that the denominator varies from $-e^2/a_o$ to $-3e^2/4a_o$, one can write

$$E(R) \simeq 2E_{1s}^o - \frac{a_o}{e^2}\left[\sum_k < 0|\mathcal{H}_{dip}|k >< k|\mathcal{H}_{dip}|0 >- < 0|\mathcal{H}_{dip}|0 >< 0|\mathcal{H}_{dip}|0 >\right] =$$

$$= 2E_{1s}^o - \frac{a_o}{e^2} < 0|\mathcal{H}_{dip}^2|0 > \quad . \tag{13.11}$$

Thus from Eq. 13.9,

$$E(R) \simeq 2E_{1s}^o - \frac{a_o}{e^2}\frac{e^4}{R^6}\left[4 < x_1^2 >< x_2^2 > + < y_1^2 >< y_2^2 > + < z_1^2 >< z_2^2 >\right].\tag{13.12}$$

For Hydrogen the expectation values of the square of the components x, y and z are $< r^2 > /3 = a_o^2$ and then

$$E(R) \simeq 2E_{1s}^o - \frac{6e^2 a_o^2}{R^6}a_o^3, \tag{13.13}$$

showing that an attractive interaction has arisen.

The London interaction can be depicted as related to the dipolar interaction between an instantaneous dipole in one atom and the one induced in the neighboring atom, thus explaining the role of the atomic polarizability $\alpha \propto a_o^3$ (see §4.2), as schematically described below

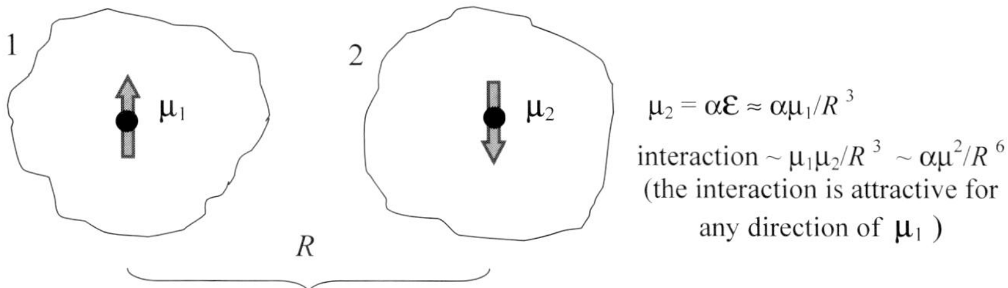

The result in Eq. 13.13 can be generalized, leading to the assumption of an attractive potential energy of the form

$$V_{att} \simeq -\frac{e^2 a_o^2}{R^6}\alpha, \tag{13.14}$$

with α proper atomic polarizability. A short-range repulsive term given by $V_{rep} = B/R^{12}$ can be heuristically added.

The **Lennard-Jones potential** between two atoms collects the concepts described above and it reads

$$V_{ij}(R) = \varepsilon\left[(\frac{\sigma}{R_{ij}})^{12} - 2(\frac{\sigma}{R_{ij}})^6\right] , \tag{13.15}$$

where ε and σ are related to the repulsion coefficient B and to the atomic polarizability α.

Note that according to the form 13.15 for the Lennard-Jones potential ε and σ are simply related to the shape of the interaction energy:

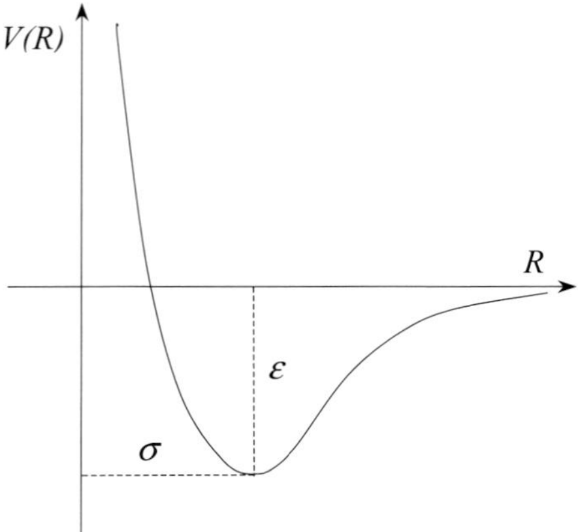

To evaluate the cohesive energy in molecular crystals one can proceed in a way similar to the one carried out in ionic crystals (§13.2.1). At variance with that case one can now limit the summation to the z first nearest neighbors for both the repulsive and the attractive terms. From the condition of minimum at $R = R_e$, one derives

$$V_T^{eq} = -\frac{Nz}{2R_e^6}e^2 a_o^2 \alpha. \qquad (13.16)$$

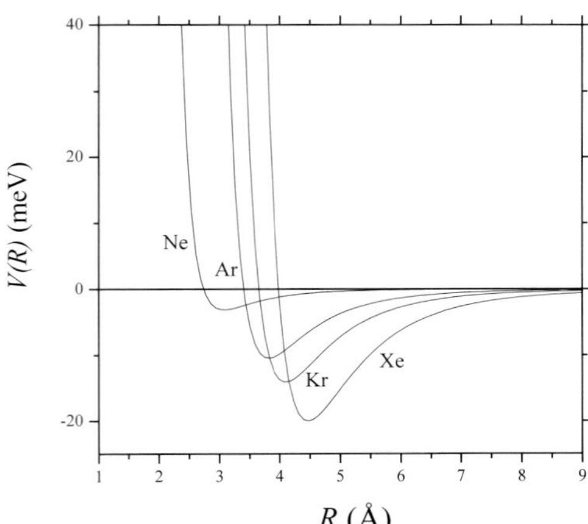

Fig. 13.1. Energy curves in crystals of inert atoms as a function of the interatomic distance.

The assumption of London interaction and of short-range repulsion as in Eq. 13.15 qualitatively justifies the cohesive energy in inert atoms crystals. In particular, through the dependence of the atomic polarizability from the third power of the "size" of the atom, Eq. 13.16 explains why the cohesive energy increases rapidly with the atomic number (see Fig. 13.1).

Problems XIII.2

Problem XIII.2.1 For ionic crystals assume that the short-range repulsive term in the interaction energy between two point-charge ions is of the form R^{-n}. Show that the cohesive energy is given by

$$E = -\frac{N\alpha e^2}{R_e}\left(1 - \frac{1}{n}\right)$$

with α Madelung constant and R_e equilibrium nearest neighbor distance. Then, from the value of the bulk modulus $B = 2.4 \cdot 10^{11}$ dyne/ cm^2, estimate n for NaCl crystal.
In KCl R_e is 3.14 Å and the cohesive energy (per molecule) is 7.13 eV. Estimate n.
   Solution:
   Modifying Eq. 13.3 we write

$$E(R) = -N\left(\frac{\alpha e^2}{R} - A\frac{1}{R^n}\right)$$

where $A = \rho z$ (z number of first nearest neighbors and ρ a constant in the repulsive term ρ/R_{ij}^n). From

$$\left(\frac{dE(R)}{dR}\right)_{R=R_e} = 0$$

$$E(R_e) = -\frac{N\alpha e^2}{R_e}\left(1 - \frac{1}{n}\right) .$$

The compressibility is defined

$$k = -\frac{1}{V}\frac{dV}{dP}$$

and from $dE = -PdV$ the **bulk modulus** is

$$B = k^{-1} = V\frac{d^2 E}{dV^2} .$$

(see Problem F.XII.5). For N molecules in the fcc Bravais lattice the volume of the crystal is $V = 2NR^3$ and then

$$\frac{d^2E}{dV^2} = \frac{dE}{dR}\frac{d^2R}{dV^2} + \frac{d^2E}{dR^2}\left(\frac{dR}{dV}\right)^2 .$$

From

$$\left(\frac{dR}{dV}\right)^2 = \frac{1}{36N^2R^4}$$

one obtains

$$k^{-1} = \frac{1}{18NR_e}\left(\frac{d^2E}{dR^2}\right)_{R=R_e} .$$

Since

$$\frac{d^2E}{dR^2} = -N\left[\frac{2\alpha e^2}{R^3} - \frac{n(n+1)A}{R^{n+2}}\right],$$

$$k^{-1} = \frac{(n-1)\alpha e^2}{18R_e^4} .$$

For NaCl $\alpha = 1.747$ while $R_e = 2.82\,\text{Å}$, therefore $n \simeq 9.4$.
For KCl, from $E(R_e) = 7.13$ eV/molecule, one obtains $n = 9$.

Problem XIII.2.2 In KBr the distance between the first nearest-neighbors is $R_e = 3.3\,\text{Å}$, while the Madelung constant is $\alpha = 1.747$. The compressibility is found $k = 6.8 \cdot 10^{-12}$ cm^2/dyne. Evaluate the constant ρ in the Born-Mayer repulsive term and the cohesive energy.

Solution:
From $V = 2NR^3$ and $dV = 6NR^2dR$ the pressure is

$$P = -\frac{1}{6R^2}\frac{dE}{dR} ,$$

where E is the cohesive energy per molecule. Then

$$\frac{dP}{dR} = -\frac{1}{6R^2}\left(\frac{d^2E}{dR^2}\right) + \frac{1}{3R^3}\frac{dE}{dR}$$

The second term being zero, the compressibility becomes

$$k = -\frac{3}{R}\frac{dR}{dP} .$$

Since

$$\frac{d^2E}{dR^2}\bigg|_{R=R_e} = \frac{18R_e}{k}$$

and

$$E = -\frac{\alpha e^2}{R}\left[1 - \frac{R\rho}{R_e^2}e^{-\frac{(R_e - R)}{\rho}}\right],$$

one finds

$$\frac{\rho}{R_e} = \left(2 + \frac{18 R_e^4}{\alpha e^2 \kappa}\right)^{-1},$$

so that

$$\frac{R_e}{\rho} \simeq 9\,,$$

and $\rho \simeq 3 \cdot 10^{-9}\, cm.$

The cohesive energy per molecule turns out
$E = -\frac{\alpha e^2}{R_e}\left(1 - \frac{\rho}{R_e}\right) \simeq -6.8$ eV.

13.3 Electron states of magnetic ions in a crystal field.

In a crystal the energy levels of partially filled d and f shells of transition metal and rare earth atoms are modified by the electric field generated by the neighboring atoms, yielding significant changes in the magnetic properties. To account for the perturbative effect two approaches can be used: the **crystal field** (CF) approximation or the **ligand field theory**.

In the first case the magnetic ion is assumed to be surrounded by point charges (with no covalency) which modify the electronic energies, in a way analogous to the Stark effect (§4.2). Thus one writes

$$\mathcal{H} = \mathcal{H}_{atom} + V_{CF}, \tag{13.17}$$

where

$$\mathcal{H}_{atom} = \sum_i \left(-\frac{\hbar^2}{2m}\nabla_i^2 - \frac{Ze^2}{r_i}\right) + \sum_{i>j}\frac{e^2}{r_{ij}} + \sum_i \xi_{nl}^{(i)}\mathbf{l}_i\cdot\mathbf{s}_i \tag{13.18}$$

The ligand field theory, at variance, takes into account the formation of co-valent bonds with the neighboring atoms, within the molecular-orbital theory.

Let us discuss a few basic aspects of the electronic states for a magnetic ion within the CF approach. As regards the order of magnitude of the V_{CF} term one can remark the following:

a) for $4d$ and $5d$ states usually one has $V_{CF} > \sum_{i>j}\frac{e^2}{r_{ij}} > \xi_{nl}$. In this **strong field limit** the CF yields splitting of the atomic levels of the order of 10^4 cm^{-1}.

b) for $3d$ states one usually has $\sum_{i>j}\frac{e^2}{r_{ij}} \geq V_{CF} > \xi_{nl}$. In this case the splitting of the atomic levels due to the CF is of the order of $10^3 - 10^2$ cm^{-1}.

c) for rare-earth atoms $\sum_{i>j} \frac{e^2}{r_{ij}} > \xi_{nl} > V_{CF}$, since the CF on the $4f$ electrons is sizeably shielded by the $5s$ and $5p$ electrons. Thus small CF splitting occurs, of the order of 1 cm^{-1}.

To understand qualitatively the role of the CF local symmetry in removing the d electron degeneracy, let us first consider the effect of point charges Ze placed at distances a from the reference ion along the x, y, z axes:

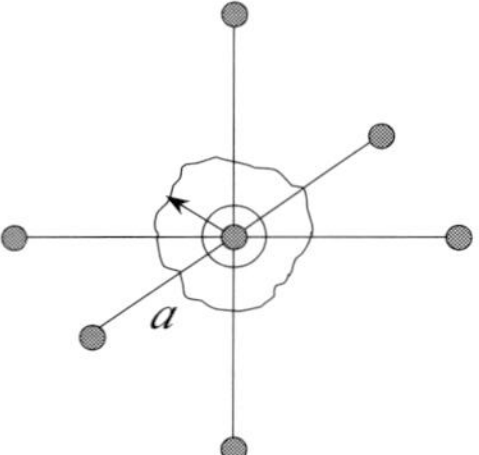

Then the perturbative potential, for instance from the charge at $(a, 0, 0)$ is

$$V_{CF} = -\frac{Ze^2}{|\mathbf{r} - a\mathbf{i}|} = -\frac{Ze^2}{\sqrt{(x-a)^2 + y^2 + z^2}} \equiv -\frac{Ze^2}{a}\frac{1}{\sqrt{1 + \frac{r^2}{a^2} - 2\frac{x}{a}}}, \quad (13.19)$$

where r is the nucleus-electron distance within the reference ion.
For $r \ll a$, by collecting the various terms and using

$$(1+x)^{-1/2} = 1 - \frac{x}{2} + \frac{3x^2}{8} - \frac{5x^3}{16} + \frac{35x^4}{128} + \dots$$

one writes

$$V_{CF} = -Ze^2\left[\frac{6}{a} + \frac{35}{4a^5}(x^4 + y^4 + z^4 - \frac{3}{5}r^4) + \dots\right]. \quad (13.20)$$

More in general, the CF potential due to the surrounding ions, on a given i-th electron is written

$$V_{CF}(\mathbf{r}_i) = -\sum_{k=1}^{N}\frac{Z_k e^2}{|\mathbf{R}_k - \mathbf{r}_i|}, \quad (13.21)$$

with Z_k the charge of the ion at $\mathbf{R}_k$. Since $r_i \ll R_k$, the validity of the Laplace equation $\nabla^2 V(\mathbf{r}_i) = 0$ is safely assumed. Then the CF potential can be expanded in terms of Legendre polynomials P_l (see Problem II.2.1):

$$V_{CF}(\mathbf{r}_i) = -e^2\sum_{k=1}^{N} Z_k \sum_{l=0}^{\infty}\frac{r_i^l}{R_k^{(l+1)}}P_l(cos\Omega_{ki}), \quad (13.22)$$

with Ω_{ki} angle between $\mathbf{r}_i$ and $\mathbf{R}_k$. By expressing P_l in terms of spherical harmonics

$$P_l(cos\Omega_{ki}) = \frac{4\pi}{(2l+1)} \sum_{m=-l}^{l} Y_{lm}(\theta_i, \phi_i) Y_{lm}^{*}(\theta_k, \phi_k), \qquad (13.23)$$

the CF Hamiltonian is written

$$\mathcal{H}_{CF} = \sum_{i=1}^{n} \sum_{l=0}^{\infty} \sum_{m=-l}^{l} A_l^m r_i^l Y_{lm}(\theta_i, \phi_i), \qquad (13.24)$$

with

$$A_l^m = \frac{-4\pi e^2}{(2l+1)} \sum_{k=1}^{N} \frac{Z_k Y_{lm}^{*}(\theta_k, \phi_k)}{R_k^{(l+1)}}, \qquad (13.25)$$

The coefficients A_l^m can be calculated once that the local coordination of the ion is known.

To give an example, let us consider the CF potential on one electron of a transition metal ion placed at the center of a regular octahedron formed by six negative charges Ze at distance R along the coordinate axes. In this case Eq. 13.24 reads

$$\mathcal{H}_{CF} = Ze^2 \left[\frac{6}{R} + \frac{7\sqrt{\pi} r^4}{3R^5} \left(Y_4^0 + \sqrt{\frac{5}{14}} (Y_4^4 + Y_4^{-4}) \right) \right] + \dots \qquad (13.26)$$

resembling Eq. 13.20.

Now one has to look for the effects of this perturbative hamiltonian on the degenerate d states. The electron wavefunction has to be of the form

$$\phi = c_o \phi_o + c_1 \phi_1 + c_{-1} \phi_{-1} + c_2 \phi_2 + c_{-2} \phi_{-2}, \qquad (13.27)$$

where $\phi_{0, \pm 1, \pm 2}$ are eigenfunctions of the unperturbed Hamiltonian.

One can notice that the matrix elements $< \phi_{0,\pm1,\pm2}|\mathcal{H}_{CF}|\phi_{0,\pm1,\pm2} >$ are all of the form nDq, with n an integer, $D = Ze^2/6R^5$ and $q \propto < r^4 >$.

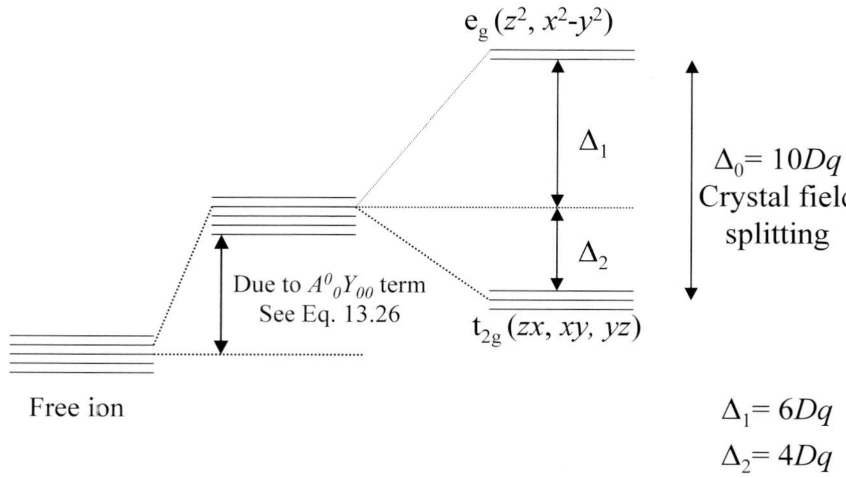

Fig. 13.2. Crystal field splitting of the $3d$ electron levels in regular octahedral coordination. The elongation of the octahedron along the z axis would cause the further splitting of the upper e_g levels.

The secular equation becomes

$$\begin{pmatrix} Dq - E & 0 & 0 & 0 & 5Dq \\ 0 & -4Dq - E & 0 & 0 & 0 \\ 0 & 0 & 6Dq - E & 0 & 0 \\ 0 & 0 & 0 & -4Dq - E & 0 \\ 5Dq & 0 & 0 & 0 & Dq - E \end{pmatrix} = 0 \quad (13.28)$$

with solutions $E_1 = E_2 = 6Dq$ and $E_3 = E_4 = E_5 = -4Dq$, in correspondence to the eigenfunctions $\phi_1' \equiv \phi_0 \equiv d_{z^2}$, $\phi_2' = (1/\sqrt{2})(\phi_2 + \phi_{-2}) \equiv d_{x^2-y^2}$, $\phi_3' = (1/\sqrt{2})(\phi_1 + \phi_{-1}) \equiv d_{xz}$, $\phi_4' = (-i/\sqrt{2})(\phi_1 - \phi_{-1}) \equiv d_{yz}$ and $\phi_5' = (-i/\sqrt{2})(\phi_2 - \phi_{-2}) \equiv d_{xy}$.

The structure of the energy levels is shown in Fig. 13.2.

The core of high-temperature superconductors is an octahedron of oxygen atoms surrounding the Cu^{2+} $3d^9$ ion, yielding the splitting of the $3d$ levels depicted in Fig. 13.2 (it should be reminded that the CF levels for a single hole in the $3d$ sub-shell are equivalent to the ones for a single electron).

The case of one p electron in a perturbative CF due to an octahedron of ions is discussed in Problem F.XIII.3, including the effect of an external magnetic field.

13.4 Simple picture of the electric transport

Let us first recall a few introductory remarks based on the **Drude model**, basically classical considerations for a free electron gas, which help to grasp some aspects of electrical conductivity in solids.

In analogy to the molecular collisions in classical gases, for the electrons colliding with impurities or with the ions (oscillating around their equilibrium positions, see Chapter 14) one can define a **mean free path** λ. This is the average distance covered by an electron between two collisions, while it is moving with an average velocity $< v >$. This average velocity can be related to the Fermi energy E_F by referring to the average energy $< E > \simeq 3E_F/5$ (see §12.7.1): $< v > = \sqrt{< \mathbf{v}^2 >} \sim \sqrt{E_F/m} \simeq 10^8$ cm/s.

An external electric field $\mathcal{E}$ modifies the random motions of the electrons in such a way that a charge flow opposite to the field arises, with a neat drift velocity $\mathbf{v}_d$. The drift velocity is estimated as follows. After a collision a given electron experiences an acceleration $\mathbf{a} = e\mathcal{E}/m$, for an average time $\lambda/<v>$. Then $v_d = a\lambda/<v> = -e\mathcal{E}\lambda/m<v>$, which is usually much smaller than $<v>$. Then, indicating with n the electron density, the current density turns out

$$\mathbf{j} = -ne\mathbf{v}_d = \frac{ne^2\mathcal{E}\lambda}{m<v>}. \tag{13.29}$$

This equation corresponds to the Ohm law, where the resistivity is $\rho = \mathcal{E}/j$.

The **mobility** μ, defined by the ratio $|\mathbf{v}_d|/|\mathcal{E}|$, is thus given by $\mu = e\lambda/m<v>$ and the conductivity σ is

$$\sigma = ne\mu. \tag{13.30}$$

For totally filled bands the conductivity is zero, as it will be emphasized subsequently. When a band is **almost filled** an expression for the conductivity due to positive charges (**holes**) can be considered. A contribution to the conductivity analogous to Eq. 13.30 can then be written: $\sigma_h = n_h e_h \mu_h$.

It should be noticed that due to the opposite sign of their charges and of their drift velocities, both electron and hole conductivities contribute with the same sign to the electric transport.

In the **Drude** model for metallic conductivity all the free electrons contribute to the current, a situation in contradiction to the Pauli principle. In fact, the electron at energy well below E_F cannot acquire energy from the field, the states at higher energy being occupied. Furthermore the temperature dependence of the conductivity (which around room temperature goes as

$\sigma \propto T^{-1}$ is not explicitly taken into account in Drude-like descriptions, the ions being considered immobile.

The quantum mechanical description of the current flow would require solving Schrödinger equation in the spatially periodic lattice potential in the presence of an electric field. Here we shall limit to a semi-classical picture in order to better clarify the phenomenological concepts given above, taking into account the band structure and resorting to the wave-packet-like properties of the electrons.

In the semiclassical approach the motion of the electron (see §12.3 and §12.6) is based on the equation for the increase of energy δE in a time δt, due to the force associated with the electric field $\boldsymbol{\mathcal{E}}$:

$$\delta E = -e\boldsymbol{\mathcal{E}}.\mathbf{v}\delta t. \tag{13.31}$$

Here $\mathbf{v}$ represents the group velocity of the Bloch wave-packet describing the electron:

$$\mathbf{v} = \boldsymbol{\nabla}_{\mathbf{k}}\omega(\mathbf{k}) \equiv \frac{1}{\hbar}\boldsymbol{\nabla}_{\mathbf{k}}E(\mathbf{k}). \tag{13.32}$$

It is recalled that in order to have particle properties, still retaining the required wave-like structure, an electron cannot have a precise definite momentum but must possess a range of $\mathbf{k}$ values.

From Eqs. 13.31 and 13.32 the equation of motion

$$\hbar\dot{\mathbf{k}} = -\boldsymbol{\mathcal{E}}e \tag{13.33}$$

describes how the wave-vector and hence the state of the electron, changes. From Eqs. 13.32 and 13.33 the effective mass m^*, reflecting the effect of the crystal field included in $E(\mathbf{k})$, was obtained (§12.6). From the components of the acceleration

$$\dot{v}_\alpha = \frac{1}{\hbar}\frac{d}{dt}(\boldsymbol{\nabla}_{\mathbf{k}}E)_\alpha = \frac{1}{\hbar}\sum_\beta \frac{\partial^2 E}{\partial k_\alpha \partial k_\beta}\dot{k}_\beta = \frac{1}{\hbar^2}\sum_\beta \frac{\partial^2 E}{\partial k_\alpha \partial k_\beta}(-e\mathcal{E}_\beta)$$

the components of the effective mass tensor turn out (see Eq. 12.24 and Prob. XII.3.3)

$$(m^*)^{-1}_{\alpha\beta} = \frac{1}{\hbar^2}\frac{\partial^2 E(\mathbf{k})}{\partial k_\alpha \partial k_\beta}. \tag{13.34}$$

As already discussed at §12.6, the effective mass concept is useful to describe the effect of the lattice in regards of the response of the electrons to external forces. It has already been emphasized how the effective mass changes along a given band $E(\mathbf{k})$, so that the electrons can move along the direction of the electric field or along the opposite direction.

By extending Eq. 13.32 and considering that the density of $\mathbf{k}$ states is $Nv_c/8\pi^3$, the current density (Eq. 13.29) can be written

$$\mathbf{j} = \frac{-e}{8\pi^3\hbar} \int_{BZ} \nabla_\mathbf{k} E_\mathbf{k} d\mathbf{k}, \qquad (13.35)$$

where the integration is over all states occupied by electrons, within the Brillouin zone. For a fully occupied band the integral extends over all the BZ.

It must be remarked that for each electron with velocity $\mathbf{v}(\mathbf{k})$ there is another electron at - $\mathbf{k}$ for which

$$\mathbf{v}(-\mathbf{k}) = \frac{1}{\hbar}\nabla_\mathbf{k} E(-\mathbf{k}) = -\frac{1}{\hbar}\nabla_{-\mathbf{k}} E(-\mathbf{k}) = -\frac{1}{\hbar}\nabla_\mathbf{k} E(\mathbf{k}) = -\mathbf{v}(\mathbf{k}) \quad (13.36)$$

(since $E(\mathbf{k}) = E(-\mathbf{k})$, due to the inversion symmetry). Thus the current associated with a full band is **zero**, as it was anticipated. The crystal is an insulator, if no thermal excitation to the upper empty band is considered.

For a partially filled band, according to Eq. 13.33 the electric field redistribute the electrons, so that the distribution is no longer symmetric around $k = 0$. Therefore for a certain time interval there is no cancellation of the contributions to the drift and an electronic current flow along $-\mathcal{E}$ occurs, as sketched below in a one-dimensional reciprocal space:

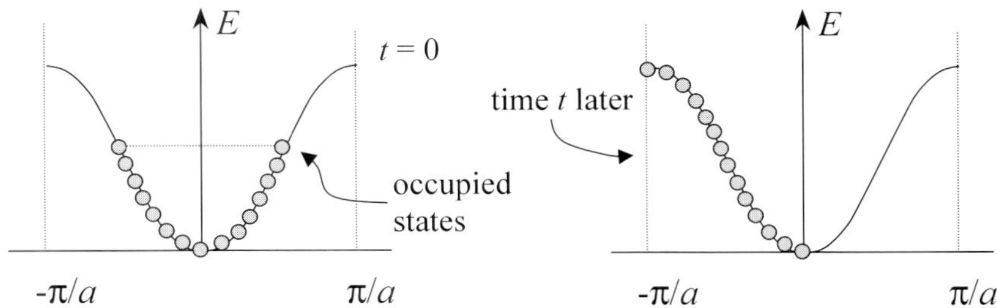

By extending to the band what has been derived for a single electron at Problem XII.6.1, one realizes that after some time the distribution in $\mathbf{k}$-space changes. The states at positive $\mathbf{k}$ are refilled, as sketched below (for the moment, as in Problem XII.6.1, **no scattering** process is assumed to occur).

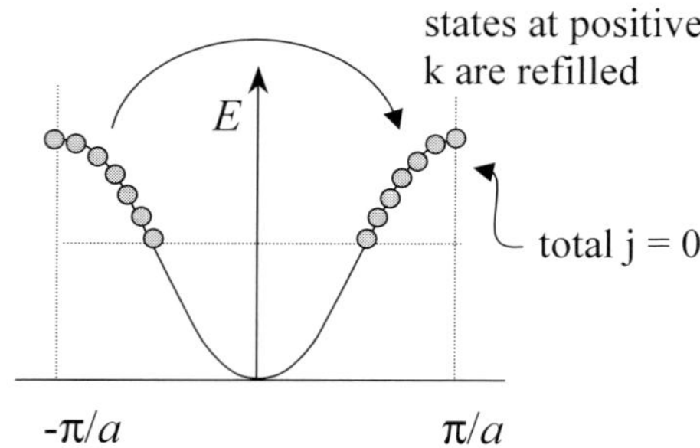

Then, at this moment the current flows due to the regions at positive and at negative **k**'s compensate each other. Later on a neat flow in opposite direction (see Eq. 13.36) should occur. Therefore, as a whole, an oscillating current should be expected upon application of a constant electric field (the so-called **Bloch oscillations**, see Problem XII.6.1 for a single electron). However, we have to take into account the inelastic collisions of the electrons with impurities or oscillating ions. In a simple description one can imagine that after each collision the entire group of electrons is forced to re-take the equilibrium thermal distribution over the **k**-states. Then, for frequent collisions, only the evolution of the system in the first time interval mentioned above is practically effective. The net effect of the field can be thought to generate a stationary distribution skewed in the opposite direction of the field:

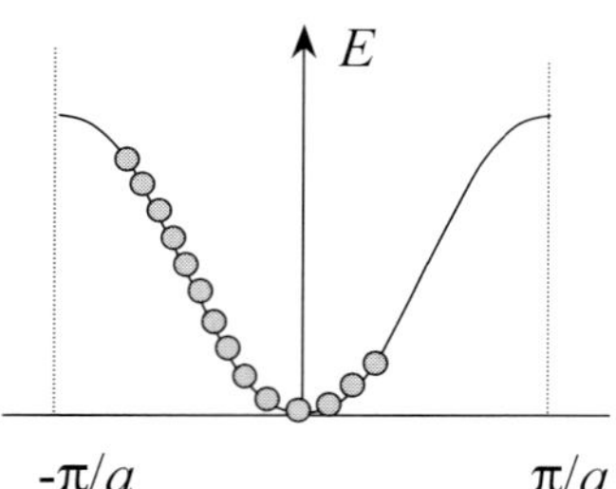

yielding a net flow of current.

For almost totally filled bands a description in terms of **pseudo-particles** (the **holes**) occupying the empty states can be given, as anticipated. In fact, the integral in Eq. 13.35 extends only over the occupied states. Therefore for the current density one can write

$$\mathbf{j} = \frac{-e}{8\pi^3}\left[\int_{BZ}\mathbf{v}(\mathbf{k})d\mathbf{k} - \int_{empty}\mathbf{v}(\mathbf{k})d\mathbf{k}\right] = +\frac{e}{8\pi^3}\int_{empty}\mathbf{v}(\mathbf{k})d\mathbf{k} \ . \qquad (13.37)$$

Thus the current has been formally transformed to a current of **positive particles** occupying **empty electron states**. To those quasi-particles Eqs. 13.31-13.35 and the related concepts do apply.

At the thermal equilibrium the holes are usually confined to the **k** states in the upper part of the band, where the electron effective mass is usually negative. Thus the holes behave as positive charges with a positive effective mass m_h^* moving along the electric field direction.

These concepts are particularly useful in intrinsic semiconductors, where the thermal excitations promote a limited number of electrons from the valence band (fully occupied at $T = 0$) to the conduction band (fully empty at $T = 0$). Since the holes in the valence band and the electrons at the bottom of the conduction band move along opposite directions and have opposite charges, the neat effect is that the electron and hole conductivities sum up.

Appendix XIII.1 Magnetism from itinerant electrons

The magnetic properties associated with **localized magnetic moments**, therefore of crystals with magnetic ions, have been addressed at Chapter 4. At §12.7 and Problem XII.7.6 the paramagnetic susceptibility of the Fermi gas has been described.

The issue of the magnetic properties associated with an ensemble of delocalized electrons, with no interaction (Fermi gas) or in the presence of electron-electron interactions is much more ample. In this Appendix we first recall the diamagnetism due to free electrons (**Landau diamagnetism**). Then some aspects of the magnetic properties of interacting delocalized electrons (**ferromagnetic** or **antiferromagnetic metals**) are addressed, in a simplified form.

The conduction electrons in metals are responsible of a **negative susceptibility**, associated with orbital motions under the action of external magnetic field. To account for this effect one has to refer to the generalized momentum operator (see Eq. 1.26) $-i\hbar\boldsymbol{\nabla}+(e/c)\mathbf{A}$, with $\mathbf{A} = (0, Hx, 0)$ (**second Landau gauge**) [1], for a magnetic field $\mathbf{H}$ along the z axes.

Then the Schrodinger equation takes the form

$$-\frac{\hbar^2}{2m}\left[(\frac{\partial}{\partial x})^2 + (\frac{\partial}{\partial y} + \frac{ieHx}{\hbar c})^2 + (\frac{\partial}{\partial z})^2\right]\psi = E\psi \qquad \text{(A.XIII.1.1)}$$

Since $-i\hbar\boldsymbol{\nabla}_{y,z}$ describe constants of motion with eigenvalues $\hbar k_{y,z}$ one can rewrite this equation in the form

$$\left[-\frac{\hbar^2}{2m}\frac{\partial^2}{\partial x^2} + \frac{1}{2}\frac{e^2 H^2}{mc^2}(x - \frac{\hbar k_y c}{eH})^2 + \frac{\hbar^2 k_z^2}{2m}\right]\psi = E\psi \qquad \text{(A.XIII.1.2)}$$

---

[1] This gauge is translationally invariant along the y-axis, with eigenstates of the y-component of the momentum.

where the first two terms represent the Hamiltonian for a displaced linear oscillator, with characteristic frequency

$$\omega_c = \frac{eH}{mc} = \frac{2\mu_B H}{\hbar} = 2\omega_L \qquad \text{(A.XIII.1.3)}$$

(ω_L Larmor frequency, see Prob.III.2.4). ω_c is the **cyclotron frequency**, while $x_o = \hbar c k_y / eH$ is the center of the oscillations.

Therefore, from Eq. A.XIII.1.2 the eigenvalues turn out

$$E_{nk_z} = \frac{\hbar^2 k_z^2}{2m} + (n + \frac{1}{2})\hbar\omega_c \qquad \text{(A.XIII.1.4)}$$

where the quantum number n labels the **Landau levels**.

The one-electron eigenfunctions in the presence of the magnetic field are plane waves along one direction (dependent on the choice of the gauge for $\mathbf{A}$) multiplied by the wavefunctions for the harmonic oscillator.

The semiclassical view of the result given at Eq. A.XIII.1.4 is that under the Lorenz force $\mathbf{F}_L = -(e/c)\mathbf{v}_g \times \mathbf{H}$ (with $\mathbf{v}_g$ the group velocity) the evolution of the crystal momentum $\hbar d\mathbf{k}/dt = \mathbf{F}_L$ induces a cyclotron rotational motion in the xy plane while the electron propagates along the z direction (see Problem XII.3.2).

It is noticed that each Landau level is degenerate, the degeneracy depending on the number of possible values for x_o. For a crystal of volume $V = L_x.L_y.L_z$, then $0 \leq x_0 \leq L_x$, while one has $0 \leq k_y \leq L_x eH/\hbar c \equiv k_y^{max}$. Then, k_y being quantized in steps $\Delta k_y = 2\pi/L_y$, the degeneracy of each Landau level, given by the number of oscillators with origin within the sample, is

$$N_L(H) = \frac{k_y^{max}}{\Delta k_y} = L_x L_y H \frac{e}{hc} = \frac{\Phi(H)}{\Phi_o} \, , \qquad \text{(A.XIII.1.5)}$$

where $\Phi(H)$ is the flux of the magnetic field across the crystal and $\Phi_o = hc/e \simeq 4 \times 10^{-7}$ Gauss cm^2 is the **flux quantum** [2].

It is observed that the degeneracy, the same for all the n levels, increases linearly with H. Hence, by increasing H one can vary the population of each level and eventually when H is very high (and for moderate electron densities) all electrons will occupy just the first $n = 0$ level. Accordingly, on increasing H different Landau levels will cross the Fermi energy.

By resorting to the results outlined above one can calculate the energy of the electrons $E(H)$ in presence of the field and then the magnetization. One can conveniently distinguish two regimes, for $k_B T$ large or small compared to $\hbar\omega_c$. For $k_B T \ll \hbar\omega_c$ an oscillatory behaviour of $E(H)$ is observed. The

---

[2] It can be noticed that the flux quantum here is by a factor 2 larger than the superconducting fluxon $\Phi_{SC} = hc/2e$, since in the latter case a Cooper pair, of charge $2e$, is involved.

oscillations occur when the Landau level pass through the Fermi surface and cause changes in the energy of the conduction electrons, namely for

$$(n + \frac{1}{2})\hbar\omega_c = E_F \quad , \qquad\qquad \text{(A.XIII.1.6)}$$

Characteristic oscillations in the magnetization, known as **De Haas-Van Alphen oscillations** can be detected.

For $k_B T \gg \hbar\omega_c$ the discreteness of the Landau levels is no longer effective and the energy increases with H^2:

$$E(H) \propto \hbar\omega_c[\hbar\omega_c D(E_F)] \quad ,$$

corresponding to an increase by $\hbar\omega_c$ of the energy for all the $\hbar\omega_c D(E_F)$ electrons in a Landau level ($D(E_F)$ density of states at the Fermi level, see §12.7.1). Therefore, the susceptibility turns out

$$\chi_L = -\frac{1}{12}(\frac{e\hbar}{mc})^2 \frac{D(E_F)}{N v_c} = -\frac{1}{12\pi^2} \frac{e^2}{mc^2} k_F \quad , \qquad\qquad \text{(A.XIII.1.7)}$$

k_F being the Fermi wave vector. From the Pauli susceptibility χ_P (see Problem XII.7.6) one can write

$$\chi_L = -\frac{1}{3}\chi_P \quad . \qquad\qquad \text{(A.XIII.1.8)}$$

Modifications in χ_L (as well as in χ_P) have to be expected when the effective mass m^* of the electrons is different from m_e. For instance, when $m^* \ll m_e$ (as for example in bismuth, where $m^* \sim 0.01 m_e$) the metal can become diamagnetic. In fact, the total susceptibility for non-interacting delocalized electrons has to be written

$$\chi_{total} = \mu_B^2 D(E_F)[1 - \frac{1}{3}(\frac{m_e}{m^*})^2] \equiv \chi_P[1 - \frac{1}{3}(\frac{m_e}{m^*})^2]$$

For further insights on the behaviour of the Fermi gas in the presence of constant magnetic field, Chapter XV in the book by **Grosso** and **Pastori Parravicini** (quoted in the Preface) should be read.

In transition metals, with partially occupied d bands, the electrons involved in the magnetic properties are itinerant with relevant **many-body correlation** effects. The Fermi-gas picture for the conduction electrons is no longer adequate and significant modifications to the Pauli susceptibility have to be expected, including the possibility of the transition to an ordered state. In these cases one often speaks of **ferro** (or **antiferro)magnetic metals**. For example, an experimental evidence of a particular itinerant ferromagnetism is iron metal: the magnetic moment per atom is found around $2.2\mu_B$. This value cannot be justified in terms of localized moments on Fe^{2+} ion, in the $^5D_4$ state (see §3.2.3).

The simplest model to account for the correlation effects on the magnetic properties of itinerant electrons is the one due to **Stoner** and **Hubbard**. In this model the electron-electron Coulomb interaction is replaced by a constant repulsive energy U between electrons on the same site, with opposite spins according to Pauli principle. Then the total Hamiltonian is written

$$\mathcal{H} = \sum_{\mathbf{k}} E(\mathbf{k})(n_{\mathbf{k},\uparrow} + n_{\mathbf{k},\downarrow}) + U \sum_m p_{m,\uparrow}p_{m,\downarrow} \qquad \text{(A.XIII.1.9)}$$

where the first term is the usual free electron kinetic Hamiltonian, while the second term describes the repulsive on-site interaction, with the sum running over all lattice sites.

The total magnetization can be derived in a way analogous to the one used for the Pauli susceptibility (Problem XII.7.6), by estimating the numbers of electrons with spin up and spin down, following the application of the magnetic field. For N electrons per cubic cm, in the conduction band of width larger than U, $N_\uparrow$ and $N_\downarrow$ are the numbers of electrons of spin up and spin down respectively. Then the energy for spin-up electrons turns out

$$E(\mathbf{k})_\uparrow = E(\mathbf{k}) + Un_\downarrow + \mu_B H \qquad \text{(A.XIII.1.10)}$$

while for electrons with spin-down

$$E(\mathbf{k})_\downarrow = E(\mathbf{k}) + Un_\uparrow - \mu_B H \ . \qquad \text{(A.XIII.1.11)}$$

where $n_{\uparrow,\downarrow} = N_{\uparrow,\downarrow}/N$.

The decrease of the energy of the spin-down band with respect to the spin-up band yields an increase in the population of spin-down electrons and a non-zero magnetization. Since (see again Problem XII.7.6) for $N_{\uparrow,\downarrow}$ one writes

$$N_\downarrow = \frac{1}{2} \int_{Un_\uparrow - \mu_B H}^{\infty} f(E)D(E - Un_\uparrow + \mu_B H)dE \simeq$$

$$\simeq \frac{1}{2} \int_0^{\infty} f(E)D(E)dE + \frac{1}{2}(\mu_B H - Un_\uparrow)D(E_F) \quad \text{(A.XIII.1.12)}$$

while

$$N_\uparrow \simeq \frac{1}{2} \int_0^{\infty} f(E)D(E)dE - \frac{1}{2}(\mu_B H + Un_\downarrow)D(E_F) \ . \qquad \text{(A.XIII.1.13)}$$

The magnetization (per unit volume) becomes

$$M = \mu_B \frac{(N_\downarrow - N_\uparrow)}{V} \simeq \frac{\mu_B U D(E_F)}{2N}(N_\downarrow - N_\uparrow) + \mu_B^2 D(E_F)H \quad \text{(A.XIII.1.14)}$$

(V the reference volume). Therefore the magnetic susceptibility becomes

$$\chi = \frac{M}{H} = \frac{\mu_B^2 D(E_F)}{1 - \frac{UD(E_F)}{2N}} = \frac{\chi_P}{1 - (U\chi_P/2\mu_B^2 N)} \, , \qquad \text{(A.XIII.1.15)}$$

with χ_P Pauli susceptibility (for bare electrons) and $D(E_F)$ the density of states per unit volume.

It is noted that when $UD(E_F)/2N \to 1$ (**Stoner criterium**) the susceptibility diverges and ferromagnetic order is attained.

Even if the Stoner condition is not fulfilled, Eq. A.XIII.1.15 shows that the susceptibility is significantly modified with respect to the one for bare free-electrons. Eq. A.XIII.1.15 can be considered a particular case of Eq. 4.33, where the enhancement factor corresponds to the mean field acting on a particular electron due to the interaction with all the others. Stoner criterium rather well justifies the ferromagnetism in metals like Fe, Co and Ni, as well as the enhanced susceptibility (about $5\ \chi_P$) measured in Pt and Pd metals.

Finally a few words are in order about the magnetic behaviour of itinerant electrons when the concentration n is reduced (**diluted electron fluid in the presence of electron-electron interaction**). As shown in Problem F.XIII.2 the Coulomb repulsive energy of the electrons goes as $< E_C >\propto e^2 n^{1/D}$ (D the dimensionality), while for the kinetic energy (for $T \to 0$) one has $< T >\propto n^{2/D}$. Thus the electron dilution causes a decrease of the average kinetic energy which is more rapid than the one for the average Coulomb repulsion. Eventually, below $n_{3D} = 1.4 \times 10^{-3}/a_o^3$ and below $n_{2D} = 0.4/a_o^2$, when $< E_C >$ becomes dominant, a spontaneous "crystallization" should occur, in principle (**Wigner crystallization**).

Monte Carlo simulations predict a three-dimensional crystallization at densities below 2×10^{18} cm^{-3}, while at densities below 2×10^{20} cm^{-3} the Coulomb interaction should be strong enough to align all spins, according to the Stoner criterium. Charge or spin ordering are hard to be experimentally tested, mainly because of the difficulty of the physical realization of the electron fluid at low density sufficiently free from impurities and/or defects.

Problems F.XIII

Problem F.XIII.1 Silver is a monovalent metal, with density $10.5\,\text{g/cm}^3$ and fcc structure. From the values of the resistivity at $T = 20$K and $T = 295$K given by $\rho_{20} = 3.8 \cdot 10^{-9}\,\Omega\,\text{cm}$ and $\rho_{295} = 1.6 \cdot 10^{-6}\,\Omega\,\text{cm}$, estimate the mean free paths of the electrons.

Solution:

The Fermi wavevector turns out $k_F = 1.2 \cdot 10^8\ cm^{-1}$ and then the Fermi energy is $E_F = 63800$ K.

From $\rho = m/ne^2\tau$ and $<v> = e\mathcal{E}\lambda/mj = (e\lambda m)\rho$, with $\lambda = <v>\tau$ and $<v> \sim \sqrt{E_F/m}$ (see the *incipit* at §13.4), one derives

$$\lambda = 5.2 \cdot 10^{-12}\,\text{cm}\quad\text{at}\ 295\,K\quad\text{and}\quad \lambda = 0.2 \cdot 10^{-8}\,\text{cm}\quad\text{at}\ 20\,K\,.$$

Problem F.XIII.2 For three-dimensional and for two-dimensional metals, in the framework of the free-electron model and for $T \to 0$, evaluate the electron concentration n at which the average kinetic energy coincides with the average Coulomb repulsion (which can be assumed $U = e^2/d$, with d the average distance between the electrons).

Solution:
In 3D $d = 1/(4\pi n/3)^{1/3}$, while in 2D $d = 1/n^{1/2}$. Thus

$$U^{3D} = e^2\left(\frac{4\pi}{3}\right)^{1/3}n^{1/3}\quad\text{and}\quad U^{2D} = e^2 n^{1/2}$$

The average kinetic energy per electron (for $T \to 0$) is $<E> = \int_0^{E_F} D(E)E\,dE$, with $D(E)^{3D} = (3/2)E^{1/2}/E_F^{3/2}$ and $D(E)^{2D} = 1/E_F$.

Then $<E>^{3D} = (3/5)E_F$ and $<E>^{2D} = (1/2)E_F$.
The average kinetic energy coincides with the Coulomb repulsion for $n^{3D} = 1.4 \times 10^{-3}/a_0^3$ and $n^{2D} = 0.4/a_0^2$, with a_0 Bohr radius.

Problem F.XIII.3 A magnetic field is applied on an atom with a single p electron in the crystal field at the octahedral symmetry (§13.3), with six charges Ze along the $\pm x$, $\pm y$, $\pm z$ axes. Show that without the distortion of the octahedron (namely $a = b$, with a the distance from the atom of the charges in the xy plane and b the one along the z axis) only a shift of the p levels would occur. Then consider the case $b \neq a$ and discuss the effect of the magnetic field (applied along the z axis). Prove that quenching of the angular momentum occurs and derive the eigenvalues.

Solution:
By summing the potential due to the six charges, analogously to the case described at §13.3, for $r \ll a$ the crystal field perturbation turns out

$$V_{CF} = -Ze^2\left\{\left(\frac{1}{a^3} - \frac{1}{b^3}\right)r^2 + 3\left(\frac{1}{b^3} - \frac{1}{a^3}\right)z^2\right\} + ... = A(3z^2 - r^2) + \text{const}$$

where $A \neq 0$ only for $b \neq a$.
From the unperturbed eigenfunctions the matrix elements of V_{CF} are

$$<\phi_{p_x}|V_{CF}|\phi_{p_x}> = A\int r^2|\mathcal{R}(r)|^2 r^2 dr \int \sin^2\theta \cos^2\phi(3\cos^2\theta - 1)\sin\theta\,d\theta\,d\phi$$
$$= -A<r^2>\tfrac{8\pi}{15} = <\phi_{p_y}|V_{CF}|\phi_{p_y}>$$

while
$$< \phi_{p_z} |V_{CF}| \phi_{p_z} >= A < r^2 > \frac{16\pi}{15} \; .$$

In the absence of magnetic field the energy levels are

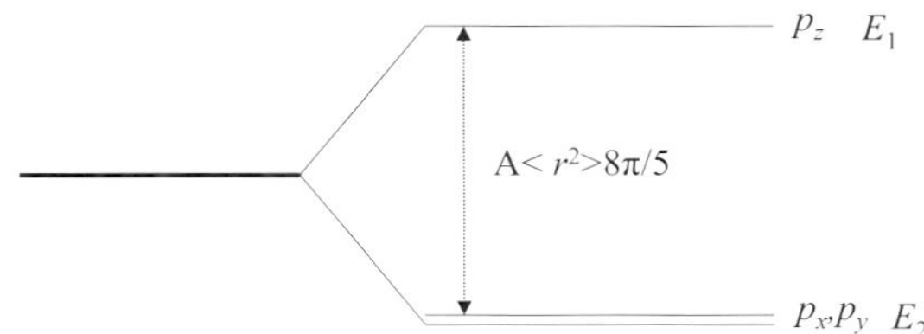

In the presence of the field, the total perturbative Hamiltonian becomes $V_{CF} + \mu_B l_z H$.
The diagonal matrix elements of l_z in the basis of the unperturbed eigenfunctions are zero. In fact,

$$< \phi_{2p_x}|l_z|\phi_{2p_x} >= +i\hbar \int_0^\infty f(r)dr \int_0^\pi \sin^3\theta \, d\theta \int_0^{2\pi} \sin\phi\cos\phi \, d\phi = 0$$

(Problems F.IV.1 and F.IV.2) and, analogously,

$$< \phi_{2p_y}|l_z|\phi_{2p_y} >=< \phi_{2p_z}|l_z|\phi_{2p_z} >= 0 \; .$$

The non-diagonal matrix elements are $< \phi_{2p_y}|l_z|\phi_{2p_x} >= i\hbar = - < \phi_{2p_x}|l_z|\phi_{2p_y} >$.
The secular equation becomes

$$\begin{vmatrix} E_0 - E & -i\mu_B H & 0 \\ i\mu_B H & E_0 - E & 0 \\ 0 & 0 & E_1 - E \end{vmatrix} = 0$$

yielding $E' = E_1$ and $E'' = E_0 \pm \mu_B H$, i.e.:

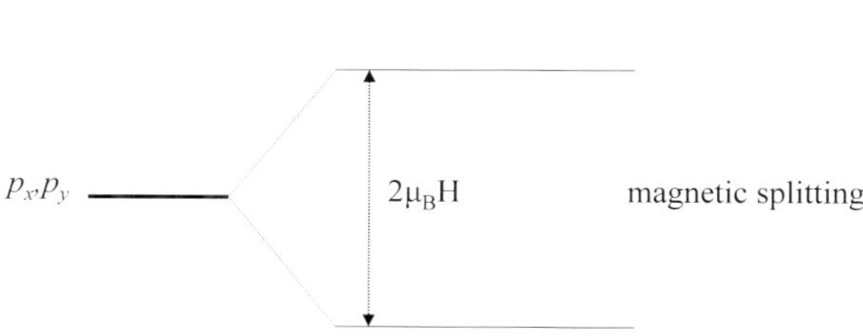

14

Vibrational motions of the ions and thermal effects

Topics

Elastic waves in crystals
Acoustic and optical branches
Debye and Einstein models
Phonons
The melting temperature
Mössbauer effect

14.1 Motions of the ions in the harmonic approximation

Hereafter we shall afford the problem of the motions of the ions around their equilibrium positions in an ideal (disorder- and defect-free) crystal. The motions are called **lattice vibrations**. The Born-Oppenheimer separation and the adiabatic approximation (§7.1) will be implicit and the concepts involved in the description of the normal modes (§10.6) in the harmonic approximation will be used. In fact, the crystal cell will be considered as a molecular unit: its normal modes propagate along the crystal with a phase factor, in view of the spatial periodicity.

According to the definitions sketched below,

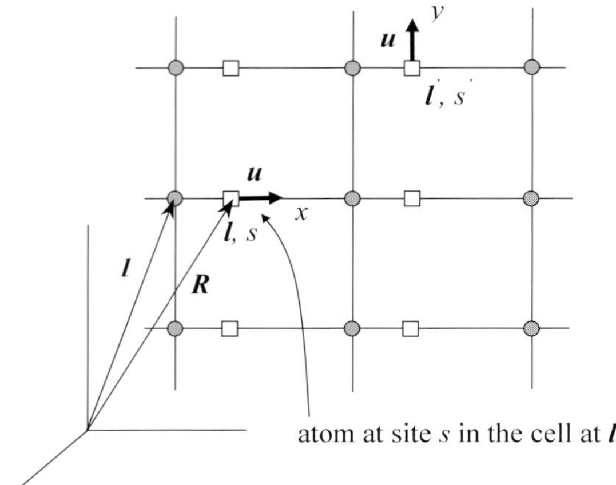

within the harmonic approximation the potential energy will be written

$$V_2 = \frac{1}{2} \sum_{\mathrm{l},s,\alpha} \sum_{\mathrm{l}',s',\beta} \left(\frac{\partial^2 V}{\partial \alpha(\mathrm{l}, s) \partial \beta(\mathrm{l}', s')} \right)_o \mathbf{u}_\alpha(\mathrm{l}, s) \mathbf{u}_\beta(\mathrm{l}', s')$$

$$\equiv \sum_{\mathrm{l},s,\alpha} \sum_{\mathrm{l}',s',\beta} \Phi^{(\alpha,\beta)}_{\mathrm{l},s,\mathrm{l}',s'} \mathbf{u}_\alpha(\mathrm{l}, s) \mathbf{u}_\beta(\mathrm{l}', s') \ , \qquad (14.1)$$

where $\Phi^{(x,y)}_{\mathrm{l},s,\mathrm{l}',s'}$ involves the force along the x direction on the ion at site s of the l-th cell when the ion at site s' in the l' cell is displaced by the unit length along the y direction. From Eq. 14.1 the equations of motion turn out

$$m_s \frac{d^2 \mathbf{u}_{l,s}}{dt^2} = -\frac{\partial V_2}{\partial \mathbf{u}_{l,s}} = -\sum_{\mathrm{l}',s'} \Phi_{\mathrm{l},s,\mathrm{l}',s'} \mathbf{u}_{\mathrm{l}',s'}, \qquad (14.2)$$

namely $3SN$ coupled equations (S number of atoms in each cell).

Recalling the normal modes in the molecules (§10.6) it is conceivable that due to the translational invariance, the motion of the atom at site s in a given cell differs only by a phase factor with respect to the one in another cell (this is the analogous of the Bloch orbital condition for the electron states). Therefore the displacement of the $(\mathbf{l}, s)$ atom along a given direction is written in terms of plane waves propagating the normal coordinates within a cell:

$$u_\alpha^{(\mathbf{q})}(\mathbf{l}, s) = U_\alpha(s, \mathbf{q}) e^{i\mathbf{q} \cdot \mathbf{R}(\mathbf{l}, s)} e^{-i\omega_\mathbf{q} t}, \qquad (14.3)$$

where $\mathbf{q}$ are the wavevectors defined by the boundary conditions (the analogous of the electron wavevector $\mathbf{k}$, §12.4).

From Eqs. 14.3 and 14.2 for each $\mathbf{q}$, by taking $\mathbf{h} = \mathbf{l} - \mathbf{l}'$, one has

$$m_s \omega_{\mathbf{q}}^2 U_\alpha(s, \mathbf{q}) = \sum_{\beta, s'} U_\beta(s', \mathbf{q}) M_{\alpha,\beta}(s, s', \mathbf{q}), \qquad (14.4)$$

where

$$M_{\alpha,\beta}(s, s', \mathbf{q}) \equiv \sum_{\mathbf{h}} \Phi_{\mathbf{l},s,\mathbf{l}',s'}^{(\alpha,\beta)} e^{i\mathbf{q}\cdot\mathbf{h}} \qquad (14.5)$$

is the **dynamical matrix**, namely the Fourier transform of the elastic constants.

14.2 Branches and dispersion relations

For a given wave-vector Eq. 14.4 can be rewritten in the compact form

$$\omega^2 m \mathbf{U} = \mathbf{MU} \qquad (14.6)$$

where $\mathbf{M}$ is a square matrix of $3S$ degree, m is a diagonal matrix and $\mathbf{U}$ is a column vector. As for the normal modes in molecules (see Eq. 10.53) the condition for the existence of the normal coordinates is

$$|\mathbf{M} - \omega^2 m| = 0 \ . \qquad (14.7)$$

For each wavevector $\mathbf{q}$ Eq. 14.7 yields $3S$ angular frequencies $\omega_{\mathbf{q},j}^2$. Here j is a **branch index**. $3S - 3$ branches are called **optical** since, as it will appear at §14.3.2, they can be active in infrared spectroscopy, while 3 branches are called **acoustic**, since in the limit $\mathbf{q} \to 0$ the crystal must behave like an **elastic continuum**, where $\omega_{\mathbf{q}} = v_{sound}\mathbf{q}$. At variance, for the optical branches (see §14.3.2) for $q = 0$ one has $\omega_{\mathbf{q},j} \neq 0$.

The $\mathbf{q}$-dependence of $\omega_{\mathbf{q},j}$ is called **dispersion relation**. In analogy to the density of $\mathbf{k}$-states for the electrons (§12.5), one can define a density of $\mathbf{q}$ values in the reciprocal space: $D(\mathbf{q}) = N v_c/8\pi^3$. One also defines the **vibrational spectrum** $D_j(\omega)$ for each branch, with the sum rule $\sum_{j=1}^{3S} \int D_j(\omega)d\omega = 3NS$.

In the next Section illustrative examples of vibrational spectra will be given.

14.3 Models of lattice vibrations

In this Section the classical vibrational motions of the ions within the harmonic approximation will be addressed for some model systems.

14.3.1 Monoatomic one-dimensional crystal

Let us refer to a linear chain of identical atoms, for simplicity by considering only the longitudinal motions along the chain direction:

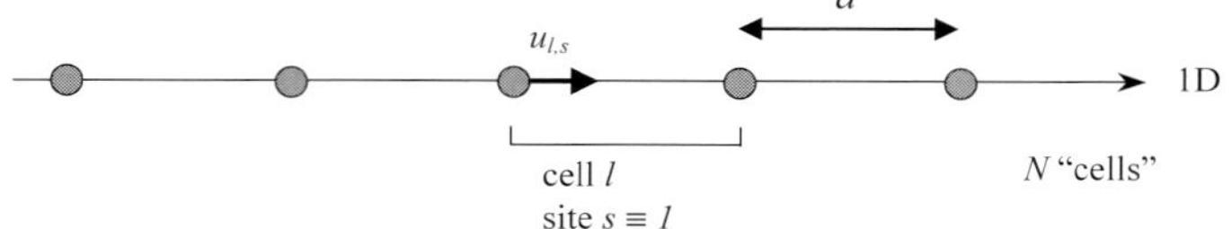

The equations of motions are of the form Eq. 14.2, the index s being redundant. One first selects in the reciprocal space a wavevector $q = n_1 2\pi/Na$, with $-N/2 \leq n_1 \leq N/2$. Then one writes the $u_{l,s}$ displacement as due to the superposition of the ones caused by the waves propagating along the chain, for each q (correspondent to Eq. 14.3). From Eq. 14.2 and 14.4 one writes

$$m_s \omega_{\mathbf{q}}^2 U(s,\mathbf{q}) = \sum_{s'} U(s',\mathbf{q}) M(s,s',\mathbf{q}), \qquad (14.8)$$

where

$$M(s,s',\mathbf{q}) \equiv \sum_{\mathbf{h}} \Phi_{1,s,1',s'} e^{i\mathbf{q}\cdot\mathbf{h}} \qquad (14.9)$$

is the **collective force constant**, representing the Fourier transform of the elastic constants. Eqs. 14.8 and 14.9 describe the propagation of the normal modes of the "cell" along the chain.

By limiting the interaction to the nearest neighbors,

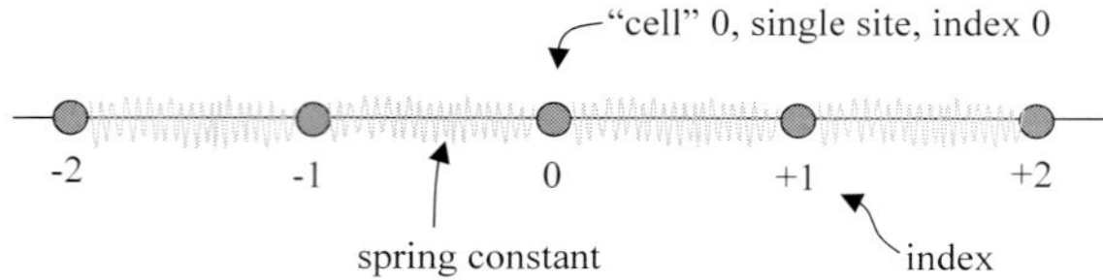

the equation of motion for the atom in the cell at the origin ($l = 0$) turns out

$$m\frac{d^2 u_0}{dt^2} = -2ku_0 + ku_1 + ku_{-1} \qquad (14.10)$$

implying $\Phi(0,0) = 2k$ and $\Phi(\pm 1, 0) = -k$.

The dynamical matrix (Eq. 14.5) is reduced to

$$M = \Phi(0,0) + \sum_{n=\pm 1} \Phi(n,0) e^{iqna}$$

and Eq. 14.8 takes the form

$$m\omega_q^2 U_q = (2k - 2k\cos(qa)) U_q, \qquad (14.11)$$

namely the one for a single **normal oscillator**, with an effective elastic constant taking into account the coupling to the nearest neighbors.

The solubility condition (Eq. 14.7) corresponds to

$$\omega_q^2 = \frac{2k}{m}(1 - cos(qa)),\tag{14.12}$$

yielding the dispersion relation

$$\omega_q = 2\sqrt{\frac{k}{m}}\,sin(qa/2)\tag{14.13}$$

sketched below:

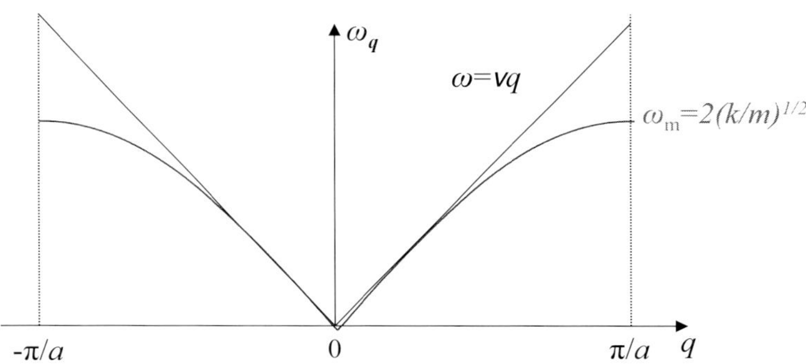

The vibrational spectrum, or density of states $D(\omega) = D(q)dq/d\omega$, with $D(q) = Na/2\pi$, turns out

$$D(\omega) = (2N/\pi\sqrt{\omega_m^2 - \omega^2}),\tag{14.14}$$

reported below

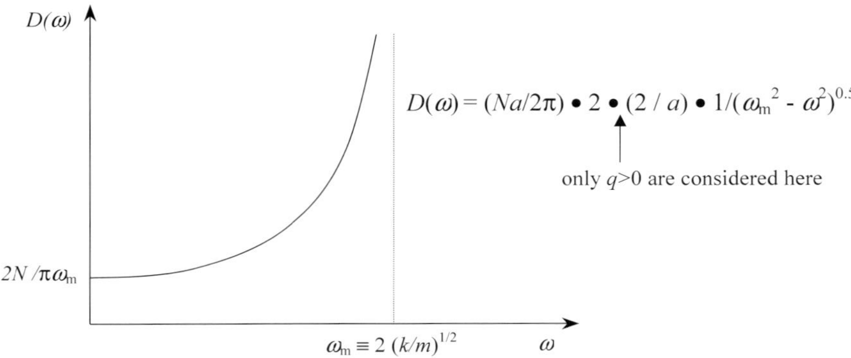

The situation arising at the zone boundary, where $\omega_{q=\pi/a} \equiv \omega_m$, is equivalent to the one encountered at the critical points of the electronic states (see §12.5).

14.3.2 Diatomic one-dimensional crystal

For a chain with two atoms per unit cell, with mass m_1 and m_2 ($m_1 > m_2$), again considering the longitudinal modes and assuming a single elastic constant and nearest neighbour interactions,

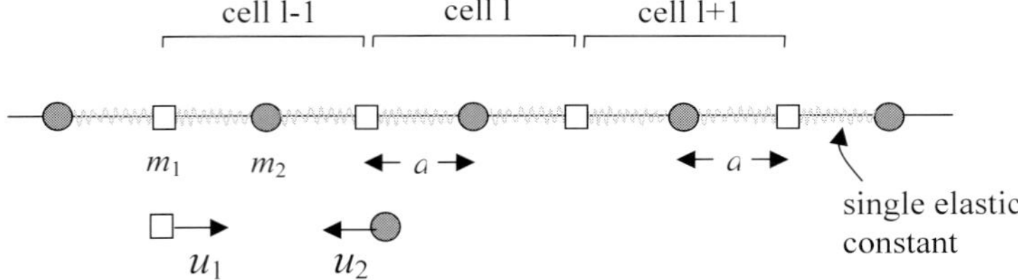

the equations of motions for the atoms at sites $s = 1$ and $s = 2$, within the l-th cell, are

$$m_1 \frac{d^2 u_{l,1}}{dt^2} = -2ku_{l,1} + ku_{l,2} + ku_{l-1,2}$$

$$m_2 \frac{d^2 u_{l,2}}{dt^2} = -2ku_{l,2} + ku_{l,1} + ku_{l+1,1} \tag{14.15}$$

Again resorting to solutions of the form

$$u(l,1) = U_1 e^{iq2la} e^{-i\omega_q t}$$

and

$$u(l,2) = U_2 e^{iq(a+2la)} e^{-i\omega_q t}$$

(the index q in $U_{1,2}$ is dropped here), one has

$$(\frac{2k}{m_1} - \omega^2)U_1 - \frac{k}{m_1}(e^{iqa} + e^{-iqa})U_2 = 0$$

$$-\frac{k}{m_2}(e^{iqa} + e^{-iqa})U_1 + (\frac{2k}{m_2} - \omega^2)U_2 = 0 \ . \tag{14.16}$$

The dynamical matrix is

$$M = \begin{pmatrix} \dfrac{2k}{} & -k(e^{iqa} + e^{-iqa}) \\ -k(e^{-iqa} + e^{iqa}) & 2k \end{pmatrix}$$

and the solubility condition

$$\begin{pmatrix} 2k - m_1\omega^2 & -2k\cos(qa) \\ -2k\cos(qa) & 2k - m_2\omega^2 \end{pmatrix} = 0$$

leads to

$$\omega_q^2 = k(\frac{1}{m_2} + \frac{1}{m_1}) \pm k\left[(\frac{1}{m_2} + \frac{1}{m_1})^2 - \frac{4}{m_1 m_2}\sin^2(qa)\right]^{\frac{1}{2}} \ . \tag{14.17}$$

The dispersion relations are shown in Fig.14.1, with μ reduced mass.

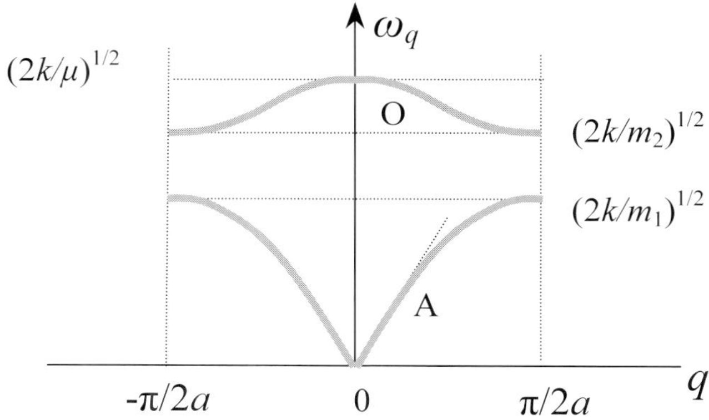

Fig. 14.1. Frequencies of the acoustic (A) and optical (O) longitudinal modes in one-dimensional diatomic crystal, according to Eq. 14.17.

At the boundaries of the Brillouin zone ($q = \pm\pi/2a$) the frequencies of the acoustic and optical modes are $\omega^{\mathrm{A}} = \sqrt{2k/m_1}$ and $\omega^{\mathrm{O}} = \sqrt{2k/m_2}$, respectively.

It is noted that when $m_1 = m_2$ the two frequencies coincide, the gap at the zone boundary vanishes: the situation of the monoatomic chain is restored, once that the length of the lattice cell becomes a instead of $2a$.

For a given wavevector one can obtain the atomic displacements induced by each normal mode. For instance, by choosing $q = 0$ for the acoustic branch one derives $U_{\mathrm{A}}(0,1) = U_{\mathrm{A}}(0,2)$, the same displacement for the two atoms, corresponding to the translation of all the crystal. For the optical mode, again for $q = 0$ one has $m_1 U_{\mathrm{O}}(0,1) = -m_2 U_{\mathrm{O}}(0,2)$, keeping fixed the center of mass. As for the diatomic molecule (see §10.6) the difference of the two displacements corresponds to the normal coordinate.

In a similar way one can derive the displacements associated with the zone boundary wavevectors (Fig. 14.2, where also the transverse modes are schematized).

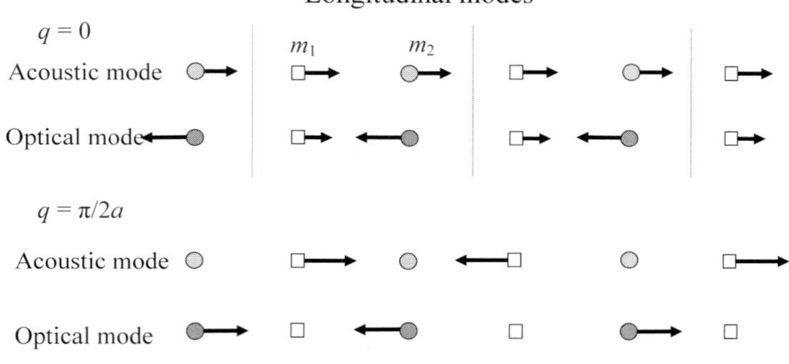

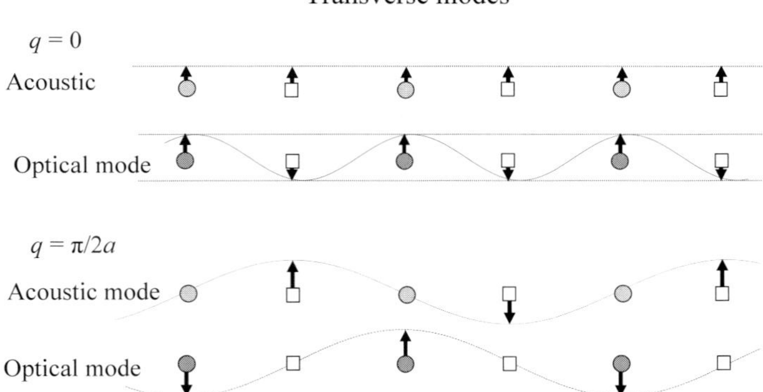

Fig. 14.2. Atomic displacements associated with the $q = 0$ and the $q = \pi/2a$ acoustic (A) and optical (O) modes, for one-dimensional diatomic crystal.

From the dispersion relations (Eq. 14.17) the vibrational spectra reported in Fig. 14.3 are derived.

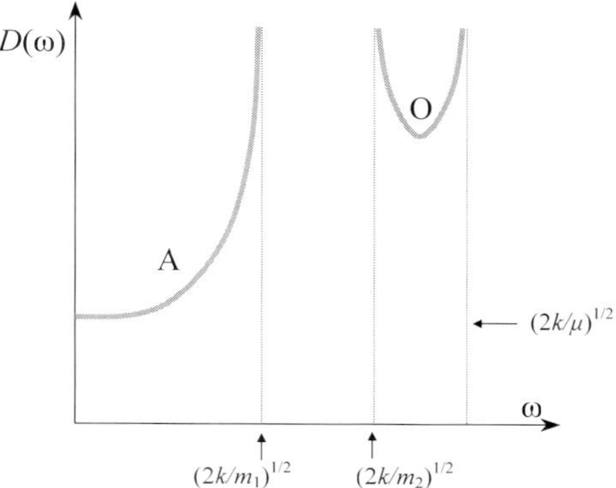

Fig. 14.3. Vibrational spectra for the longitudinal acoustic (A) and optical (O) branches in one-dimensional diatomic crystal.

Up to now only longitudinal modes have been considered. To describe the transverse vibrations the elastic constants for the displacements perpendicular to the chain should be considered. In this way, for a given wave-vector, 3 vibrational branches would be obtained for the monoatomic chain and 6 branches for the diatomic one, at longitudinal (L) and transverse (T) optical and acoustic characters (see Fig. 14.2).

Finally one should observe that the interaction with electromagnetic waves requires the presence of **oscillating electric dipole** within the cell. To grant energy and momentum conservation, the absorption process should occur in correspondence to the **photon momentum** $q = \hbar\omega/c$, which for typical values of the frequencies ($\omega \sim 10^{13} - 10^{14}$ rad s^{-1}) is much smaller than $h/2a$. For $q \to 0$, at the center of the Brillouin zone, the acoustic modes do not yield any dipole moment. Therefore only the optical branches, implying in general oscillating dipoles (as schematized in Fig. 14.2), can be active for the absorption of the electromagnetic radiation, similarly to the case described for the molecules.

14.3.3 Einstein and Debye crystals

The phenomenological models due to Einstein and to Debye are rather well suited for the approximate description of specific properties related to the lattice vibrations in real crystals.

The **Einstein crystal** is assumed as an ensemble of independent atoms elastically connected to equilibrium positions. The interactions are somewhat reflected in a vibrational constant common to each oscillator, yielding a characteristic frequency ω_E. As regards the dispersion curves, one can think that for each $\mathbf{q}$ there is a threefold degenerate mode at frequency ω_E. Thus, the vibrational spectrum could be schematized as below:

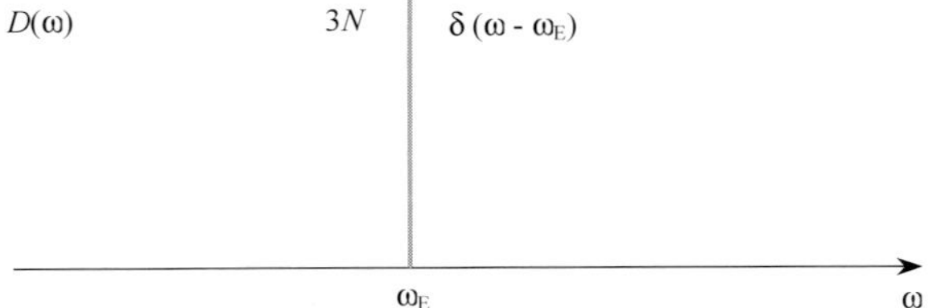

Although introduced to justify the low-temperature behavior of the specific heat (see §14.5), the Einstein model is often applied in order to describe the properties of the optical modes in real crystals. In fact, the optical modes are often characterized by weakly q-dependent dispersion curves with a narrow $D(\omega)$, not too different from the delta-like vibrational spectrum of the Einstein model heuristically broadened, as sketched below:

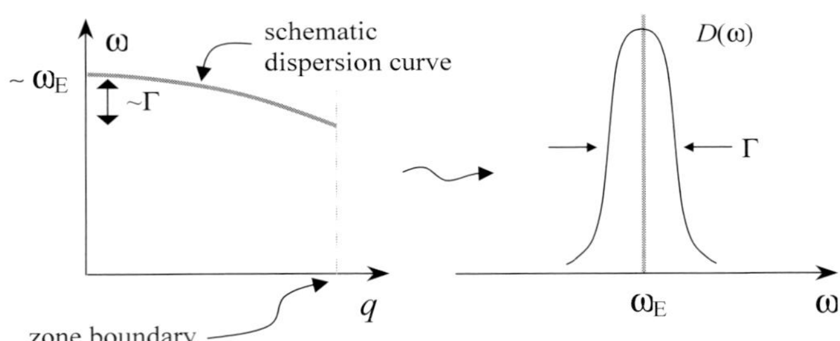

In the **Debye model** it is assumed that the vibrational properties are basically the ones of the **elastic** (and sometimes isotropic) **continuum**, with *ad hoc* conditions in order to take into account the discrete nature of any real crystal. In particular:

i) the Debye model describes rather well the acoustic modes of any crystal, since for $\mathbf{q} \to 0$ the dispersion curves of the acoustic branches practically coincide with the ones of the continuum solid, the wavelength of the vibration being much larger than the lattice step.

ii) the model cannot describe the vibrational contribution from optical modes.

iii) one has to introduce a cutoff frequency ω_D in the spectrum in order to keep the number of modes limited to $3N$ (for N atoms).

iv) only 3 branches have to be expected, with dispersion relations of the form $\omega_q^j = \mathbf{v}_{sound}^j \mathbf{q}$, where the sound velocity can refer to transverse or to longitudinal modes.

For a given branch, in the assumption of isotropy, the vibrational spectrum turns out

$$D_j(\omega) = \frac{Nv_c}{8\pi^3}\,d\mathbf{q} = \frac{Nv_c}{8\pi^3}\,4\pi q^2 dq = \frac{Nv_c}{8\pi^3}\,\frac{4\pi\omega^2}{v_j^3}. \tag{14.18}$$

One can introduce an average velocity v and again in the isotropic case, $3/v^3 = 2/v_T^3 + 1/v_L^3$. Therefore

$$D(\omega) = \frac{Nv_c}{8\pi^3}\,\frac{12\pi\omega^2}{v^3} = \frac{Nv_c}{v^3}\,\frac{3}{2\pi^2}\omega^2, \tag{14.19}$$

the typical vibrational spectrum characteristic of the continuum.

Now a cutoff frequency ω_D (known as **Debye frequency**) has to be introduced. The role of ω_D in the dispersion relation and in the vibrational spectrum $D(\omega)$ is illustrated below:

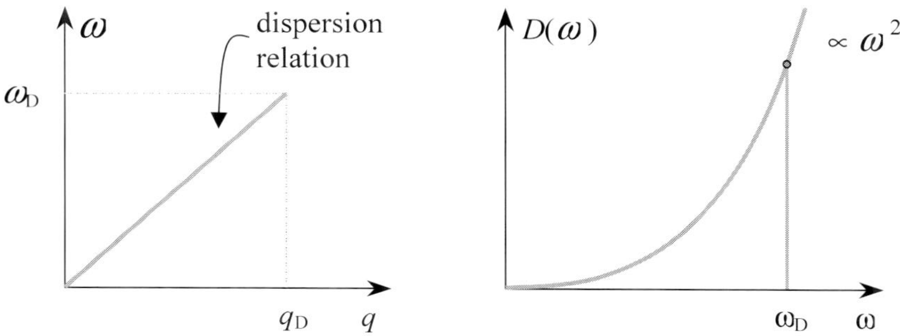

ω_D can be derived from the condition $\int D(\omega)d\omega = 3N$ or, equivalently, by evaluating the Debye radius q_D of the sphere in the reciprocal space which includes the N allowed wavevectors.

Thus $(Nv_c/8\pi^3)(4\pi q_D^3/3) = N$ and then

$$q_D = \left(\frac{6\pi^2}{v_c}\right)^{\frac{1}{3}} \tag{14.20}$$

and

$$\omega_D = vq_D = v\left(\frac{6\pi^2}{v_c}\right)^{\frac{1}{3}}. \tag{14.21}$$

In real crystals detailed description of the vibrational modes are difficult. One can recall the following. In the $\mathbf{q} \to 0$ limit one can refer to the conditions of the continuum and the acoustic branches along certain symmetry directions can be discussed in terms of effective elastic constants. These constants are usually derived from **ultrasound propagation** measurements.

The frequencies of the various branches can become equal in correspondence to certain wavevectors, implying degeneracy. Although the optical branches have non-zero frequency even for $q = 0$ they are not always optically active, since do not always imply oscillating electric dipoles. For instance, in diamond, although the optical modes cause the vibration of the two sublattices (see §11.3) against each other, no electric dipole is induced and no interaction with the electromagnetic waves can occur.

The dispersion curves are usually obtained by inelastic **neutron spectroscopy**. The schematic structure of a triple axes spectrometer is reported below:

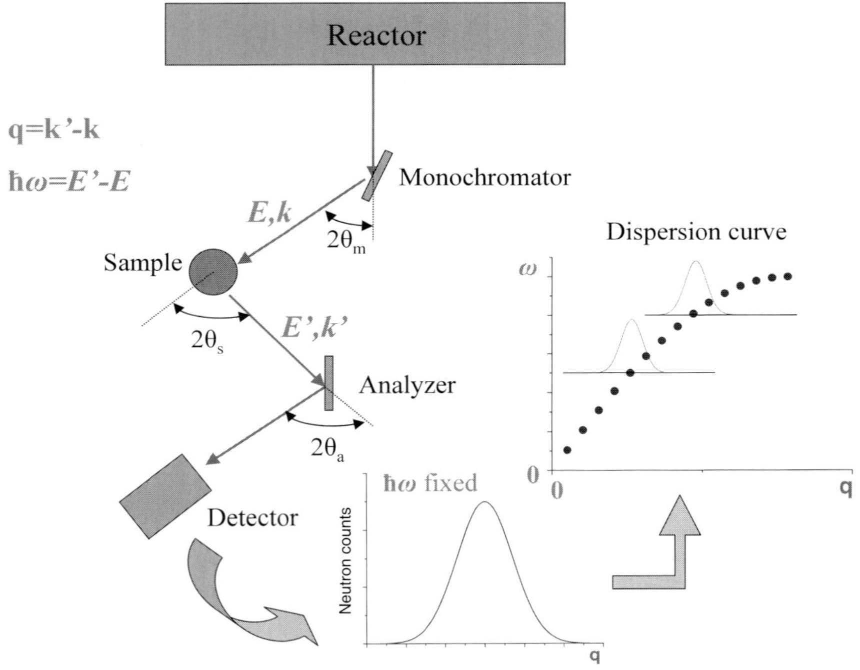

14.4 Phonons

While discussing the normal modes in molecules (§10.6) it was shown how a non-normal Hamiltonian (in terms of local coordinates) could be transformed into a normal one by writing the local displacements as a superposition of excitations, each one associated to a normal oscillator. The collective normal coordinate was shown to be a linear combination of the local ones. The treatment given at §10.6 can be extended to the displacements of the atoms around their equilibrium positions in a crystal. Thus, returning to Eq. 14.3, for each branch (j) we write the displacement in the form

$$\mathbf{u} = \sum_{\mathbf{q}} \mathbf{U_q} e^{i\mathbf{q}\cdot\mathbf{R}} e^{-i\omega_{\mathbf{q}}t} \tag{14.22}$$

Therefore the problem is reduced to the evaluation of the normal coordinates $\mathbf{Q}_{\mathbf{q}}^{(j)}$ of the crystal cell, that one can build up from the amplitudes $\mathbf{U_q}$ by including the masses and the normalization factors. The translational invariance of the crystal implies the propagation of the normal excitations of the cell with phase factor $e^{i\mathbf{q}\cdot\mathbf{R}}$.

Hence, one can start from Hamiltonians of the form $\mathcal{H} = \sum_j \mathcal{H}_j[Q^j(\mathbf{q})]$, for each wavevector $\mathbf{q}$ of a given branch j. By indicating with $\mathbf{Q}$ the group of the normal coordinates and with $\phi(\mathbf{Q})$ the related wavefunction, one expects

$$\phi(\mathbf{Q}) = \prod_{\mathbf{q},j} \phi_{\mathbf{q}}^{(j)}(Q^j(\mathbf{q})) \ . \tag{14.23}$$

In the harmonic approximation $\phi_{\mathbf{q}}^{(j)}$ is the eigenfunction of single normal oscillator, characterized by quantum number $n_j(\mathbf{q})$ and eigenvalues

$$E_{\mathbf{q}}^{(j)} = \hbar\omega_{\mathbf{q}}^{(j)}\left[1/2 + n_j(\mathbf{q})\right].$$

The total energy is

$$E_T = \sum_j \sum_{\mathbf{q}} (n_j(\mathbf{q}) + \frac{1}{2})\hbar\omega_{\mathbf{q}}^{(j)}. \tag{14.24}$$

Therefore the vibrational state of the crystal is defined by the set of $3SN$ numbers $|..., ..., n_j(\mathbf{q}), ... >$ that classify the eigenfunctions of the normal oscillators. At $T = 0$, the ground-state is labelled $|0, 0, 0... >$ and the wavefunction is the product of Gaussian functions (see §10.3.1).

At finite temperature one has to take into account the thermal excitations to excited states, for each normal oscillator. Two different approaches can be followed:

A) - the **normal oscillators** are **distinguishable** and the numbers $n_j(\mathbf{q})$ select the stationary states for each of them. Then the Boltzmann statistics holds and for a given oscillator with characteristic frequency ν the average energy is

$$\overline{E} = \sum_v p_v E_v, \tag{14.25}$$

with

$$p_v = \frac{e^{-E_v/k_B T}}{\sum_v e^{-E_v/k_B T}}$$

and

$$E_v = (v + 1/2)h\nu \quad v = 0, 1, 2, \ldots.$$

For each normal mode the average energy $\overline{E}$ is found as shown at Problem F.I.2 for photons (Planck derivation), here having to include the zero-point energy:

$$\overline{E} = h\nu\left(\frac{1}{2} + \frac{1}{e^{\frac{h\nu}{k_B T}} - 1}\right) \tag{14.26}$$

The energy turns out the one for the quantum oscillator, provided that an **average excitation number**

$$< v >= \frac{1}{e^{\frac{h\nu}{k_B T}} - 1} \tag{14.27}$$

is introduced.

The total thermal energy of the crystal is obtained by summing Eq. 14.26 over the various modes, for each branch.

B) - the crystal is considered as an assembly of **indistinguishable pseudo-particles**, each of energy $\hbar\omega_{\mathbf{q},j}$ and momentum $\hbar\mathbf{q} = (\hbar\omega_{\mathbf{q},j}/v_{j,\mathbf{q}})\hat{q}$. These quasi-particles are the quanta of the elastic field and are called **phonons** in analogy with the photons for the electromagnetic field.

Then the total energy has to be written

$$< \overline{E} >= \sum_{\mathbf{q},j}(\overline{n}_{\mathbf{q},j} + \frac{1}{2})\hbar\omega_{\mathbf{q},j}, \tag{14.28}$$

where the average number of pseudo-particles is given by the Bose-Einstein statistics, i.e.

$$\overline{n}_{\mathbf{q},j} = \frac{1}{e^{\frac{\hbar\omega_{\mathbf{q},j}}{k_B T}} - 1}, \tag{14.29}$$

for a given branch j.

The two ways A and B to conceive the quantum aspects of the lattice vibrations give equivalent final results, as it can be seen by comparing Eq. 14.26 (summed up to all the single oscillators) and Eq. 14.28. The derivation of some thermal properties (§14.5) will emphasize the equivalence of the two ways to describe the quantum aspects of the vibrational motions of the ions.

14.5 Thermal properties related to lattice vibrations

As usual, all the thermodynamical properties related to the vibrational state of the crystal can be derived from the total partition function $Z_{TOT} = \prod_{\mathbf{q},j} Z_{\mathbf{q},j}$, with

$$Z_{\mathbf{q},j} = \sum e^{\frac{-E(\mathbf{q},j)}{k_B T}} \tag{14.30}$$

where the sum is over all energy levels, for each $\mathbf{q}$-dependent oscillator of each branch.

The thermal energy can be directly evaluated by resorting to the vibrational spectra $D(\omega)$, in the light of Eqs. 14.28 and 14.29, by writing

$$U = \int \hbar\omega \left(\frac{1}{2} + \frac{1}{e^{\frac{\hbar\omega}{k_B T}} - 1}\right) D(\omega) d\omega. \tag{14.31}$$

For instance, for Einstein crystals where $D(\omega) = 3N\delta(\omega - \omega_E)$ one derives

$$U = 3N\hbar\omega_E[1/2 + 1/(e^{\frac{\hbar\omega_E}{k_B T}} - 1)].$$

The molar $(N = N_A)$ specific heat for $T \gg \Theta_E \equiv \hbar\omega_E/k_B$ turns out $C_V \simeq 3R$. For $T \ll \Theta_E$ one has

$$C_V \simeq 3R\left(\frac{\Theta_E}{T}\right)^2 e^{\frac{-\Theta_E}{T}} \tag{14.32}$$

For Debye crystals, from Eq. 14.31 by resorting to Eq. 14.19, one writes

$$C_V = \frac{\partial}{\partial T}\left\{\int_0^{\omega_D} D(\omega)\hbar\omega \frac{1}{e^{\frac{\hbar\omega}{k_B T}} - 1} d\omega\right\}$$

and then

$$C_V = 9R\left(\frac{T}{\Theta_D}\right)^3 \int_0^{\Theta_D/T} \frac{z^4 e^z}{(e^z - 1)^2} dz, \tag{14.33}$$

with $z = \hbar\omega/k_B T$.

For $T \gg \Theta_D$, with $\Theta_D \equiv \hbar\omega_D/k_B$ (known as **Debye temperature**), one again finds the classical result $C_V \to 3R$.

In the low temperature range $(\Theta_D/T \to \infty)$ Eq. 14.33 yields $C_V \simeq (12\pi^4/5)R(T/\Theta_D)^3$.

The temperature behavior of the molar specific heat in the framework of Einstein and Debye models is sketched below:

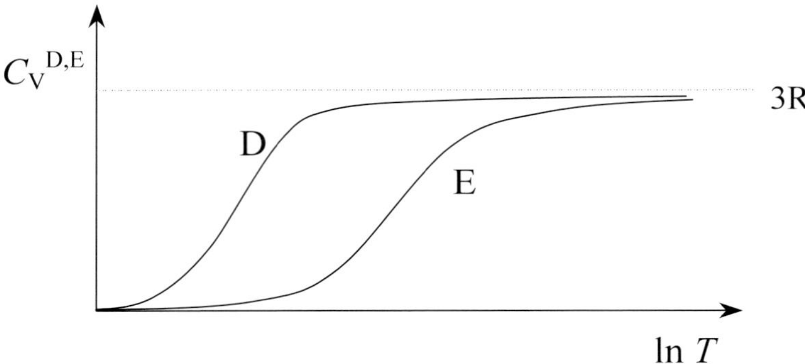

For $T \to 0$ the Debye specific heat C_V^D vanishes less rapidly than the Einstein C_V^E. The different behavior of C_V^D for $T \to 0$ originates from the fact that the vibrational spectrum in the Debye model includes oscillatory modes with energy separation of the order of $k_B T$, even at low temperature. On the contrary in the Einstein crystal in the low-temperature range one has $\hbar\omega_E \gg k_B T$.

In the Table below the Debye temperatures of some elements are reported.

Li	Be	Debye temperature in Kelvin (estimated at low temperature)											B	C	N	O	F	Ne
344	1440												...	2230 (diamond)	...	...	...	75
Na	Mg												Al	Si	P	S	Cl	Ar
158	400												428	645	...	...	...	92
K	Ca	Sc	Ti	V	Cr	Mn	Fe	Co	Ni	Cu	Zn	Ga	Ge	As	Se	Br	Kr
91	230	360	420	380	630	410	470	445	450	343	327	320	374	282	90	...	72
Rb	Sr	Y	Zr	Nb	Mo	Tc	Ru	Rh	Pd	Ag	Cd	In	Sn	Sb	Te	I	Xe
56	147	280	291	275	450	...	600	480	274	225	209	108	200	211	153	...	64
Cs	Ba	La	Hf	Ta	W	Re	Os	Ir	Pt	Au	Hg	Tl	Pb	Bi	Po	At	Rn
38	110	142	252	240	400	430	500	420	240	165	71.9	78.5	105	119	...	...	64

By resorting to the expression for the thermal energy in terms of the vibrational spectra, the mean square displacement of a given ion as a function of temperature can be directly derived. According to the extension of Eq. 14.3 to include all the normal excitations, the mean square vibrational amplitude of each atom around its equilibrium position is written

$$< |\mathbf{u}|^2 > = \sum_{\mathbf{q},j} |\mathbf{U}_{\mathbf{q},j}|^2 \ . \tag{14.34}$$

By recalling that for each oscillator the mean square displacement can be related to the average energy $m\omega^2 < u^2 >=< E >$, then for a given branch j one can write $|\mathbf{U_q}|^2 =< E_{\mathbf{q}} > /Nm\omega_{\mathbf{q}}^2$. Hence,

$$< u^2 >= \frac{1}{mN} \sum_{\mathbf{q},j} \frac{< E_{\mathbf{q},j} >}{\omega_{\mathbf{q},j}^2} = \frac{\hbar}{mN} \int \left[\frac{1}{2} + \frac{1}{e^{\frac{\hbar\omega}{k_B T}} - 1} \right] \frac{D(\omega)}{\omega} d\omega. \quad (14.35)$$

For Debye crystals, at temperatures $T \gg \Theta_D$, from Eq. 14.19 one obtains

$$< u^2 >\simeq \frac{9k_B T}{m\omega_D^2}. \quad (14.36)$$

It can be remarked that $< u^2 >$ controls the temperature dependence of the strength of the elastic component in scattering processes, through the **Debye-Waller** factor $e^{-4\pi<u^2>/\lambda^2}$, with λ wavelength of the radiation (see §14.6 for the derivation of this result).

According to the **Lindemann criterium** the crystal melts when the mean square displacement $< u^2 >$ reaches a certain fraction ξ of the square of the nearest neighbor distance R, $< u^2 >= \xi R^2$.

Empirically it can be devised that ξ is around 1.5×10^{-2} ($\sqrt{< u^2 >} \simeq 0.12R$). This criterium allows one to relate the melting temperature T_m to the Debye temperature. From Eq. 14.36 one writes $\xi R^2 = 9k_B T_m/m\omega_D^2$ and then

$$T_m = \xi\Theta_D^2 \frac{mk_B R^2}{\hbar^2}. \quad (14.37)$$

Problems XIV.5

Problem XIV.5.1 Derive the vibrational entropy of a crystal in the low temperature range $(T \ll \Theta_D)$.

Solution:
From C_V^D (Eq. 14.33) in the low temperature limit, by recalling that

$$S = \int_0^T \frac{C_V^D}{T} dT$$

the molar entropy is $S(T) = [12R\pi^4/(15\Theta_D^3)]T^3$. The contribution from optical modes can be neglected.

Problem XIV.5.2 Derive the temperature dependence of the vibrational contribution to the Helmoltz free energy and to the entropy, for Einstein crystals.

Solution:
For N oscillators the total partition function is $Z_T = Z^N$, with

$$Z = e^{-\hbar\omega_E/2k_BT} \sum_v e^{-\hbar\omega_E v/k_BT} = \frac{e^{-\hbar\omega_E/2k_BT}}{1 - e^{-\hbar\omega_E/k_BT}}$$

(remind that $\sum x^n = 1/(1-x)$, for $x < 1$).
Then the total free energy turns out

$$F = -Nk_BT\ln Z = N\{\frac{\hbar\omega_E}{2} + k_BT\ln(1 - e^{-\hbar\omega_E/k_BT})\}$$

and the entropy is

$$S = -(\frac{\partial F}{\partial T})_V = -Nk_B\{\ln(1 - e^{-\hbar\omega_E/k_BT}) - \frac{\hbar\omega_E}{k_BT}\frac{1}{e^{\hbar\omega_E/k_BT} - 1}\} \quad .$$

Problem XIV.5.3 Evaluate the specific heat per unit volume for Ag crystal (fcc cell, lattice step $a = 4.07$ Å) at $T = 10$ K, within the Einstein model (the elastic constant can be taken $k = 10^5$ dyne/cm) and within the Debye model, assuming for the sound velocity is $v \simeq 2 \times 10^5$ cm/s.

Solution:
The Einstein frequency $\omega_E \simeq \sqrt{k/M_{Ag}}$, corresponds to Einstein temperature $\Theta_E \simeq 180$ K. In the unit volume (1 cm^3) there are $n = 1/(N_A v_c)$ moles, with $v_c = a^3/4$ the primitive cell volume. Then, since $T = 10$K $\ll \Theta_E$, from Eq. 14.32 one derives $C_V^E \simeq 112$ erg/K.

The Debye frequency can be estimated from Eq. 14.21 and the corresponding Debye temperature turns out $\Theta_D \simeq 230$ K$\gg 10$ K. Then

$$C_V^D \simeq \frac{12\pi^4 k_B}{5v_c}(\frac{T}{\Theta_D})^3 \simeq 3 \times 10^4 \text{erg/K} \ .$$

Problem XIV.5.4 Specific heat measurements in copper (fcc cell, lattice step $a = 3.6$ Å, sound velocity $v = 2.6 \times 10^5$ cm/s) show that C_V/T (in 10^{-4}Joule/ mole K^2) is linear when reported as a function of T^2, with extrapolated value (C_V/T) for $T \to 0$ given by about 7 and slope about 6. Derive:
i) Fermi temperature, ii) Debye temperature, iii) the temperature at which the electronic and vibrational contributions to the specific heat are about the same.

Solution:
From the specific mass $\rho = 9.018$ g/cm^3 the number of electrons per cm^3 is $n = 8.54 \cdot 10^{22}$cm^{-3}. Then the Fermi temperature turns out $T_F = \frac{\hbar^2}{2mk_B}\left(3\pi^2 n\right)^{2/3} = 8.2 \cdot 10^4$K.

The Debye temperature, for the primitive cell of volume v_c is $\theta_D = \frac{\hbar v}{k_B}\left(\frac{6\pi^2}{v_c}\right)^{1/3} = 343$K.

From

$$\frac{\pi^2}{2} n k_B \frac{T^*}{T_F} = 32\frac{1}{v_c} k_B \frac{4\pi^4}{5}\left(\frac{T^*}{\theta_D}\right)^3$$

the temperature T^* at which the electronic and vibrational contributions are the same is obtained: $T^* = \sqrt{5 v_c n}(\Theta_D)^{3/2}/(16\pi\sqrt{T_F}) \simeq 1$ K.

Problem XIV.5.5 Write the zero-point vibrational energy of a crystal in the Debye model and derive the bulk modulus for $T \to 0$.

Solution:
The zero-point energy is $E_0 = \frac{1}{2}\int_0^{\omega_D} \hbar\omega D(\omega)d\omega$ (Eq. 14.31). From Eq. 14.19 one derives $E_0 = 9N\hbar\omega_D/8$.

At low temperature the bulk modulus is ($B \simeq V\partial^2 E_0/\partial V^2$). Then, by writing ω_D in terms of the volume $V = N v_c$ one finds

$$B = \frac{1}{2}\frac{N}{V}\hbar\omega_D \ .$$

14.6 The Mössbauer effect

The recoil-free emission or absorption of γ-ray (for the first time experimentally noticed by Mössbauer in 1958) is strictly related to the vibrational properties of the crystals. Meantime it allows one to recall some aspects involving the interaction of radiation with matter.

Let us consider an atom, or a nucleus, ideally at rest, emitting a photon due to the transition between two electronic or nucleonic levels. At the photon energy $h\nu$ is associated the momentum $(h\nu/c)$. Then in order to grant the momentum conservation the atom has to recoil during the emission with kinetic energy $E_R = (h\nu/c)^2/2M$, with M the atomic mass.

Because of the energy conservation the emission spectrum (from an assembly of many atoms) displays a Lorentzian shape,

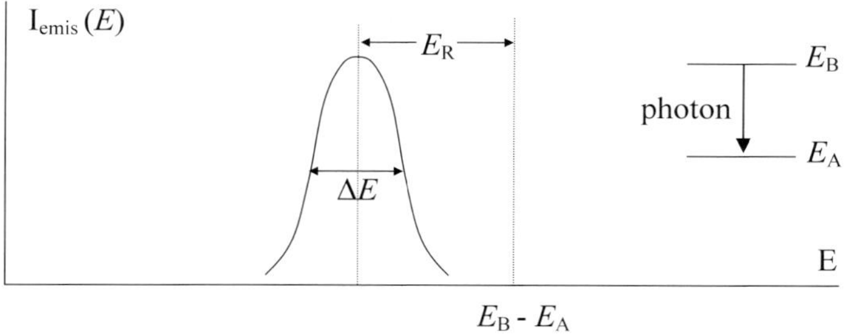

at least with the line broadening ΔE related to the **life-time** of the level (the inverse of the spontaneous emission probability, see Prob. F.I.1). Another source of broadening arises from the thermal motions of the atoms and the emission line usually takes a Gaussian shape, with width related to the distribution of the Doppler modulation in the emitted radiation (see Problem F.I.7).

Let us suppose to try the **resonance absorption** of the same emitted photon from an equivalent atom (or nucleus). Again, by taking into account the energy and momentum conservation in the absorption process, the related spectrum must have an energy distribution of Gaussian shape, centered at $E = (E_{\rm B} - E_{\rm A}) + E_{\rm R}$:

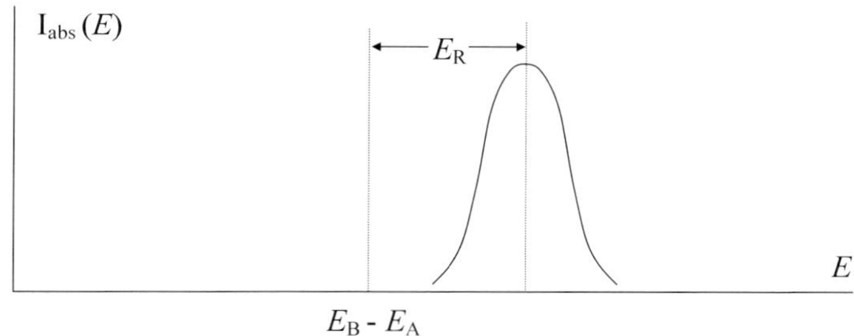

From the comparison of the emission and absorption spectra one realizes that the fraction of events that grant the resonance absorption is only the one corresponding to the energy range underlying the emission and absorption lines.

In atomic spectroscopy, where energy separations of the order of the eV are involved, the condition of resonant absorption is well verified. In fact, the recoil energy is $E_R \sim 10^{-8}$ eV, below the broadening $\Delta E \sim 10^{-7}$ eV typically associated with the life time of the excited state. At variance, when the emission and the absorption processes involve the γ-rays region, with energies around 100 keV, the recoil energy increase by a factor of the order of 10^{10}. Since the lifetime of the excited nuclear levels is of the same order of the one for electronic levels, only a limited number of resonance absorption processes can take place, for free nuclei.

In crystals, in principle, one could expect a **decrease** in the fraction of resonantly absorbed γ-rays upon cooling the source (or the absorber), due to the decrease of the broadening induced by thermal motions. Instead, an **increase** of such a fraction was actually detected at low temperature. This phenomenon is due to the fact that in solids a certain fraction f of emission and absorption processes occurs **without recoil**. Thus the spectrum schematically reported below

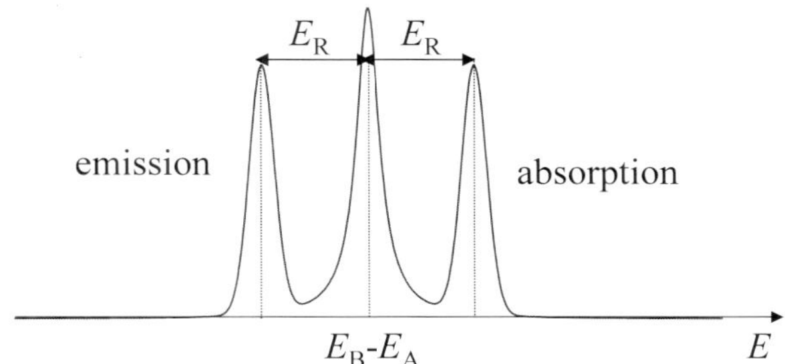

can be conceived, with a sizeable superposition of events around the energy difference $(E_B - E_A)$.

The momentum conservation is anyway granted, since the recoil energy goes to the whole crystal, with negligible subtraction of energy to the emitted or absorbed photons. The reason for the recoilless processes can be grasped by referring to the Einstein crystal, with energy $\hbar\omega_E$ larger than E_R. It is conceivable that when the quantum of elastic energy cannot be generated, then the crystal behaves **as rigid**.

Another interpretation (not involving the quantum character of the vibrational motions) is based on the classical consideration of the spectrum emitted by a source in motion. For a sinusoidal motion with frequency ω_S, the emitted spectrum has Fourier components at $\omega_i, \omega_i \pm \omega_S, ...$, so that a component at the intrinsic frequency ω_i should remain.

The fraction f of recoilless processes can be evaluated by considering, in the framework of the time dependent perturbation theory used in App.I.3, the emitting system as one nucleus imbedded in the crystal, looking for the transition probability between states having the same vibrational quantum numbers, while the nuclear state is changed. Since the long wave-length approximation cannot be retained, the perturbation operator reads $\sum_i \mathbf{A}_i . \mathbf{\nabla}_i$ (the sum is over all nucleons) (see Eq. A.I.3.3).

Let us refer to an initial state corresponding to the vibrational ground-state $|0, 0, 0, ... >$, by writing the amplitude of the time-dependent perturbative

Hamiltonian $\sum_i e^{i\mathbf{k}\cdot\mathbf{R}_i}$. Expressing $\mathbf{R}_i$ in terms of the nucleon coordinates with respect to the center of mass, the effective perturbation term entering the probability amplitude $f^{1/2}$ is of the form $e^{i\mathbf{k}\cdot\mathbf{u}}$, with $\mathbf{u}$ the displacement of the atom from its lattice equilibrium position: $f^{1/2} \propto\, <0,0,0...|e^{i\mathbf{k}\cdot\mathbf{u}}|0,0,0...>$.

The proportionality factor includes the matrix element of the variables and spins of the nucleons as well as the mechanism of the transition.

The vibrational ground-state (see Eq. 14.23) for a given branch is $||0,0,0...>= \prod_\mathbf{q} e^{-Q_\mathbf{q}^2/4\Delta_\mathbf{q}^2}$. The displacement $\mathbf{u}$ can be written as a superposition of the normal modes coordinates: $\mathbf{u} = \sum_\mathbf{q} \alpha_\mathbf{q}\mathbf{Q_q}$ ($\alpha_\mathbf{q}$ normalizing factors which include the masses). Then, by referring to the component along the direction of the γ-rays, one writes

$$f^{1/2} \propto \int_{-\infty}^{+\infty} \prod_\mathbf{q} e^{\frac{-Q_\mathbf{q}^2}{2\Delta_\mathbf{q}^2}} e^{ik\alpha_\mathbf{q}Q_\mathbf{q}} dQ_\mathbf{q} \propto \prod_\mathbf{q} e^{\frac{-\alpha_\mathbf{q}^2\Delta_\mathbf{q}^2 k^2}{2}} = e^{-\frac{1}{2}\sum_\mathbf{q}\alpha_\mathbf{q}^2\Delta_\mathbf{q}^2 k^2}$$

The mean square displacement turns out

$$<0,0...|u_x^2|0,0...> \,\equiv\, <0,0...|\sum_{\mathbf{q},\mathbf{q}'}\alpha_\mathbf{q}Q_\mathbf{q}\alpha_{\mathbf{q}'}Q_{\mathbf{q}'}|0,0...> =$$

$$= \sum_\mathbf{q}\alpha_\mathbf{q}^2 <0,0...|Q_\mathbf{q}^2|0,0...> = \sum_\mathbf{q}\alpha_\mathbf{q}^2\Delta_\mathbf{q}^2$$

and then

$$f \propto e^{-k^2<u_x^2>} = e^{-k^2<\mathbf{u}^2>/3}.$$

Since for $k = 0$ one can set $f = 1$, one has

$$f = e^{-k^2<\mathbf{u}^2>/3}. \tag{14.38}$$

For $T \to 0$ f depends from the particular transition involved in the emission process (through k^2) and from the spectrum of the crystal through the zero-point vibrational amplitude $<u^2(T=0)>$.

The temperature dependence of f originates from the one for $<u^2>$. f is also known as the **Debye-Waller** factor, since it controls the intensity of X-ray and neutron diffraction peaks. The Bragg reflections, in fact, do require elastic scattering and therefore recoilless absorption and re-emission.

By evaluating $<|\mathbf{u}|^2>$ for the Debye crystal, for instance, (see Eq. 14.36) for $T \ll \Theta_D$ one has

$$f = e^{-\frac{3E_R}{2k_B\Theta_D}}. \tag{14.39}$$

The typical experimental setup for Mössbauer absorption spectroscopy is sketched below

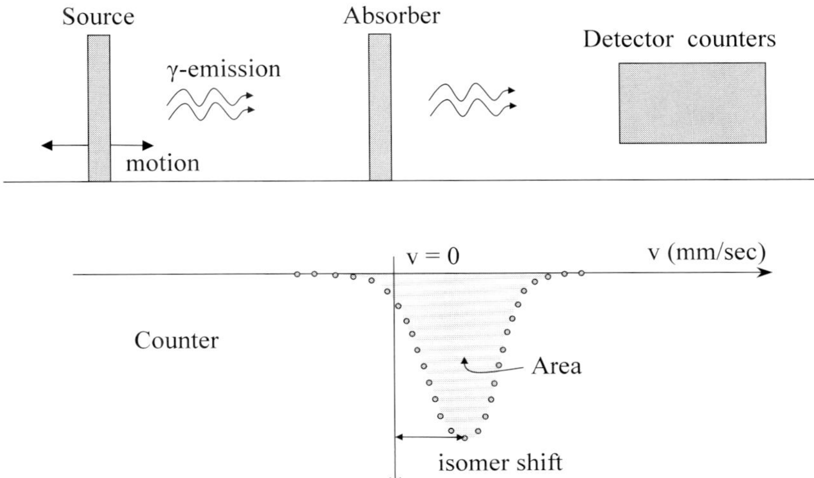

The source (or the absorber) is moved at the velocity v in order to sweep through the resonance condition. As a function of the velocity, one observes the Mössbauer absorption line, the area being proportional to the recoilless fraction f.

The shift with respect to the zero-velocity condition, **isomer shift**, is related to the finite volume of the emitting and absorbing nuclei (try to understand the shift by returning to Problem I.4.6 and F.V.16).

Since the motions do not affect the linewidth, the resolution of the Mössbauer line in principle depends only on the intrinsic lifetime of the level. Typically, for ~ 100 keV γ-rays, a resolution around 10^{-14} can be achieved. Therefore, the Mössbauer spectroscopy can be used in solid state physics to investigate the magnetic and electric hyperfine splitting of the nuclear levels. It has been used also in order to detect subtle relativistic effects (see Problem F.XIV.9).

Problems F.XIV

Problem F.XIV.1 Show that an approximate estimate of the Debye temperature in a monoatomic crystal can be obtained from the specific heat, by looking at the temperature at which $C_V \simeq 23 \cdot 10^7$ erg/mole K.

Solution:

From Eqs. 14.31 and 14.19

$$U = \int_0^{\omega_D} \frac{\hbar\omega}{e^{\frac{\hbar\omega}{k_B T}} - 1} D(\omega)d\omega = \frac{3}{2\pi^2} \frac{N v_c}{v^3} \int_0^{\omega_D} \frac{\hbar\omega^3}{e^{\frac{\hbar\omega}{k_B T}} - 1} d\omega \, ,$$

(having neglecting the zero-point energy which does not contribute to the thermal derivatives). v is the sound velocity (an average of the ones for longitudinal and transverse branches). The specific heat turns out

$$C_V = 9R \left[4 \left(\frac{T}{\theta_D} \right)^3 \int_0^{\frac{\theta_D}{T}} \frac{z^3}{e^z - 1} dz - \frac{\theta_D}{T} \frac{1}{e^{\frac{\theta_D}{T}} - 1} \right].$$

For $T = \theta_D$

$$C_V(T = \theta_D) \simeq 36R \left[\int_0^1 \frac{z^3}{e^z - 1} dz - \frac{1}{1.72} \right]$$

and then

$$C_V(T = \theta_D) \simeq 2.856R \simeq 23.74 \cdot 10^7 \, \frac{\text{erg}}{\text{mole K}} \, .$$

Problem F.XIV.2 In a linear diatomic chain of alternating Br^- and Li^+ ions (lattice step $a = 2$ Å) the sound velocity is $v = 2.7 \cdot 10^5 \, cm/s$. Derive the effective elastic constant under the assumption used at §14.3.2 and the gap between the acoustic and optical branches.

Solution:

From Eq. 14.7 the sound velocity is

$$v = \sqrt{\frac{2k}{m_1 + m_2}} a \, ,$$

and the elastic constant turns out

$$k = \frac{1}{2}(m_1 + m_2) \left(\frac{v_s}{a} \right)^2 \simeq 10^5 \text{ dyne/cm} \, .$$

The gap covers the frequency range from $\omega_{min} = (2k/m_1)^{1/2}$ to $\omega_{max} = (2k/m_2)^{1/2}$, with

$$\omega_{min} = 0.15 \cdot 10^{14} \mathrm{rad\,s}^{-1} \quad \text{and} \quad \omega_{max} = 0.989 \cdot 10^{14} \mathrm{rad\,s}^{-1}.$$

Problem F.XIV.3 For a cubic crystal, with lattice step a, show that within the Debye model and for $T \ll \Theta_D$, the most probable phonon energy is $\hbar\omega_p \simeq 1.6 k_B T$ and that the wavelength of the corresponding excitation is $\lambda_p \simeq a\Theta_D/T$.

Solution:

In view of the analogy with photons (see Problem F.I.2) the number of phonons with energy $\hbar\omega$ is given by

$n(\omega) = D(\omega)/(e^{\hbar\omega/k_B T} - 1)$.

From Eq. 14.19 and from $dn(\omega)/d\omega = 0$, one finds

$$\frac{\hbar\omega_p}{k_B T} e^{\hbar\omega/k_B T} = 2(e^{\hbar\omega/k_B T} - 1)$$

and then $\hbar\omega_p/k_B T \simeq 1.6$.

Since $\lambda_p(\omega_p/2\pi) = v$ one has $\lambda_p \simeq 2\pi v\hbar/1.6 k_B T$. For cubic crystal $\Theta_D = (v\hbar/k_B a)(6\pi)^{1/3}$, and then $\lambda_p \simeq a\Theta_D/T$.

Problem F.XIV.4 Evaluate the root-mean squared amplitude of the atomic displacement within the Debye model, at low and at high temperatures.

Solution:

From Eq. 14.35 for $T \gg \hbar\omega_D/k_B$ one has

$$< u^2 > \simeq \frac{9 k_B T}{m\omega_D^2} ,$$

while at low temperature $< u^2 > \simeq \frac{9\hbar}{4 m\omega_D}$.

Problem F.XIV.5 Show that in a Debye crystal at high temperature the thermal energy U is larger than the classical one by a factor going as $1/T^2$.

Solution:

From

$$U = 9 N k_B T \left(\frac{T}{\Theta_D}\right)^3 \int_0^{x_D} \left(\frac{x^3}{e^x - 1} + \frac{x^3}{2}\right) dx$$

with $x_D = \Theta_D/T$ and $x = \hbar\omega/k_B T$, for $x \to 0$ and after series expansion of the integrand

$$\int_0^{x_D} \left(\frac{x^3}{e^x - 1} + \frac{x^3}{2}\right) dx \simeq \int_0^{x_D} \left(\frac{x^3}{x + \frac{x^2}{2} + \frac{x^3}{6} + \dots} + \frac{x^3}{2}\right) dx \simeq \int_0^{x_D} x^2 \left(1 + \frac{x^2}{12} - \dots\right) dx,$$

one can write

$$U = 9Nk_BT(\frac{T}{\Theta_D})^3 \left(\frac{1}{3}(\frac{\Theta_D}{T})^3 + \frac{1}{60}(\frac{\Theta_D}{T})^5 + ... \right).$$

The specific heat turns out

$$C_V \simeq 3R\left(1 - \frac{1}{20}(\frac{\Theta_D}{T})^2 - .... \right).$$

Problem F.XIV.6 In the Figures below

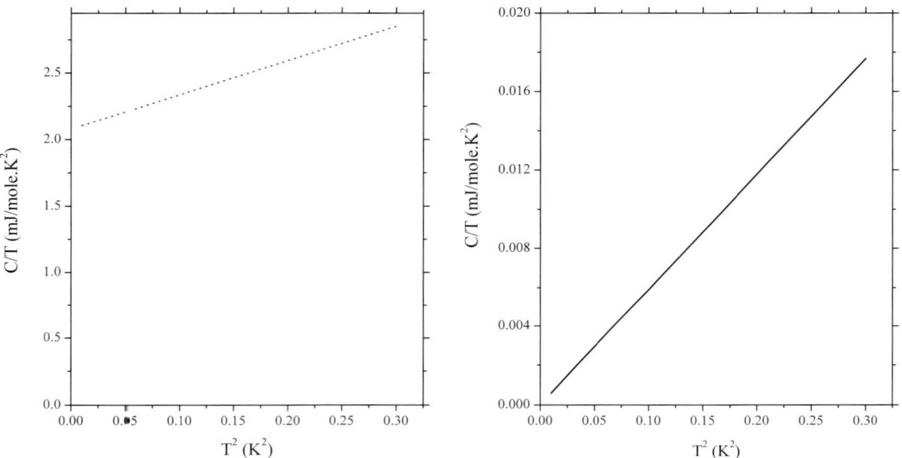

the low temperature specific heats of two crystals are reported. Are they metals or insulators?

Estimate the Debye temperatures and the Fermi energy.

Solution:
From $C_V/T = A + BT^2$, $A = R(\pi^2/3)D(E_F)$ is the term associated with the free-electron contribution (see §12.7.1), while $B = (12\pi^4/5)(R/\Theta_D^3)$ originates from the phonon contribution. Hence the Figure on the left refers to a metal while the one on the right to an insulator.

From the data on the left $A \simeq 2.1 \times 10^4 \text{erg/K}^2\text{mole}$ one finds $E_F \simeq 2.1$ eV. From $B \simeq 2.6 \times 10^4 \text{erg/K}^4\text{mole}$, then $\Theta_D \simeq 90$ K. From the data on the right $B \simeq 590 \text{ erg/K}^4\text{mole}$, yielding $\Theta_D \simeq 320$ K.

Problem F.XIV.7 Derive the vibrational contribution to the specific heat for a chain, at high and low temperatures, within the Debye and the Einstein approximations. Compare the results with the exact estimates obtained in the harmonic approximation and nearest-neighbor interactions (§14.3.1).

Solution:
Within the Debye model $D(\omega) = Na/(2\pi v)$ and then

$$U_D = \frac{N}{2}\hbar\omega + \frac{N}{\omega_D}\int_0^{\omega_D} \frac{\hbar\omega}{e^{\beta\hbar\omega} - 1}d\omega \ .$$

The specific heat turns out $C_V \simeq R$ for $T \gg \Theta_D$ and $C_V \simeq 2IR(T/\Theta_D)$ for $T \ll \Theta_D$, with $I = \int_0^\infty x/(e^x - 1)dx$.

Within the Einstein model $D(\omega) = N\delta(\omega - \omega_E)$ and results independent from the dimensionality are obtained (see Eq. 14.32).

In the harmonic approximation with nearest neighbors interactions the density of states is $D(\omega) = (2N/\pi)(1/\sqrt{\omega_m^2 - \omega^2})$ for $\omega \leq \omega_m$, while it is zero for $\omega > \omega_m$ (see Eq. 14.14). Then

$$U = \frac{N}{2}\hbar\omega + \frac{2Nk_BT}{\pi}\int_0^{x_m} \frac{1}{\sqrt{x_m^2 - x^2}}\frac{x}{e^x - 1}dx$$

with $x = \beta\hbar\omega$ and $x_m = \beta\hbar\omega_m$. For $T \gg \Theta_m = \hbar\omega_m/k_B$ one has

$$U \simeq \frac{N}{2}\hbar\omega + \frac{2Nk_BT}{\pi}\left(\frac{\pi}{2} - \frac{x_m}{2} + ...\right)$$

and $C_V \simeq R$. For $T \ll \hbar\omega_m/k_B$

$$U \simeq \frac{N}{2}\hbar\omega + \frac{2N(k_BT)^2}{\pi\hbar\omega_m}I$$

so that

$$C_V \simeq \frac{4I}{\pi}R\frac{T}{\Theta_m}$$

showing that the Debye approximation yields the same low temperature behavior.

Problem XIV.8 A diatomic crystal has two types of ions, one at spin $S = 1/2$ and $g = 2$ and one at $S = 0$. The Debye temperature is $\Theta_D = 200\,K$. Evaluate the entropy at $T = 20\,K$ in zero external magnetic field and for magnetic field $H = 1\,\text{kGauss}$. Assume no interaction among the magnetic moments and the same atomic masses.

Solution:
The vibrational entropy is

$$S_{vib} = \int_0^T \frac{C_V(T')}{T'}dT'$$

where (per ion and in k_B units)

$$C_V(T') = \frac{4}{5}\pi^5 \left(\frac{T'}{\theta_D}\right)^3 .$$

Then, for $T' = 20$ K

$$S_{vib} = \frac{4}{15}\pi^5 \left(\frac{T'}{\theta_D}\right)^3 = 0.078 .$$

The magnetic partition function is

$$Z_{mag} = \exp\left(-\frac{1}{2}y\right) + \exp\left(\frac{1}{2}y\right)$$

with

$$y = \frac{\mu_B g H}{k_B T} \simeq \frac{0.9 \cdot 10^{-20}}{1.38 \cdot 10^{-16}} g\frac{H}{T} = 6.72 \cdot g\frac{H}{T} \cdot 10^{-5}.$$

Then, from

$$F = -k_B T \ln Z \qquad \text{and} \qquad S = -\frac{\partial F}{\partial T}$$

one has

$$S_{mag}(T') \simeq \ln 2$$

$$S = S_{vib} + \frac{1}{2}S_{mag} = 0.078 + 0.34 = 0.42 ,$$

namely

$$S = 0.42\, k_B/\text{ion} .$$

Problem XIV.9 The life time of the ^{57}Fe excited state decaying through γ emission at 14.4 keV is $\tau \simeq 1.4 \times 10^{-7}$ s (see Problems F.I.1, F.I.7 and F.III.6). Estimate the height at which the γ-source should be placed with respect to an absorber at the ground level, in order to evidence the gravitational shift expected on the basis of Einstein theory.

Assume that a shift of 5 % of the natural linewidth of Mössbauer resonant absorption can be detected [in the real experiment by **Pound** and **Rebka** (Phys. Rev. Lett. 4, 337 (1960)) by using a particular experimental setup resolution of the order of $10^{-14} - 10^{-15}$ could be achieved, with a fractional full-width at half-height of the resonant Lorentzian absorption line of 1.13×10^{-12}]. Try to figure out why the source-absorber system has to be placed in a liquid He bath.

Solution:
On falling from the height L the energy of the γ photon becomes

$$h\nu(0) = h\nu(L)[1 + \frac{gL}{c^2}]$$

where mgL/mc^2 (mass independent and therefore valid also for photons) is the ratio of the gravitational potential energy to the intrinsic energy. The natural linewidth of the Mössbauer line is $2\hbar/\tau$. Therefore, to observe a 5% variation

$$\frac{2\hbar}{20\tau} = h\nu(L)\frac{gL}{c^2}$$

and then

$$L = \frac{\hbar c^2}{10g\tau 14.4\text{keV}} = 284\text{m}$$

(in the real experiment the height of the tower was about 10 times smaller!). Note that the natural linewidth, when sweeping with velocity v the absorber (or the source) corresponds to a velocity width

$$\Delta v = \frac{2\hbar c}{h\nu\tau} \simeq 0.2 \text{ mm/s}$$

(the actual full-width at half height in the experiment by Pound and Rebka was 0.43 mm/s).

A difference in the temperatures of the source and the absorber of 1 K could prevent the observation of the gravitational shift because of the temperature-dependent **second-order** Doppler shift resulting from lattice vibrations, since $< v^2 > \sim k_B T/M$. Low temperature increases the γ-recoilless fraction f.

Index

Printing in March 2007